# Advances in Intelligent Systems and Computing

Volume 1400

The series "Advances in Intelligent Systems and Computing" contains publications on theory, applications, and design methods of Intelligent Systems and Intelligent Computing. Virtually all disciplines such as engineering, natural sciences, computer and information science, ICT, economics, business, e-commerce, environment, healthcare, life science are covered. The list of topics spans all the areas of modern intelligent systems and computing such as: computational intelligence, soft computing including neural networks, fuzzy systems, evolutionary computing and the fusion of these paradigms, social intelligence, ambient intelligence, computational neuroscience, artificial life, virtual worlds and society, cognitive science and systems, Perception and Vision, DNA and immune based systems, self-organizing and adaptive systems, e-Learning and teaching, human-centered and human-centric computing, recommender systems, intelligent control, robotics and mechatronics including human-machine teaming, knowledge-based paradigms, learning paradigms, machine ethics, intelligent data analysis, knowledge management, intelligent agents, intelligent decision making and support, intelligent network security, trust management, interactive entertainment, Web intelligence and multimedia.

The publications within "Advances in Intelligent Systems and Computing" are primarily proceedings of important conferences, symposia and congresses. They cover significant recent developments in the field, both of a foundational and applicable character. An important characteristic feature of the series is the short publication time and world-wide distribution. This permits a rapid and broad dissemination of research results.

Indexed by DBLP, INSPEC, WTI Frankfurt eG, zbMATH, Japanese Science and Technology Agency (JST).

All books published in the series are submitted for consideration in Web of Science.

More information about this series at http://www.springer.com/series/11156

Juan José Gude Prego ·
José Gaviria de la Puerta ·
Pablo García Bringas · Héctor Quintián ·
Emilio Corchado
Editors

# 14th International Conference on Computational Intelligence in Security for Information Systems and 12th International Conference on European Transnational Educational (CISIS 2021 and ICEUTE 2021)

Springer

*Editors*
Juan José Gude Prego
Department of Computing, Electronics
and Communication Technologies
University of Deusto
Bilbao, Spain

José Gaviria de la Puerta
Department of Computing, Electronics
and Communication Technologies
University of Deusto
Bilbao, Spain

Pablo García Bringas
Department of Mechanics, Design
and Industrial Management
University of Deusto
Bilbao, Spain

Héctor Quintián
Department of Industrial Engineering
University of A Coruña
Ferrol, Spain

Emilio Corchado
BISITE Research Group
University of Salamanca
Salamanca, Spain

ISSN 2194-5357 ISSN 2194-5365 (electronic)
Advances in Intelligent Systems and Computing
ISBN 978-3-030-87871-9 ISBN 978-3-030-87872-6 (eBook)
https://doi.org/10.1007/978-3-030-87872-6



This Springer imprint is published by the registered company Springer Nature Switzerland AG
The registered company address is: Gewerbestrasse 11, 6330 Cham, Switzerland

# Preface

This volume of *Advances in Intelligent and Soft Computing* contains accepted papers presented at the *14th International Conference on Computational Intelligence in Security for Information Systems* (CISIS 2021) and the *12th International Conference on EUropean Transnational Education* (ICEUTE 2021). These conferences were held in the beautiful seaside city of Bilbao, Spain, in September 2021.

The aim of the CISIS 2021 conference is to offer a meeting opportunity for academic and industry-related researchers belonging to the various, vast communities of computational intelligence, information security, and data mining. The need for intelligent, flexible behavior by large, complex systems, especially in mission-critical domains, is intended to be the catalyst and the aggregation stimulus for the overall event.

After a through peer-review process, the CISIS 2021 International Program Committee selected 23 papers, written by authors from 16 different countries, which are published in these conference proceedings. In this edition, one special session including six papers was organized: Building Trust in Ecosystems and Ecosystem Components.

The aim of ICEUTE 2021 conference is to offer a meeting point for people working on transnational education within Europe. It provides a stimulating and fruitful forum for presenting and discussing the latest works and advances on transnational education within European countries.

In the case of ICEUTE 2021, the International Program Committee selected 17 papers (from seven countries), which are also published in these conference proceedings. One special session with five papers was organized related to the relevant topic: Sustainable personal goals: engaging students in their learning process.

The selection of papers was extremely rigorous in order to maintain the high quality of the conference, and we would like to thank the members of the Program Committees for their hard work in the reviewing process. This is a crucial process to the creation of a high-standard conference, and the CISIS and ICEUTE conferences would not exist without their help.

CISIS 2021 and ICEUTE 2021 enjoyed outstanding keynote speeches by distinguished guest speakers: Prof. Javier del Ser, who is a principal researcher in data analytics and optimization at Tecnalia, Spain, is a part-time lecturer at the University of the Basque Country, Spain; Prof. Concha Bielza, who is a Full Professor in the Department of Artificial Intelligent, Polytechnic University of Madrid, Spain; and Prof. Enrique Zuazua, who holds a Chair in Applied Analysis at Friedrich Alexander University, Germany, and a Chair of Computational Mathematics at University of Deusto, Spain.

For this CISIS 2021 edition, as a follow-up of the conference, we anticipate further publication of selected papers in a special issue of the prestigious *Logic Journal of the IGPL*, published by Oxford Journals.

Particular thanks go as well to the conference main sponsors, Startup Ole, Department of Education and Universities of the Basque Government, Logistar Project of DeustoTech, and University of Deusto, who jointly contributed in an active and constructive manner to the success of this initiative.

We would like to thank all the special session organizers, contributing authors, as well as the members of the Program Committees and the Local Organizing Committee for their hard and highly valuable work. Their work has helped to contribute to the success of the CISIS 2021 and ICEUTE 2021 events.

The editors

September 2021

Juan José Gude Prego
José Gaviria de la Puerta
Pablo García Bringas
Héctor Quintián
Emilio Corchado

# CISIS 2021

International Conference on CISIS Computational Intelligence in Security for Information Systems

## Organization

### General Chair

| | |
|---|---|
| Emilio Corchado | University of Salamanca, Spain |

### Local Chair

| | |
|---|---|
| Pablo García Bringas | University of Deusto, Spain |

### Local Co-chairs

| | |
|---|---|
| Hugo Sanjurjo González | University of Deusto, Spain |
| Alberto Tellaeche Iglesias | University of Deusto, Spain |

### Program Committee Chairs

| | |
|---|---|
| Emilio Corchado | University of Salamanca, Spain |
| Pablo García Bringas | University of Deusto, Spain |
| Héctor Quintián | University of A Coruña, Spain |

### Program Committee

| | |
|---|---|
| Adam Wójtowicz | Poznan University of Economics and Business, Poland |
| Agusti Solanas | Rovira i Virgili University, Spain |
| Agustin Martin Muñoz | CSIC, Spain |
| Aitor Martinez Seras | University of Deusto, Spain |
| Alberto Peinado | Universidad de Malaga, Spain |
| Alberto Tellaeche | University of Deusto, Spain |
| Alvaro Michelena Grandío | University of A Coruña, Spain |

| | |
|---|---|
| Alessandra De Benedictis | University of Naples Federico II, Italy |
| Amparo Fuster-Sabater | C.S.I.C., Spain |
| Andreea Vescan | Babes-Bolyai University, Romania |
| Andrés Blazquez | University of Salamanca |
| Andrysiak Tomasz | University of Technology and Life Sciences UTP, Poland |
| Angel Arroyo | University of Burgos, Spain |
| Angel Martin Del Rey | University of Salamanca, Spain |
| Antonio J. Tomeu-Hardasmal | University of Cadiz, Spain |
| Borja Sanz Urquijo | University of Deusto, Spain |
| Bruno Baruque | University of Burgos, Spain |
| Carlos Pereira | ISEC, Portugal |
| Carmen Benavides | University of León, Spain |
| Cataldo Basile | Politecnico di Torino, Italy |
| Ciprian Pungila | West University of Timisoara, Romania |
| Cosmin Sabo | Technical University of Cluj-Napoca, Romania |
| Cristina Alcaraz | University of Malaga, Spain |
| Daniel Urda | University of Burgos, Spain |
| David Alvarez Leon | University of León, Spain |
| David Arroyo | C.S.I.C., Spain |
| David Buján Carballal | University of Deusto, Spain |
| David García-Retuerta | University of Salamanca, Spain |
| David Zamora Arranz | University of Deusto, Spain |
| Eduardo Solteiro Pires | UTAD University, Portugal |
| Elena Hernández Nieves | University of Salamanca, Spain |
| Eneko Osaba Icedo | Tecnalia Research & Innovation, Spain |
| Enrique Onieva | University of Deusto, Spain |
| Esteban Jove | University of A Coruña, Spain |
| Evgenia Adamopoulou | NTUA/ICCS, Greece |
| Fernando Tricas | University of Zaragoza, Spain |
| Francisco Martínez Álvarez | Pablo de Olavide University, Spain |
| Francisco Zayas Gato | University of A Coruña, Spain |
| Guillermo Morales-Luna | CINVESTAV-IPN, Mexico |
| Hugo Sanjurjo González | University of Deusto, Spain |
| Hugo Scolnik | University of Buenos Aires, Argentina |
| Igor Santos Grueiro | Mondragon University, Spain |
| Iker Pastor López | University of Deusto, Spain |
| Isaias Garcia | University of León, Spain |
| Javier Nieves | Azterlan, Spain |
| Javier Prieto | University of Salamanca, Spain |
| Jesús Díaz-Verdejo | University of Granada, Spain |
| Jon Ander Garamendi | University of Deusto, Spain |
| José Gaviria de la Puerta | University of Deusto, Spain |
| Jose A. Onieva | University of Malaga, Spain |
| José F. Torres | Pablo de Olavide University, Spain |

| | |
|---|---|
| Jose Luis Calvo-Rolle | University of A Coruña, Spain |
| José Luis Casteleiro-Roca | University of A Coruña, Spain |
| Jose Luis Imana | Complutense University of Madrid, Spain |
| Jose M. Molina | University Carlos III de Madrid, Spain |
| Jose Manuel Lopez-Guede | University of the Basque Country, Spain |
| Josep Ferrer | Universitat de les Illes Balears, Spain |
| Juan Caubet | EURECAT, Spain |
| Juan Jesús Barbarán | University of Granada, Spain |
| Juan José Gude Prego | University of Deusto, Spain |
| Juan Pedro Hecht | University of Buenos Aires, Argentina |
| Lidia Sánchez-González | University of León, Spain |
| Luis Hernandez Encinas | C.S.I.C., Spain |
| Luis Alfonso Fernández Serantes | FH-Joanneum University of Applied Sciences, Austria |
| Manuel Castejón-Limas | University of León, Spain |
| Manuel Graña | University of the Basque Country, Spain |
| Marc Ohm | Rheinische Friedrich-Wilhelms-University of Bonn, Germany |
| Michal Choras | ITTI Ltd., Poland |
| Michal Wozniak | Wroclaw University of Technology, Poland |
| Oscar Llorente-Vazquez | University of Deusto, Spain |
| Pau Figuera Vinué | University of Deusto, Spain |
| Paulo Moura Oliveira | UTAD University, Portugal |
| Rafael Alvarez | University of Alicante, Spain |
| Rafael Corchuelo | University of Seville, Spain |
| Ramón-Ángel Fernández-Díaz | University of León, Spain |
| Raúl Durán | University of Alcalá, Spain |
| Robert Burduk | Wroclaw University of Technology, Poland |
| Roberto Casado-Vera | University of Salamanca, Spain |
| Roland Jones | Blue Jellyfish, USA |
| Ruxandra F. Olimid | University of Bucharest, Romania |
| Salvador Alcaraz | Miguel Hernandez University, Spain |
| Sorin Stratulat | Université de Lorraine, France |
| Urko de la Torre | Azterlan, Spain |
| Wenjian Luo | Harbin Institute of Technology, China |
| Wilson Rojas | El Bosque University, Colombia |
| Zita Vale | ISEP/IPP, Portugal |

# CISIS 2021: Special Sessions

## Building Trust in Ecosystems and Ecosystem Components

### Program Committee

| | |
|---|---|
| José Barata (Organizer) | NOVA University of Lisboa, Portugal |
| Oliviu Matei (Organizer) | North University of Baia Mare, Romania |
| Petrica Pop (Organizer) | Technical University of Cluj-Napoca, North University Center at Baia Mare, Romania |
| Anca Avram | Technical University of Cluj-Napoca, Romania |
| Andre Rocha | NOVA University of Lisboa, Portugal |
| Emilia Cioroaica | Fraunhofer IESE, Germany |
| Ileana Buhan | Radboud University, Netherlands |
| Ioana Zelina | Technical University of Cluj-Napoca, Romania |
| Ioannis Sorokos | Fraunhofer IESE, Germany |
| Ovidiu Cosma | Technical University of Cluj-Napoca, Romania |
| Ricardo Peres | NOVA University of Lisboa, Portugal |
| Rudolf Erdei | Holisun SRL, Romania |
| Sanaz Nikghadam | NOVA University of Lisboa, Portugal |

## CISIS 2021 Organizing Committee Chairs

| | |
|---|---|
| Emilio Corchado | University of Salamanca, Spain |
| Pablo García Bringas | University of Deusto, Spain |
| Héctor Quintián | University of A Coruña, Spain |
| Hugo Sanjurjo González | University of Deusto, Spain |

## CISIS 2021 Organizing Committee

| | |
|---|---|
| David Buján Carballal | University of Deusto, Spain |
| José Gaviria De la Puerta | University of Deusto, Spain |
| Juan José Gude Prego | University of Deusto, Spain |
| Iker Pastor López | University of Deusto, Spain |
| Borja Sanz Urquijo | University of Deusto, Spain |
| Alberto Tellaeche Iglesias | University of Deusto, Spain |

# ICEUTE 2021

## Organization

### General Chair

| | |
|---|---|
| Emilio Corchado | University of Salamanca, Spain |

### Local Chair

| | |
|---|---|
| Pablo García Bringas | University of Deusto, Spain |

### Local Co-chairs

| | |
|---|---|
| José Gaviria de la Puerta | University of Deusto, Spain |
| Hugo Sanjurjo González | University of Deusto, Spain |

### Program Committee Chairs

| | |
|---|---|
| Emilio Corchado | University of Salamanca, Spain |
| Pablo García Bringas | University of Deusto, Spain |
| Héctor Quintián | University of A Coruña, Spain |

### Program Committee

| | |
|---|---|
| Aitor Martinez Seras | University of Deusto, Spain |
| Alberto Fraile | University of Salamanca |
| Alessandra Raffaetà | Universita Ca'Foscari Venezia, Italy |
| Ana María Lara Palma | University of Burgos |
| Ana Rosa Pereira Borges | ISEC, Portugal |
| Andreea Vescan | Babes-Bolyai University, Romania |
| Andrés Blázquez | University of Salamanca |
| Angel Arroyo | University of Burgos, Spain |

| | |
|---|---|
| Monika Cielkiewicz | Complutense University of Madrid, Spain |
| Oscar Llorente-Vazquez | University of Deusto, Spain |
| Paola Clara Leotta | University of Catania, Italy |
| Pau Figuera Vinué | University of Deusto, Spain |
| Paulo Moura Oliveira | UTAD University, Portugal |
| Rafael Corchuelo | University of Seville, Spain |
| Ramona Mihaila | Dimitrie Cantemir Christian University, Bucharest |
| Ramón-Ángel Fernández-Díaz | University of León, Spain |
| Richard Duro | University of A Coruña, Spain |
| Sorin Stratulat | Université de Lorraine, France |
| Urko de la Torre | Azterlan, Spain |
| Viorel Negru | West University of Timisoara, Romania |

# ICEUTE 2021: Special Sessions

## Sustainable Personal Goals: Engaging Students in their Learning Process

## Program Committee

| | |
|---|---|
| Marián Queiruga-Dios (Organizer) | Francisco de Vitoria University, Spain |
| Pedro Mauricio Acosta Castellanos (Organizer) | Santo Tomás University, Colombia |
| Araceli Queiruga-Dios | University of Salamanca |
| David Rodriguez de Vera Ríos | Francisco de Vitoria University, Spain |
| Farrah Kristel Batista | University of Salamanca, Spain |
| Juan José Bullón Pérez | University of Salamanca, Spain |
| María Angeles Alonso Navales | Francisco de Vitoria University, Spain |
| Maria Fernanda Gambarini | Francisco de Vitoria University, Spain |
| María Jesús Santos | University of Salamanca, Spain |
| Miguel Angel Queiruga-Dios | University of Burgos, Spain |
| Soraya Muñoz Pérez | Francisco de Vitoria University, Spain |
| Yuddy Alejandra Castro Ortegon | Santo Tomás University, Colombia |

## ICEUTE 2021 Organizing Committee Chairs

| | |
|---|---|
| Emilio Corchado | University of Salamanca, Spain |
| Pablo García Bringas | University of Deusto, Spain |
| Héctor Quintián | University of A Coruña, Spain |
| Hugo Sanjurjo González | University of Deusto, Spain |

## ICEUTE 2021 Organizing Committee

| | |
|---|---|
| David Buján Carballal | University of Deusto, Spain |
| José Gaviria De la Puerta | University of Deusto, Spain |
| Juan José Gude Prego | University of Deusto, Spain |
| Iker Pastor López | University of Deusto, Spain |
| Borja Sanz Urquijo | University of Deusto, Spain |
| Alberto Tellaeche Iglesias | University of Deusto, Spain |

# Contents

## SPECIAL SESSION: Building Trust in Ecosystems and Ecosystem Components



## ICEUTE 2021

### General Track

# General Track

# A Fast Heterogeneous Approach to Enhanced Blockchain Attack Resilience and Mitigation

Ciprian Pungila(✉) and Viorel Negru

Faculty of Mathematics and Informatics, West University of Timisoara, Timisoara, Romania
{ciprian.pungila,viorel.negru}@e-uvt.ro

**Abstract.** We are introducing a new computational model for heterogeneous architectures for performing enhanced security validations and processing of transactions on-chain, a model applicable to both permissioned and permissionless blockchains. By proposing an efficient storage model for the Merkle trees involved in the transaction and security validation processing, and combining it with an efficient heterogeneous approach for processing it, we are outlining the improved computational advantages to ensuring attack resilience on-chain, which surpasses the performance of traditional CPU-based approaches by a significant factor.

**Keywords:** Blockchain · Heterogeneous architecture · High-performance computing · Consensus validation · Efficient storage · Merkle trees · GPGPU processing

## 1 Introduction

As of today, blockchain technology is slowly growing to become a part of day-to-day technology. Strong focus falls on how such technology can be put to good use in various use-cases and industries, ranging from education, digital certification and supply management, to fintech and security tokens. As Bitcoin [1] has grown to become a strong point of interest to financial investors, and with more than 2,000 altcoins [2] in existence today, the market share for crypto-assets has grown spectacularly in the past few years. Blockchains however have practical purpose outside the crypto world as well, with multiple solutions emerging in the world of fintech [3], health systems [4], manufacturing [5], road safety [6], disaster prevention [7], as well as many more. While blockchains today each have advantages and disadvantages, a common way to validate blockchain transactions involves the use of cryptography approaches such as those offered through Merkle [9] trees, which are very common in the Bitcoin protocol, as well as many other blockchains using a Proof-of-Work (PoW) approach (whether directly derived from Bitcoin or otherwise).

We are proposing an approach to efficiently store, process and enhance, in a heterogeneous architecture combining both CPU and GPGPUs, the Merkle


J. J. Gude Prego et al. (Eds.): CISIS 2021/ICEUTE 2021, AISC 1400, pp. 3–12, 2022.
https://doi.org/10.1007/978-3-030-87872-6_1

tree validation process, an essential part of the consensus process in all major proof-of-work blockchains nowadays. We discuss related work in Sect. 2 of the paper, describe our approach, with data structures, algorithm and methodology in Sect. 3, while Sect. 4 provides the necessary experimental results achieved during our testing.

## 2 Related Work

Merkle trees, as primary data structures used to validate authenticity and consistency of transactions in major blockchains nowadays, are a security measure to confirm and validate transactions, as well as prevent injection of false transactions. Whilst common blockchains today are slow (for example, Bitcoin processes about 7 transactions per second (tx/s) on average, and Ethereum [8] 20 tx/s), several newer blockchains (e.g. TRON [11], EOS [10], etc.) promise to introduce better, more competitive performance, better suited for real-world use-cases. In Proof-of-Work (PoW) systems such as Bitcoin and Ethereum, to name the most popular ones (but also applicable to thousands of other coins in existence nowadays), the principle of consensus validation is fundamental to the way blockchains function. In most such approaches, Merkle trees are being used as structures for efficient, secure validation of digital data in large amounts of information, as an essential mechanism to verify the consistency and authenticity of that content.

A Merkle tree is a structure that performs a security check of all the transactions included in a block of the blockchain, enabling user to verify whether or not a transaction is a part of a block. In Fig. 1, transactions A and B are hashed through a SHA-256 mechanism into a corresponding Merkle node, producing Hash A and Hash B, and then again both of these are hashed consecutively into a parent node, Hash AB. Following on the same logic for transactions C and D, and hashing in the final step nodes Hash AB with Hash CD, we obtain the Merkle root of the tree, with the unique property that should any of the underlying hashes (or transactions) change, at any depth, the root also changes adequately. The Merkle tree validation process helps prevent malicious nodes from a blockchain from injecting fake or invalid transactions into the blockchain, while also helping prevent double-spend [13] and denial-of-service (BDoS [14]) attacks. Resilience to BDoS attacks is a critical security aspect for ensuring that data can be processed correctly and on-time, as part of a blockchain. In a BDoS attack, a node is becoming so busy processing new malicious blocks and their associated transactions, that is taken out of the functional list of nodes, sometimes with a rather low amount of resources required to make this happen, even for widely spread blockchains such as Bitcoin (in [12], authors outline how it would be possible with just 17% mining power to achieve this).

### 2.1 The CUDA Framework

The CUDA framework [15] was introduced back in 2007, and has opened up the possibility of offloading the CPU, to allow certain parallel processing to happen on the GPU directly. However, not all algorithms are parallelizable, and often finding adequate computational models can be a daunting task.

A warp in CUDA is a group of 32 threads, which is the minimum size of the data processed in SIMD (Single Instruction Multiple Data) fashion by a CUDA multiprocessor. The CUDA architecture works with blocks that can contain 64 to 512 threads, with every graphics card having an optimal number of threads per block, which often depends on the use-case. Blocks are organized into grids.

Parallel portions of an application are executed on the device (GPU) as kernels, with one kernel being executed at a time, while many threads may execute each kernel - this aspect being an essential part of the architecture required to achieve the highest throughputs therefore. Parallelism (and throughput, consequently) however greatly depends on several limiting factors, such as the number of CUDA cores, the memory bandwidth, the efficient use of memory and lack of data coalescence. A requirement in order to process the data, is to transfer it from RAM to video memory as a first step. This transfer is done nowadays through the use of the PCI Express architecture, whose theoretical throughput exceeds 64Gbps in the PCI Express 3.0 standard. Once the data is transferred in video memory, the CUDA parallel processing can begin through the use of kernels, with threads being scheduled internally. A data parallel approach usually works best with CUDA.

## 2.2 Challenges in Heterogeneous Implementations

Data on a GPU can only be processed from video memory (V-RAM), which requires a transfer of all necessary information from RAM to V-RAM. While the transfers are handled nowadays through the PCI Express architecture at high processing speeds, transferring many small data structures (such as individual nodes of a tree) could easily slow down the process due to the increased latencies involved between having to keep RAM and V-RAM synchronized. V-RAM is also generally smaller in size than RAM memory sizes as of today.

In addition, implementing trees usually involves internal memory fragmentation, as the classic/default routines of memory allocation for nodes will insert padding areas to cover for potential dynamic expansion or retraction of memory - and while that helps achieve faster run-time performance in the average-case scenarios on x86 Von Neumann machines, it does create a big discrepancy over expected memory usage and real memory usage. In particular, gaps are appearing between consecutive node allocations, which double or even triple the memory requirements often, creating an unnecessary memory overhead for high-performance computations. Eventually, such gaps can even double [16,17] the memory required to store such large trees.

While a CPU's job nowadays in a blockchain node varies greatly from processing data received from peers, to validating transactions, it is still desirable that the validation stage happens as fast as possible, with an immediate consequence being accommodation of as many transactions as possible in a short timespan. The Merkle tree helps prevent injection of false transactions into the blockchain, a common attack technique for attempting to tamper with the blockchain's integrity, while at the same time a common technique for denial-of-service/flood attacks (as mentioned before, primarily aimed at reducing the processing capacity of a node, through injection of fake transactions in one or more blocks, in order to undermine capacity or processing capabilities). The focus of our study therefore lies on accelerating the validation process, increasing the attack resilience of the blockchain nodes overall to such attacks, and study the approach's feasibility to real-world use-cases.

## 3 Implementation

Our proposed approach uses a memory compression algorithm, for a much more efficient storage of the Merkle tree in video memory, as well as for enhanced parallel processing. Throughout the blockchain world, a block consists of several transactions forged together. As an extension to that approach, we propose a memory model that can serialize multiple blocks, while keeping every block serialized as well. So individual blocks of transactions are serialized, after which blocks are concatenated in a serial manner just as well. Every node is compacted through the use of a child offset in the list of nodes, that points the location where the first child resides in the stack and the hash as an array of 256-bits, totaling only 36 bytes per node (detailed in Fig. 1) when using 4 bytes to store the number of nodes (allowing up to $2^{32}$ nodes per tree, which we consider to be sufficient). The first step represents constructing the Merkle tree in RAM, using a bottom-up approach (starting from the maximum depth until the root). This process nonetheless is very fast and efficient, taking only 136 ms for 8 million nodes in our experiments with real-world data, as outlined in the experimental section. After this, the memory-compact tree is computed according to the serialization approach below:

```
void serializeByDepth(merkleNode, depth)
{
    serializedNodes[pos].offset = lastIndexAt[depth+1]
    serializedNodes[pos].data = merkleNode.data
    old = lastIndexAt[depth+1]
    lastIndexAt[depth+1] += 2
    if (merkleNode.hasChild(left))
        serializeByDepth(merkleNode.leftChild, old);
    if (merkleNode.hasChild(right))
        compactMerkleTree(merkleNode.rightChild, old+1);
}

void processDepths(maxDepth)
{
        for (i = maxDepth-1, i > 0, i--)
          kernelCall <<lastIndexAt[i-1], lastIndexAt[i]>>
}
```

**Listing 1.1.** Memory-compact preprocessing of the Merkle tree

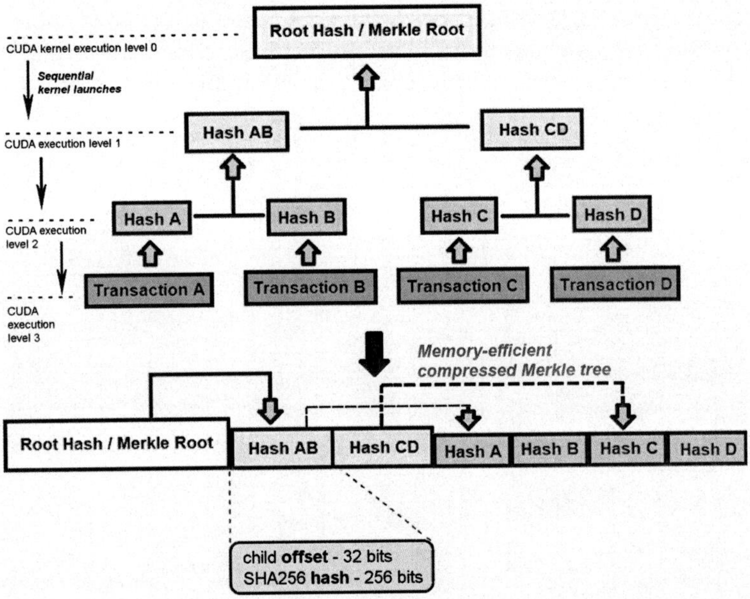


**Fig. 1.** Proposed methodology for processing the Merkle tree in a heterogeneous architecture.

The result is a stack of nodes which can be processed in a data-parallel manner by our proposed CUDA processing architecture. The child offset at every node indicates the starting offset (in the stack of nodes) of the two children of the binary Merkle tree, as the resulting compressed memory layout always places children sequentially in the compressed stack. Additionally, all nodes from a certain depth, are stored sequentially in the stack of nodes, a very important process that allows processing depths sequentially through consecutive kernel calls, with every individual depth being handled separately in a CUDA kernel. Another advantage of this approach is that storage requirements are always deterministic at $O(T \times ((2^N - 1) \times (P + |\Sigma|)))$, where $T$ is the total number of Merkle trees to be verified, $N$ is the average number of transactions in each Merkle tree being validated, $P$ is the size of a pointer in the architecture used (for x86, commonly 4 or 8 bytes) and $\Sigma$ is the alphabet used for the hashing methodology employed (for SHA256, this would be the entire ASCII spectrum, so $|\Sigma| = 256$).

The system architecture used for our approach to real-time tree processing is presented in Fig. 2. The initialization step ensures that data is processed real-time as it is coming through the blockchain's supporting backend, after which it is processed into the memory-compact storage model proposed. The next processing step involves moving the updated data into the video memory and initiating the kernel calls through individual, lightweight GPGPU threads, where the hash validation is performed for a single node through its siblings. Once validation is completed, the process is resumed with the next batch of available information. The validation process could be used as intermediary step towards blacklisting malicious nodes that are repeatedly sending invalid information, given a specific metric, to further enhance processing capabilities.

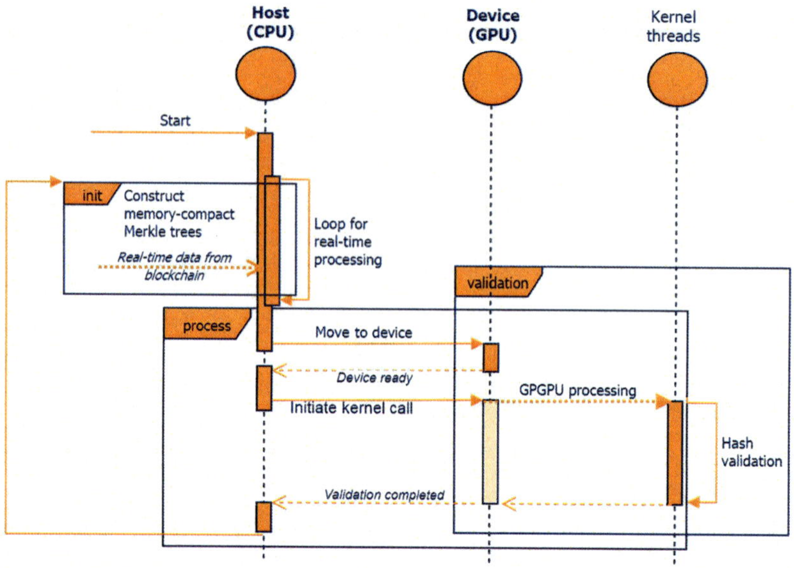


**Fig. 2.** Overall architecture workflow for our proposed heterogeneous approach.

The data parallel approach for processing nodes is exemplified with more details in Fig. 1. The biggest advantage of the approach is that, while it would be very time-consuming to move a big tree such as the ones used in this example, from RAM to V-RAM (in our experiments with real-world data, taking more than 2 h for 2 million Merkle tree nodes, when using individual node transfers through PCI Express), the new compact memory model can be transferred in one single burst, in a full-duplex manner, benefiting of the full bandwidth offered by the PCI Express architecture. At the same time, this represents a major step towards achieving real-time, polymorphic (dynamic) Merkle trees, as now entire sets of trees can be moved back and forth between RAM and V-RAM, with minimal latency, on a per-need basis, in simple bursts.

After one such Merkle tree is compacted as detailed, another serialized compacted tree follows it, with multiple such trees being concatenated in video memory, depending on how many blocks are being processed by the blockchain node. In practice, throughout the most popular blockchains with respect to their market cap, every tree corresponds to a block of transactions, with the transactions representing the last level (maximum depth) of the actual compacted tree (where sequential processing begins, as nodes at depth $N-1$ depend on the nodes found at depth $N$).

To allow validation of the tree in the GPU, as part of our heterogeneous processing architecture and model, we create and assign a single GPU thread to each specific node of the Merkle tree, for validation purposes, in a data parallel manner. Because of the way the compact tree is constructed, data parallel processing works very well with the CUDA capabilities of nVIDIA cards.

The kernel execution in CUDA, the core of the GPGPU architectures, finally ensures that a node can easily validate, or compute, the resulting SHA256 hash of its child nodes, and determine if a match is found or not. Because the process is very simple and has a bottom-up approach to compute the hashes, starting with the underlying transactions submitted in a block and ending with the root of the hash tree, the process is fast and straightforward, significantly reducing data coalescence issues usually introduced through internal memory fragmentation, and directly contributing to increased throughput overall in the heterogeneous architecture model proposed.

## 4 Experimental Results

We have performed our testing using an i7 6700HQ CPU, backed up by 64 GB of RAM, and for heterogeneous processing we have used an nVIDIA Quadro M1000M graphics card with 2 GB of DDR5 V-RAM. The throughputs for the CPU were measured on a single-core.

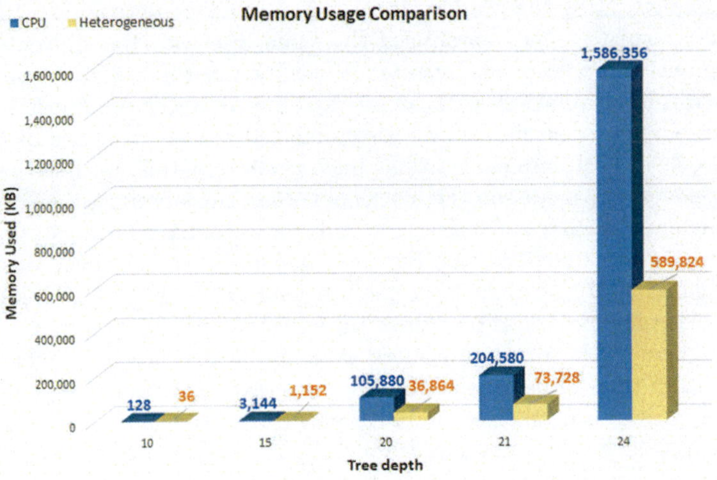


**Fig. 3.** Memory usage analysis of both the classic (CPU) processing model and our proposed heterogeneous (CPU + GPGPU) processing.

To perform the experiments, we have used real-world data from the current Bitcoin blockchain, with real transaction information, and created the corresponding Merkle trees for various different scenarios, each with a purposely high number of transaction. We started with a Merkle tree depth of 10 (totaling $2^{10} = 1024$ transactions), all the way up to a depth of 24 (totaling $2^{24} = 16.7$ million transactions). While they may not be achievable in practice at this very moment by most public blockchains (such as Bitcoin or Ethereum), there are numerous others that do claim a very high number of transactions - and we fully expect this side of the blockchain technology to mature even further in the years to come.

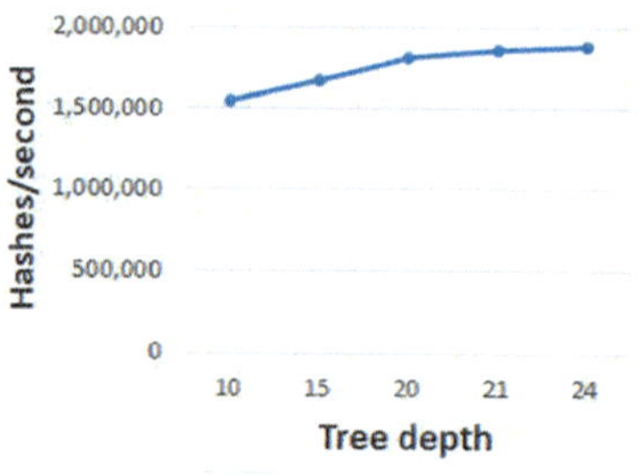


Fig. 4. CPU processing.

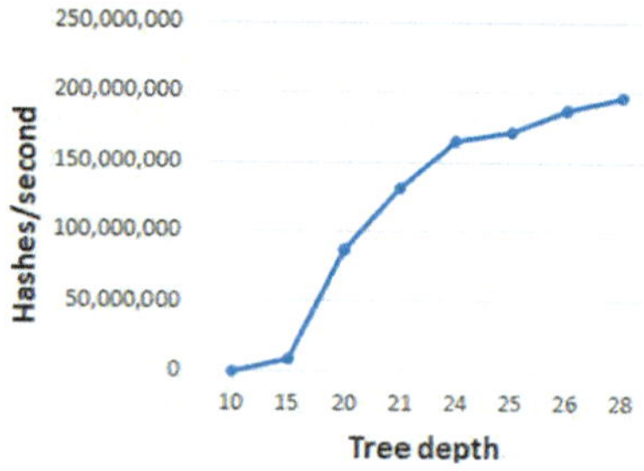


Fig. 5. Heterogeneous processing.

We tested the efficiency of the storage model proposed, as that is heavily affecting performance throughput and storage requirements. The first test was done by implementing the Merkle tree using the classic approach, on the CPU, while the second test had reproduced the same tree with the efficient storage model we outlined before. The results are outlined in Fig. 3 and present significant storage improvements, with an average improvement of almost 3× lower storage in our proposed model. We also analyzed the performance of the proposed model, in both implementations once again. For this, we have employed the execution of our model on the blocks of transactions proposed. We analyzed the throughput, in SHA256 hashes/second, for each model, and outlined the results in Fig. 4 for the CPU, and Fig. 5 for our proposed heterogeneous model. At a rate of 312,000 tx/day in the Bitcoin network as seen on the 22nd of April 2020 (this number varies, often times by a significant margin), our proposed model would be able to validate more than 21 months worth of transactions (totaling more than 92,000 blocks, at Bitcoin's rate of 1 block every 10 min) in the Bitcoin network in one second, with an average consumer GPU. In Fig. 6, the resulting storage efficiency speed-ups range an average of 2.9× less storage in our proposed heterogeneous model (with a more asymptotic behavior rounding up at about 2.7× less storage for higher tree depths).

The run-time performance speed-ups are in line with the GPGPU bandwidth saturation observation made earlier. At the same time, Fig. 7 outlines that the GPU bandwidth begins to saturate only for much higher tree depths, for example at about a tree depth of 28. In essence, while it underperforms for low tree depths of up to around 14 (making it less suitable for a low number of transactions per second, and therefore potentially less feasible), it only achieves its true potential on a high-number of transactions. The advantage of the model however lies in the fact that with our approach, multiple modified Merkle trees can be compacted without internal memory fragmentation issues, stacked one upon each other without external memory fragmentation gaps, therefore making a parallel computation not only possible, but desirable.

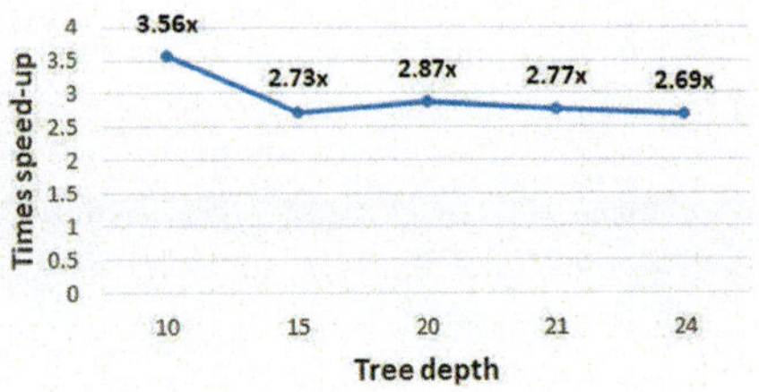


**Fig. 6.** Effective storage speed-up (Heterogeneous vs CPU).

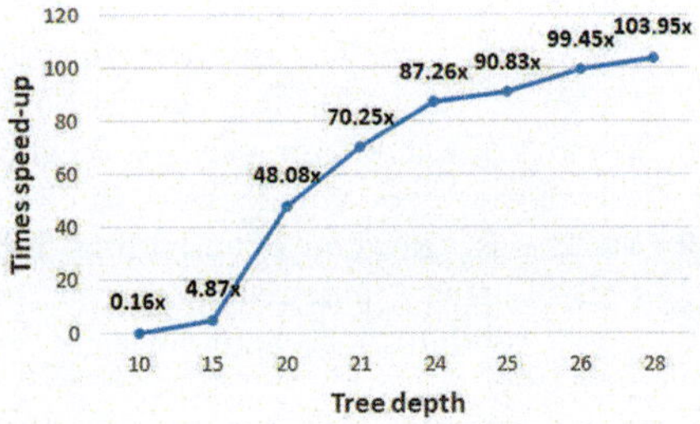


**Fig. 7.** Run-time performance speed-up (Heterogeneous vs CPU).

## 5 Conclusion

We have outlined a new and efficient heterogeneous computational model for performing faster Merkle tree validation, an essential process for security validation purposes and transaction processing in multiple popular blockchains nowadays. Through the use of an efficient storage model, which ensures full-duplex transfers between RAM and V-RAM at full throughput speeds offered by the PCI-Express architecture, alongside an enhanced processing model for computation which allows concurrent processing of multiple Merkle trees in-memory directly, as well as a real-time approach to processing transaction validation, we have provided a way to process Merkle trees in video memory directly, benefiting of the massive computational capabilities offered by GPGPU computing, and paving the way to real-time Merkle tree processing in heterogeneous architectures overall.

**Acknowledgment.** This work was supported by the West University of Timişoara, BloxHIVE [18] and MOISE project number 240/2020, ID POC/398/1/1, financed by EU and Romanian government. The views expressed in this paper do not necessarily reflect those of the corresponding project's consortium members.

## References

1. Nakamoto, S.: Bitcoin whitepaper (2009). https://bitcoin.org/bitcoin.pdf
2. CoinMarketCap. All Cryptocurrencies. https://coinmarketcap.com/all/views/all/. Accessed 15 Feb 2019

3. Zile, K., Renate, S.: Blockchain use cases and their feasibility. Appl. Comput. Syst. **23**(1), 12–20 (2018)
4. Holbl, M., et al.: A systematic review of the use of blockchain in healthcare. Symmetry **10**(10), 470 (2018)
5. Ko, T., et al.: Blockchain technology and manufacturing industry: real-time transparency and cost savings. Sustainability **10**, 4274 (2018)
6. PwC Report: Blockchain: the next innovation to make our cities smarter (2018). https://www.pwc.in/assets/pdfs/publications/2018/blockchain-the-next-innovation-to-make-our-cities-smarter.pdf
7. DH Network: Blockchain for the humanitarian sector: future opportunities (2016)
8. Wood, G.: Ethereum: a secure decentralized generalised transaction ledger (2015). https://gavwood.com/paper.pdf
9. Merkle, R.: Method of providing digital signatures. Patent, Patent US4309569A (1979). https://patents.google.com/patent/US4309569
10. EOS: EOSIO: blockchain software architecture. https://eos.io/. Accessed 12 Feb 2020
11. TRON: TRON—Decentralize the Web. https://tron.network/. Accessed 12 Feb 2020
12. Mirkin, M., et al.: https://arxiv.org/ftp/arxiv/papers/1912/1912.07497.pdf. Accessed 23 Apr 2020
13. Double-spending: Double-spending digital currency. https://en.wikipedia.org/wiki/Double-spending. Accessed 12 Feb 2020
14. Blockchain Denial of Service: BDoS: Blockchain Denial of Service. https://medium.com/coinmonks/bdos-blockchain-denial-of-service-385c8c56b401. Accessed 12 Feb 2020
15. NVIDIA: NVIDIA CUDA Compute Unified Device Architecture Programming Guide, version 4.1. http://developer.download.nvidia.com/compute/DevZone/docs/html/C/doc/CUDA_C_Programming_Guide.pdf
16. Pungila, C., Negru, V.: A highly-efficient memory-compression approach for GPU-accelerated virus signature matching. In: Information Security Conference (ISC) (2012)
17. Pungila, C., Negru, V.: Real-time polymorphic Aho-Corasick automata for heterogeneous malicious code detection. In: Herrero, Á., et al. (eds.) International Joint Conference SOCO'13-CISIS'13-ICEUTE'13. Advances in Intelligent Systems and Computing, vol. 239, pp. 439–448. Springer, Cham (2014). https://doi.org/10.1007/978-3-319-01854-6_45
18. BloxHIVE: The BloxHIVE Project. https://bloxhive.com

# Estimating Re-identification Risk by Means of Formal Conceptualization

Gonzalo A. Aranda-Corral[1(✉)], Joaquín Borrego-Díaz[2], and Juan Galán-Páez[2]

[1] Department of Information Technology, Universidad de Huelva, Huelva, Spain
gonzalo.aranda@dti.uhu.es

[2] Department of Computer Science and Artificial Intelligence, Universidad de Sevilla, Sevilla, Spain
{jborrego,juangalan}@us.es

**Abstract.** Risk-based methodologies for de-identification provide solutions to ensure privacy. These are based on the availability of sound metrics to estimate the risk of re-identification. Two issues associated with classical risk estimation are, on the one hand, the adequacy of the metric and, on the other hand, its static nature -that is, any change in the database to reduce the risk could imply recomputing the metrics, for example, by removing compromised data (data with a high probability of re-identification). This paper presents a semantic-based risk estimation -by means of Formal Concept Analysis- that allows to estimate *a priori* the risk of compromised data deletion.

**Keywords:** Privacy · Re-identification · Formal concept analysis · Risk metrics

## 1 Introduction

The anonymization (or de-identification) is the problem of avoiding individual identification from databases [7]. As opposed to re-identification, that concerns manipulating databases, in order to determine the identity of individuals whose information appears within a de-identified database. Both problems deals with taking into account the purpose of anonymization (mainly for data sharing) as well as the nature of agents that will attempt the re-identification [7]. The attacking agents range from a particular scientific group to anyone with access to Open Data (e.g. open health data about COVID-19 pandemic). Different security scenarios induce several specific types of attacks against the de-identified dataset [8]. Also, it allows illustrating how to manage other privacy concerns [14].

The privacy issue has been a major topic of study in Healthcare, from which Data Science has inherited a number of good practices, guides and methodologies. Among them, are those addressing the re-identification attack problem [8]. Considering that, as a general rule, the sharing of an anonymized dataset does not need the express consent of the individuals represented, de-identification has


J. J. Gude Prego et al. (Eds.): CISIS 2021/ICEUTE 2021, AISC 1400, pp. 13–22, 2022.
https://doi.org/10.1007/978-3-030-87872-6_2

become a requirement for sharing information in sensitive fields such as the aforementioned area or within Security, and the potential interest in any Engineering area [4]. To this practical reason, other scientific (reproducibility of experiments), ethical, transparency policy, reusability, etc. reasons should be added. Returning to the case of Healthcare, the Safe Harbor (SH) method [13] provides a first approach to the problem, detailing which fields should be eliminated or anonymized through pseudonyms. Moreover, other standards for de-identification have been published [2,9,11].

The problem of attribute disclosure -that is, the attacker learns that an attribute or set of attributes provides him with information-rich enough to know who one of the individuals is- could be the main problem in the area. Those attributes are not necessarily some of the ones listed in SH. For this problem, it persists the need to consider different techniques, according to the Closed World scenario (in which only a restricted set of agents have access and there is little or no linkability of the data to other external data) or the Open World one. In this second scenario, data is linkable to external data, or the set of agents with access is very large or open to anyone). For the latter case, other techniques can -and should- enhance the traditional ones, such as Differential Privacy [5,6].

The fundamental requirement for any Risk-Based Methodology (RBM) is the need for risk quantification. That is, a metric for estimating the danger of re-identification, and how good is the solution offered [7]. In this way, the RBM will try, on the one hand, to keep a small risk value (such value is relative to the working scenario) by considering sound thresholds. On the other hand, the techniques used to reduce risk should not cause an unacceptable loss of information that would make the dataset unusable. Among these techniques are those applied to variables or variable sets that are direct identifiers (e.g. masking or elimination of variables) or to others for which, if some combination of values is known, re-identification is very likely (quasi-identifiers). Finding such kind of variables represents the first stage in RBM. This paper focuses on such a task, proposing a cognitive and semantic technique where it aims to propose new risk metrics.

The structure of the paper is as follows. The next section recalls the fundamental elements of both, risk-based metrics and Formal Concept Analysis. Section 3 is devoted to establish the formal framework that will allow introducing the FCA-based, risk-based metrics (Sect. 3.2). These are semantic in nature (related to the concepts mined from the dataset). The metrics also allow estimating a priori the accumulated risk, by removing potentially re-identifying entries from the dataset, based on their impact on the concept structure (Sect. 3.3).

## 2 Background

### 2.1 Risk-Based Methodologies for Preserving Privacy

An example, from [12] (shown in Fig. 1), has been chosen in order to illustrate the RBM concepts introduced in the paper. In this table, a direct identifier, the

| | Quasi-identifiers | | Other Variables | |
|---|---|---|---|---|
| ID | Sex | Year of Birth | Lab Test | Lab Result |
| 1 | Male | 1959 | Albumin, Serum | 4.8 |
| 2 | Male | 1969 | Creatine Kinase | 86 |
| 3 | Female | 1955 | Alkaline Phosphatase | 66 |
| 4 | Male | 1959 | Bilirubin | Negative |
| 5 | Female | 1942 | BUN/Creatinine Ratio | 17 |
| 6 | Female | 1975 | Calcium, Serum | 9.2 |
| 7 | Female | 1966 | Free Thyroxine Index | 2.7 |
| 8 | Female | 1987 | Globulin, Total | 3.5 |
| 9 | Male | 1959 | B-type Natriuretic Peptide | 134.1 |
| 10 | Male | 1967 | Creatine Kinase | 80 |
| 11 | Male | 1968 | Alanine Aminotransferase | 24 |
| 12 | Female | 1955 | Cancer Antigen 125 | 86 |
| 13 | Male | 1967 | Creatine Kinase | 327 |
| 14 | Male | 1967 | Creatine Kinase | 82 |
| 15 | Female | 1966 | Creatinine | 0.78 |
| 16 | Female | 1955 | Triglycerides | 147 |
| 17 | Male | 1967 | Creatine Kinase | 73 |
| 18 | Female | 1956 | Monocytes | 12 |
| 19 | Female | 1956 | HDL Cholesterol | 68 |
| 20 | Male | 1978 | Neutrophils | 83 |
| 21 | Female | 1966 | Prothrombin Time | 16.9 |
| 22 | Male | 1967 | Creatine Kinase | 68 |
| 23 | Male | 1971 | White Blood Cell Count | 13.0 |
| 24 | Female | 1954 | Hemoglobin | 14.8 |
| 25 | Female | 1977 | Lipase, Serum | 37 |
| 26 | Male | 1944 | Cholesterol, Total | 147 |
| 27 | Male | 1965 | Hematocrit | 45.3 |

NOTE: BUN = blood urea nitrogen; HDL = high-density lipoprotein

| | Quasi-identifiers | | Probability of Re-identification |
|---|---|---|---|
| ID | Sex | Decade of Birth | |
| 1 | Male | 1950-1959 | 0.33 |
| 2 | Male | 1960-1969 | 0.125 |
| 3 | Female | 1950-1959 | 0.167 |
| 4 | Male | 1950-1959 | 0.33 |
| 5 | Female | 1940-1949 | 1 |
| 6 | Female | 1970-1979 | 0.33 |
| 7 | Female | 1960-1969 | 0.33 |
| 8 | Female | 1980-1989 | 1 |
| 9 | Male | 1950-1959 | 0.33 |
| 10 | Male | 1960-1969 | 0.125 |
| 11 | Male | 1960-1969 | 0.125 |
| 12 | Female | 1950-1959 | 0.167 |
| 13 | Male | 1960-1969 | 0.125 |
| 14 | Male | 1960-1969 | 0.125 |
| 15 | Female | 1960-1969 | 0.33 |
| 16 | Female | 1950-1959 | 0.167 |
| 17 | Male | 1960-1969 | 0.125 |
| 18 | Female | 1950-1959 | 0.167 |
| 19 | Female | 1950-1959 | 0.167 |
| 20 | Male | 1970-1979 | 1 |
| 21 | Female | 1960-1969 | 0.33 |
| 22 | Male | 1960-1969 | 0.125 |
| 23 | Male | 1970-1979 | 0.33 |
| 24 | Female | 1950-1959 | 0.167 |
| 25 | Female | 1970-1979 | 0.33 |
| 26 | Male | 1940-1949 | 1 |
| 27 | Male | 1960-1969 | 0.125 |

**Fig. 1.** Example from [12]. The application will focus on these two quasi-identifiers.

patient's name, has already been masked by an ID variable whose correspondence is securely stored by the dataset sponsor. Please note that the set of IDs $\{10, 13, 14, 17, 22\}$ are the individuals sharing the values $\{Male, 1967\}$ for the quasi-identifier $\{Sex, Year-of-Birth\}$. This type of dataset is referred to as equivalence class [7]. The variable $Year-of-Birth$ can be modified to lose precision, augmenting such a set. For example, by taking decades as the time scale. Thus the equivalence class for the modified quasi-identifier $\{Male, 1960-1959\}$ increases: $\{2, 10, 11, 13, 14, 17, 22, 27\}$.

**Definition 1.** *Let $D$ be a dataset and $Q = \{v_1, \ldots, v_k\}$ a variable subset (the quasi-identifier) of $D$.*

- *The* **re-identification risk** *of an entry (individual) $e$ with respect to a set of values $v_1 = q_1, \ldots, v_k = q_k$ is defined as*

$$r(e, q_1, \ldots, q_k) = \frac{1}{s(q_1, \ldots, q_k)}$$

*where $s(q_1, \ldots, q_k)$ is the number of individuals in the dataset that share such values.*

- *An entry is a* **unique record** *with respect to $A$ if there exist values $v_1 = q_1, \ldots, v_k = q_k$ such that $r(e, q_1, \ldots, q_k) = 1$*

For example, $r(14, sex, Year-of-birth) = 0.2$. A unique record has maximum re-identification risk. If the attacker knew the values of these variables, then he or she could extract the remaining information. For example, in the dataset of Fig. 1, if the attacking agent knows the identity of the man who was born in 1965, then attacker may also know the result of his clinical results since $r(27, Male, 1965) = 1$.

Risk metrics are useful for estimating risk when the adversary knows someone in the real world and is trying to find his entry in the dataset, or has chosen a record from the dataset and is trying to find out the identity of the individual [7,12].

**Definition 2.** *Under the above conditions,*

- *The maximum risk of $D$ with respect to $q_1, \ldots, q_k$ is:*

$$r_M(D, v_1, \ldots, v_k) := \max\{r(e, q_1, \ldots, q_k) \ : \ e \in D, \ q_i \ \textit{value for} \ v_i, 1 \leq i \leq k\}$$

- *The average risk of $D$ with respect to $q_1, \ldots, q_k$, $r_m(D, q_1, \ldots, q_k)$ is the average risk of the records,*

$$r_m(D, v_1, \ldots, v_k) = \mathbb{E}[r(., q_1, \ldots, q_k)]$$

For the running example, it has $r_M(D, Sex, Year - of - Birth) = 1$, whilst $r_m(D, Sex, Year - of - Birth) = 0.59$.

It is usual to select the maximum risk when the data sponsor aims to share data within an open environment (Open Data). The reason is that it is reasonable to think that the adversary will focus on the records of maximum risk. The opposite case is when the access to data is restricted, and under contract is forbidden the re-identification. In this case, medium risk is selected. The reason is that the problem will be that inadvertent re-identification may occur. Moreover, unique records represent an unacceptable risk. In general terms, an average risk greater than 0.5 is unacceptable. The deletion of unique records relieves both risks. For example, after applying unique records suppression in the example of Fig. 1, a database $D'$ with $r_m(D') = 0.33$ would be obtained. It is generally considered acceptable if the minimum size of the equivalence classes $s(e, q_1, \ldots, q_k)$ is 5 (i.e., $r_M(D, v_1, \ldots, v_k) = 0.2$, although for some studies higher risks are adopted, such as 0.33 (used by US CDC in HIV studies). There are even some cases where much larger sizes are used (such as 20).

## 2.2 Formal Concept Analysis

The Formal Concept Analysis (FCA) [10] mathematizes the philosophical understanding of a *concept* as a unit of thought, comprising its extent and its intent. The extent covers all objects belonging to the concept, and the intent comprises all common attributes valid for all the objects under consideration. FCA provides the extraction and analysis of concepts from data. FCA provides algorithms to extract, from data, all units of knowledge with meaning in the sense of the above-mentioned concept notion, as well as it also allows the computation of concept hierarchies from data tables.

The information format used in FCA is organized in an object–attribute table called *Formal Context*. It is a three elements set $\mathbb{K} = (G, M, I)$, where $G$ is a non-empty set of *objects* (events), $M$ is a non-empty set of *attributes*, and $I \subseteq G \times M$ is a (object–attribute) binary relation. Figure 2 (left) shows a

| Formula 1 | River | Coast | Sea |
|---|---|---|---|
| Carp | × | | |
| Escatofagus | × | × | |
| Bream | | × | × |
| Sparus | | × | × |
| Eel | × | × | × |

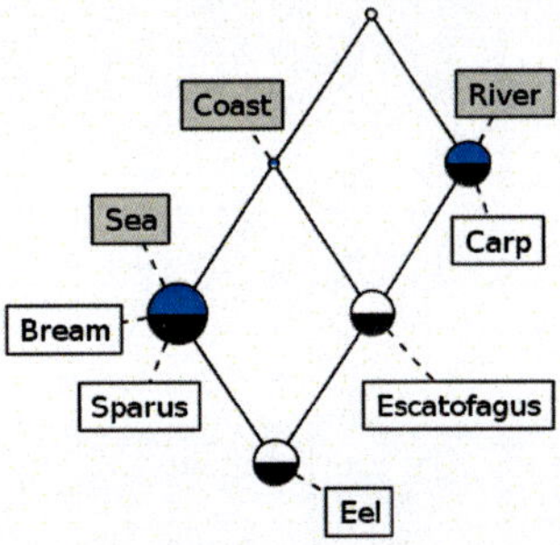


**Fig. 2.** Formal context on fishes, and its associated concept lattice

formal context describing fishes (objects) living on different aquatic ecosystems (attributes) is shown.

A formal context $\mathbb{K} = (G, M, I)$ induces a pair of operators, which we will call both *derivation operators*. Given $A \subseteq G$ and $B \subseteq M$, they are defined by

$$A' = \{m \in M \mid (g, m) \in I \text{ for all } g \in A\}$$

(that is, the set of attributes shared by all the objects in $A$), and reciprocally

$$B' = \{g \in G \mid (g, m) \in I \text{ for all } m \in B\}$$

(the objects that have all the attributes of $B$). In terms of FCA, and identifying the variable values with attributes, the extent of the quasi-identifier would correspond to its extension.

**Proposition 1.** *Both derivation operators on formal contexts enjoy the following properties:*

1. $X_1 \subseteq X_2 \Longrightarrow X_2' \subseteq X_1'$
2. $X \subseteq X''$
3. $X' = X'''$

**Definition 3.** *A* **formal concept** *is a pair $C = (A, B)$ of object and attribute sets (called the extent, $Ext(C) = A$, and the intent, $Int(C) = B$ of the concept, respectively) verifying that $A' = B$ and $B' = A$.*

By using the properties described in Proposition 1, it is possible to characterize both, the smallest concept containing a given set of objects and the largest concept containing, in its intent, a given set of attributes:

**Proposition 2.** *Given $\mathbb{K} = (G, M, I)$,*

- *The smallest concept that contains $X \subseteq G$ is $(X'', X')$*
- *The biggest concept that contains $Y \subseteq M$ is $(Y', Y'')$*

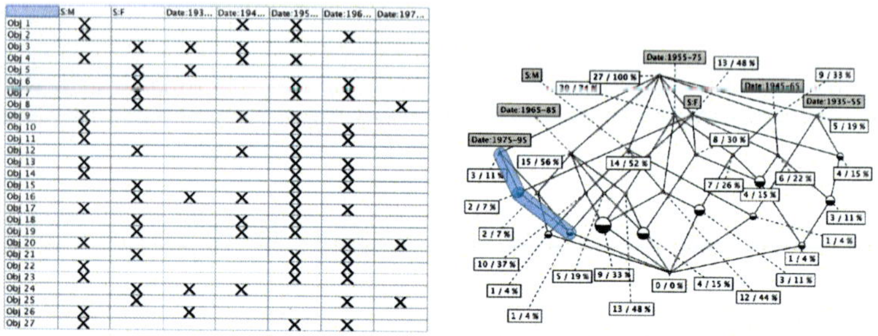

**Fig. 3.** Transformation of the dataset from Fig. 1 in a formal context by means of scales, obtaining $\{S : M, S : F, Date : 1935 - 55, \ldots Date : 1965 - 85\}$

$Con(g)$ denotes the smallest concept containing $g \in G$. That is, $Con(g) := (\{g\}'', \{g\}')$.

The set of concepts of $\mathbb{K}$ a context, $\mathfrak{B}(\mathbb{K})$, can be endowed with the mathematical structure of lattice, by means of the *subconcept* relationship, $\leq$. For example, the Hasse diagram of the concept lattice associated with the formal context on fishes of Fig. 2 left, is shown in Fig. 2, right. It actually represent the Hasse diagram (a graph representing a partial order from bottom to top) induced by the partial order relation among concepts, where $C_1 \subset C_2$ *and there is not any intermediate concept* (which will be denoted by $C_1 \prec C_2$).

In this representation each node is a concept, and its intent (extent resp.) is formed by the set of attributes (objects resp.) included along the path to the top (bottom resp.) concept. For example, the bottom concept

$$(\{eel\}, \{Coast, Sea, River\})$$

is a concept that could be interpreted as *euryhaline-fish* (this is not a term of the language represented by attribute set, is something new). This is an example of how FCA can be used as knowledge discovering tool.

## 3 Estimating Risk Using Formal Concepts

A well-known way to transform a dataset $D$ into a formal context $\mathbb{K}$ is by employing scales [3,10]. Scales specify values or value ranges for the dataset variables, transforming a multivalued variable into an attribute (i.e., a Boolean attribute). To transform the example into a more interesting one, we have chosen the so-called nominal scale for the variable $Sex$ and another one composed of (non-disjoint) time intervals for $Year - of - Birth$. In this way, the risk may increase, since the value for an entry belongs to two different intervals, thus more information is available. The result is shown in Fig. 3, left. The object in the formal context $\mathbb{K}$ corresponding to a record $e \in D$ will be denoted by $e^{\mathbb{K}}$. The concept lattice of the formal context appears in Fig. 3, right.

Please note that when working with the formal context, the set of individuals sharing a given set of values has been endowed with semantic meaning and is

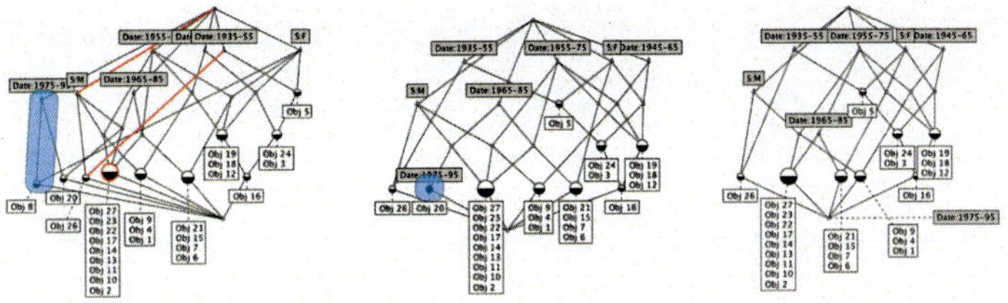

**Fig. 4.** Effect of sequential record suppression from identifier chain

related to others, as shown in the concept lattice. The number $s(e^{\mathbb{K}}, q_1, \ldots, q_k)$ is the size of the big concept that contains attributes $\{q_1, \ldots, q_k\}$.

## 3.1 Semantic Interpretation of Risk Metrics

In order to illustrate our proposal for semantics-based risk metric, the simplest case will be introduced first, that is, unique records. With respect to FCA, this would imply that there exists a concept with extent size 1. In other words, there exists $e^{\mathbb{K}} \in G$ such that $r_M(e^{\mathbb{K}}) = 1$. Note that this means $|\{e^{\mathbb{K}}\}''| = 1$. But also -and this is a new feature that the classic metric does not possess- one can also determine, in the same concept lattice, the effect of eliminating the unique record can produce other unique records.

**Definition 4.** *Let $\mathbb{K}$ be the formal context associated to the database $D$. A chain $\mathcal{C} = C_1 \prec C_2 \prec \cdots C_k$ with $C_j = (X_j, Y_j) \in \mathfrak{B}(\mathbb{K})$ $(1 \leq j \leq k)$ is called a* **identifier chain** *if $|X_1| = 1$ and $X_{j+1} \backslash X_j| = 1$ for any $j < k$.*

In Fig. 1, right, a maximal identifier chain is highlighted in blue. The interest of detecting this type of chain lies in that it allows predict the effect on the risk metric of deleting an unique record (it can create another one, the next concept in the chain). Following the example, an identifier chain will be:

$$
\begin{aligned}
&(\{Obj25\}, \{Date: 1975-95, Date: 1965-85, S: F\}) \prec \\
&\qquad \prec (\{Obj25, Obj28\}, \{Date: 1975-95, S: F\}) \prec \\
&\qquad \prec (\{Obj25, Obj28, Obj20\}, \{Date: 1975-95\})
\end{aligned}
$$

The effect of the sequential suppression of unique records on the concept lattice, showing the apparition of new unique records, is shown in Fig. 4.

Along the rest of the paper, a threshold $u$ for risk metric is supposed to be fixed. The generalization of the identifier chain notion is the following. The definition expresses the idea that eliminating a concept with high risk will cause a risk increment to the next concept of the chain.

**Definition 5.** *Let $\mathbb{K}$ be the formal context associated to the database $D$.*

- *A chain* $C = C_1 \prec C_2 \prec \cdots C_k$ *with* $C_j - (X_j, Y_j)\mathfrak{B}(\mathbb{K})$ $(1 \leq j \leq k)$ *is called a* chain with identification risk *if:*
  - $|X_1| \leq 1/u$
  - *For any* $j \leq k$ $|X_{j+1} \backslash X_j| \leq 1/u$
- *The risk associated with the chain* $C$, $r_C : \bigcup X_i \to \mathbb{R}$ *is defined as*

$$r_C(e^{\mathbb{K}}) = \begin{cases} \frac{1}{|X_1|} & \textit{if } e^{\mathbb{K}} \in X_1 \\ \frac{1}{|X_i \backslash X_{i-1}|} & \textit{if } e^{\mathbb{K}} \in X_i \backslash X_{i-1} \end{cases}$$

### 3.2 Semantic Risk

It is feasible to define a new risk on the concepts by recursion on $\prec$, without dependence on a particular chain. To simplify the notation, the quasi-identifiers are omitted in the definitions.

**Definition 6.** *Let $\mathbb{K}$ be the formal context associated to the database $D$. Let $e$ be a record of $D$. The* **basic semantic risk** *of $e$ is defined by*

$$r_{BS}(e, \mathbb{K}) = \max\{\frac{1}{|\{e^{\mathbb{K}}\}'' \backslash ext(C)|} \ : \ C \in K, \ C \prec Con(e^{\mathbb{K}})\}$$

The basic semantic risk estimates the recalculated risk under the assumption that there has been a deletion of records at risk on some subconcept of $Con(e^{\mathbb{K}})$. Please note that the basic semantic risk depends on $Con(e^{\mathbb{K}})$:

**Lemma 1.** $\{e_1^{\mathbb{K}}\}'' = \{e_2^{\mathbb{K}}\}'' \Longrightarrow r_{BS}(e_1^{\mathbb{K}}, \mathbb{K}) = r_{BS}(e_2^{\mathbb{K}}, \mathbb{K})$

The following metric picks up the idea of not assuming the suppression of the elements of a subconcept but using the probability of the occurrence of that event. It is also well defined on the partial order $\prec$.

**Definition 7.** *The semantic risk of $e$ is defined by* $r_S(e, K) := r_L(Con(e^{\mathbb{K}}))$ *where* $r_S(.)$ *defined by*

$$r_L(C, \mathbb{K}) = \begin{cases} \frac{1}{|ext(C)|} & \textit{if } \{e^{\mathbb{K}}\}'' \textit{ has not non-empy subconcepts} \\ \max\{\frac{1}{|ext(C) \backslash ext(D)|} \cdot r_L(D, \mathbb{K}) \ : \ D \prec C\} & \textit{in other case} \end{cases}$$

### 3.3 Semantic Risk Management

To recap, the procedure proposed in this paper for risk estimation extends the classic one by carrying the following tasks:

1. The risk associated with a particular record is computed if one wants to study specific individuals. The concepts containing such records at risk are located.

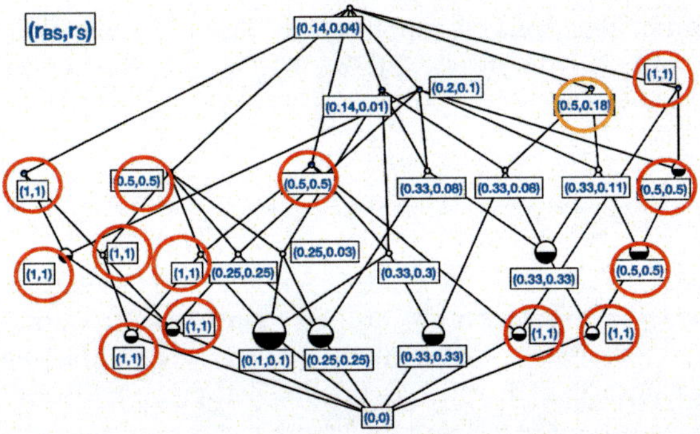


**Fig. 5.** Basic semantic risk and semantic risk for the example with threshold $u = 0.35$, the records belonging to highlighted concepts have to be discarded. Note that such discarding does not affect to risks of superconcepts. Concept highlighted in orange has $r_{BS} > u$. However, $r_S < u$ indicates that it is very unlikely that records located in that concept and not in the subconcepts are reidentifiable, so it would be acceptable.

2. The elimination process is analyzed by simulating the effect on the lattice. The possible strategies for dealing with the records involved, are also studied.
3. Finally, if it is not desirable to eliminate particular individuals at risk, noise (e.g. using Differential Privacy) should be introduced in the values of the original dataset and go back to step 1.

The contexts of Fig. 4 has been simulated using the tool Conexp[1]. It has been centered on the chain discussed above. It has recomputed the lattice after each element deletion. Fig. 5 shows both, the basic semantic risk and the semantic risk for the running example.

## 4 Conclusions and Future Work

In this paper, we propose a risk metric based on the formal concept consideration of value sets of quasi-identifiers. It enriches the basic definition of risk metric by exploiting the lattice structure of the concepts associated with different sets of variables. In this way, new data at risk can be detected by exploring the lattice. Moreover, the impact on the risk of eliminating records can be estimated. Other approaches address the problem from statistical modeling (such as the one based on copulas [15]), These approaches differ from our proposal in that ours is semantic in nature. The proposed approach is concerned with identifying the quasi-identifiers that put privacy at risk and assessing the knock-on effect of eliminating such risk.

The future work will be devoted to change the use of the concept lattice by association rules, for which bases can be obtained [10]. The interest lies in the fact that we have efficient reasoning methods for association rules that allow us to reduce the number of attributes to focus the reasoning on a particular set of quasi-identifiers [1].

[1] http://conexp.sourceforge.net/.

**Acknowledgments.** This work is supported by State Investigation Agency (Agencia Estatal de Investigación), project PID2019-109152GB-I00/AEI/10.13039/501100011033 and Universidad de Huelva, project UHU-1266216.

## References

1. Aranda-Corral, G.A., Borrego-Díaz, J., Galán-Páez, J., Caballero, A.T.: On experimental efficiency for retraction operator to stem basis. In: Cornejo, M.E., Kóczy, L.T., Medina, J., De Barros Ruano, A.E. (eds.) Trends in Mathematics and Computational Intelligence. SCI, vol. 796, pp. 73–79. Springer, Cham (2019). https://doi.org/10.1007/978-3-030-00485-9_8
2. Canadian Institute for Health Information: Best practice guidelines for managing the disclosure of de-identified health information. Technical report, Canadian Institute for Health Information (2011)
3. Cole, R., Eklund, P., Walker, D.: Constructing conceptual scales in formal concept analysis. In: Wu, X., Kotagiri, R., Korb, K.B. (eds.) PAKDD 1998. LNCS, vol. 1394, pp. 378–379. Springer, Heidelberg (1998). https://doi.org/10.1007/3-540-64383-4_33
4. Dodds, L., L'Hénaff, P., Maddison, J., Yates, D.: A manifesto for increasing access to data in engineering. Data-Cent. Eng. **1**, 11 (2020). https://doi.org/10.1017/dce.2020.3
5. Dwork, C.: A firm foundation for private data analysis. Commun. ACM **54**(1), 86–95 (2011)
6. Dwork, C., McSherry, F., Nissim, K., Smith, A.: Calibrating noise to sensitivity in private data analysis. J. Priv. Confident. **7**(3), 17–51 (2017)
7. El Emam, K., Arbuckle, L.: Anonymizing Health Data: Case Studies and Methods to Get You Started, 1st edn. O'Reilly Media, Inc. (2013)
8. El Emam, K., Jonker, E., Arbuckle, L., Malin, B.: A systematic review of re-identification attacks on health data. PLoS One **6**(12), 1–12 (2011)
9. Federal Committee on Statistical Methodology: Report on Statistical Disclosure Limitation Methodology. Technical report 12, Federal Committee on Statistical Methodology (2005)
10. Ganter, B., Wille, R.: Formal Concept Analysis: Mathematical Foundations, 1st edn. Springer, New York (1997)
11. ICO: Anonymisation: managing data protection risk code of practice. Technical report, UK's Information Commissioner's Office (2012)
12. Institute of Medicine: Sharing Clinical Trial Data: Maximizing Benefits. Minimizing Risk. The National Academies Press, Washington, DC (2015)
13. Office for Civil Rights: Guidance Regarding Methods for De-identification of Protected Health Information in Accordance with the Health Insurance Portability and Accountability Act (HIPAA) Privacy Rule. Technical report, U.S. Department of Health & Human Services, Washington, DC (2012)
14. Ribeiro-Navarrete, S., Saura, J.R., Palacios-Marqués, D.: Towards a new era of mass data collection: Assessing pandemic surveillance technologies to preserve user privacy. Technol. Forecast. Soc. Change **167**, 120,681 (2021)
15. Rocher, L., Hendrickx, J., de Montjoye, Y.A.: Estimating the success of re-identifications in incomplete datasets using generative models. Nat. Commun. **10**, 1–9 (2019)

# Beta Hebbian Learning for Intrusion Detection in Networks of IoT Devices

Álvaro Michelena Grandío[1], María Teresa García Ordás[2], David Yeregui Marcos del Blanco[2], José Aveleira-Mata[2], Francisco Zayas-Gato[1], Esteban Jove[1], José-Luis Casteleiro-Roca[1], Héctor Quintián[1](✉), Héctor Alaiz-Moretón[2], and José Luis Calvo-Rolle[1]

[1] Department of Industrial Engineering, University of A Coruña, CTC, CITIC, Ferrol, A Coruña, Spain
{alvaro.michelena,f.zayas.gato,esteban.jove,jose.luis.casteleiro, hector.quintian,jlcalvo}@udc.es

[2] Department of Electrical and Systems Engineering, University of León, León, Spain
{m.garcia.ordas,dmarcb01,jose.aveleira,hector.moreton}@unileon.com

**Abstract.** This research paper is focused on security in IoT devices network, by providing a visual tool based on Beta Hebbian Learning (BHL) to easily identify attacks in the network to human experts. Contrary to Artificial Intelligent-driven solutions based on supervised learning BHL does not require labelled information. A testing environment of IoT devices and web clients is created and attack by using a Sybil Attack type, recording all traffic information in separating the fields of the captures by the most relevant, these fields are taken from "Wireshark Display Filter Reference" tool. Results obtained by BHL algorithm provide clear projections where most of the attacks can be easily identified by human expert through visual inspection in real time, proving a powerful tool to be easily implemented in IoT environments.

**Keywords:** Beta Hebbian Learning · IoT · MQTT · Cyberattack

## 1 Introduction

Nowadays, cybersecurity represent an important change in the field of mobile phones, due to the rapidly growing of number of this devices in the last decade. Despite the fact that there are currently an important industry dedicate to protect mobile devices from cyberattacks, in case of Internet of Things (IoT) devices, they are easy to hack and compromise, due to their hardware limitations, in terms of storage, computational power, etc., making them more vulnerable to be attack than other devices such as mobile ones [13].

Currently, it is estimated that a total of 12 billions of IoT connected devices. Despite the popularity of IoT concept, the expected growth of number of connected devices were slower than expected [21], due, among other causes, the security and privacy issues. However, due to COVID-19 pandemic the adoption


J. J. Gude Prego et al. (Eds.): CISIS 2021/ICEUTE 2021, AISC 1400, pp. 23–32, 2022.
https://doi.org/10.1007/978-3-030-87872-6_3

of IoT devices in some sectors such as healthcare, transport, smart buildings, smart cities, and industrial IoT has experimented a rapidly growth [1,20].

Among all available data protocols used for IoT devices, MQTT is one of the most popular and used due to its lightweight message protocol used for sending data between sensor and applications/middlewares. This protocol is for communications M2M (machine to machine) and is base on TCP-IP by using a publish-subscribe pattern. Main actors in MQTT are publishers, broker and subscribers. In this protocol, publishers and subscriber never communicate directly, and all messages are passed trough broker which distribute the publisher message to those devices subscribed to publisher, following a star topology with a central node acting as a server or broker.

Some of the most common attacks in IoT networks can be classified in 9 groups [7]:

- Traffic Analysis Attacks: Network messages are intercepted and analysed to get information of the network.
- Spoofing: An attacker capture information by eavesdropping the communication channel or by phishing.
- Cloning: Attacker copies credentials of one of the network devices being able to modify the data transmitted in the network.
- Unauthorized Access: Due to the low or null security in most of the IoT devices authentication mechanisms, tags can be easily accessed, and therefore an attacker modify and even delete data in network nodes.
- Sinkhole Attack: Nodes of the network are attacked inside the network by other node previously compromised by an external attacker, sending fake routing information to its neighboring nodes.
- Man in the Middle Attacks: In these attacks the attacker interposing himself between two points of communication so that attacker is able to know and modify the messages he intercepts.
- Denial of Service: Attacker floods the broker (in case of MQTT protocol) with large traffic data, more than broker can handle, in order to disable services of subscribers.
- Routing Information Attacks: Attacker modify the routing information which can cause routing loops, error messages, or even partitioning the network.
- Sybil Attack: The attacker impersonate another client of the system, being able to see all the broker's messages in such a way that he can send false information and see the content of the messages managed by the broker.

Most of the Artificial Intelligent-driven solutions have proposed supervised learning and only until recently, appropriate methodologies for applying and validating such methods have been released [6,10,11,15]. Additionally, AI visualization methods have been also applied to Intrusion Detection (ID) [9,14,17–19], as well as to other problems such as the detection of Android malware [22]. As opposed to supervised methods, visualization ones do not decide which data can be considered as "normal" or "anomalous". Instead, they visualize data in such an easy and intuitive way that the visualization itself reveals the anomalous nature of the patterns. As a result, anyone could easily identify such abnormality without requiring previous knowledge about the applied method.

In keeping with this idea, the present paper proposes visualization for Intrusion Detection by applying a novel neural method called Beta Hebbian Learning (BHL). The analysed real-life data comes from a test environment developed to make attacks on the MQTT protocol and capture the generated traffic, the environment consists of several IoT devices and several applications that interact with them. This dataset has been used in previous studies for detection of attacks with supervised methods, obtaining good results but with a lower performance in the detection of sybil attack with respect to other attack types [3,8]. In this research the Sybil Attack has been applied in the created network, by capturing the messages and sending false information to different sensor in the network.

The remaining sections of this study are structured as follows: Sect. 2 presents the case of study to later in Sect. 3 describe the dataset used in this research. Section 4 introduce the algorithm use to analysed the dataset and detail in Sect. 5 the experiments and results obtained. Finally, the conclusions of this study are discussed in Sect. 6.

## 2 Case Study

In order to generate the dataset, the first step is to define and establish a test context, comprised of several IoT devices and application interacting among them by means of the previously presented MQTT protocol.

It has been configured a testing environment with the aim to test different type of attacks under MQTT protocol, by capturing the traffic data. This network is composed by several IoT devices that control lighting room and presence through relays and sonar sensors. Each node is composed by one low cost NodeMCU device with WiFi connection through ESP8266 chipset connect to sensors and actuators through GPIO inputs.

As broker one server developed in node.js has been used and web clients have been created in angular.js in order to connect to the broker as another MQTT client. Each web client has user interface to show the status of the different clients and allows to interact with them in real time.

As connection point between all devices, it has been used a TP-LINK router model TL-WR1043ND with openWRT firmware, that allows to capture the traffic data in the network.

All the captured traffic is stored in a PCAP file through OpenWRT. Subsequently, it is dissected considering every field in the MQTT protocol. Then, the different frames are marked depending on whether they are under attack depending on the timestamps related to the offensive. By doing so, the resulting CSV file includes the dataset of the DoS attacks.

The structure of the IoT environment is presented in Fig. 1.

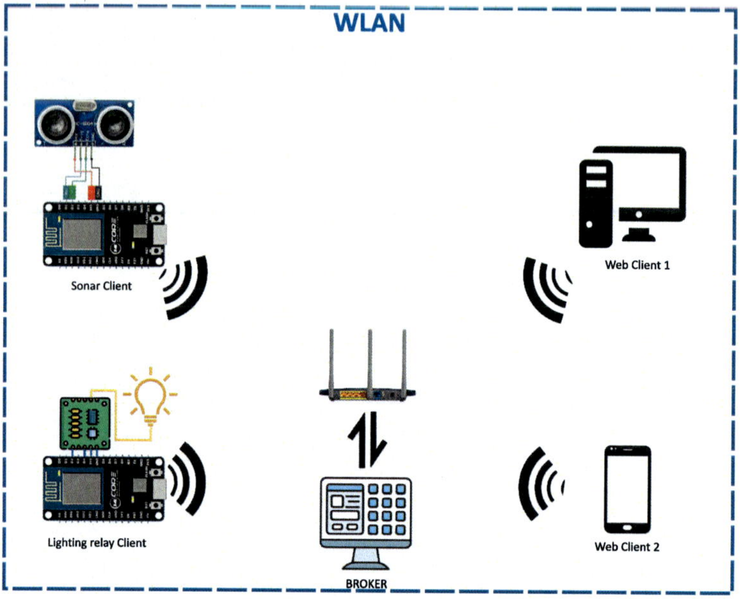


**Fig. 1.** MQTT network environment

## 3 Dataset

Considering that the well-known port of the MQTT protocol is 1883, an attacker could scan the networks to detect which ones use this protocol and access them with a not legitimate client [2,12]. In the testing environment the Sybil Attack type has been used. In this situation MQTT broker has been configured with no authentication criteria, being susceptible to being attacked by intruders posing as one more client of the system, and with "#" special character can capture all the messages in the system and sending false information.

During attacks the attacker capture the messages sent by lighting relay and sonar clients to web clients and he also send false information to these nodes to provoke unwanted actions in the controlled systems by NodeMCU devices.

During experiments all the traffic information is captured and stored in a .csv file including a tag as "attack" or "normal" for each sample stored.

The dataset consist on 80893 samples with a total of 65 variables, which represent 1898 attacks and 78995 normal samples. Information stored in the 41 variables is obtained from a software specialized in separating the fields of the captures by the most relevant, these fields are taken from "Wireshark Display Filter Reference" tool. Variables stored are shown in Table 1:

After an initial analysis of the original dataset, there were identified a total of 21 variables with constant values and therefore remove from the original dataset, finally obtaining a dataset with 41 variables. In all cases categorical variables has been coded as numeric ones.

**Table 1.** Variables used as inputs

| | | |
|---|---|---|
| frame.time_delta | frame.time_delta_displayed | |
| frame.time_epoch | frame.time_invalid | frame.time_relative |
| ip.src | ip.dst | tcp.srcport |
| tcp.dstport | eth.src | eth.dst |
| frame.cap_len | frame.coloring_rule.name | frame.coloring_rule.string |
| frame.comment | frame.comment.expert | frame.encap_type |
| frame.file_off | frame.ignored | frame.incomplete |
| frame.interface_id | frame.interface_name | frame.len |
| frame.link_nr | frame.marked | frame.md5_hash |
| frame.number | frame.offset_shift | mqtt.clientid |
| mqtt.clientid_len | mqtt.conack.flags | mqtt.conack.flags.reserved |
| mqtt.conack.flags.sp | mqtt.conack.val | mqtt.conflag.cleansess |
| mqtt.conflag.passwd | mqtt.conflag.qos | mqtt.conflag.reserved |
| mqtt.conflag.retain | mqtt.conflag.uname | mqtt.conflag.willflag |
| mqtt.conflags | mqtt.dupflag | mqtt.hdrflags |
| mqtt.kalive | mqtt.len | mqtt.msg |
| mqtt.msgid | mqtt.msgtype | mqtt.passwd |
| mqtt.passwd_len | mqtt.proto_len | mqtt.protoname |
| mqtt.qos | mqtt.retain | mqtt.sub.qos |
| mqtt.suback.qos | mqtt.topic | mqtt.topic_len |
| mqtt.username | mqtt.username_len | mqtt.ver |
| mqtt.willmsg | mqtt.willmsg_len | mqtt.willtopic |
| mqtt.willtopic_len | | |

## 4 Beta Hebbian Learning Algorithm as Projection Model for CyberActtacks Visualization

Among the range of applications of unsupervised artificial neural networks, data projection or visualization is the one that facilitates, human experts, the analysis of the internal structure of a dataset. This can be achieved by projecting data on a more informative axis or by generating maps that represent the inner structure of datasets. This kind of data visualization can usually be achieved with techniques such as Exploratory Projection Pursuit (EPP) [4,5,16] which project the data onto a low dimensional subspace, enabling the expert to search for structures through visual inspection.

The use of projectionist techniques for intrusion detection networks has been successfully applied in previous researches [17–19,22], as it provides an usefully tool not only for detecting anomalous situations in a network but also for understanding these attacks as it provides a clear visualization of the internal structure of the data in a network that can be interpreted by human experts.

In spite other previous EPP algorithm has obtained good results in this field, recently the novel Beta Hebbian Learning technique (BHL) [16] has reported a clear improvement in the results compared with the most used ones (PCA, MLHL, CMLHL, etc.)

The Beta Hebbian Learning technique (BHL) [16] is an Artificial Neural Network belonging to the family of unsupervised EPP, which uses Beta distribution as part of the weight update process, for the extraction of information from high dimensional datasets by projecting the data onto low dimensional (typically 2 dimensional) subspaces. This technique is better than other exploratory methods in that it provides a clear representation of the internal structure of data.

BHL uses Beta distribution to update its learning rule to match the Probability Density Function (PDF) of the residual (e) with the dataset distribution, where the residual is the difference between input and output feedback through the weights (4).

Thus, the optimal cost function can be obtained if the PDF of the residuals is known. Therefore, the residual ($e$) can be expressed by 5 in terms of Beta distribution parameters ($B(\alpha\ and\ \beta)$):

$$p(e) = e^{\alpha-1}(1-e)^{\beta-1} = (x - Wy)^{\alpha-1}(1 - x + Wy)^{\beta-1} \tag{1}$$

where $\alpha$ and $\beta$ control the PDF shape of the Beta distribution, $e$ is the residual, $x$ are the inputs of the network, $W$ is the weight matrix, and $y$ is the output of the network.

Finally, gradient descent can be used to maximize the likelihood of the weights (Eq. 2):

$$\begin{aligned} \frac{\partial p_i}{\partial Wij} &= (e_j^{\alpha-2}(1-e_j)^{\beta-2}(-(\alpha-1)(1-e_j) + e_j(\beta-1))) \\ &= (e_j^{\alpha-2}(1-e_j)^{\beta-2}(1-\alpha+e_j(\alpha+\beta-2))) \end{aligned} \tag{2}$$

Therefore, BHL architecture can be expressed by means the following equations:

$$Feed-forward: y_i = \sum_{j=1}^{N} W_{ij}x_j, \forall i \tag{3}$$

$$Feedback: e_j = x_j - \sum_{i=1}^{M} W_{ij}y_i \tag{4}$$

$$Weightupdate: \Delta W_{ij} = \eta(e_j^{\alpha-2}(1-e_j)^{\beta-2}(1-\alpha+e_j(\alpha+\beta-2)))y_i \tag{5}$$

where $\eta$ is the learning rate

## 5 Experiments and Results

Over previously described dataset, a normalization of each variable between the range $-1$ to 1 has been applied to guarantee the stability of the BHL network

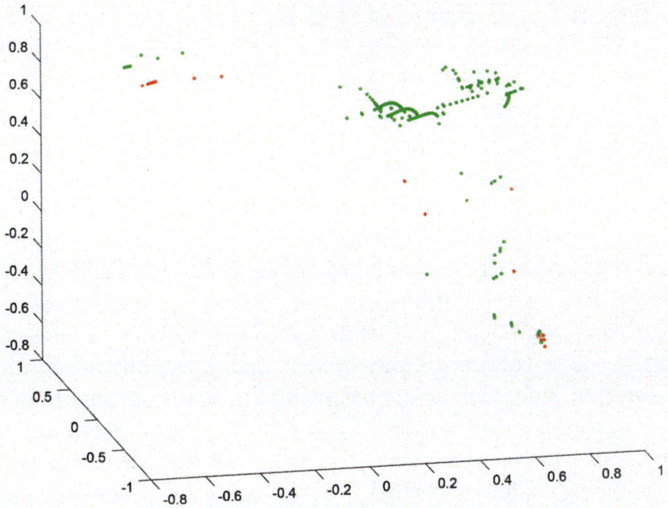


**Fig. 2.** 3D BHL 3 first components projection

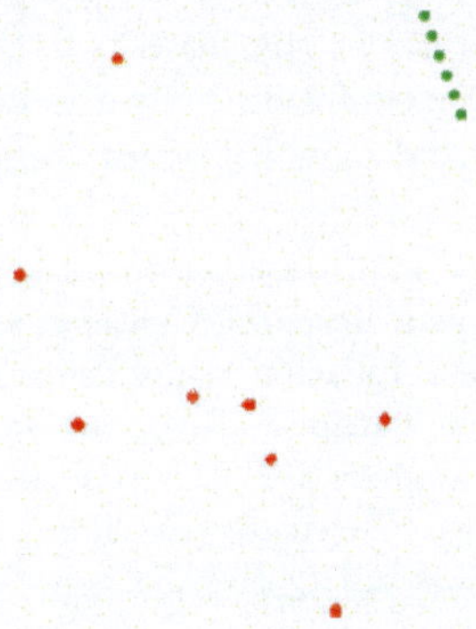

**Fig. 3.** 2D BHL 3 first components projection, rotated and zoom

during the training process [16]. Finally best projections are presented and normal and attacks are presented in different colors.

Figure 2 shows the first 3 components of the dataset projected over the new subspace representing in green normal samples and in red the attacks. BHL parameters used to obtain this projections were: iters = 5000, lrate = 0.001, $\alpha = 4$, $\beta = 3$.

In this 3D projection normal samples are clearly separated from attacks except a few isolated samples and one zone in the right down corner of the image, where it is not easy to differentiate both types. However, by means of a rotation of the view and by zooming, it can be clearly separated both types as it is shown in Fig. 3.

## 6 Conclusions and Future Works

Security in IoT devices is becoming more relevant nowadays, therefore the use of a tool that can present visually the internal structure and behaviors of networks can be an advantage as new types of attacks can be identified and classified. Especially, industrial processes and communication protocols such as MQTT are quite vulnerable to be attacked. Therefore, early attack detection is crucial to guarantee the robustness of the process.

In the present research, a novel algorithm BHL has been applied over a dataset of Sybil Attack type applied over a testing IoT environment that uses MQTT protocol communication. Results obtained provide interesting projections of the internal structure of the dataset being able to clearly differentiate attacks from normal behavior in the network. The main advantage of these projections is that provides an easy tool for a human expert to extract knowledge of the behavior of the network when it is attacked, and therefore new types of attacks can be identified and classified, without any type of previous information.

Therefore, BHL is a powerful tool that can be used in the early identification and characterization of cyberattacks.

Future works include testing the novel BHL algorithm over new and more complex datasets with new types of cyberattacks in the testing environment, and its application to anomaly detection in industrial sectors. It is also contemplated the combination with supervised learning algorithms to provide a more robust system.

**Acknowledgements.** CITIC, as a Research Center of the University System of Galicia, is funded by Consellería de Educación, Universidade e Formación Profesional of the Xunta de Galicia through the European Regional Development Fund (ERDF) and the Secretaría Xeral de Universidades (Ref. ED431G 2019/01).

Spanish National Cybersecurity Institute (INCIBE) and developed Research Institute of Applied Sciences in Cybersecurity (RIASC).

## References

1. Juniper research: IoT platform revenue to grow 20% in 2020, despite global covid-19 pandemic (2020). https://www.juniperresearch.com/press/press-releases/iot-platform-revenue-to-grow20-in-2020
2. Andy, S., Rahardjo, B., Hanindhito, B.: Attack scenarios and security analysis of MQTT communication protocol in IoT system. In: 2017 4th International Conference on Electrical Engineering, Computer Science and Informatics (EECSI), pp. 1–6 (2017). https://doi.org/10.1109/EECSI.2017.8239179
3. Aveleira-mata, J., et al.: Prototipo de ids para detección de intrusiones con modelos de machine learning en sistemas IoT de la industria 4.0 ids prototype for intrusion detection with machine learning models in IoT systems of the industry 4.0. Dyna **96**, 1–11 (2021)
4. Berro, A., Larabi Marie-Sainte, S., Ruiz-Gazen, A.: Genetic algorithms and particle swarm optimization for exploratory projection pursuit. Ann. Math. Artif. Intell. **60**, 153–178 (2010). https://doi.org/10.1007/s10472-010-9211-0

5. Corchado, E., Fyfe, C.: Connectionist techniques for the identification and suppression of interfering underlying factors. IJPRAI **17**, 1447–1466 (2003). https://doi.org/10.1142/S0218001403002915
6. Crespo-Turrado, C., et al.: Comparative study of imputation algorithms applied to the prediction of student performance. Log. J. IGPL **28**(1), 58–70 (2020)
7. Deogirikar, J., Vidhate, A.: Security attacks in IoT: a survey. In: 2017 International Conference on I-SMAC (IoT in Social, Mobile, Analytics and Cloud)(I-SMAC), pp. 32–37. IEEE (2017)
8. García-Ordás, M.T., Aveleira-Mata, J., Casteleiro-Roca, J.-L., Calvo-Rolle, J.L., Benavides-Cuellar, C., Alaiz-Moretón, H.: Autoencoder latent space influence on IoT MQTT attack classification. In: Analide, C., Novais, P., Camacho, D., Yin, H. (eds.) IDEAL 2020. LNCS, vol. 12490, pp. 279–286. Springer, Cham (2020). https://doi.org/10.1007/978-3-030-62365-4_27
9. Jove, E., Casteleiro-Roca, J., Quintián, H., Méndez-Pérez, J., Calvo-Rolle, J.: Anomaly detection based on intelligent techniques over a bicomponent production plant used on wind generator blades manufacturing. Rev. Iberoamericana Autom. Inform. Ind. **17**(1), 84–93 (2020)
10. Jove, E., et al.: Missing data imputation over academic records of electrical engineering students. Log. J. IGPL **28**(4), 487–501 (2020)
11. Jove, E., et al.: Modelling the hypnotic patient response in general anaesthesia using intelligent models. Log. J. IGPL **27**(2), 189–201 (2019)
12. Kant, D., Johannsen, A., Creutzburg, R.: Analysis of IoT security risks based on the exposure of the MQTT protocol. IEEE Access **3**(6), 1–8 (2021)
13. Khan, M.A., Salah, K.: IoT security: review, blockchain solutions, and open challenges. Futur. Gener. Comput. Syst. **82**, 395–411 (2018)
14. Luis Casteleiro-Roca, J., Quintián, H., Luis Calvo-Rolle, J., Méndez-Pérez, J.A., Javier Perez-Castelo, F., Corchado, E.: Lithium iron phosphate power cell fault detection system based on hybrid intelligent system. Log. J. IGPL **28**(1), 71–82 (2020)
15. Magán-Carrión, R., Urda, D., Díaz-Cano, I., Dorronsoro, B.: Towards a reliable comparison and evaluation of network intrusion detection systems based on machine learning approaches. Appl. Sci. **10**(5), 1775 (2020)
16. Quintián, H., Corchado, E.: Beta hebbian learning as a new method for exploratory projection pursuit. Int. J. Neural Syst. **27**(6), 1–16 (2017). https://doi.org/10.1142/S0129065717500241
17. Quintián, H., et al.: Beta-hebbian learning for visualizing intrusions in flows. In: Herrero, Álvaro., Cambra, Carlos, Urda, Daniel, Sedano, Javier, Quintián, Héctor., Corchado, Emilio (eds.) CISIS 2019. AISC, vol. 1267, pp. 446–459. Springer, Cham (2021). https://doi.org/10.1007/978-3-030-57805-3_42
18. Sánchez, R., Herrero, Á., Corchado, E.: Clustering extension of MOVICAB-IDS to distinguish intrusions in flow-based data. Log. J. IGPL **25**(1), 83–102 (2017). https://doi.org/10.1093/jigpal/jzw047
19. Sedano, J., González, S., Chira, C., Herrero, Á., Corchado, E., Villar, J.R.: Key features for the characterization of android malware families. Log. J. IGPL **25**(1), 54–66 (2017). https://doi.org/10.1093/jigpal/jzw046
20. Singh, R.P., Javaid, M., Haleem, A., Suman, R.: Internet of things (IoT) applications to fight against COVID-19 pandemic. Diabetes Metab. Syndr.: Clin. Res. Rev. **14**(4), 521–524 (2020)

21. Umair, M., Cheema, M., Cheema, O., Li, H., Lu, H.: Impact of COVID-19 on IoT adoption in healthcare, smart homes, smart buildings, smart cities, transportation and industrial IoT (2021)
22. Vega, R.V., Quintián, H., Calvo-Rolle, J.L., Herrero, Á., Corchado, E.: Gaining deep knowledge of android malware families through dimensionality reduction techniques. Log. J. IGPL **27**(2), 160–176 (2019). https://doi.org/10.1093/jigpal/jzy030

# Attacks Detection on Sampled Netflow Traffic Through Image Analysis with Convolutional Neural Networks (CNN)

Alberto Fernández de Retana(✉), Alberto Miranda-García, Ángel Manuel Guerrero, and Camino Fernández-Llamas

University of Leon, León, Spain
{afernf,amirg,am.guerrero,camino.fernandez}@unileon.es

**Abstract.** The interest in attacks detection has increased significantly in recent years together with the internet traffic and connections. Due to the big amount of packages, it is not feasible to analyze the payload of every packet that goes through the network. In order to have a statistical solution, the NetFlow protocol was designed. The payload of the packets is not included in the information stored by this protocol, making the detection of malicious attacks more challenging. Furthermore, to alleviate the performance penalty generated by the NetFlow on the routers, the Sampled NetFlow was developed. Sampled NetFlow allows the system administrators to define the interval in which these flows are going to be gathered. In the literature, there are several approaches that make use of traditional Machine Learning methods like KNN or SVM. To the best of our knowledge, there is currently no study attempting to probe Convolutional Neural Network on Sampled NetFlow. In this paper, we present the results obtained using Convolutional Neural Network on flows of Sampled NetFlow v5 to fill this gap. Our approach was able to obtain 94.15% of accuracy on sampling rate of 500. Additionally the limitations of this technique are going to be discussed if the interval of the Sampled Netflow is greather than 500.

**Keywords:** Attack detection · Sampled Netflow data · Convolutional Neural Network

## 1 Introduction

The NetFlow technology was developed in an attempt to collect information about IP traffic in a simple way, thus being able to track the flows. NetFlow, which allows us to obtain several characteristics of the packets that pass through network devices, has become an industry standard protocol. Nowadays, it is used by the majority of routers in the world. This protocol has several versions but the most widely used standard versions are versions 5 and 9, which gather data such as the source IP, the destination IP or the source and destination ports, among others. One of the consequences of the design of this protocol is that we


J. J. Gude Prego et al. (Eds.): CISIS 2021/ICEUTE 2021, AISC 1400, pp. 33–40, 2022.
https://doi.org/10.1007/978-3-030-87872-6_4

are unable to inspect the payload of the packets, called deep packet inspection (DPI). Additionally, in order to avoid the saturation of the routers the Sampled NetFlow is used. This allows the system administrators to define the threshold in which the flows of Netflow are gathered.

As of today, due to the impossibility of exhaustively analyzing each and every one of the network packets, the interest in gathering information based on the network flows is growing dramatically. One of the first topics to be studied was the traffic classification [7]. The accuracy obtained without sampling was 96.67% [2] and the accuracy with sampling was 51.02% [4]. Nowadays, the researchers go further trying to recognize worm attacks [1] or building generic detection systems [16].

One of the latest topics in this research area is the detection of network attacks, such as DDoS or Ports scanning. For this purpose, a large number of papers using machine learning, e.g. KNN, SVM, have been published [3,5,8,12–15]. However, there are few published research [9] working with the increasingly popular Convolutional Neural Networks, in particular, with the well known architecture called ResNet. That paper describes a methodology with better results than some traditional methods described before, reaching a 95.86% accuracy. Despite that, the paper was written ambiguously, not describing the version of NetFlow used nor the columns selected on the research, among other examples. This method of Convolutional Neural Networks applied in Netflow for detection networks attacks is a new field opened by this group of researchers. Our work continues the research done in this new field working on Sampled Neflow.

In summary, we present results applying Convolutional Neural Network method on Sampled Netflow (version 5). Achieving 94.15% of accuracy in 500 of sampling rate. Furthermore, the limitations generated by the sampling are discussed getting a low accuracy with a big value of interval in sampling. This is the main contribution due to the fact that sampling is used in real environments.

This paper is structured as follows: Sect. 2 explains the different datasets used in the project and describes both image generation and model training. Additionally, Sect. 3 describes the results obtained by our models and their time consumption. Finally, Sect. 4 presents our conclusions and discusses about the further work in the area and the limitations.

## 2 Materials and Methods

In this section, the core of the research is explained. The different parts of the project are explained separately below, from the generation of images, to the training of the network. We follow a three-step methodology (Fig. 1). First, data gathering of NetFlow flows is done by using DOROTHEA, a NetFlow dataset generator explained below. Next, in the image generation phase NetFlow data (1D) is converted to images (2D). Finally, the model of CNN is trained, and later validated.

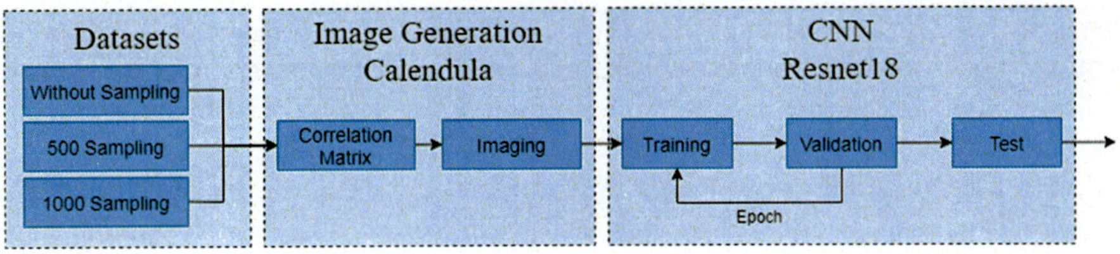


**Fig. 1.** Project architecture

## 2.1 Data Gathering

In this section, we explain the datasets used in our research. In our work, we have a different datasets for each sampling. The datasets were obtained using DOROTHEA, an open-source tool to gather netflow datasets [3]. We used three types of dataset, the first one without sampling, the next one with a sampling rate of 500 and the last one with 1000. All of them are NetFlow version 5. Datasets were taken in similar duration periods, but as a consequence of sampling, the size of datasets are smaller. The first dataset is used in order to probe that we have a similar accuracy to previous research [9], verifying that the solution of CNN on NetFlow works. We have used a dataset with a sampling rate of 1000 as it resembles a real production environment due to the fact that it is usually the most common sampling rate on big networks. Finally, we chose a sampling rate of 500 because it is a middle value between the real environment and the NetFlow without sampling. In the Table 1, the total number of images generated for the different datasets are shown. The dataset without sampling is the bigger one with a total of 1M images separated in training and validation phases. The dataset with a sampling rate of 500 has about 10k images. Finally, the dataset with a sampling rate of 1000, that imitates a real environment, has about 5000 images. All the datasets have a total of 12GB of disk usage. The dataset is publicly available with the project.

**Table 1.** Datasets of images used to train the models.

| Sampling rate | Training | | Validation | |
|---|---|---|---|---|
| | Attack | Normal | Attack | Normal |
| – | 241.910 | 236.855 | 303.907 | 303.907 |
| 500 | 2742 | 5506 | 2727 | 2727 |
| 1000 | 1428 | 1428 | 1292 | 1292 |

## 2.2 Image Generation

Imaging is one of the most important aspects of this research field. For the training and testing of the network, it is very important to be able to correctly represent all the flow data, in such a way that the network can be able to catalog the flows. For this task, we have used the parallel computing cluster of Calendula. Calendula is the supercomputer of SCAYLE (Supercomputing Center of Castilla and León) with a total of 397 TFlops which has allowed us to generate a large number of images in a reduced time for the generation of the model. This phase is one of the most critical ones, because it is a process that requires a big amount of time. On our approach, we decided to run 36 processes simultaneously. The creation of the images from the datasets took a few weeks executing on Calendula.

To carry out the task of representing the values properly, only numerical values have been used, discarding other NetFlow features with different types of values. Regarding the eliminated characteristics, in addition to the non-numerical ones, we have also discarded the ones which are related to the exact time in which the flow was extracted, the next hop of the packet or the IP from which the flow was extracted, since they do not provide any type of relevant information for the image. It should be noted, that some columns, e.g. *src_as*, *dst_as*, *src_mask* or *dst_mask*, may include relevant information on datasets retrieved from real infrastructures. The source IP and the destination IP are converted to decimal values before the image is generated. Finally, NetFlow v5 features used in this research are: *dpkts*, *doctets*, *first*, *last*, *srcaddr*, *dstaddr*, *input*, *output*, *srcport*, *dstport*, *prot*, *tos* and *tcp_flags*. Therefore, the official NetFlow v5 columns discarded due to the lack of information are: *unix_secs*, *unix_nsecs*, *sysuptime*, *exaddr*, *engine_type*, *engine_id*, *nexthop*, *src_mask*, *dst_mask*, *src_as*, *dst_as*.

First of all, a correlation matrix is generated with all the characteristics defined before. In order to create the correlation matrix, the Pandas library [10] is used. After that, each of the columns is surrounded by their eight more correlated values making a matrix of $3 \times 3$. Finally, all the $3 \times 3$ matrices are joined on a big matrix called Surrounding Correlation matrix (SC matrix). This methodology was presented by a group of researchers [9] from Purdue University. Finally, this SC matrix values are replaced by each of the flows, generating an image similar to Fig. 2.

### 2.3 Training the CNN

The training phase was done using the popular library PyTorch [11]. This tool is a machine learning library which is focused on usability and speed, allowing us to train ResNet-18. Our approach was to use the model offered by PyTorch, freezing all the layers except the final one. This methodology allowed us to train the model with a maximum performance. The Convolutional Neural Network model used was ResNet-18 [6]. This architecture is a reliable model used in the previous work in this field by Purdue's researchers. ResNet-18 is well-known to be used for classification [6].

Before feeding the model with data, the images need to be resized to $224 \times 224$ in order to fit the requirements of the network. After that, the last layer of the model is trained using a defined number of epochs in order to get the best

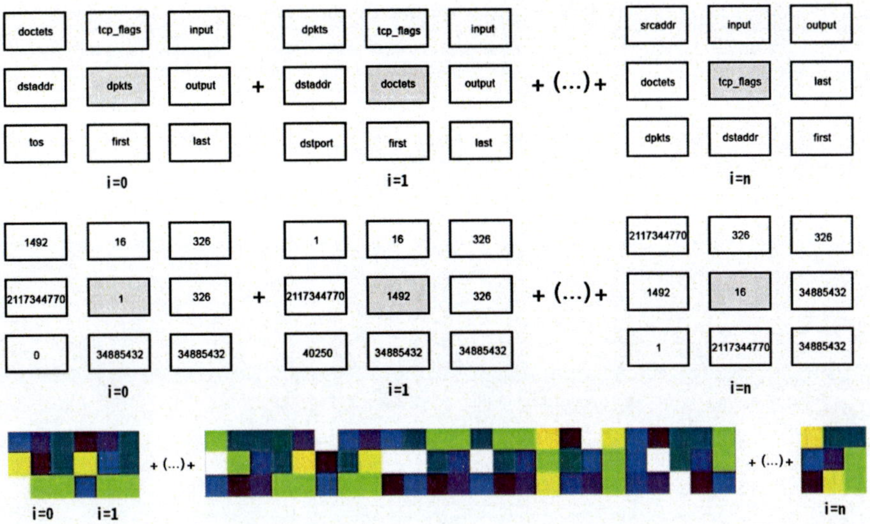


**Fig. 2.** Netflow matrix numeric representation

accuracy possible. In each of the epochs there are two phases defined, training and then validation phase. This technique lets us save the best model.

## 3 Results

In this section, a detailed overview of the results are presented. The results are obtained using the methodology explained above. First of all, the accuracy of Netflow without sampling is 96.58%. This accuracy, that is greater than the previous work [9], demonstrates that Convolutional Neural Network (CNN) could be used on Netflow with good results. Moreover, with this dataset, we have double-checked the previous research. This dataset is the bigger one, so the training of the model was the largest in terms of time. In other words, the training with this dataset delays approximately 11 h. The accuracy of Convolutional Neural Network (CNN) is similar to any other method used in the literature, e.g. KNN, SVM [3,5,8,12–15]. In the Table 2 the accuracies of other classifiers are shown.

The second dataset contains the data with a sampling rate of 500. Like we explained before, this sampling rate is a middle value between the actual literature and the real environments. Using this second dataset, the accuracy was 94.15%, decreasing a 2.52% below the results without sampling. This value shows that Convolutional Neural Network (CNN) could be applied successfully on Sampled NetFlow.

The last one is the simulation of a real environment with a sampling rate of 1000. This is the sampling rate that is normally applied in real routers to avoid the degradation of the performance due to their saturation. The accuracy obtained by the model with this dataset is slightly disappointing, getting a

**Table 2.** Accuracy from the literature.

| Classifier | Accuracy |
|---|---|
| CNN | 96.58% |
| KNN | 96.41% |
| LR | 94.83% |
| SGD | 92.10% |
| OvR | 93.23% |
| CART | 90.18% |
| RF | 90.18% |
| AB | 90.18% |
| BRBM | 78.22% |
| QDA | 51.71% |
| LDA | 51.71% |
| NB | 51.71% |
| BC | 51.71% |

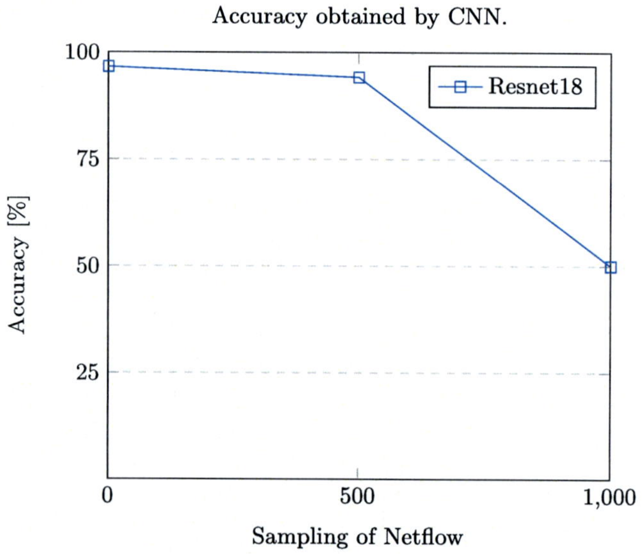


**Fig. 3.** Accuracy obtained by CNN

50.11%. This supposes a 48.12% decrease below the first dataset and a 46.78% below the second one with 500 of sampling rate. Although the results are not encouraging, they open an opportunity to research new methodologies in this way.

In summary, the accuracy decreased drastically when the rate of Sampled NetFlow is greater than 500. The reason for that is the amount of information that is lost when the NetFlow is sampled. In the Fig. 3 the accuracies of the datasets are shown.

## 4 Conclusions and Further Work

This paper presents the first work using Convolutional Neural Networks on Sampled Netflow. Moreover, in this research we describe the limitations that Convolutional Neural Networks have on Sampled NetFlow. This is an issue that also appear in other similar approaches [4]. Using Convolutional Neural Networks we obtain a 94.15% of accuracy with a sampling rate of 500. We believe these insights can be the foundation for more systematic future works in this field. Additionally, we have made the project[1] publicly available, allowing other researchers to double-check and improve our findings.

Our work shows that Convolutional Neural Networks (CNN) accuracy decreases when the interval of Sampled Netflow increases. Furthermore, with a simulated real environment dataset with a sampling rate of 1000, the accuracy is greatly decreased. However, there are smaller networks where the sampling rate is 500 or less in which this work could be applied. On a future work, the model is going to be tested with real data to compare the accuracy obtained. Besides, another line of research is the study of the information added to the model for each of the features of NetFlow. In this line of work, different architectures of CNN or different methods of training could be studied. Furthermore, the performance of the image creation process is critical in order to implement this methodology on a real environment, so this could be a future limitation of this new approach.

**Acknowledgement.** The research described in this article has been partially funded by Instituto Nacional de Ciberseguridad de España (INCIBE), under the grant "ADENDA 4: Detección de nuevas amenazas y patrones desconocidos (red Regional de Ciencia y Tecnología)", addendum to the framework agreement INCIBE–Universidad de León, 2019–2021; and by the Spanish Ministry of Science, Innovation, and Universities RTI (RTI2018-100683-B-I00) grant.

## References

1. Abdulla, S., Ramadass, S., Taha, A., Amer, N.: Setting a worm attack warning by using machine learning to classify netflow data. Int. J. Comput. Appl. **36**, 49–56 (2011)
2. Bakhshi, T., Ghita, B.: On internet traffic classification: a two-phased machine learning approach. J. Comput. Netw. Commun. **2016**, 1–21 (2016)
3. Campazas-Vega, A., Crespo-Martínez, I.S., Guerrero Higueras, A.M., Fernández Llamas, C.: Flow-data gathering using netflow sensors for fitting malicious-traffic detection models. Sensors **20**(24), 7294 (2020)

[1] https://niebla.unileon.es/AlbertoMGV/cnn/.

4. Carela-Español, V., Barlet-Ros, P., Cabellos-Aparicio, A., Solé-Pareta, J.: Analysis of the impact of sampling on netflow traffic classification. Comput. Netw. **55**(5), 1083–1099 (2011)
5. Casas, P., Mazel, J., Owezarski, P.: Unsupervised network intrusion detection systems: detecting the unknown without knowledge. Comput. Commun. **35**(7), 772–783 (2012)
6. He, K., Zhang, X., Ren, S., Sun, J.: Deep residual learning for image recognition. In: 2016 IEEE Conference on Computer Vision and Pattern Recognition (CVPR) (2016)
7. Jiang, H., Moore, A.W., Ge, Z., Jin, S., Wang, J.: Lightweight application classification for network management. In: Proceedings of the 2007 SIGCOMM workshop on Internet network management - INM 07 (2007)
8. Kanda, Y., Fontugne, R., Fukuda, K., Sugawara, T.: ADMIRE: anomaly detection method using entropy-based PCA with three-step sketches. Comput. Commun. **36**(5), 575–588 (2013)
9. Liu, X., Tang, Z., Yang, B.: Predicting network attacks with CNN by constructing images from netflow data. In: 2019 IEEE 5th International Conference on Big Data Security on Cloud (BigDataSecurity), IEEE International Conference on High Performance and Smart Computing, (HPSC) and IEEE International Conference on Intelligent Data and Security (IDS) (2019)
10. Mckinney, W.: Pandas: a foundational python library for data analysis and statistics. Python High Perform. Sci. Comput. **14**, 1–9 (2011)
11. Paszke, A., et al.: PyTorch: an imperative style, high-performance deep learning library (2019)
12. Shone, N., Ngoc, T.N., Phai, V.D., Shi, Q.: A deep learning approach to network intrusion detection. IEEE Trans. Emerg. Top. Comput. Intell. **2**(1), 41–50 (2018)
13. Tran, Q.A., Jiang, F., Ha, Q.M.: Evolving block-based neural network and field programmable gate arrays for host-based intrusion detection system. In: 2012 Fourth International Conference on Knowledge and Systems Engineering (2012)
14. Tran, Q.A., Jiang, F., Hu, J.: A real-time netflow-based intrusion detection system with improved bbnn and high-frequency field programmable gate arrays. In: 2012 IEEE 11th International Conference on Trust, Security and Privacy in Computing and Communications (2012)
15. Winter, P., Hermann, E., Zeilinger, M.: Inductive intrusion detection in flow-based network data using one-class support vector machines. In: 2011 4th IFIP International Conference on New Technologies, Mobility and Security (2011)
16. Zhenqi, W., Xinyu, W.: Netflow based intrusion detection system. In: 2008 International Conference on MultiMedia and Information Technology (2008)

# Analysis of the Performance of Different Accountability Strategies for Autonomous Robots

Laura Fernández-Becerra, Ángel Manuel Guerrero-Higueras(✉), Francisco Javier Rodríguez-Lera, and Camino Fernández-Llamas

Robotics Group, University of León, Campus de Vegazana S/N, 24071 León, Spain
inflfb00@estudiantes.unileon.es
{am.guerrero,fjrod,camino.fernandez}@unileon.es
http://robotica.unileon.es

**Abstract.** Providing an efficient logging engine to determine the causes of an event is needed to enforce humans trust in computing systems. This fact becomes increasingly challenging in robotics, where unexpected decisions can have unpredictable consequences in sensitive environments, endangering human interests. In order to analyze the causes of an incident or to anticipate future behaviors, it is necessary to record every cause of a given event. Although this task is usually developed through logging systems, characteristics such as incomplete recording of events, the inclusion of unhelpful data, or the impact on robot performance, make it necessary to conceive new accountability solutions. This work presents a generic approach to developing an accountability system that consists of four main components: a system event logger, a message producer, a distributed event streaming platform, and a database. The proposed solution has been tested in several assessment scenarios in order to define the best strategy to reduce the impact of auditing and logging tasks on robot performance.

**Keywords:** Accountability · Auditing · Explicability · Security · Robotics · Experimental research

## 1 Introduction

Traditional logging systems usually lack fundamental features to achieve an efficient journal. Unstructured and disorganized information, useless data, or incomplete records of events are some of the obstacles found during their analysis. New requirements should arise when identifying the reasons behind an incident apart from traditional preventive security measures based on protocols or authorization mechanisms enforcement.

Accountability implies that the causes that trigger an event are traceable, allowing us to determine them. The lack of accountability systems signifies a


J. J. Gude Prego et al. (Eds.): CISIS 2021/ICEUTE 2021, AISC 1400, pp. 41–51, 2022.
https://doi.org/10.1007/978-3-030-87872-6_5

waste of resources and effort in identifying whether a system event is caused by an attack or by an erroneous software or hardware component [14].

Robotics is one of the research fields where accountability is an important issue [2]. Providing explanations to robots' actions and the reasons guiding their behaviors has become necessary to encourage humans to interact with them in an effective way [1]. A primary concern in service robotics is to allow robots to ease the interaction with non-expert users in open-ended environments. Any faults or errors should be quickly detected, avoiding potential threats.

To this end, every conduct that takes place should involve a record of events that allows rebuilding the set of steps that have triggered it, with the view to providing a way to analyze the causes of an incident or helping to anticipate future robot's actions.

Nearly all autonomous robots deployed in real-world environments do not have accountability systems for auditing, and if they do, the logging usually implies a drop in performance [15]. The most common methods for error analysis in autonomous robots are usually terminal output or logs. While the first one is fundamentally volatile memory whose content is lost after being closed, avoiding further analysis, ordinary logs impact the autonomous agent performance. Furthermore, the absence of structure or links between related data avoids efficient query development [4].

Robot Operating System (ROS) is the most commonly used framework in robotics [10]. However, difficulties to satisfy real-time embedded systems requirements and the need to improve its multi-platform capabilities promoted a significant upgrade to ROS2, which major change lies in using the Data Distribution Service (DDS) [6]. Both versions include the logging engine as its main functionality for describing and understanding robot behavior in addition to other tools that allow assessing robot performance, such as topic statistics.

The use of these tools to provide accountability services results in high computer resource demand, negatively impacting on robot's execution [11]. In order to minimize this drawback, Niemueller et al. in [7,8] introduced MongoDB for logging purposes in robotics by open-sourcing the mongodb_log library. After that, several researchers have promoted the use of a NoSQL database to store data from the sensors or logging messages produced by robots middlewares [9,13].

Although these works could improve ROS and ROS2 logging engine impact, they do not scale to changes in the system, in the framework, or the events triggered by both frameworks apart from the ones written in the logs by default.

Based on these facts, we propose a generic accountability system whose primary purpose is to support debugging for software developers and the robotic system by using a system event logger, a real-time messaging system, and a NoSQL database. The prototype system has been successfully implemented on different virtual machines used to test our solution in several assessment scenarios. The comparison of ROS and ROS2 default logging engine with our proposed system not only has been checked in the host-based version but also in its containerized deployment. Finally, we have evaluated system performance when deploying our approach in a single server and a distributed architecture with MongoDB Atlas as a database management system.

The remainder of this paper is organized as follows: Sect. 2 describes the generic system architecture of the proposed accountability system. Applying the results of the above analysis, Sect. 3 details the technical elements of the accountability system that has been built. Section 4 presents and discusses the results obtained when testing our accountability solution. Section 6 discussed the obtained results. Finally, Sect. 7 summarizes the conclusions and outlines the steps for future work.

## 2 System Architecture

This section details our approach to release an accountability system. Our solution aims to identify the causes that have triggered a set of specific events, making them traceable and auditable.

A generic solution for the accountability system that we propose consists of (i) a system event logger responsible for capturing the relevant system information for a given purpose, (ii) a message producer needed to send previous filtered information to (iii) a distributed event streaming platform that handles producer batches of messages and (iv) a database where previous data is finally stored. Figure 1 represents a generic architecture of our solution.

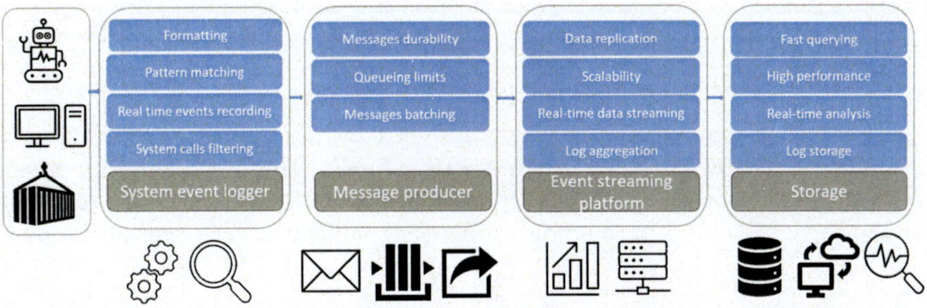


**Fig. 1.** System architecture.

## 3 System Application Using Sysdig, Librdkafka, Apache Kafka and MongoDB

Several tools can be used to implement the system architecture that has been described in Sect. 2. In the present section, we detail the technologies we have chosen to build a solution based on it.

### 3.1 Sysdig

System events loggers such as DTrace, Strace, Linux Auditd, Event Tracing for Windows or Sysdig can be used to generate logs. Regarding features such as performance overhead, required technical knowledge to use, and flexibility, we have selected Sysdig for capturing all the syscalls and events of interest that are executed in our assessment scenarios. Sysdig includes functionalities from other tools such as Strace, Tcpdump, Htop, Iftop, or Lsof. Furthermore, it is scriptable in Lua. This feature has been used to perform the pattern matching, the captured information optimization and output conversion to a JSON (Javascript Object Notation) representation.

### 3.2 Librdkafka Message Producer

Many programming languages provide client libraries for event streaming platforms. In order to develop our message producer, we have chosen Librdkafka, a C++ library implementation of the Apache Kafka protocol used for our message producer implementation. Its design enforces message delivery reliability, high performance and low latency, getting rates of about 1 million msgs/second for the producer.

This library also includes characteristics such as TLSv1.3 support, compression and statistics metrics, features that have allowed us to develop an efficient client for the chosen event streaming platform.

### 3.3 Apache Kafka

Message systems such as Apache ActiveMQ, RabbitMQ or ZeroMQ can handle asynchronous communication between systems by managing the transference of queues of records. However, these solutions need some improvements when are used in distributed systems [12]. Apache Kafka includes a more efficient storage format and high throughput, processing millions of messages per second with limited resources. These facts make it to be commonly used for real-time data streaming. Its architecture consists of producers, consumers, and brokers distributed over several servers to improve the performance and fault-tolerance of the system [5]. This characteristic provokes that Apache Kafka would be usually used in log aggregation. Furthermore, Kafka relies on the underlying operating system to lazily flush the data to disk instead of flushing to it immediately, improving performance. The risk of data loss is reduced by configuring more than one replica in each topic partition. This way, data will be in several brokers.

### 3.4 MongoDB

NoSQL databases are highly scalable and efficient when storing a large amount of data. They are usually categorized into four main types: key-value stores, graph databases, vast column stores, and document stores. Previous researches have found that the document base is the most suitable type to achieve high performance. Therefore, they provide better flexibility, performance, and scalability when the data is in JSON format [3].

Accordingly, MongoDB was chosen as a good candidate database for evaluation since it is a NoSQL document-oriented database. Apart from storing all data in BSON and being optimized for fast BSON-based queries, since version 3.6, it also supports JSON Schema validation.

## 4 Experimentation

To evaluate our proposal, we conducted an experimental study comparing the performance of an accountability system built according to Sect. 3. In this solution, data is collected from ROS and ROS2 logging engine by Sysdig. Next, Sysdig's output in JSON format is gathered by the Librdkafka client, which sends batch messages to an Apache Kafka Server topic. Thanks to the MongoDB Kafka Sink Connector, data from Chatter and Rosout topics will be saved in MongoDB.

### 4.1 Software Artifacts

Our accountability solution has been tested in three Linux virtual machines, each with one 2.60 GHz core and 8 GB of memory. The first one includes ROS Kinetic over a Ubuntu 16.04 OS in its host-based and containerized version. The second server contains ROS 2 Foxy over Ubuntu 20.04. Both of them hold Sysdig (version 0.27.1), Librdkafka (version 1.6), Apache Kafka (version 2.7.0), Openssl (version 1.1.1g), JDK (version 11.0.11-ea) and MongoDB (version 4.4.5). The third machine includes Apache Kafka and its related libraries to test our solution in a distributed architecture that saves the data in MongoDB Atlas.

We have developed three Lua scripts for Sysdig to perform the ROS, ROS2 and containerized ROS syscalls filtering, pattern matching and output conversion to JSON format. Regarding the filtering functionality, the processes filter by Sysdig are shown in Table 1.

**Table 1.** Processes filtered by Sysdig

| Audited framework | *.py | amcl | bash | catkin_find | gzclient | gzserver | map_server | python | roscore | rosmaster | rosout | rostopic | rosversion | rviz | ros2 | rviz2 |
|---|---|---|---|---|---|---|---|---|---|---|---|---|---|---|---|---|
| 1. ROS | ✓ | ✓ | ✓ | ✓ | ✓ | ✓ | ✓ | ✓ | ✓ | ✓ | ✓ | ✓ | ✓ | ✓ | | |
| 2. ROS 2 | ✓ | | | | ✓ | ✓ | | | | | | | | | ✓ | ✓ |
| 3. Containerized ROS | ✓ | ✓ | ✓ | ✓ | ✓ | ✓ | ✓ | ✓ | ✓ | ✓ | ✓ | ✓ | ✓ | ✓ | | |

The processes system calls that are monitored by Sysdig in our solution are clone(), connect(), accept(), sendto(), setsockopt(), recvfrom() and execve(). The capture of these functions when being executed by processes included in Table 1 is enough to gather all the information contained in the ROS logs generated by its default engine. Information related to sockets establishment performed by

ROS frameworks is also saved in our approach. Referring to the Librdkafka message producer, parameters related to the number of messages batched and the established time to wait for messages in the producer queue are set. The configuration of these variables is necessary to get a compromise between throughput and message delivery delay. The messages sent by the Librdkafka producer to Apache Kafka are ciphered with TLS 1.3 protocol. The communication with the database uses TLS 1.3 in the MongoDB local installation and TLS 1.2 in the Atlas option. All the mentioned code is available online on GitHub[1].

## 5 Evaluation

The evaluation of our proposal has been developed through a set of experiments in three different assessment scenarios. Performance comparison of our solution and ROS default logging engine has been made in all of them. Next, we will provide a detailed description of each other.

### 5.1 Assessment Scenarios

The assessment of the proposed accountability system have been made in three main scenarios:

i. ROS and ROS2 logging engine. This scenario consists of the default accountability engine in ROS and ROS2 frameworks that stores log data in files. It includes all the information related to the nodes and the topics.
ii. Sysdig, Apache Kafka and MongoDB (Local). In this scenario, all the components of our accountability solution have been deployed in one single server.
iii. Sysdig, Apache Kafka and MongoDB Atlas (Distributed). This scenario employs the same components as (ii). Sysdig and the Apache Kafka client are deployed in one server, and the Apache Kafka server is in a second machine. MongoDB Atlas, a cloud alternative of MongoDB, is used as a storage system.

The performance comparison of our solution and ROS default logging engine has been evaluated through a set of experiments. Each consists of running a simple publisher and a listener that interchange information by using a topic repeatedly. Both Talker and Listener nodes of the experiments have been developed for ROS and ROS2 frameworks [10]. A containerized deployment of ROS has been assessed in the same way. Nodes execution time has been settled to 3 min, enough time to gain an overall vision of the performance and impact related to our proposal.

Regarding the sent and received message, additional experiments have been done with the default ROS or ROS2 publisher message (54 and 83 bytes, respectively), a 1 KB message, and an 8 KB message to assess the performance of our solution when handling significant log entries.

[1] https://github.com/inflfb00/accountability-sysdig-kafka.

Taking this into consideration, the developed experiments are:

i. Default publisher message sending and ROS logging engine (1_1).
ii. Default publisher message sending and Sysdig, Apache Kafka and MongoDB (Local) for logs recording (2_1).
iii. Default publisher message sending and Sysdig, Apache Kafka and MongoDB (Distributed) for logs recording (3_1).
iv. 1K publisher message sending and ROS logging engine (1_2).
v. 1K publisher message sending and Sysdig, Apache Kafka and MongoDB (Local) for logs recording (2_2).
vi. 1K publisher message sending and Sysdig, Apache Kafka and MongoDB (Distributed) for logs recording (3_2).
vii. 8K publisher message sending and ROS logging engine (1_3).
viii. 8K publisher message sending and Sysdig, Apache Kafka and MongoDB (Local) for logs recording (2_3).
ix. 8K publisher message sending and Sysdig, Apache Kafka and MongoDB (Distributed) for logs recording (3_3).

### 5.2 Metrics

Several metrics have been used to evaluate the impact of the accountability system in the chosen scenarios. Firstly, we have considered logs size. Regarding the system performance, we have studied variables such as CPU, memory or disk usage. Writing logs to the filesystem from ROS and ROS2 frameworks could not be disabled due to the terminal buffer limit of 4096 bytes restricts our message size when capturing logs entries from the screen only.

We use Dstat to collect the mentioned data. This tool covers vmstat, iostat, netstat and ifstat functionalities. This fact makes it to be frequently used in systems performance monitoring.

## 6 Results and Discussion

This section includes the results of our experiments to compare the cost of service resources in the selected scenarios.

### 6.1 Log Benchmarking

Table 2 presents the overall size for experiments described in Sect. 5.1 on the selected scenarios. Although the ROS2 framework improves logs storage requirements considerably in comparison to ROS, the best performance is achieved in scenario iii. Furthermore, our solution gets better results when handling bigger messages thanks to the MongoDB compression properties. This fact highlights the ability of our proposal to handle messages from different applications without causing a remarkable performance impact in the system.

Differences between scenario ii and iii regarding size storage are related to Sysdig behaviour when system load average grows. In this situation, Sysdig drops incoming events to avoid slowing down the monitored system.

**Table 2.** Overall size of log files in MB

| Experiment | ROS | | ROS2 | | Containerized ROS | |
|---|---|---|---|---|---|---|
| | Storage size | Total document size | Storage size | Total document size | Storage size | Total document size |
| 1_1 | 57,0 | | 8,1 | | 41,1 | |
| 2_1 | 6,0 | 42,6 | 4,1 | 29,1 | 5,3 | 37,7 |
| 3_1 | 7,1 | 50,2 | 4,4 | 30,8 | 6,1 | 41,9 |
| 1_2 | 437,0 | | 93,0 | | 466,0 | |
| 2_2 | 7,9 | 72,2 | 6,5 | 60,1 | 9,7 | 69,7 |
| 3_2 | 8,4 | 74,4 | 8,4 | 78,0 | 9,3 | 77,5 |
| 1_3 | 1126,0 | | 741 | | 1128,8 | |
| 2_3 | 24,7 | 87,0 | 50,9 | 133,0 | 20,1 | 80,3 |
| 3_3 | 26,6 | 95,7 | 51,0 | 159,2 | 21,6 | 86,7 |

## 6.2 System Benchmarking

Results obtained with the Dstat tool related to CPU performance, memory and disk used are described next. Concerning CPU usage, Dstat provides CPU usage by user processes, system processes and the number of idle and waiting processes. As the Table 3 shows, most experiments developed show better performance results when being executed in scenarios ii and iii. Nevertheless, the more significant difference in average values of all situations does not exceed three per cent. These results mean that not only our solution does not introduce CPU impact, but also, in most of the experiments, distributed architecture of our solution improves CPU consumption.

**Table 3.** CPU usage percentage of user processes in ROS framework

| Experiment | ROS | | | ROS2 | | | Containerized ROS | | |
|---|---|---|---|---|---|---|---|---|---|
| | Average | Std. Deviation | Median | Average | Std. Deviation | Median | Average | Std. Deviation | Median |
| 1_1 | 83,12 | 8,05 | 84,00 | 86,20 | 11,89 | 88,20 | 82,64 | 5,97 | 83,80 |
| 2_1 | 84,63 | 9,74 | 86,20 | 86,10 | 12,27 | 88,40 | 78,26 | 14,17 | 81,80 |
| 3_1 | 82,97 | 8,95 | 84,37 | 85,12 | 12,15 | 87,17 | 79,27 | 8,06 | 80,40 |
| 1_2 | 87,52 | 10,71 | 89,38 | 89,55 | 12,37 | 91,60 | 86,40 | 7,77 | 88,18 |
| 2_2 | 88,41 | 10,50 | 90,16 | 89,16 | 12,45 | 91,57 | 88,43 | 8,56 | 89,62 |
| 3_2 | 86,70 | 10,73 | 88,42 | 86,56 | 12,47 | 88,58 | 88,15 | 9,88 | 89,72 |
| 1_3 | 87,99 | 11,59 | 89,96 | 92,57 | 13,49 | 94,78 | 90,31 | 12,18 | 92,57 |
| 2_3 | 90,72 | 10,25 | 92,37 | 90,12 | 12,97 | 92,37 | 89,60 | 11,31 | 91,58 |
| 3_3 | 90,03 | 10,53 | 91,60 | 90,32 | 12,98 | 92,59 | 88,88 | 11,54 | 90,74 |

Memory usage remarkably changes when introducing Apache Kafka in the evaluated system. Experiments developed in scenario ii show a significant difference in RAM usage terms. Nevertheless, average values are similar in scenarios i and iii. Memory consumption impact is owned to the Apache Kafka java process that achieves values over 1 GB when sending 8K messages in scenario ii when evaluating ROS framework (Table 4).

Concerning disk write operations, scenario iii improve i and ii scenarios notably when evaluating the ROS framework. ROS 2 logging engine improvements achieve similar or even better results than those from scenarios ii and iii.

Nevertheless, this circumstance reverts again in the containerized ROS assessment. Considering that our approach has not overwritten ROS and ROS2 logging engine, these results show that our solution provides similar or better values regarding disk usage. This information is shown in Table 5.

**Table 4.** Memory usage (MB)

| Experiment | ROS | | | ROS2 | | | Containerized ROS | | |
|---|---|---|---|---|---|---|---|---|---|
| | Average | Std. Deviation | Median | Average | Std. Deviation | Median | Average | Std. Deviation | Median |
| 1_1 | 1194,52 | 2,75 | 1194,48 | 1026,69 | 5,15 | 1027,45 | 1510,73 | 2,39 | 1510,34 |
| 2_1 | 2180,39 | 14,13 | 2184,34 | 2052,33 | 32,98 | 2038,62 | 2795,33 | 85,34 | 2750,79 |
| 3_1 | 1163,29 | 2,70 | 1163,31 | 996,07 | 5,44 | 997,12 | 1581,27 | 4,98 | 1581,30 |
| 1_2 | 1197,10 | 5,97 | 1196,72 | 1027,45 | 5,34 | 1028,32 | 1532,71 | 6,34 | 1533,78 |
| 2_2 | 2952,13 | 47,64 | 2969,56 | 2801,71 | 37,34 | 2804,82 | 2521,70 | 61,78 | 2519,54 |
| 3_2 | 1204,97 | 4,56 | 1205,46 | 1000,57 | 5,88 | 1002,06 | 1582,12 | 3,39 | 1582,35 |
| 1_3 | 1340,27 | 8,73 | 1340,13 | 1154,42 | 18,70 | 1168,51 | 1331,03 | 10,60 | 1333,27 |
| 2_3 | 2921,92 | 50,28 | 2932,32 | 2339,18 | 11,52 | 2341,16 | 2555,34 | 46,65 | 2555,21 |
| 3_3 | 1230,70 | 6,76 | 1231,14 | 1112,47 | 7,46 | 1115,29 | 1332,27 | 12,19 | 1335,45 |

**Table 5.** Disk write usage (KB/s)

| Experiment | ROS | | | ROS2 | | | Containerized ROS | | |
|---|---|---|---|---|---|---|---|---|---|
| | Average | Std. Deviation | Median | Average | Std. Deviation | Median | Average | Std. Deviation | Median |
| 1_1 | 755,51 | 2757,22 | 41,60 | 41045,00 | 249135,61 | 34,00 | 238,10 | 376,12 | 27,20 |
| 2_1 | 768,62 | 498,93 | 535,20 | 44785,08 | 269612,41 | 305,00 | 764,72 | 615,49 | 405,60 |
| 3_1 | 222,95 | 292,14 | 32,00 | 45245,46 | 274725,03 | 43,00 | 242,97 | 477,10 | 24,00 |
| 1_2 | 2317,14 | 3001,20 | 71,20 | 41547,84 | 249745,22 | 31,00 | 3285,21 | 4447,40 | 49,60 |
| 2_2 | 2501,38 | 2395,66 | 1804,80 | 42756,24 | 253693,58 | 445,00 | 2600,24 | 2630,38 | 651,20 |
| 3_2 | 1676,69 | 2155,36 | 54,40 | 45696,94 | 273819,17 | 40,00 | 1809,02 | 4450,45 | 28,80 |
| 1_3 | 5513,66 | 6994,40 | 117,60 | 59361,62 | 335376,71 | 257,00 | 6820,46 | 8369,42 | 574,52 |
| 2_3 | 4070,01 | 4490,15 | 1625,00 | 68450,49 | 395743,99 | 990,00 | 3936,86 | 5730,19 | 719,20 |
| 3_3 | 3339,89 | 4700,83 | 94,40 | 71195,73 | 422959,75 | 58,00 | 3177,95 | 5475,32 | 52,80 |

## 7 Conclusions

Efficient logging systems are needed to identify the circumstances behind an event. Nevertheless, the information usually recorded by this systems is not always complete or meaningful to the problem needed to be solved. Accountability solutions aim to determine the causes that trigger a given action, improve the quality of the recorded information and optimise non-expert users' efforts when dealing with an unexpected system or application behaviour. This work proposes a generic and decentralised accountability solution where the monitored application and systems are analysed without disturbing their real-time execution. Our approach enforces logging and auditing mechanisms as a mean to increase trust and security in computing systems. The developed experimental study demonstrates the efficiency and effectiveness of the proposed system when using a distributed architecture regarding robotics. This design avoids performance issues caused by the robot default logging engine. Nevertheless, future work such as scaling the event streaming platform to additional nodes and using new actors to test our solution could be helpful to get a more profound vision of our approach.

**Acknowledgments.** The research described in this article has been partially funded by Instituto Nacional de Ciberseguridad de España (INCIBE), under the grant "ADENDA 4: Detección de nuevas amenazas y patrones desconocidos (red Regional de Ciencia y Tecnología)", addendum to the framework agreement INCIBE-Universidad de León, 2019–2021; and by the Spanish Ministry of Science, Innovation, and Universities RTI2018-100683-B-I00 grant.

## References

1. Anjomshoae, S., Najjar, A., Calvaresi, D., Främling, K.: Explainable agents and robots: results from a systematic literature review. In: Proceedings of the 18th International Conference on Autonomous Agents and MultiAgent Systems, AAMAS 2019, pp. 1078–1088. International Foundation for Autonomous Agents and Multiagent Systems, Richland, SC (2019)
2. Guerrero-Higueras, Á.M., Rodríguez-Lera, F.J., Martín-Rico, F., Balsa-Comerón, J., Matellán-Olivera, V.: Accountability in mobile service robots. In: Fuentetaja Pizán, R., García Olaya, Á., Sesmero Lorente, M., Iglesias Martínez, J., Ledezma Espino, A. (eds.) Advances in Physical Agents. WAF 2018. Advances in Intelligent Systems and Computing, vol. 855, pp. 242–254. Springer, Cham (2019). https://doi.org/10.1007/978-3-319-99885-5_17
3. Gupta, A., Tyagi, S., Panwar, N., Sachdeva, S., Saxena, U.: NoSQL databases: critical analysis and comparison. In: 2017 International Conference on Computing and Communication Technologies for Smart Nation (IC3TSN), pp. 293–299 (2017)
4. Han, Z., Allspaw, J., Norton, A., Yanco, H.A.: Towards a robot explanation system: a survey and our approach to state summarization, storage and querying, and human interface (2019)
5. Kreps, J., Corp, L., Narkhede, N., Rao, J., Corp, L.: Kafka: a distributed messaging system for log processing. In: NetDB 2011 (2011)
6. Maruyama, Y., Kato, S., Azumi, T.: Exploring the performance of ROS2. In: Proceedings of the 13th International Conference on Embedded Software, EMSOFT 2016. Association for Computing Machinery, Inc. (2016). https://doi.org/10.1145/2968478.2968502
7. Niemueller, T., Abdo, N., Hertle, A., Lakemeyer, G., Burgard, W., Nebel, B.: Towards deliberative active perception using persistent memory. Technical report (2013)
8. Niemueller, T., Lakemeyer, G., Srinivasa, S.S.: A generic robot database and its application in fault analysis and performance evaluation. In: IEEE International Conference on Intelligent Robots and Systems, pp. 364–369 (2012)
9. Oliveira, M., Lim, G.H., Lopes, L.S., Kasaei, S.H., Tome, A.M., Chauhan, A.: A perceptual memory system for grounding semantic representations in intelligent service robots. In: IEEE International Conference on Intelligent Robots and Systems, pp. 2216–2223. Institute of Electrical and Electronics Engineers Inc. (2014)

10. Quigley, M., et al.: ROS: an open-source Robot Operating System. In: ICRA Workshop on Open Source Software (2009). http://stair.stanford.edu
11. Rodríguez-Lera, F.J., Guerrero-Higueras, Á.M., Martín-Rico, F., Gines, J., Sierra, J.F.G., Matellán-Olivera, V.: Adapting ROS logs to facilitate transparency and accountability in service robotics. In: Silva, M., Luís Lima, J., Reis, L., Sanfeliu, A., Tardioli, D. (eds.) Robot 2019: Fourth Iberian Robotics Conference. ROBOT 2019. Advances in Intelligent Systems and Computing, vol. 1093, pp. 587–598. Springer, Cham (2020). https://doi.org/10.1007/978-3-030-36150-1_48
12. Wang, Z., et al.: Kafka and its using in high-throughput and reliable message distribution. In: Proceedings - 8th International Conference on Intelligent Networks and Intelligent Systems, ICINIS 2015, pp. 117–120. Institute of Electrical and Electronics Engineers Inc. (2016)
13. Winkler, J., Tenorth, M., Bozcuoğlu, A.K., Beetz, M.: CRAMm – memories for robots performing everyday manipulation activities (2014)
14. Xiao, Y.: Flow-net methodology for accountability in wireless networks. IEEE Netw. **23**(5), 30–37 (2009)
15. Yoon, M.K., Shao, Z.: ADLP: accountable data logging protocol for publish-subscribe communication systems. In: 2019 IEEE 39th International Conference on Distributed Computing Systems (ICDCS), pp. 1149–1160. IEEE (2019)

# Analysis of NetFlow Features' Importance in Malicious Network Traffic Detection

Adrián Campazas-Vega[1(✉)], Ignacio Samuel Crespo-Martínez[1], Ángel Manuel Guerrero-Higueras[1], Claudia Álvarez-Aparicio[1], and Vicente Matellán[2]

[1] Robotics Group, University of León, Campus de Vegazana S/N, 24071 León, Spain
{acamv,icrem,am.guerrero,calvaa}@unileon.es
[2] Supercomputación Castilla y León (SCAyLE), Campus de Vegazana S/N, 24071 León, Spain
vicente.matellan@scayle.es
http://robotica.unileon.es

**Abstract.** Malicious traffic detection allows for preventing cybersecurity-related threats. Machine learning algorithms are commonly used to detect such traffic in computer networks by analyzing packets. In wide-area networks, such as RedCAYLE (Red de Ciencia y Tecnología de Castilla y León), it is not possible to analyze every packet routed. So we pose that in such networks sampled flow data may be used to provide malicious traffic detection. This work presents the analysis carried out of the relevance that every NetFlow feature has in the K-Nearest Neighbors (KNN) algorithm in order to detect malicious traffic. Validation of the model has been carried out with real network data from RedCAYLE. Results show that it is necessary to train the models with sampled flow data. They also show that the *nexthop* feature has a negative influence on malicious traffic detection in wide-area networks such as RedCAYLE.

**Keywords:** Netflow features analysis · K-Nearest Neighbors (KNN) · Network traffic · Machine learning · Network security · Malicious traffic detection

## 1 Introduction

Cybercriminals have been professionalized using different techniques to achieve profit through the theft of victims' data. They use a large number of techniques, such as port scanning, malware distribution, lateral network movements,

The research described in this article has been partially funded by Instituto Nacional de Ciberseguridad de España (INCIBE), under the grant "ADENDA 4: detección de nuevas amenazas y patrones desconocidos (red Regional de Ciencia y Tecnología)", addendum to the framework agreement INCIBE-Universidad de León, 2019–2021; the Spanish Ministry of Science, Innovation, and Universities RTI2018-100683-B-I00 grant; and the regional Government of Castilla y León under the grant BDNS (487971).


J. J. Gude Prego et al. (Eds.): CISIS 2021/ICEUTE 2021, AISC 1400, pp. 52–61, 2022.
https://doi.org/10.1007/978-3-030-87872-6_6

privilege escalation, zero-day attacks, and social engineering. Most attacks generate malicious traffic. Such traffic may be used to detect and stop them. In the literature, various techniques have been used to detect malicious traffic. For instance, in [18], the authors present a payload-based anomaly detector for intrusion detection. In [19], the authors characterize the benign network traffic so that anomalous traffic patterns may be detected. Different works show that it is possible to detect malicious traffic using machine learning models. In [16] authors propose an anomaly detection system using supervised learning algorithms that have previously been trained with the UNSW-NB15 dataset. The authors obtained 94.36% accuracy using the AODE algorithm, 92.70% using Bayesian networks, and 75.73% using the NB algorithm. This work was validated by the same authors in [17], in which, using the same dataset, they tried to verify which algorithms were the best in detecting malicious traffic in terms of accuracy and efficiency. The results showed that the AODE algorithm obtained the best results with an accuracy of 97.26% and a running time of approximately 7 s.

In order to detect malicious traffic, in [23] a hybrid model is proposed which implements bagging and tree rotation techniques. To test the efficiency of the generated models, the authors used the NSL-KDD and UNSW-NB15 datasets, obtaining a precision of 85.8%. Finally, in [22] the authors carried out a review of the literature in which they collected what are the machine learning algorithms and datasets most used to detect malicious traffic. Although it is possible to detect malicious traffic using machine learning models, all the above proposals used network packets as input for the different algorithms. In wide-area networks, it is not possible to analyze every packet that the network routes. That is why this type of infrastructure uses flow-based technologies.

A flow is defined as a set of packets that pass through an observation point in the network during a specific time interval. All packets belonging to a particular flow have a common set of properties, such as source and destination IP addresses, and source and destination port numbers [6,8]. Flow data do not store the payload of network packets, reducing the computational power required to process flows versus complete packages. In addition to those already mentioned, some of the features that flow data gathers are the number of packets that the flow contains, the IP protocol, the TCP flags, and the timestamp. One of the most widely used flow-based technologies is NetFlow. NetFlow was designed by Cisco as a network protocol for collecting network statistics [5]. NetFlow is deployed on most commercial routers.

However, the network traffic volume that some routers manage is so large that even using NetFlow it is not possible to use every packet to generate flow data. So they need to sample the packets before generating flows. For instance, RedCAYLE, which manages traffic belonging to universities and research centers in Castilla y León, manages an average amount of traffic of 6.7 Gb/s. To decrease the computational load of gathering flow data on the routers, a sampling threshold is applied. The sampling threshold used by RedCAYLE routers is 1 packet out of 1000.

This work comes from the results obtained in [4]. In this work, the authors developed a framework for collecting flow datasets using NetFlow. With such datasets, the authors fitted supervised learning models, obtaining that the best classifier to detect port scans was KNN.

This work presents an analysis of NetFlow's features to determine which are most relevant when training the KNN algorithm. The training has been carried out with two different datasets, the first one gathering sampled flow data and the second one gathering unsampled flow data. Validation has been carried out against real data obtained from RedCAYLE in order to double-cleck if the fitted model is able to detect attacks in a real environment and how the features used to fit it influence the detection of malicious traffic.

The remainder of the paper is organized as follows: Sect. 2, describes the materials and tools used in this job, as well as the methodology used to evaluate the importance of NetFlow features to detect malicious traffic; Sect. 3 shows the results obtained in the experiment and discusses the results. Finally, the conclusions are presented in the Sect. 4.

## 2 Materials and Methods

Section enumerates the materials used, presents the experiments carried out, and describes the methods used to evaluate the results obtained in the experiments.

### 2.1 NetFlow

NetFlow [7] is a protocol developed by Cisco Systems to collect information about the network traffic. The usage of NetFlow has become so popular in recent years that manufacturers such as Juniper and Enterasys Switches support this technology. NetFlow was introduced as a new feature of Cisco's routers to provide IP traffic gathering allowing administrators to have a global vision of what is happening in the infrastructure they manage.

NetFlow has multiple versions: NetFlow V1,V5 and V9. RedCAYLE uses NetFlow V5. For such release, the features that flow data gather are enumerated in Table 1.

### 2.2 RedCAYLE

RedIRIS [21], is a Spanish research network that provides communications services to the scientific and university community. RedCAYLE manages RedIRIS network traffic belonging to the region of Castilla y León. It is managed from SCAYLE. Network traffic management is carried out through two main routers located in León and Valladolid, and a series of auxiliary switches and routers distributed throughout the region.

RedCAYLE provides to its affiliated institutions (educational centers, universities, hospitals, scientific facilities, etc.) a high-capacity communications backbone infrastructure, which allows for accessing both the resources of the research

**Table 1.** NetFlow features

| Feature | Description |
|---|---|
| sysuptime | Current time in milliseconds since the export device started |
| unix_secs | Current count of seconds since 0000 UTC 1970 |
| unix_nsecs | Residual nanoseconds since 0000 UTC 1970 |
| engine_type | Flow switching motor type |
| engine_id | Slot number switching engine flow |
| exaddr | Flow exporter Ip |
| srcaddr | Source IP address |
| dstaddr | Destination IP address |
| nexthop | IP address of the next hop router |
| input | SNMP index of the input interface |
| output | SNMP index of the exit interface |
| dpkts | Number of packets contained in the flow |
| doctets | Total number of bytes of layer 3 in the packets of the flow |
| first | Sysuptime at start of flow |
| last | Sysuptime when the the last packet in the flow was received |
| srcport | TCP/UDP source port number |
| dstport | TCP/UDP destination port number |
| tcp_flags | TCP flags |
| prot | IP type of protocol (Por ejemplo, TCP = 6; UDP = 17) |
| tos | IP type of service (ToS) |
| src_as | Autonomous system number of the source, either source or pair |
| dst_as | Autonomous system number of the destination, either source or pair |
| src_mask | Source address prefix mask bits |
| dst_mask | Destination address prefix mask bits |

network and the Internet. Currently, RedCAYLE offers a wide variety of services: 1 Gbps point-to-point transport service, internet connection, IP addressing, incident management, and service supervision. The RedCAYLE monitoring service allows its affiliated institutions for analyzing and diagnosing the status of their network services. The usage of NetFlow provides a network analysis based on statistics. As mentioned above, to decrease the computational load of gathering flow data, a sampling threshold of 1 packet out of 1000 is applied.

### 2.3 MoEv

MoEv [10] is an open-source tool developed in Python [20] for building machine learning models from datasets. The latest release accepts as input both CSV files and images. Some of the MoEv features are data cleaning, normalization, dimensionality reduction, hyperparameter tuning (using the Grid-SearchCV method), and parallel running (using DASK). It evaluates several machine learning algorithms by considering accuracy, recall, precision, and F1 score. MoEv also allows for carrying out validation tests to ensure the optimal generalization of its models. Once a model, or a set of models, has been fitted, it is validated by using a different dataset.

MoEv has been used successfully in different research areas, such as in the detection of jamming and spoofing attacks in real-time location systems [11], or the prediction of academic success in educational institutions [9,12,13]. Furthermore, in [4] the authors successfully replicated the experiment proposed by [1], thus demonstrating the validity of MoEv in generating machine learning models for the detection of network attacks through flows.

### 2.4 DOROTHEA

DOROTHEA is a Docker-based solution that allows for creating virtual network topologies to generate and collect flow data. It uses a NetFlow sensor that builds flows from the packets that pass through a network interface. DOROTHEA is configurable and scalable, allowing, for example, to select the number of attack and victim nodes.

The framework offers two testbeds. The first one allows for simulating the generation of benign traffic. In order to get realistic responses, it is connected to the Internet. The second testbed allows for carrying out distributed network attacks in an isolated Internet environment. Isolation ensures that all the collected traffic corresponds to attack flows, thus avoiding noise when generating datasets. Testbed separation allows for the empirical labeling of the network traffic.

In both testbeds, all traffic goes through a router that incorporates a sensor that generates the NetFlow's flows from the packets it routes *ipt_NetFlow* [14]. The sensor processes network traffic and provides flow data in NetFlow V5, V9, and IPF formats. Flow data are sent to a warehouse every 2 min.

In addition to packet routing and flow data generation, the router performs packet sampling. As with commercial routers, DOROTHEA allows you to select the packet sampling rate to be used when generating flows. This feature allows for simulating the behavior of real routers, allowing researchers to generate datasets under the same conditions as big networks deployed in production. Finally, DOROTHEA returns a CSV file with the generated network flows.

### 2.5 Data Gathering

To carry out the experiments, two datasets have been created using DOROTHEA. Both datasets contain benign traffic and malicious traffic. Specifically, malicious traffic comes from port scanning attacks. The first dataset (aka $\mathcal{D}_1$) has been collected without packet sampling. The second dataset (aka $\mathcal{D}_2$) has been collected with a sampling threshold of 1 packet out of 1000, simulating the conditions in RedCAYLE routers.

Both datasets contain approximately 50% benign traffic. Benign flow data were labeled '0'. Malicious flow data generated were labeled '1'. Both datasets are available online under an open-access license with the following DOIs: 10.5281/zenodo.4106730 and 10.5281/zenodo.4600638.

The benign traffic was generated using three scripts developed in Python. These scripts are also available online in [3]. The first script simulates web browsing. It carries out queries in different search engines generating HTTP and HTTPS traffic. The second script simulates the sending of emails using the SMTP protocol. Finally, the third script simulates SSH connections.

Malicious traffic has been generated using the Nmap tool. Different types of port scans have been launched on both TCP and UDP ports. The scans performed were: TCP SYN; TCP Connect; UDP; TCP NULL, FIN, and Xmas; TCP ACK; TCP Window; and TCP Maimon scanning [15]. The attacks were executed by 100 machines distributed over the 65,536 ports of the 200 attacked nodes. The Python script used to carry out the attacks is available online at [2].

The range of network addresses in the training datasets was 182.168.1.1/24 for the benign traffic simulators and attack nodes, and 126.52.30.0/24 for victim nodes.

To validate the previously fitted models, a third dataset has been generated (aka $\mathcal{D}_3$), with flow data collected in the RedCAYLE.

Table 2 shows the volume of the datasets used in this work.

**Table 2.** Number of Flows

| Dataset | #Benign flows | # Malicious flows | #Total flows |
|---|---|---|---|
| $\mathcal{D}_1$ | 1.429.038 | 1.178.992 | 2.608.030 |
| $\mathcal{D}_2$ | 1447 | 1447 | 2894 |
| $\mathcal{D}_3$ | 460 | 460 | 920 |

### 2.6 Evaluation

To fit the KNN models that have been used in the experiments, the tool MoEv and the datasets $\mathcal{D}_1$ and $\mathcal{D}_2$ have been used. Then, an analysis of the features of each model has been carried out in order to know which NetFlow features have the highest importance. In order to do so, the technique permutation feature importance has been used. This procedure breaks the relationship between the feature and the target, thus the drop in the model score is indicative of how much the model depends on the feature, obtaining, as a result, the weight that each one has in the training of the KNN model.

Once the study was carried out, the detection rate of the model was verified with the $\mathcal{D}_3$ dataset. In order to know how the features affect the detection of malicious traffic in a production environment, the models have been re-fitted and obtained their accuracy removing the most relevant features according to the previous study. The accuracy of the model has been calculated as shown in the Eq. 1, where TP is true-positive rate, TN is the true-negative rate, FP is the false-positive rate and FN is the false-negative rate.

$$Accuracy = \frac{T_P + T_N}{T_P + F_P + T_N + F_N} \tag{1}$$

## 3 Results and Discussion

Table 3 shows the importance of NetFlow features, described in Table 1. The features that are not shown in the new table with respect to the older one, have an importance of 0 for the model.

**Table 3.** Weight for the features of the model KNN for $\mathcal{D}_1$ and $\mathcal{D}_2$

| Feature | Weight $\mathcal{D}_1$ | Weight $\mathcal{D}_2$ |
|---|---|---|
| nexthop | 0.266652 | 0.400000 |
| srcaddr | 0.196473 | 0.116930 |
| dstaddr | 0.160707 | 0.073476 |
| last | 0.107621 | 0.005079 |
| first | 0.107598 | 0.005079 |
| sysuptime | 0.107593 | 0.004515 |
| unix_secs | 0 | 0.004402 |
| unix_nsecs | 0 | 0.000451 |

As can be seen in the Table 3 the three most relevant features are nexthop, srcaddr, and dstaddr for both models, regardless of the dataset on which they have been trained. In the case of the trained model without sampling, dataset $\mathcal{D}_1$, the last, first and sysuptime features also take relevance. In the case of the model trained with sampling, dataset $\mathcal{D}_2$, in addition to the previous features, unix_secs and unix_nsecs also take relevance. The difference that we can observe in the weight of the features is that for the model trained with the dataset $\mathcal{D}_2$ the nexthop has greater weight than in the model trained with the dataset $\mathcal{D}_1$ which distributes the weight of the features more evenly between nexthop, srcaddr, dstaddr, last, first, and sysuptime.

In Table 4 we can see the accuracy of the model KNN validated with the dataset $\mathcal{D}_3$ and that has previously been trained with the datasets $\mathcal{D}_1$ and $\mathcal{D}_2$.

As can be seen in Table 4 the model trained without sampling ($\mathcal{D}_1$), cannot detect any malicious flow regardless of the features with which it has been trained. In all cases, the model detects benign traffic, but it does not detect any flow classified as attack, so the accuracy of the model is 50% in all cases. From the analysis of these data, we can concluded that the KNN model trained with $\mathcal{D}_1$ is not functional in detecting malicious network traffic in networks that perform packet-level sampling.

**Table 4.** True Negative, True Positive and Accuracy of KNN classifier trained with $\mathcal{D}_1$ and $\mathcal{D}_2$ based on features removed and validated with $\mathcal{D}_3$

| Features removed | $T_N$ $\mathcal{D}_1$ | $T_N$ $\mathcal{D}_2$ | $T_P$ $\mathcal{D}_1$ | $T_P$ $\mathcal{D}_2$ | Accuracy $\mathcal{D}_1$ | Accuracy $\mathcal{D}_2$ |
|---|---|---|---|---|---|---|
| nexthop | 100% | 96.52% | 0% | 100% | 50% | 98.26% |
| srcaddr | 100% | 91.09% | 0% | 0% | 50% | 45.54% |
| dstaddr | 100% | 100% | 0% | 0% | 50% | 50% |
| nexthop, srcaddr | 100% | 91.09% | 0% | 0% | 50% | 45.54% |
| nexthop, dstaddr | 100% | 100% | 0% | 0% | 50% | 50% |
| srcaddr, dstaddr | 100% | 0% | 0% | 100% | 50% | 50% |
| nexthop, srcaddr, dstaddr | 100% | 0% | 0% | 100% | 50% | 50% |

On the contrary, we can see that if we eliminate the nexthop feature, the model trained $\mathcal{D}_2$ obtains an accuracy of 98.26% when we validated it with the dataset $\mathcal{D}_3$. The rest of the combinations of features made when training the model, show an accuracy of less than 50 % discarding its validity.

From the analysis of the Table 4 we can affirm, on the one hand, that it is necessary to train models with sampled traffic to be able to detect traffic on networks such as RedCAYLE. On the other hand, the use of the nexthop feature affects negatively of the detection of malicious traffic in this type of network, therefore it is necessary to eliminate this feature when training the model. In addition, it is necessary to maintain both the source IP address and the destination IP address when training the model to increase its accuracy.

If we observe the nexthop field of the $\mathcal{D}_1$ and $\mathcal{D}_2$ datasets, we can see that in the flows classified as benign, the nexthop is always the router's laboratory that will route the packet to the Internet, however if we observe the malicious traffic flows, we can see that the nexthop field is 0, since due to the architecture of the attack laboratory, it does not have an Internet connection and therefore the jump from the attacking machine to the victim is direct.

On the other hand, if we look at the $\mathcal{D}_3$ dataset collected in RedCAYLE, the nexthop field has a much greater variance than the datasets generated with DOROTHEA since they correspond to real IP addresses and more complex network architectures. That is why it is necessary to remove this field when training models for them to work in a real environment.

## 4 Conclusions

The detection of malicious traffic is a good indicator in the detection of various threats. It has been shown in the literature that malicious traffic can be detected using machine learning algorithms. The difficulty in detecting these types of threats increases when it is not possible to analyze all the traffic on a network. This occurs in networks that manage a large amount of traffic per second, making it impossible to analyze all the packets that it route. To know the state of the network, this type of infrastructure uses flow-based protocols such as Netflow. Networks as large as RedCAYLE do not have enough computing power to analyze all packets, even using NetFlow. For this reason, these networks perform packet sampling before generating flows, in this specific case RedCAYLE selects 1 packet out of every 1000.

In this work, three different datasets have been generated, $\mathcal{D}_1$, $\mathcal{D}_2$, and $\mathcal{D}_3$. The first two datasets have been generated using DOROTHEA. Both contain approximately 50% of benign traffic and 50% of port-scanning attack traffic. The difference between both datasets is that $\mathcal{D}_2$ has been generated with a sampling threshold. Finally, $\mathcal{D}_3$ has been generated with flow data collected from RedCAYLE.

With the $\mathcal{D}_1$ and $\mathcal{D}_2$ datasets the KNN model has been fitted in order to demonstrate if it is possible to detect real malicious traffic coming from RedCAYLE. The KNN model has been selected because it is the model that has shown the best results in detecting malicious network traffic in the literature [4,22].

Once the models were trained, an analysis of the importance of NetFlow features in the generated models was carried out. This analysis reported that the features nexthop, srcaddr, and dstaddr are the most relevant when training the KNN model. With the information obtained in the analysis, the models have been re-trained eliminating the most relevant features. Once the models have been re-fitted, they have been validated with the flows obtained from RedCAYLE ($\mathcal{D}_3$).

From the validation with the $\mathcal{D}_3$ dataset, it has been shown that the results of the models trained with $\mathcal{D}_1$ do not exceed 50% accuracy regardless of the features with which are train the model. On the other hand, the model trained with the $\mathcal{D}_2$ dataset provides an accuracy of 98.26% if the nexthop feature is eliminated when training the model.

Two clear conclusions can be drawn from the experiments carried out. On the one hand, it is necessary to train models with sampled traffic to be able to detect traffic on networks such as RedCAYLE. On the other hand, the nexthop feature negatively affects the detection of malicious traffic in wide-area networks, and it is necessary to remove this feature in the training phase of the model.

## References


1. Boukhamla, A., Coronel, J.: CICIDS 2017 dataset: performance improvements and validation as a robust intrusion detection system testbed. Int. J. Inform. Comput. Secur. **9** (2018)
2. Campazas-Vega, A., Crespo-Martínez, I.: Source code DOROTHEA attacks generation. https://niebla.unileon.es/cybersecurity/dorothea/-/tree/master/labs/lab_attacks/attacks. Accessed 13 Mar 2021
3. Campazas-Vega, A., Crespo-Martínez, I.: Source code DOROTHEA normal traffic generation. https://niebla.unileon.es/cybersecurity/dorothea/-/tree/master/labs/lab_normal/generator/generate-traffic. Accessed 13 Mar 2021
4. Campazas-Vega, A., Crespo-Martínez, I.S., Guerrero-Higueras, Á.M., Fernández-Llamas, C.: Flow-data gathering using netflow sensors for fitting malicious-traffic detection models. Sensors **20**(24), 7294 (2020)
5. Cisco: About Cisco (2021). https://www.cisco.com/. Accessed 13 Mar 2021

6. Claise, B., Zander, S.: Network working group J. Quittek Request for Comments: 3917 nec europe ltd. category: Informational t. zseby fraunhofer fokus (2004)
7. Claise, B., Sadasivan, G., Valluri, V., Djernaes, M.: Cisco systems netflow services export version 9 (2004)
8. Claise, B., Trammell, B., Aitken, P.: Specification of the IP flow information export (ipfix) protocol for the exchange of flow information. RFC 7011 (Internet Standard), Internet Engineering Task Force, pp. 2070–1721 (2013)
9. Fernández, A.G., et al.: Evaluación del resultado académico de los estudiantes a partir del análisis del uso de los sistemas de control de versiones. RIED. Rev. Iberoamericana Educación a Dist. **23**(2), 127–145 (2020)
10. Guerrero-Higueras, Á.M., Campazas-Vega, A., Crespo-Martínez, I.S.: Module evaluator (moev). Technical report, Robotics group, Universidad de León (2020). https://doi.org/10.5281/zenodo.4114127
11. Guerrero-Higueras, Á.M., DeCastro-García, N., Matellán, V.: Detection of cyberattacks to indoor real time localization systems for autonomous robots. Robot. Auton. Syst. **99**, 75–83 (2018)
12. Guerrero-Higueras, Á.M., DeCastro-García, N., Rodriguez-Lera, F.J., Matellán, V., Conde, M.Á.: Predicting academic success through students' interaction with version control systems. Open Compu. Sci. **9**(1), 243–251 (2019)
13. Guerrero-Higueras, Á.M., Fernández Llamas, C., Sánchez González, L., Gutierrez Fernández, A., Esteban Costales, G., González, M.Á.C.: Academic success assessment through version control systems. Appl. Sci. **10**(4), 1492 (2020)
14. ipt_NetFlow: Source code ipt_NetFlow. https://github.com/aabc/ipt-NetFlow. Accessed 13 Mar 2021
15. Lyon, G.F.: Nmap network scanning: the official Nmap project guide to network discovery and security scanning. Insecure (2009)
16. Nawir, M., Amir, A., Lynn, O.B., Yaakob, N., Ahmad, R.B.: Performances of machine learning algorithms for binary classification of network anomaly detection system. In: Journal of Physics: Conference Series. vol. 1018, p. 012015 (2018)
17. Nawir, M., Amir, A., Yaakob, N., Lynn, O.B.: Effective and efficient network anomaly detection system using machine learning algorithm. Bull. Electr. Eng. Inform. **8**(1), 46–51 (2019)
18. Parekh, J.J., Wang, K., Stolfo, S.J.: Privacy-preserving payload-based correlation for accurate malicious traffic detection. In: Proceedings of the 2006 SIGCOMM Workshop on Large-Scale Attack Defense, pp. 99–106 (2006)
19. Pena, E.H.M., Barbon, S., Rodrigues, J.J.P.C., Proença, M.L.: Anomaly detection using digital signature of network segment with adaptive ARIMA model and paraconsistent logic. In: 2014 IEEE Symposium on Computers and Communications (ISCC), pp. 1–6 (2014). https://doi.org/10.1109/ISCC.2014.6912503
20. Python: About python (2021). https://www.python.org/about/. Accessed 13 Mar 2021
21. RedIRIRS: About redirirs (2021). https://www.rediris.es/rediris/index.html.es. Accessed 13 Mar 2021
22. Sobrín-Hidalgo, D., Campazas Vega, A., Guerrero Higueras, Á.M., Rodríguez Lera, F.J., Fernández-Llamas, C.: Systematic mapping of detection techniques for advanced persistent threats. In: Herrero, Á., Cambra, C., Urda, D., Sedano, J., Quintián, H., Corchado, E. (eds.) CISIS 2019. AISC, vol. 1267, pp. 426–435. Springer, Cham (2021). https://doi.org/10.1007/978-3-030-57805-3_40
23. Tama, B.A., Comuzzi, M., Rhee, K.H.: TSE-IDS: a two-stage classifier ensemble for intelligent anomaly-based intrusion detection system. IEEE Access **7**, 94497–94507 (2019). https://doi.org/10.1109/ACCESS.2019.2928048

# Localizing Worst-Parent Rank Attack Using Intelligent Edges of Smart Buildings

G. V. K. Sasirekha(✉), V. Bhanu Prakash, Jyotsna Bapat, and Debabrata Das

Networking and Communication Research Lab, International Institute of Information Technology – Bangalore (IIIT-B), Bangalore, Karnataka, India
{sasirekha,jbapat,ddas}@iiitb.ac.in, Bhanu.Prakash@iiitb.org

**Abstract.** Routing Protocol for Low power and lossy networks (RPL) is the underlying protocol of 6LoWPAN, a core communication standard for the Internet of Things (IoT). The concept of rank is used in RPL for routing and its optimization. However, the rank provides an avenue for "Rank Attacks (RA)" like worst parent, increased rank, and decreased rank, which are internal to 6LowPAN. In this paper, a smart building scenario with intelligent edges is considered for detecting the worst-parent attacks of RPL. Novelty of our approach is that, in addition to detecting the attack, the impact of the malicious node location on the average statistics of control messages is used to identify its locality, using Machine Learning (ML) classification at the edge of the network. An "impact factor" has been defined and analyzed in terms of the forwarding load of the node. Simulation results obtained for a given static node topology, show a localization accuracy of 0.67 using Support Vector Method (SVM), while the accuracy obtained using eXtreme Gradient Boost (XGBoost) is 0.88. Discussion on how the technique is applicable to any generalized network topology, to reduce the number of packet level inspections for malicious node detection, is also presented.

**Keywords:** Rank attacks · Smart buildings · Azure edge · Intrusion detection sytems · Smart Homes · XGBoost · SVM · Intelligent edge · IoT

## 1 Introduction

### 1.1 Background

IPv6 over Low-Power Wireless Personal Area Network [1] (6LoWPAN) is a low cost and low power communication network defined for IoT, which uses RPL as networking layer, a 6LowPAN adaptation layer, and IEEE 802.15.4 protocol at the link and physical layer. RPL is a tree-based routing protocol defined by IETF as the standard routing protocol for Low power Lossy Networks (LLNs) [2]. RPL organizes the network topology in a Destination-Oriented Graph (DAG), which is composed of one or more Destination-Oriented Directed Acyclic Graphs (DODAGs). Each DODAG represents a routing tree created by a root node, also known as a sink node. Establishment and maintenance of the routing is done by control messages i) DIO (DODAG Information Object) for setting and updating the topology. DIO contains DODAG-ID and the rank information


J. J. Gude Prego et al. (Eds.): CISIS 2021/ICEUTE 2021, AISC 1400, pp. 62–72, 2022.
https://doi.org/10.1007/978-3-030-87872-6_7

to allow other nodes to discover the DODAG. ii) DAO (DODAG Destination Object) for propagating destination information upwards during route update progress, and iii) DIS (DODAG Information Solicitation) for a new node to ask for topology information before joining the network [2]. DIO, DAO and DAO Acknowledgements (DAO-ACKs) messages create the downward routes.

RPL is vulnerable to a large variety of attacks like Rank, Neighbor, Version Number, Hello-Flooding, Selective-Forwarding, Sybil, Wormhole, Blackhole attacks, etc. [3]. In the case of rank attack, an attacker manipulates the rank value to degrade the quality of service or deny the service. These attacks generate loops in the network, exhaust node resources and congest the network. Rank attacks have variants namely, decreased-rank attack, worst-parent attack, and increased-rank attack [4, 5]. When a malicious node decreases its rank value, it causes neighboring nodes to choose it as a preferred parent. Increase of the rank by the malicious node in turns repels the neighbors and starts a process of searching for another preferred parent, which will consume considerable network resources. In the case of the worst-parent attack, the malicious node disobeys the rank rule of selecting the best parent (least rank). It chooses the worst one which delays the packet transmission and increases the Expected Transmission Count routing metric (ETX) of all descendants of the malicious node [4, 5]. This can cause the topology around the malicious node to change, leading to generation of more DIO messages, which in turn leads to more control overhead as well as more packet collisions.

In general, Intrusion Detection Systems (IDS) for IoT have been well investigated over several years. An IDS can be signature-based, anomaly-based, specification-based or hybrid type [5]. Signature-based systems are search for stored known patterns to detect intrusion. Alerts are generated in case of a match. Anomaly based systems observe the deviation from normal and generate an alert whenever the deviation exceeds the defined threshold. In specification-based systems any deviation from the defined specification monitored and the alert is generated accordingly. An IDS is called a hybrid one, when it is based on both anomaly and signature. Both signature and specification based are computationally intense involving packet header inspections, finite state machines, etc. Anomalies are more generic in nature and can be statistics based.

In this paper, an anomaly-based system for smart buildings is proposed, where in the detection of the attack along with localization information, in terms of the locality (subset of network nodes) of the malicious node is provided. This localization information helps in narrowing down of the packets for further inspection. The contributions of this paper are organized as follows. The attack detection architecture based on edges for smart buildings discussed in Sect. 2. Section 3 discusses the attack and its impact. Section 4 describes the simulation of attack using Cooja simulator. Section 5 describes the implementation and performance results of ML classifiers used for attack locality classification. The final section has the conclusions.

### 1.2 Related Work

Literature survey shows significant contributions towards IoT attacks, edge-based solution architectures and specifically, RPL attacks. The highlights of few relevant papers are discussed in this sub-section. Work in [6], discusses the criticality of IoT attacks in the healthcare vertical. An anomaly detection system is proposed in a smart hospital IoT system for detecting events of network intrusions, which is integrated with health-related alerts. Such integrated approach ensures the accuracy of the patient related decisions. [7] proposes a Secure RPL Routing Protocol (SRPL-RP) for rank and version number attacks, which detects, mitigates, and isolates attacks in RPL networks, based on a comparison of the rank strategy. The mitigation process uses threshold and attack status tables.

The effect of the DAO flooding attack on the network performance is discussed in [8]. In a DAO attack, an attacker node transmits periodically DAO messages to its preferred parent, which in turn will forward the received DAOs to its own parent, and this continues till the DAOs reach the DODAG root. The simulation results show how the attack can damage the network performance, by significantly increasing the DAO overhead and power consumption. [9] describes the neighbor attack and DIS attacks. It presents an anomaly-based lightweight IDS, based on threshold values for detecting attacks on the RPL protocol. Each normal node in specific time intervals calculates the average and standard deviation in the current time window, and then generates a threshold value. Whenever the statistics exceed the threshold, an attack is indicated. The nodes identified as attackers are temporarily blocked.

To the best of our knowledge from literature survey, none of the earlier papers discuss localization techniques based on impact of worst-parent attack.

## 2 System Description

Figure 1 shows the scenario of a smart building. The building is managed by the Building Management System (BMS), which has the IoT sub-systems for water management, Heating Ventilation and Air Conditioning (HVAC), lighting, etc. The ‘things’ of the IoT sub-systems use the 6LowPAN protocol stack involving RPL, and thus are prone to RPL attacks. The usage of direct IP devices is also common as shown for the security camera sub-system of Fig. 1.

The sensor data of each sub-system is sent via the sink node of the 6LowPAN to a corresponding edge device. Each edge device is assumed to be an intelligent one, capable of processing sensor data and making decisions to control the actuators. This has the advantages of low latency and low network bandwidth. The intelligent edges send the status and alerts of the BMS to the remote cloud-based clients. An azure IoT is considered for the ease of deployment [10]. It consists of a collection of managed and platform services across edge and cloud that connect, monitor, and control billions of IoT assets. It mainly consists of the IoT devices, IoT Hub, storage, stream processing, and User Interface (UI) comprising of command line and visualization tools. IoT business applications can be quickly developed and deployed using Azure IoT. The information is collected by IoT devices. This information is stored and analysed in the Azure cloud, thus, providing an insight into the information, thus, information is transformed to knowledge. This knowledge is used for the business.

Each of the IoT sub-system has monitoring nodes as suggested in [11], capable of monitoring its neighbors and reporting the statistics periodically every $T$ time units to the edge via the sink node. A separate management communication network is assumed between the monitoring nodes and the sink or edge. That is, each monitoring node has 2

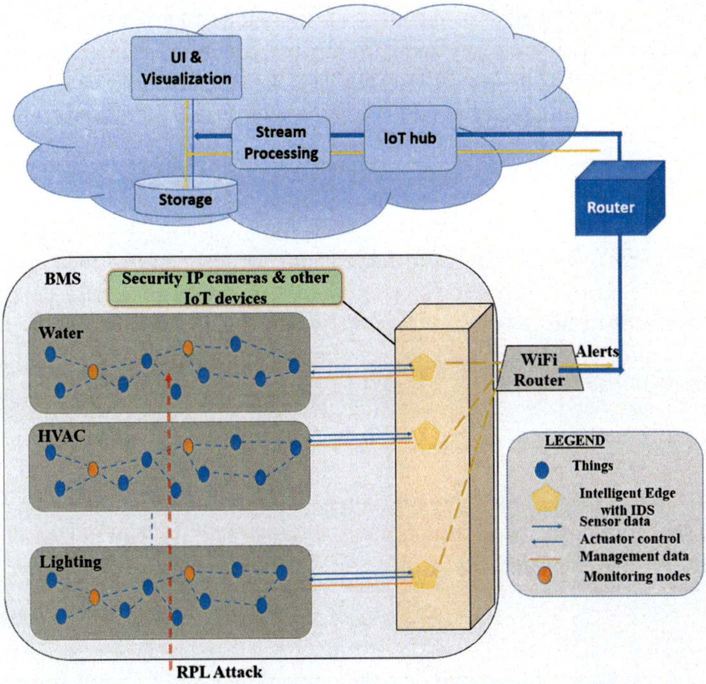


**Fig. 1.** Smart building with Building Management System & Intelligent edges

communication channels, one for sensor data and one for management information transmission [12]. The ML algorithms considered are the Support Vector Machine (SVM) and XGBoost, which are detailed in Sect. 5.

The management information consists of the extracted statistics of the RPL control message packets of the neighbors of the monitoring node of the sub-system. These statistics are aggregated by each edge device. For a given topology of the 6LowPAN, the statistics for normal, i.e., without attack situation, and with attack situation at each locality, are collected to train and test the ML models at the edge. In live (real-time) mode, the aggregated management data at each node is applied to the model to detect the attack in a locality. Thus, localized information of the presence of a malicious node is provided. This alert is sent as telemetry from the Azure edge to the Azure IoT hub.

## 3 Worst-Parent Attack

Rank property plays a vital role in the creation of an optimal topology, prevention of loop formations and to manage the control overhead. This vital role of rank, makes it a target to attacks, and calls for mechanisms to detect them. The rules of RPL define that i) the rank must increase from root to the leaves of the DODAG, in the downstream direction. ii) the rank must decrease strictly in the upstream direction.

As described earlier, in a worst-parent attack, the malicious node disobeys the rank rule and chooses the worst parent, which delays the packet transmission and increases the routing metric ETX of all descendants of the malicious node. This causes changes in the topology around the malicious node, leading to generation of more DIO messages, which further effects the number of DAO and DAO_ACK messages.

Consider a fixed topology of $N$ nodes represented by a set $V$ of integers 1 to $N$.

$$v = \{i | i = 1\ to\, N\} \tag{1}$$

Each node $i$ is at a depth $d$ from the root. $D$ is the maximum depth of the topology tree. Also, each node $i$ has $n_i$ number of descendants. Descendants of node $i$ are all those nodes for which $i$ is in the path to the root. Forwarding load of a node is the the number of packets that a node needs to forward for its neighbors during the observation time. This forwarding load $L_i$ of a node $i$ in the observation time $T$, depends on the descendants' load and its own load. This can be expressed as in (2).

$$L_i = \upsilon(n_i + 1)/T \tag{2}$$

Where $\upsilon$ is the number of packets generated by a node in time $T$, assuming all nodes are generating at the same rate. The normalized depth and the normalized number of descendants of a node $i$ is defined by (3).

$$\mu_i = d_i/D \ \& \ \eta_i = \eta_i/N \tag{3}$$

In the case of a symmetric topology tree, the forwarding load of a node is proportional to the depth. A symmetric tree is defined as a tree, which is mirror image of itself about the root node. A locality $k$, Loc($k$) is defined by a set $S_k$, which is a subset of $V$, as in (4).

$$S_k = \left\{ s | s \in V \& s \notin S_j \forall j \neq k \right\} \tag{4}$$

Additionally, forwarding load is the same, for all nodes in $S_k$.

The impact of an attack at locality $k$ is proportional to the average forwarding load of the set of nodes in the locality. Thus, for a symmetric topology, the impact factor $\gamma_k$ of a locality $k$ is as in (5).

$$\gamma_k \propto (1 - \mu_k) \tag{5}$$

For any tree topology, the impact factor is as in (6), considering both descendants and self-load.

$$\gamma_k \propto (\eta_k + 1/N) \tag{6}$$

Taking an example symmetric topology tree, where $N = 20$ (excluding root), $D = 10$, $\mu_k$ and $\eta_k$ for each locality are indicated in Fig. 2a. An example of asymmetric topology tree is shown in Fig. 2b. The nodes with the same forwarding load are grouped as a locality. In the case of asymmetric topology trees, the locations could be across hops. The values in brackets at each node are the number of descendants of that node.

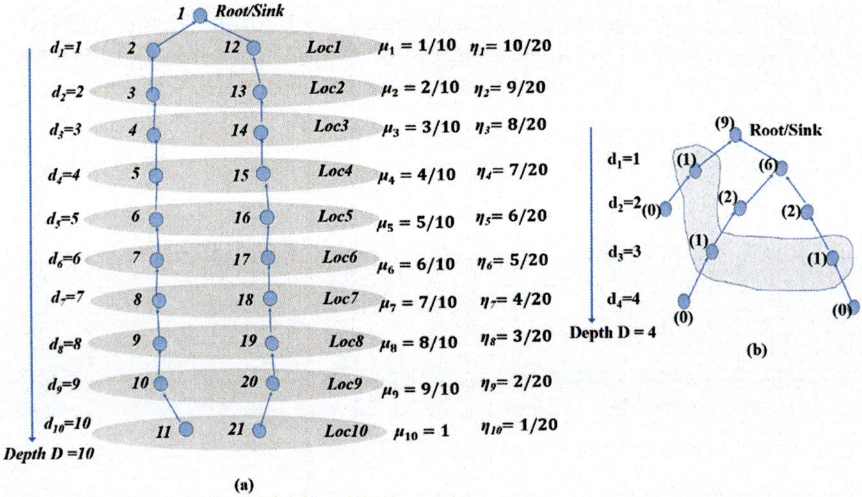

**Fig. 2.** a) Symmetric topology considered for analysis and simulation, b) Asymmetric topology with each node's number of descendants shown.

# 4 Simulation of Worst-Parent Attack

The simulation of the 6LoWPAN network nodes is performed on the Cooja simulator, running on Contiki operating system, a well-known open-source operating system for the IoT. Contiki consists of an implementation of RPL, ContikiRPL [13].

The topology set is as per Fig. 2a. The configuration parameters are set as Unit Disk Graph Medium (UDGM) Distance Loss, IEEE 802.15.4, uIPv6, 6LowPAN, UDP and Tmote Sky, OF ETX, etc. Each simulation was run for 180000 events (1 milli-second each), both with and without the attacks. The worst-parent attack has been simulated using the pseudocode given in [5]. Here, the code for the MHROF.c is modified to select the worst parent.

Each simulation has a PCAP output file. The PCAP file is parsed by a Python script, which counts the number of control packets of each type, namely DIS, DAO, DAO_ACK and DIO. Total count of control packets and attack hop number are appended to control packet counts, to form one row of the dataset. Multiple simulations, without attack and with attack at a particular hop, thus generates the dataset. Hop 0 implies no attack.

Figure 3 show box plots of the DIS, DIO, DAO and DAO_ACK messages of the network under worst-parent attack at different localities (1- 10 on x-axis). A boxplot is a standardized way of depicting the distribution of data. Five values i) minimum, ii) first quartile (Q1), iii) median, iv) third quartile (Q3), and v) maximum, are indicated in the plot. Circles depict the outliers [14]. A single malicious node is considered. The DIS packets do not show any trend, since DIS messages are generated only when a new node joins the network. The number of other control messages decrease, as the attack is launched at the lower depths of the network tree. This trend is a result of the control message packets generated, being proportional to the forwarding load of the node, which is higher as the node's location is closer to the root.

On the x-axis, 0 indicates the normal (no attack case). An attack at the leaf nodes, Loc10 is insignificant as the forwarding load is minimal. The median of the control message packet count in this case is as in normal conditions.

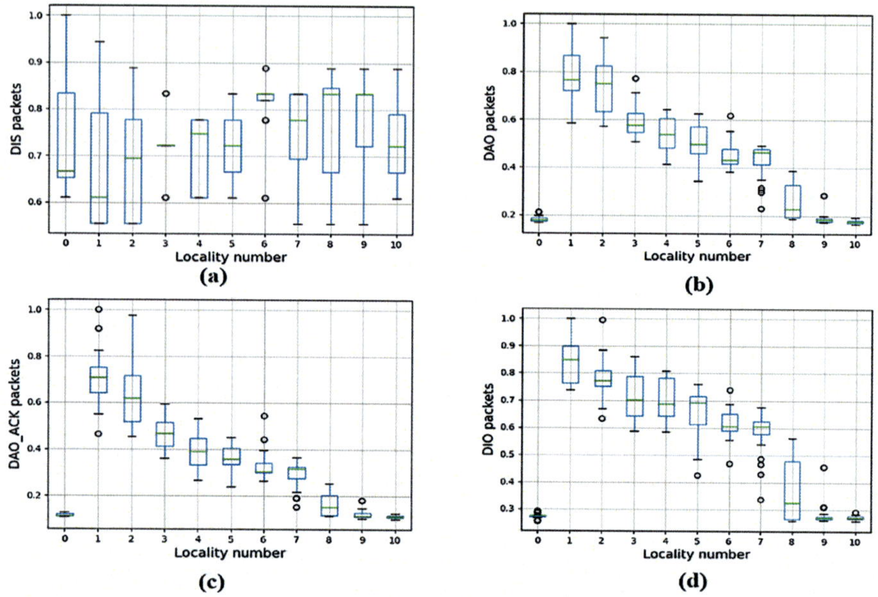


**Fig. 3.** Box plot of normalized count of control messages.

The impact factor computed from the simulated control message packets is defined as (7).

$$\text{Impact Factor} = \frac{T_{Attack\_k} - T_{Normal}}{T_{Normal}} \tag{7}$$

Where, $T_{Normal}$ is total control messages without attack, and $T_{Attack\_k}$ is total control messages with attack at locality $k$. The impact factor adjusted by empirical constant values, 3.5 & 5 in (5), (6), and simulated values computed as in (7) are plotted in Fig. 4.

# 5 Results and Discussion

## 5.1 Preprocessing and Machine Learning

After normalizing the data and filtering out DIS message packets, as their count shows no correlation with the attack location, the dataset consisting of the average DAO, DIO DAO_ACK along with the total number of packets, is used for the localization of the attack, using the HOP as the target label. Python script is used to implement the two classification models SVM and XGBoost.

SVM is a supervised learning method used or classification, regression and outliers detection [15]. SVM is more accurate as compared to other classifiers, such as logistic regression, and decision trees. SVM classifier separates data points using an optimal hyperplane, which helps in classifying new data points. XGBoost stands for "Extreme, Gradient Boosting" [17]. Gradient boosting is a machine learning technique.

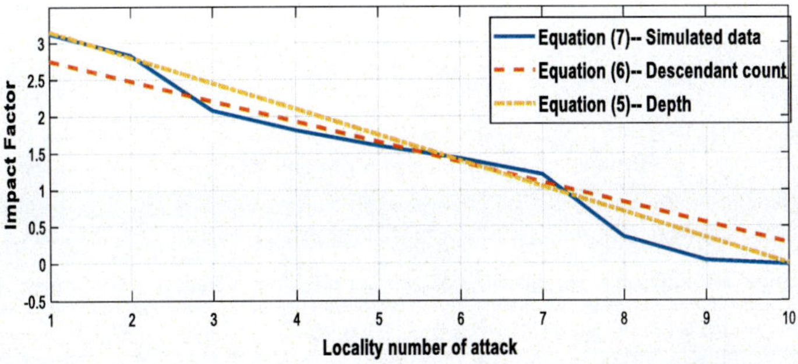


**Fig. 4.** Impact Factor computed analytically and using simulation data.

For regression and classification problems, which produces a prediction model in the form of an ensemble of weak prediction models, typically decision trees. Boosting is an ensemble technique, where new models are sequentially added to correct the errors of prior models. This is iterated till no further improvements are observed. Performance metrics of the classifiers are the accuracy, precision, recall, F-score, etc., derived from the confusion matrix.

From the various ML algorithms available, SVM was chosen because of its performance for smaller datasets, while XGBoost is the latest high performance low latency algorithm. The choice of classification algorithm depends on the available datasets (size and balance), performance requirements (accuracy and latency), and the computing power available.

The training and validation data is obtained by splitting the dataset into 2 halves, with random choice of data points.

### 5.2 Results

The results of classification are seen in Table 1. The number of classes are 11, Normal, that is, without attack and 10 localities. The accuracy with SVM is 0.67, while that with of XGBoost is 0.88. XGBoost classifier is more powerful than other two approaches, and thus is more confident, which is indicated by higher probabilities assigned to a class. However, the ML algorithms need to be evaluated for larger datasets.

### 5.3 Discussion

After training and evaluating the model, the real time application of localization is as in Fig. 5. Every $T$ units of time, the data collected by the monitoring nodes, which consists of the counts of the control message packets, is received by the corresponding edge and the data point is applied to the model. The output of classifier is normal or attack at Loc($k$). The nodes of the location under attack are further analysed at packet level, which is not in the scope of this paper. This architecture reduces the packet level inspection from $N$ nodes to the number of elements in the Loc($k$), set $S_k$. The Computation Reduction Factor (CRF) at packet level, can be expressed as in (8).

$$CRF = |S_k|/N \tag{8}$$

**Table 1.** Performance of SVM and XGBoost.

| | SVM | | | XGBoost | | | |
|---|---|---|---|---|---|---|---|
| Attack Locality | Precision | Recall | F1-Score | Precision | Recall | F1-Score | Support |
| Normal | 0.48 | 0.71 | 0.58 | 0.90 | 0.90 | 0.90 | 42 |
| Loc1 | 0.83 | 1.00 | 0.91 | 0.91 | 1.00 | 0.95 | 29 |
| Loc2 | 0.65 | 0.80 | 0.72 | 0.90 | 0.90 | 0.90 | 30 |
| Loc3 | 1.00 | 0.36 | 0.53 | 0.75 | 0.43 | 0.55 | 14 |
| Loc4 | 1.00 | 0.37 | 0.54 | 0.82 | 0.74 | 0.78 | 19 |
| Loc5 | 0.75 | 0.90 | 0.82 | 0.77 | 1.00 | 0.87 | 30 |
| Loc6 | 0.79 | 0.70 | 0.74 | 1.00 | 0.76 | 0.86 | 37 |
| Loc7 | 0.72 | 0.78 | 0.75 | 0.83 | 0.93 | 0.88 | 27 |
| Loc8 | 0.62 | 0.71 | 0.67 | 0.87 | 0.93 | 0.90 | 14 |
| Loc9 | 0.53 | 0.57 | 0.55 | 0.90 | 0.93 | 0.91 | 40 |
| Loc10 | 0.00 | 0.00 | 0.00 | 1.00 | 1.00 | 1.00 | 21 |
| Accuracy | | | 0.67 | | | 0.88 | 303 |
| Macroavg | 0.67 | 0.63 | 0.62 | 0.89 | 0.86 | 0.88 | 303 |
| Weighted avg | 0.65 | 0.67 | 0.64 | 0.88 | 0.86 | 0.86 | 303 |

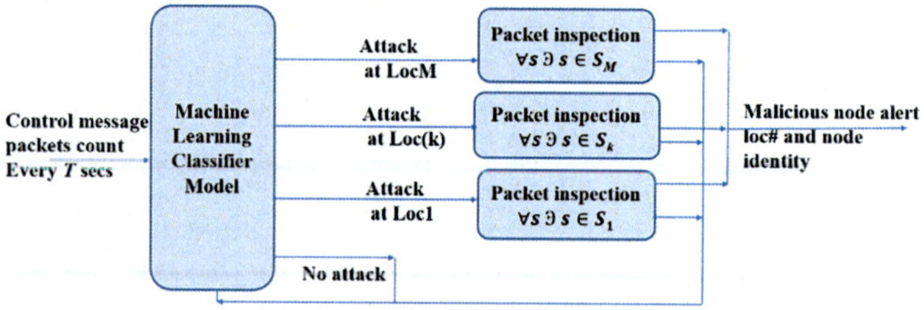


**Fig. 5.** Localization using SVM/ XGBoost in Real-time application.

## 6 Conclusions

In this paper, a smart building scenario with intelligent edges for detecting the locality of a worst parent attack is discussed. An impact factor of an attack locality has been defined and analyzed in terms of the forwarding load of the node, both for symmetric and asymmetric topologies. Simulation results show the accuracy of 0.67 using SVM and 0.88 using XGBoost. It is to be noted that the number of nodes $N$ of the network plays an important role in the accuracy obtained. $N$ defines the depth and number of descendants of a node. This in turn, defines the number of classes of the classification problem. More the number of classes, the dataset needs to be larger as the accuracy depends on the number of data points in each class. However, this would not cause any deviation from the proposed architecture, and the impact factor-based localization concept. Discussions on how the technique is applicable to any generalized network topology to reduce the number of packet level inspections, is also presented. Further studies on the effect of multiple malicious nodes, effect of mobility of nodes, and effect of using Load Balancing Objective Function (LBOF) need to be conducted.

## References

1. Shelby, Z., Bormann, C.: 6LoWPAN- The Wireless Embedded Internet, vol. 43. Wiley, New York (2011)
2. RPL: IPv6 Routing Protocol for Low Power and Lossy Networks, RFC Standard 6550, Mar (2012)
3. Verma, A., Ranga, V.: Evaluation of network intrusion detection systems for RPL based 6LoWPAN networks in IoT. Wireless Pers. Commun. **108**(3), 1571–1594 (2019)
4. Boudouaia, M.A., Ali-Pacha, A., Abouaissa, A., Lorenz, P.: Security Against Rank Attack in RPL Protocol. IEEE Network **34**(4), 133–139 (2020)
5. Le, A., Loo, L., Lasebae, A., Vinel, A., Chen, Y., Chai, M.: The impact of rank attack on network topology of routing protocol for low-power and lossy networks. IEEE Sens. J. **13**(10), 3685–3692 (2013). https://doi.org/10.1109/JSEN.2013.2266399
6. Khraisat, A., Alazab, A.: A critical review of intrusion detection systems in the internet of things techniques, deployment strategy, validation strategy, attacks, public datasets, and challenges. Cybersecur **4**, 18 (2021). https://doi.org/10.1186/s42400-021-00077-7A
7. Said, A.M., Yahyaoui, A., Abdellatif, T.: Efficient anomaly detection for smart hospital IoT systems. Sensors **21**(4), 1026 (2021)
8. Almusaylim, Z.A., Jhanjhi, N., Alhumam, A.: Detection and mitigation of rpl rank and version number attacks in the internet of things: SRPL-RP. Sensors **20**(21), 5997 (2020)
9. Wadhaj, I., Ghaleb, B., Thomson, C., Al-Dubai, A., Buchanan, W.J.: Mitigation mechanisms against the dao attack on the routing protocol for low power and lossy networks (RPL). IEEE Access **8**, 43665–43675 (2020)
10. Farzaneh, B., Montazeri, M.A., Jamali, S.: An anomaly-based IDS for detecting attacks in rpl-based internet of things. In: 5th International Conference on Web Research (ICWR), pp. 61–66 (2019)
11. https://thenewstack.io/azure-iot-edge-a-technology-primer/, Accessed 1.06.2021
12. Mayzaud, A.: Monitoring and Security for the RPL-based Internet of Things. Cryptography and Security [cs.CR]. Université de Lorraine, (2016)
13. Le, A., Loo, J., Luo, Y., Lasebae, A.: Specification-based IDS for securing RPL from topology attacks. In: IFIP Wireless Days (WD), pp. 1–3 (2011)

14. http://anrg.usc.edu/contiki/index.php/Cooja_Simulator. Accessed 1 June 2021.
15. https://towardsdatascience.com/understanding-boxplots-5e2df7bcbd51. Accessed 1 June 2021
16. https://www.datacamp.com/community/tutorials/svm-classification-scikit-learn-python. Accessed 1 June 2021
17. Chen, T., Guestrin, C.: XGBoost: A Scalable Tree Boosting System. In: Proceedings of the 22nd ACM SIGKDD International Conference on Knowledge Discovery and Data Mining, pp. 785–794 (2016)

# Towards a Fog Based Computing Conceptual Framework for Biometric Digital Identification Systems

Alphanso Murray, Ronald Brown, Sean Thorpe(✉), Maurice Edwards, Everton Walker, David Kelly, Tamra Brown, and Raj Patil

University of Technology, Kingston, Jamaica
David.Kelly@utech.edu.jm

**Abstract.** The contribution posited in this paper is the provision of a conceptual framework for the use of Biometric facial recognition camera systems as a form of digital identification within your internet of things (IOT) integrated - data cloud environments, and in of itself an overlay to a data fog network layer. The authors conduct quantitative quasi experimental assessments that seek to answer the following research questions (i) To what extent can the characteristics of an Internet of things (IOT) biometric facial recognition camera system be used to generate false alerts. The false alerts seek to determine veracity of the collected and enrolled image templates stored within our test Microsoft Azure cloud environments and compared with non-cloud deployments to determine the veracity. The characteristics referred to in question # 1 are treated as the independent variables within the study and the false alerts as the dependent variable. We assume a threshold on all alerts to be a maximum of 0.1. Our experiments show the delta change between the non-cloud and the cloud based setup was identical for matching templates that were collected. The second research question seek to determine (ii) what is a suitable unifying framework to address question # 1 raised. In case of question # 2, the overlay of using the NIST biometric standards, NIST, Cyber-Security, and the NIST digital identity provided a strong terms of reference of this work such that a unified conceptual framework based on all three were integrated. The overall results of our study based on the two research questions provide a comprehensive baseline for which benchmark testing on biometric facial recognition identification technologies located within data clouds, and situated on edge fog networks can be further studied and our work represents one of the first anywhere in literature that speaks to these application study approaches described in this paper. While we grounded our benchmark simulation studies within the University environment, future work seeks to explore enterprise wide operational use cases for supporting performance evaluations within small to medium size production environments.

**Keywords:** Biometric · Cloud · Digital · Fog · Identity · Internet of things


J. J. Gude Prego et al. (Eds.): CISIS 2021/ICEUTE 2021, AISC 1400, pp. 73–83, 2022.
https://doi.org/10.1007/978-3-030-87872-6_8

# 1 Introduction

## 1.1 Motivation

The Caribbean Community (CARICOMM) was established to promote economic integration and development among its member countries. A vision to build an ICT-enabled borderless environment to strengthen integration among its Caribbean citizens was adopted as early as February 2017 [1]. However, in order for this vision to become a reality, a robust digital identity system that allows for secure and safe interconnectedness throughout the digital domain (including an IOT cloud-based environment) must be implemented. Therefore, this study is driven by the need to develop biometric passport profiles for its citizens where this passport is the singular identity used to recognize the citizens as they travel throughout the CARICOMM. And hence the approach of a federated digital identity encapsulated as a one biometric passport dispenses with the need for disparate functional IDs such as passports, electoral ID, and driver's license, together with a tax registration number (TRN), are used to authenticate citizen's identity. Those means of identification were not intended to verify identities in other contexts, or by third parties, but rather to meet the objectives of issuing organizations. The current ID systems are not only limited in their coverage and, for some, costly to obtain, but there is a lack of interconnectivity; that is, each system's collection, processing, storage, and management of identity data is isolated from each other.

Against the background of the above, the introduction of a National Identity card (NIC) which is IOT friendly, data cloud enabled, where this IOT cloud based integration supports this biometric NIC card to run within any edge Fog computing network. As the Caribbean citizen travels globally using this card as a form of global identification, it also allows for the opportunity to use this biometric card to engage in business and trade. Therefore the use of a biometric facial recognition system managed through this NIC, provides a useful case application which has driven the reason behind this study. This work represents the first in the context of biometric facial recognition systems within an integrated IOT-cloud layered environment. Our work is further contextualized that within the University environment for which we apply our proof of context, the IOT-cloud stack for the Biometric facial recognition readers is developed to run on the edge of a data fog network. The rest of this paper is organized to look at; the background to the problem, the statement of the problem, the related work, the current state of research, the conceptual framework, an evaluation of the framework, a conclusion and future work of the study.

## 1.2 Background of the Problem

Biometric face recognition systems it is well understood have had their roots in biometric systems that date back to 5000BC for which they have been used as a form of identity passports. In recent times cloud computing is quickly becoming a common platform to offer services to governments and businesses due to its flexibility and scalability giving rise to cloud-based biometric authentication systems, also referred to biometric-as-a-service. Traditional problems in biometric security varies from facial recognition, feature mismatch, poor quality data [2]. On the matter of the cloud there are security issues relating to session hijacking, buffer overflows and denial-of-service attacks [3] thee problem becomes exacerbated when one introduces the hidden layers of concern within an Internet of Things connected data cloud environment. Therefore, when a biometric system is deployed in the cloud, we can appreciate the challenges with such an implementation. While we recognize these challenges, the pace of technology adoption is even faster such that we have to deal with the reality of deploying biometric systems within an IOT cloud base edge environment, which are now mainstream for Fog computing services.

### 1.3 Statement of Background of the Problem

Current biometric facial recognition system techniques for the Internet of things (IOT) cloud environments are not well understood, particularly from industry application use cases and the nuance of the concerns that come with same. We could argue that as it relates to biometric facial recognition systems research in traditional non-cloud environments is well published however we have not seen anywhere in the literature of its application in a fully integrated IOT-cloud stack or data fog domain. The problem becomes exacerbated when biometric facial recognition systems are deployed in an IOT cloud stack. This is a clear problem because the IOT cloud stack has become a mainstream service for the larger digital public and as if we were to run a biometric facial recognition system services within an IOT data cloud, the need to discover suitable ways to verify and identify a digital identity profile in facial recognition biometric system is a relevant problem. This specific problem is what our quantitative quasi-experimental study seeks to address by showing that we can still have a consistently higher level of reliability for facial biometric identity proofing in these lucid environments through the implementation of our novel IOT cloud-based Digital Identity framework. So in the context of this study we therefore asked two (2) research questions – (i) To what extent can the characteristics of a Biometric facial recognition profile collected from Internet of Things (IOT) facial cameras situated within a data cloud environment be used to trigger a false alert ? This false alert rate is measured in terms of acceptance and error rates as a means of verifying attributable identity of the facial images collected for template enrolment and matching (ii) What is a suitable unifying framework for evaluating question # 1 above? From question # 1, the characteristics represent the independent variables and the false alert represents the dependent variable in this study.

## 2 Related Works

While it is well documented that the history of biometric digital identity systems date back to 5000 BC, it is with the advent of the internet since the 1960's–70's that is has become ubiquitous. And hence the scope of identity systems now comprises (i) human based digital identities which includes your face, iris, speech, and gait recognition (ii) non-human identities include the identity assertions related to your IOT objects.

DNA identification has typically been used in forensic sciences, but it is now being pursued as a biometric technology. There are still technology issues such as invasive data capture and accessing time, which runs in hours and not in seconds. The definition of biometrics specifically refers to automated recognition, and although current DNA analysis [4] is not completely automated, future advances in this technology should not rule out such a possibility. It's a high level of distinctiveness among individuals makes it extremely promising technology for the future.

There are other biometric technology like retina recognition, gait recognition, scent recognition, gesture recognition, knuckle recognition and others that ae being researched by the community [5]. This paper does not discuss those other technologies beyond this point, as many have not yet transitioned to world deployment standards and most tend to be unimodal in nature. In short unimodal simply suggest that they use a single trait for recognition. The ability to use multiple biometric technologies to enhance performance and eliminate weaknesses has led to the design of multibiometric systems [6]. These systems combine multiple traits (e.g. face and finger), multiple traits of the same characteristics (e.g. different fingers) or a variety of other information sources (e.g. speech).

Specific to biometric facial authentication system research, the authors in [7] designed and implemented a stand-alone access control system based on face recognition at a distance that used local binary pattern to detect participant's face and eyes. The authors in [8] developed a facial authentication system using K-nearest Neighbor (KNN) classification method and uses Principal Component Analysis for feature extraction, while in [9] a face recognition system for gender classification was implemented, the authors in [10] designed a system where a user could use their fingerprint to log into an IOT environment in an effort to strengthen user privacy while [11] looked at how biometric authentication systems could be implemented to combat identity theft in the cloud.

While biometrics, cloud and IOT are widely studied independently, the study of the areas collectively is still an emergent space. This creates a gap in the literature, one that our own body of work seeks to fill.

## 3 Current State of Research

When we consider the current state of research on Biometric IOT cloud based digital identity systems, the areas of IOT security, cloud security and biometric security independently have had their own challenges for some time now, with biometric being the long standing. So the traditional problems in biometric security includes but not limited to facial recognition, feature mismatch and poor quality data. Whilst we do not seek to debunk the concerns in the traditional biometric spaces, we are not seeking to take on those myriad of issues we are simply looking at facial recognition systems.

With facial recognition systems the underline problems traditional have to do with the data collection process and the pattern matching process as well as verifying those patterns form the traditional systems [12]. However we have seen very little work to test the components mentioned in the previous sentence, in the cloud, and please note on the matter of the cloud there are security issues that affect that environment independently, such as session hijacks which can lead to buffer overflow attacks and man-in-the-middle-attacks [13]. Therefore how we connect biometric facial recognition systems to the cloud is that if we were to run pattern matches for data image that we collect for virtual machine sessions, our study do not take into account of the fact that the virtual machines can be session hijacked or it could be subject to a man-in-the-middle attack, memory buffer overflow attacks and even denial of service attacks. In our study we do not argue about these specific issues because they were not a focus of our study but we are very aware that those challenges that could compromise even placing the biometric image profiles and enrolled templates into the cloud.

The problem becomes compounded in that many biometric devices are now operating as internet of things (IOT) device layer, and this overlayed with the cloud stack in essence complicates the network design at the edges of a fog based network for these types of cyber physical systems deployment. So, let us take the case of internet of things connected devices such as an access control to a building or light bulbs within a building, to turn on or off those lights within the building can be controlled by a computer system which also can be hacked. They are still subjected to open web attacks (reference) like the cloud environment and the traditional internet. So being that a biometric camera also has an IP address, they are also subjected to similar attacks such as session hijacking, memory overflow attacks and man-in-the-middle attacks. Our study does not debunk that these areas are not important individual security challenges, but when we integrate everything into a data fog layer. The ability to decouple the integrated IOT-cloud stack remains an open challenge, and formed the suitable base of our own study.

## 4 A Conceptual Framework for IOT and Data Cloud Biometric Digital Identities

The suggestions for this proposed framework has strong overlay on existing frameworks namely: (i) NIST Digital identity framework [14] (ii) NIST Cyber-Security framework [15] and (iii) NIST Biometric standards [16] from which we developed a unified framework for the IOT-cloud based environments with respect to biometric facial recognition and detection. From the literature our work represents the first single unifying framework that sufficiently demonstrates how a Human identity albeit by Face recognition or otherwise can be clearly attributed over an Internet of Things- cloud stack integrated layer. The challenge is to make sure the attributes relating to the enrolled/registered templates for your image sample remain consistently true in these lucid domains.

The suggestions of building a suitable data fog based conceptual framework for Biometric identity is premised on this Internet of things-cloud stack integration. Hence one must step back to consider a generalized biometric system being viewed as a functional decomposition of its parts in this data fog environment, which includes five subsystems with an allowed signal processes that can be used to represent this newer framework are as follows:

*Data Collection or Acquisition.* This subsystem is responsible for capturing the raw biometric sample from the user. Acquisition is typically performed using a sensor that could require physical interaction with the user. Thus this subsystem represents the only point of interaction between a user and a biometric system and hence, the source of all interaction errors that are injected from the system. The biometric data that is collected at this stage can reside in a physical layer, a cloud layer and IOT layer. Importantly to note that errors introduced during the acquisition process will propagate throughout the rest of the system and increase the probability of system errors.

*Enrollment.* This subsystem is responsible for extracting unique features from the biometric sample provided by the user. This module pre-processes the sample to create a feature representation for subsequent use for either storage or for authentication. Storage is now compounded to address not only physical storage but virtual network storage. The feature representation from this subsystem is used for two purposes. For initial enrollment of a user the sample the feature representation is used as the reference sample. However for identity proofing the feature sample that is extracted is used for verification purposes.

*Data Storage.* This subsystem stores the feature representation produced by the enrollment subsystem. This feature is representation that is stored for future use can either be centralized i.e. is stored on a physical server or within data cloud servers leveraged across several internet of things devices, on a smart card or personal storage media. The framework also consist of a separate data storage responsible for storing the results from the pattern matching process.

*Authentication.* This subsystem compares two feature representations and produces a similarity score. The similarity score is the degree of confidence that the two samples are from the same individual. A biometric matching subsystem is probabilistic in nature as two samples from the same individual does not necessarily have to provide a perfect match. In comparison, password and cryptographic techniques require a perfect match in order to declare it successful. Due to the human interaction with the acquisition subsystem, successive samples from the same individual are never exactly the same. Instead of providing a binary response, a similarity score is calculated.

*Federation.* The framework also supports a federated approach, in which the user's biometric identity information and the pattern recognition function is managed by a trusted third party. In a federated setup the pattern matching process can span across several IOT cloud-based infrastructures. The results of the pattern matching process can be centralized i.e. the results are stored on a physical server or within data cloud servers leveraged across several internet of things devices.

*Pattern Recognition.* This subsystem uses the similarity score generated by the matching subsystem and compares it to a threshold value to generate success or failure decision. The threshold value indicates the variability allowed between two biometric samples for them to be considered from the same source. The threshold value is also a reflection of the acceptable level of risk of the biometric system owner or owners. The threshold value plays an important role in decision errors produced by the biometric systems. The decisions about the specific threshold value should only be taken after careful discussions among all organizational units that are affected by the biometric system.

An overview of our IOT cloud based Digital Identity framework is presented in Fig. 1.

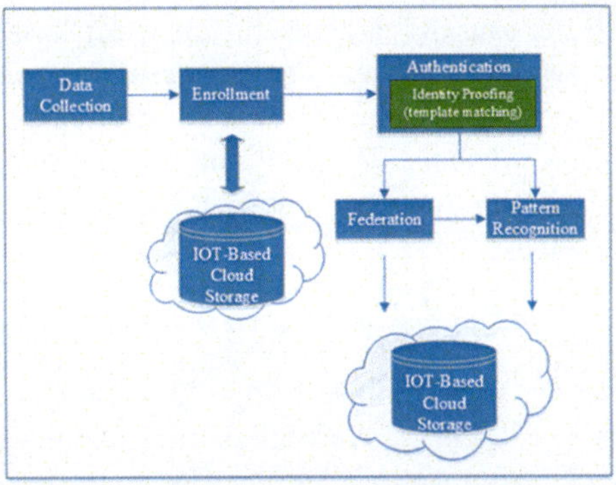


**Fig. 1.** An overview of our IOT cloud based Digital Identity Framework

## 5 Evaluation of the Conceptual Framework

To evaluate the accuracy of our proposed framework we tested a facial biometric recognition system within a pre-cloud environment and a cloud environment using four (4) participants' image databases. The accuracy of the system was determined by calculating the false acceptance rate (FAR) and the false rejection rate (FRR) for each participant's image database for both environment to determine if there was any variance in the results. The results of the matching process for the each participant's image templates was also compared in both environments to determine any variations at that granular level as a function of the pattern recognition within our framework. An identical result for the FAR, FRR and the individual template matching suggested that the template matching component of our framework of the pattern recognition was functioning with a degree of accuracy. Considering that we ran our benchmark experiments in a closed on premise Azure cloud network and compared this with a non-cloud environmental setup on a local Windows environment, with no variance or delta change on pattern match for enrolled templates, is significant for this baseline test case.

The FAR can be defined as the frequency at which a participant biometric sample from a user is incorrectly "matched" to a reference sample. "Incorrectly matched" refers to the samples belong to two different participants, while the FRR can be defined as the frequency at which a participant biometric sample is not a match to the correct reference sample. A correct reference sample suggest that both samples are from the same participant.

To calculate the FAR our four (4) participant's image databases were compared to a reference image of 5$^{th}$ participant. The FAR was calculated on a per participant basis and was found by dividing the number of images that match the reference image divided by total number of attempts. The total number of attempts is the number of images in the participant image database minus the images that failed to enroll. The formula for calculating the FAR is below:

$$FAR = \#\ of\ Matched\ Images\ /\ (\#\ of\ images - \#\ of\ Images\ failed\ to\ enroll)$$

To calculate the FRR our four (4) participant's image databases were compared to a reference image of their own. The FRR was also calculated on a per participant basis and was found by dividing the number of images that failed to match the reference image divided by total number of attempts. The total number of attempts is the number of images in the participant database minus the images that failed to enroll. The formula for calculating the FAR is below:

$$FAR = \#\ of\ Unmatched\ Images\ /\ (\#\ of\ images - \#\ of\ Images\ failed\ to\ enroll)$$

To fully stimulate our own facial biometric Fog edge network we configured our IOT cloud architecture as shown in Fig. 2. We configured a Windows Server 2019 virtual machine on the Microsoft Azure cloud platform. To stimulate our IOT based cloud storage, a 15 GB cloud data collection sandbox was configured as a file system on a physical machine using Google Backup and Sync application. The cloud data collection sandbox was also configured within the Azure Server 2019 environment. Therefore image data stored on the cloud data collection sandbox configured on physical machine automatically sync to the Azure Windows Server 2019 VM in the cloud environment. An IOT camera was configured to capture and save participant's images to the cloud data collection sandbox within the physical environment. The facial matching algorithm was configured within the Azure Windows Server 2019 environment where it ran the pattern matching to verify the identity of the images captured from the IOT camera and stored in the cloud data collection sandbox.

To evaluate the system however, the participant's image database that was tested in the pre-cloud environment was copied to the cloud data collection sandbox. The detailed automated biometric facial authentication process as described was done as a proof of concept, which is not covered in this paper.

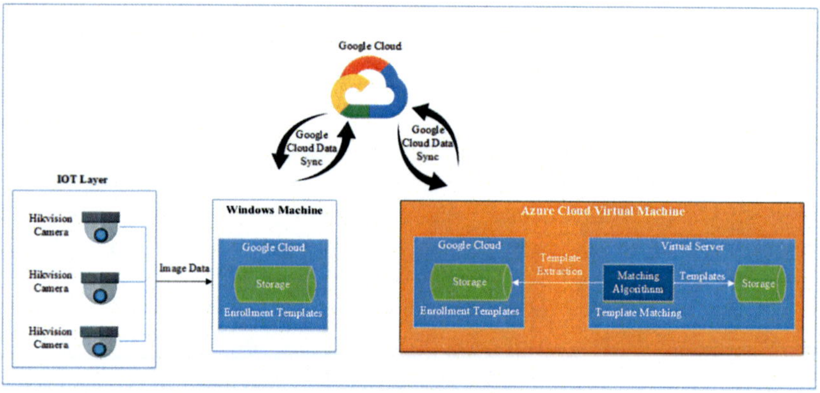


**Fig. 2.** An Overview of our IOT Cloud Architecture.

### 5.1 Results

Four image databases were prepared, 1000 images each for 3 of the participants and 847 images for the fourth participant. One reference image was prepared for each of the 4 participant including a reference image of for 5th participants.

To determine the FAR, the image databases of four of our participants were compared to the reference image of our 5th participant using a face matching algorithm. The results showed that for Participant 1, 215 templates failed to enroll, 71 templates matched the reference image and 561 templates did not match. Participant 2 had 2 templates that matched the reference image, 765 templates that did not match while 233 templates failed to enroll. Participant 3 however had 525 templates that matched the reference image while 56 templates did not match and 419 templates failed to enroll. Our final participant (Participant 4) had 7 templates that matched the reference image, 590 templates that did not match and 403 templates failed to enroll. The False acceptance rate of the system for Participant 1, Participant 2, Participant 3 and Participant 4 were 0.11, 0.002, 0.98 and 0.01 respectively. There were no variations in the FAR calculations for the pre-cloud and cloud environments for our participant's image databases. The results for the FAR test is shown in Table 1.

**Table 1.** False acceptance rate test results (Pre-cloud and Cloud)

| Match outputs | Alphanso Murray | Brictoni Piper | Davia Clarke | Ronald Brown |
|---|---|---|---|---|
| Match Found | 71 | 2 | 525 | 7 |
| No Match | 561 | 765 | 56 | 590 |
| Failed to Enroll | 215 | 233 | 419 | 403 |
| FAR | 0.11 | 0.002 | 0.90 | 0.01 |

In the FRR test the image databases of the 4 participants were compared against their own reference image. For Participant 1, 630 templates match the reference image while 2 did not match and 215 templates failed to enroll. The FRR for Participant 1 was 0.003. As it relates to Participant 2, 741 images match the reference image, 26 did not match while 233 failed to enroll. The FRR for Participant 2 was 0.033. For Participant 3 enrollment database, 580 templates matched the reference image, 26 did not match and 419 failed to enroll. The FRR for Participant 3 was 0.001. Participant 4 had 584 templates matched the reference image, 13 did not match while 403 templates failed to enroll. The FRR for Participant 4 was 0.021. Similarly to the FAR results there were no variations in the FRR results for both the pre-cloud and the cloud environments. The results are also listed in Table 2.

When we analyzed the results for the FAR and the FRR the threshold values fall with-in the NIST Biometric accuracy threshold with the exception of Participant 3 FAR results. According to the NIST Biometric Accuracy Standard, a biometric identity profile that falls outside of the 0.1 threshold is deemed unusable for identification purposes. Therefore the results are consistent with the NIST biometric accuracy standard albeit in an IOT cloud-based environment with the exception of Participant 3 FAR result (0.9),

**Table 2.** False rejection rate test results (Pre-cloud and Cloud)

| Match outputs | Alphanso Murray | Brictoni Piper | Davia Clarke | Ronald Brown |
|---|---|---|---|---|
| Match Found | 630 | 741 | 580 | 584 |
| No Match | 2 | 26 | 1 | 13 |
| Failed to Enroll | 215 | 233 | 419 | 403 |
| FRR | 0.003 | 0.033 | 0.001 | 0.021 |

which was above the accuracy threshold. This would suggest that the identity of the reference image and the image database is not one of the same person. More importantly there were no variance in the results of the cloud environment when compared to the pre-cloud which proves that it is possible to provide consistent and proactive digital identification system tracking for these lucid environments. While our data cloud testing was done on premise using a Microsoft Azure environmental setup, the assumptions do support our environment where stable. While our study does produce the merit of a benchmark testing of the technologies within the data cloud environment that was setup. However we do reconcile that an enterprise wide scenario study where we have heterogeneous IOT-data cloud stacks, the evaluation of the framework posited here could be subject to further stress testing as a further good measure of the impact of our approach and this is ongoing work by the team. Our preliminary results here seek to demonstrate merits of the benchmark testing approach on this framework only.

## 6 Conclusion and Future Work

This paper carried out a quantitative quasi experimental study that sought to evaluate the research questions as paraphrased (i) To what extent from the characteristics of a biometric facial recognition profile of images collected (within an Internet-of things - data cloud environment) will this resul in false alerts related to verifying the identity of the enrolled image(s).? (ii) What is a suitable unifying framework for making the assessment raised in question # 1? The characteristics of the images represented the independent variables within the study, while the false alerts (in terms of the acceptance and enrolment rates) represented the dependent variable in this study. The results of our study are promising and represents the first of its kind for the integrated Internet of Things – data cloud environment or what is not trending towards the edge fog layer networks. Our results revealed that we could from the experiments do suitable pattern matching within both the non- cloud and on premise private cloud based setup to verify the identity of the images collected within our sample for the same use cases. We observed that where we are guided by the standard threshold for accuracy as a 0.1 margin, there was no evident delta change in our results between the non-cloud and cloud based environments. Our work forms a suitable and novel baseline study for benchmarking biometric facial recognition profile systems testing for the edge fog environments. We must concede that a comprehensive large scale industry use case scenario study that maps detailed operational exchanges among several edge point fog devices and clouds in disparate environments still warrants a test to determine the verifiability of the enrolled templates in such situations and to track the behavior. While this work is used for biometric facial recognition systems, the experiments are reproducible to other biometric digital identification systems as a recommendation for evaluation in these new lucid domains of the fog.

## References

1. Caribbean Telecommunications Union Secretariat: Vision and Roadmap for a CARICOM Single ICT Space. 1–27 (2017). https://caricom.org/documents/15510-vision_and_roadmap_for_a_single_ict_space_-_final_version_updated.pdf. Accessed 27 Jan 2017
2. Kumar, M.: Fingerprint recognition system: issues and challenges. Int. J. Res. Appl. Sci. Eng. Techno. **6**, 556–561 (2018)
3. Ogundele, I.O., Akinade, A.O., Alakiri, H.O., Aromolaran, A.A., Uzoma, B.O.: Detection and Prevention of Session Hijacking in Web Application Management. Inte. Adv. Res. Comput. Commun. Eng. **9**(6), 1–10 (2020)
4. Jain, A.k., Ross, A.A., Nandakumar, K.: Introduction to Biometrics. In: Handbook of Biometrics, 1–22. Springer, Boston (2008)
5. ISO/IEC 19795 -1.: Information technology, Biometric performance testing
6. Siddiqui, M.F.H., Javaid, A.Y.: Multimodal facial emotion recognition framework through the fusion of speech with visible and infrared images. Multimodal Technol. Interact. **4**(3), 46 (2020)
7. Lee, H., Park, S.-H., Yoo, J.-H., Jung, S.-H., Huh, J.-H.: Face recognition at a distance for a stand-alone access control system. Sensors **20**(3), 785 (2020)
8. Wirdiani, A., Hridayami, P., Widiari, A., Rismawan, K.D., Candradinata, P.B., Jayantha, I.P.: Face identification based on K-nearest neighbor. Sci. J. Inform. **6**(2), 150–159 (2019)
9. Lin, Y., Xie, H.: Face Gender Recognition based on Face Recognition Feature Vectors. In: 2020 IEEE 3rd International Conference on Information Systems and Computer Aided Education (ICISCAE), 162–166 (2020)
10. Golec, M., Gill, S.S., Bahsoon, R., Rana, O.: BioSec: A Biometric Authentication Framework for Secure and Private Communication among Edge Devices in IoT and Industry 4.0. IEEE Consumer Electronics Magazine (2020)
11. Carmel, V.V., Akila, D.: A Survey on Biometric Authentication Systems in Cloud to Combat Identity Theft. Advance Scientific Research **7**(3), 540–547 (2019)
12. Ross, A., et al.: Some research problems in biometrics: the future beckons. In: International Conference on Biometrics (ICB), Crete (2019)
13. Baitha, A., Smitha, V.: Session hijacking and prevention techique. Int. J. Eng. Technol. **7**, 193–198 (2018)
14. NIST SP 800–63: Digital Identity Framework. https://pages.nist.gov/800-63-3/. Accessed 21 May 2021
15. NIST Cyber-security framework. https://pages.nist.gov/800-63-3/ Accessed May 21
16. M1, 4 Study Report on Biometrics in E-Authentication

# Detection of Denial of Service Attacks in an MQTT Environment Using a One-Class Approach

José Aveleira-Mata[1], Esteban Jove[2(✉)], David Yeregui Marcos del Blanco[1], María Teresa García Ordás[1], Francico Zayas-Gato[2], Álvaro Michelena[2], José-Luis Casteleiro-Roca[2], Héctor Quintián[2], Héctor Alaiz-Moretón[1], and José Luis Calvo-Rolle[2]

[1] Department of Electrical and Systems Engineering, University of León, León, Spain
{jose.aveleira,dmarcb01,m.garcia.ordas,hector.moreton}@unileon.com

[2] Department of Industrial Engineering, University of A Coruña, CTC, CITIC, Ferrol, A Coruña, Spain
{esteban.jove,f.zayas.gato,alvaro.michelena,jose.luis.casteleiro, hector.quintian,jlcalvo}@udc.es

**Abstract.** Nowadays, Internet of things (IoT) systems add connectivity to physical and common objects offering new possibilities, this systems have special features such as the low capacity of the devices and behaviour of the protocols used. These facts make cybersecurity in this kind of systems is critical. The current work uses a dataset is based on denial of service attacks over a traffic protocol used in IoT systems, called MQTT. In order to address the classification of new denial of service attacks, one-class technique is applied, obtaining good results using the Principal Component Analysis (PCA) algorithm as complement to this method.

**Keywords:** Cybersecurity · One-class classification · MQTT · Internet of Things

## 1 Introduction

The "Internet of Things" (IoT) concept denotes the addition of Internet connection to everyday objects and devices in order to provide new and/or improved functionalities. The incorporation of such connectivity to industrial environments is referred to as Industry 4.0 [20]. It provides several key upgrades such as remote access and monitoring to automated control systems, allowing real time management from any client device (PC, tablet or smartphone). Additionally, it is possible to reach a higher level of interconnection and automation by means of collected data utilization and cloud computing techniques [22]. Probably the best known IoT protocols in the link and physical layers are LORA, Zigbee, or Bluetooth; together with MQTT and CoAP for the Application layer [23, 25].

Some of the key characteristics of the devices inserted in an IoT system are [4]:


J. J. Gude Prego et al. (Eds.): CISIS 2021/ICEUTE 2021, AISC 1400, pp. 84–93, 2022.
https://doi.org/10.1007/978-3-030-87872-6_9

- Use of small, embedded systems
- Low energy needs
- A user interface is typically not required
- Light communication protocols
- Ability to operate with limited bandwidths

While the above mentioned attributes result in a reinforced versatility and flexibility of the IoT, it also introduces renewed challenges in cybersecurity due to the limited capacity of the devices to implement security primitives against recent, more sophisticated hostile actions.

One of the most relevant attacks to date was "Mirai botnet" back in 2016, in which 400.000 vulnerable IoT devices were turned into bots or zombies and used in a coordinated DDoS offensive [27]. In order to tackle the cybersecurity challenge in IoT networks, the implementation of an Intrusion Detection System or IDS is one of the most widely introduced. The two main reasons are the relatively low computing capacity required for the device as well as the adaptability of the detection rules according to the requirements of the protocol.

There are two main detection techniques: signature-based and anomaly-based:

- The former screens the network traffic, constantly comparing it with previous attack patterns to recognize a potential assault as soon as possible.
- The latter monitors the performance of the whole system in real time, comparing unusual traffic with predefined behaviors, considered to be "acceptable" or "normal".

In the case of an anomaly-based IDS, it is mandatory to develop and train a model with deep learning and machine learning techniques, together with large datasets comprised of both "normal" and "attack" patterns [11,16].

With the purpose to obtain a fully-functional, state of the art model, it is critical to use top-quality datasets such as KDD99, NSL-KDD or AWID [2,19].

In this article, the authors present the development of an intrusion detection method with the implementation of machine learning techniques to expose denial-of-service attacks (DOS) in the MQTT (Message Queuing Telemetry Transport) protocol [21], which is described in the subsequent chapter.

## 2 Case Study

The MQTT (Message Queuing Telemetry Transport) protocol operates at the application level of the TCP protocol. It is one of the most broadly implemented for new IoT system implementation [18,21]. MQTT introduces a star architecture a star topology pivoting around a central server/broker in charge of the network and message management.

Message administration follows the publication/subscription approach, in which the messages are characterized as a string implemententing a nested structure.

The separation of each hierarchy is represented by the character "/". Additionally, each client pertaining to the topic obtains the associated messages in real time with very little computation cost. Therefore IoT systems implementing MQTT can increase the number of connected devices while keeping computational requirements contained.

On the other hand, one potential vulnerability is identified in the use of the port 1883 by the vector. It could be possible that a potential attacker scans and identifies which networks are actively using the MQTT protocol and proceed with a DoS offensive [9], collapsing the broker and therefore compromising the whole system.

### 2.1 Dataset Description

In order to generate the dataset, the first step is to define and establish a test context, comprised of several IoT devices and application interacting among them by means of the previously presented MQTT protocol.

The context implements a node.js server with the Aedes [1] library acting as the broker and a web application communicating with the real time server system through websockets.

A ESP 8266 chip comprising the GPIO inputs establishes a connection with the different devices, an ultrasonic sensor HC-SR04, and another chip connected to an actuator comprised of a relay that switches a light bulb on and off.

A TP-LINK, TL-WR1043ND router manages connections among the different devices connection of the devices. In order to capture and store the traffic in the wi-fi network with the command "tcpdump -i wlan0 -w file. pcap", an openWRT firmware is installed.

As previously mentioned, the broker managing all connections is vulnerable to DoS attacks through port MQTT 1883 [3]. In this attack, a MQTT-malaria program is used by mimicking multiple clients, publishing messages of a certain size, with a predefined frequency just as quickly as possible: *malaria publish -P 1000 -n 1000 -h 192.168.1.179 -s 100 -T 100.*

The previous command generates 1000 processes, each sending 1000 messages with a fixed size of 100 characters and a frequency of 100 per second to the broker's IP 162.168.1.171 on port 1883, effectively collapsing the server temporarily.

Figure 1, the attack is made from a test environment.

All the captured traffic is stored in a PCAP file through OpenWRT. Subsequently, it is dissected considering every field in the MQTT protocol. Then, the different frames are marked depending on whether they are under attack depending on the timestamps related to the offensive. By doing so, the resulting CSV file includes the dataset of the DoS attacks.

The dataset is comprised of 28 variables containing information about the network, such as ips, tcp, and several MQTT protocol frames.

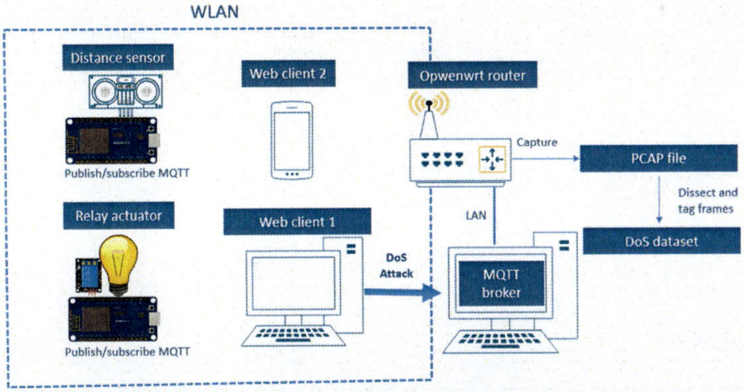


**Fig. 1.** Dataset creation environment

## 3 Classifier Approach to Detect DoS Attacks

The present section describes the classifier topology proposed to detect DoS attacks and the four one-class techniques used to implement it.

### 3.1 Classifier Approach

One of the most common issues of DoS detection is the lack of prior knowledge about the attacks. Then, a useful way to solve this problem consists of the use of one-class techniques. These techniques learn from information about the normal behaviour of a given system and, then, an attack is detected when test instances differ from the normal operation. Then, the methods are considered as semi-supervised, since only information about one class is available. The procedure to achieve a one-class classifier is depicted in Figs. 2 and 3. First, the classifier is trained to learn patterns from the normal operation. Then, the classifier is assessed.

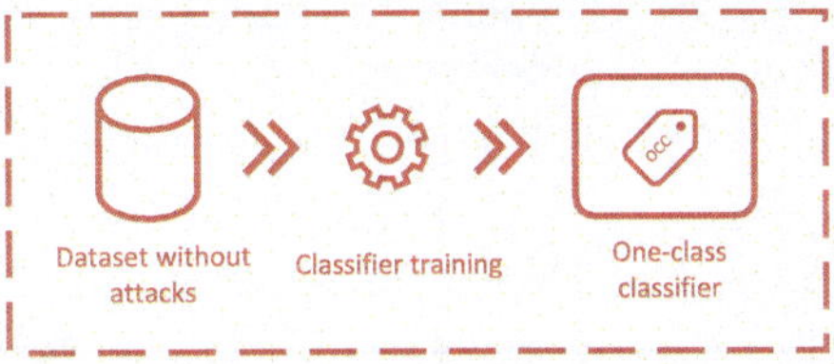


**Fig. 2.** Process to implement a one-class classifier

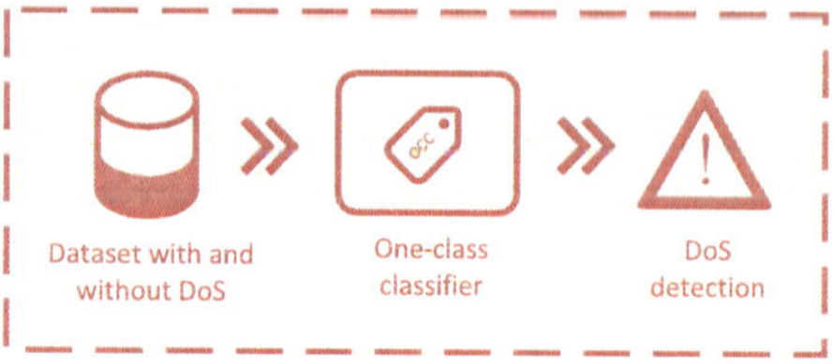


**Fig. 3.** Process to test a one-class classifier

### 3.2 Methods

Once the main concept of one-class classification is introduced and, taking into consideration the training and test phases, the main basis of four one-class techniques are described in this subsection.

**Approximate Polytope Ensemble.** A very common approach to determine the behaviour of a system whose normal behaviour consists of the delimitation of the geometric boundaries of the training set. Using this approach, the Approximate Polytope Ensemble (APE) technique has proven to be a useful boundary one-class classification technique, showing great performance has been proved over many different applications.

This technique is based on the idea of approximating the limits of a training set $S \in \mathbb{R}^n$ by means of the calculation of its convex hulls. This approach presents the disadvantage of computational cost, since the convex hull of $S$ with $n$ samples and $t$ variables has a computational cost of $O(n^{(t/2)+1})$ [6]. In this context, an alternative way to determine the approximate convex hull consists of using $\theta$ random projections on $2-D$ planes. Then, the convex hull of each projection is computed, leading to a significant reduction in the computational effort [16].

After computing the *tita* hulls of each projection, the criteria to label a new test instance takes into consideration if it is inside of all $\theta$ hulls. If the test point is outside the hull in at least one projection, it is considered anomaly (Fig. 4a). Furthermore, the adaptability of this method to different dataset shapes can be improved using a size factor. This helps to increase and decrease the different hulls, avoiding overfitting and underfitting issues. When the factor is greater than 1, the limits are widened, and the are narrowed when it is below 1.

**Non Convex Boundary over Projections.** Although APE represents a useful way to determine geometric boundaries of a target set, it presents the main weakness of dealing with non-convex sets. Then, the Non Convex Boundary over Projections (NCBoP) algorithm shown great performance with different UCI datasets [15]. Taking into consideration the APE principles, this algorithm performs significantly better in non convex datasets, since it proposes the computation of the dataset boundaries through non-convex limits, avoiding false positive classification when the anomalies are inside the convex hull.

Although this technique has significantly better performance compared to APE, its training time is higher, so its use is depends on the shape and complexity of the target set to be learnt. This technique has also the possibility of applying a size factor to avoid overfitting and underfitting [15]. An example of the goodness of NCBoP in a non-convex set is shown in Fig. 4b.

(a) (b)

**Fig. 4.** (a) Representation of APE in 3D. (b) Representation of NCBoP in 3D

**K-Means.** The K-Means algorithm has been widely used for data clustering. This unsupervised technique [7,17] groups the initial dataset in the number clusters defined by the user. To do so, the clustering error, which is the sum of distances from each point to its cluster centroid, is minimized.

Once the centroids are calculated from the training set, this technique can be used for one-class purposes taking into consideration the distance of a test data to its closest centroid. If this distance is higher than the distance of every cluster data to the centroid, the anomaly is detected [8,14].

**Principal Component Analysis.** The well known Principal Component Analysis (PCA) algorithm has been commonly applied especially for dimension reduction problems [12,13,26]. However, it offered effectiveness in one-class classification problems and anomaly detection taks.

The PCA technique calculates the directions with higher data variability by means of the eiginvectors of the covariance matrix. Once these directions, known as principal components, are calculated, they are used to perform linear projections of the original data. Consequently, the data projected has lower dimensions. The procedure to implement a PCA classifier es based on the distance between the original data and the data projected, also known as reconstruction error. If a test data presents a reconstruction error greater than the registered in the training process, the anomaly is detected.

### 3.3 Experiments Setup

The preprocessing stage, the different techniques configuration and the measures used to evaluate the performance of each classifier are described in this section.

### 3.4 Preprocessing

First, from the original dataset with 66 variables, the ones without variation where discarded since they do not contain valuable information. Furthermore, each categorical variable was converted using a natural number codification. Furthermore, with the aim of implementing the best classifier, the dataset was normalized using a 0 to 1 interval and also using the z-score method [24].

**Techniques Configuration.** The four one-class algorithms have been tested using different configurations by sweeping their hyperparameters. These configurations are summarised next:

- APE. The number of projections checked was 5, 10, 50, 100 and 500. The size parameter was set to 0.8, 1, 1.2, 1.4, 1.6, 1.8 and 2 to evaluate different restriction boundaries.
- NCBoP. The number of projections checked was 5, 10, 50, 100 and 500. The size parameter was set to 0.8, 1, 1.2, 1.4, 1.6, 1.8 and 2 to evaluate different restriction boundaries.
- K-Means. The number of clusters in which the dataset is divided varied from 1 to 10.
- PCA. The number of principal components was varied from 1 to $n-1$, being $n$ the number of variables of the training set.

**Performance Measurement.** The evaluation of the classifier performance has been done though the Area Under the Receiver Operating Characteristic Curve (AUC) [5]. This parameter, in percentage, has offered a really interesting performance, with the particular feature of being insensitive to changes in class distribution, which is a key factor in one-class classification tasks [10]. Furthermore, the time needed to train each classifier is also presented as a measure of the computational cost. The tests were validated following a $k-fold$ cross-validation method, using k = 5.

### 3.5 Results

After the preprocessing stage and the experiments detailed in the previous subsections, the obtained results are presented below. The best AUC results achieved for each size factor in APE technique are presented in Table 1. It is interesting to remark that the best result belongs to a size factor of 0.8, which means a hull contraction. The NCBoP technique has shown significantly better results than APE, with a top value of 94,82% with 100 projections and a Size Factor of 1 (Table 2). However, this configuration presents a higher train time. Regarding K-Means, the best results for each cluster configuration is shown Table 3. A very performance is achieved using this clustering algorithm, with a highest value of 94,969 %. Besides the good behaviour of these classifiers, a low training time is also a great advantage of this method. In a similar way, PCA classifiers show great performance regardless the number of components taken into consideration Table 4, with a significantly low training time.

**Table 1.** Best AUC results for APE algorithm

| APE | | | |
|---|---|---|---|
| Size factor | Proj. | AUC (%) | Train time (s) |
| 0,8 | 5 | 70,496 | 0,029 |
| 1 | 5 | 50,000 | 2,450 |
| 1,2 | 5 | 50,000 | 0,165 |
| 1,4 | 5 | 50,000 | 0,029 |
| 1,6 | 5 | 50,000 | 1,392 |
| 1,8 | 5 | 50,000 | 0,574 |
| 2 | 5 | 50,000 | 0,376 |

**Table 2.** Best AUC results for NCBoP algorithm

| NCBoP | | | |
|---|---|---|---|
| Size factor | Proj. | AUC (%) | Train time (s) |
| 0,8 | 5 | 63,184 | 8,011 |
| 1 | 100 | 94,820 | 154,492 |
| 1,2 | 5 | 58,709 | 9,240 |
| 1,4 | 5 | 50,891 | 7,188 |
| 1,6 | 10 | 56,791 | 16,993 |
| 1,8 | 5 | 50,000 | 8,494 |
| 2 | 5 | 50,000 | 7,804 |

**Table 3.** Best AUC results K-Means algorithm

| K-means | | |
|---|---|---|
| Clusters | AUC (%) | Train time (s) |
| 1 | 90,419 | 0,058 |
| 2 | 94,969 | 0,102 |
| 3 | 93,525 | 0,126 |
| 4 | 91,724 | 0,153 |
| 5 | 90,928 | 0,170 |
| 6 | 92,369 | 0,191 |
| 7 | 91,320 | 0,225 |
| 8 | 91,532 | 0,283 |
| 9 | 90,853 | 0,309 |
| 10 | 91,394 | 0,271 |

**Table 4.** Best AUC results PCA algorithm

| PCA | | |
|---|---|---|
| Components | AUC (%) | Train time (s) |
| 1 | 92,645 | 0,122 |
| 5 | 91,926 | 0,094 |
| 10 | 93,253 | 0,164 |
| 15 | 92,717 | 0,165 |
| 20 | 93,561 | 0,164 |
| 25 | 92,449 | 0,168 |
| 30 | 92,343 | 0,097 |
| 35 | 92,553 | 0,120 |
| 38 | 92,343 | 0,097 |

## 4 Conclusions and Futures Works

The results summarised in previous section show a good performance of NCBoP in one of all possible configurations, competing with techniques such as K-Means and PCA, whose good performance is achieved for all configurations in the task of detecting Denial of Service attacks. Although the three techniques presented similar behaviour, the computational cost of K-Means and PCA makes these techniques more suitable for this application. Focusing on AUC value, the best configuration is achieved using K-Means with 2 clusters. The APE results are not as good as the three other algorithms, which could be caused by the non-convex shape of the dataset.

Due to the good performance of clustering algorithms and dimensional reduction techniques, other one-class techniques based on similar idea may be applied: k-centers, k-nn, autoencoders. On the other hand, the use of dimensional reduction techniques could be considered prior to the classifier implementation in future works. In this case, reducing the data dimension could lead to better results.

**Acknowledgements.** Spanish National Cybersecurity Institute (INCIBE) and developed Research Institute of Applied Sciences in Cybersecurity (RIASC).

CITIC, as a Research Center of the University System of Galicia, is funded by Consellería de Educación, Universidade e Formación Profesional of the Xunta de Galicia through the European Regional Development Fund (ERDF) and the Secretaría Xeral de Universidades (Ref. ED431G 2019/01).

## References

1. GitHub - moscajs/aedes: Barebone MQTT broker that can run on any stream server, the node way. https://github.com/moscajs/aedes
2. KDD Cup 99 Dataset (1999). http://kdd.ics.uci.edu/databases/kddcup99/kddcup99.html
3. Andy, S., Rahardjo, B., Hanindhito, B.: Attack scenarios and security analysis of MQTT communication protocol in IoT system. In: 2017 4th International Conference on Electrical Engineering, Computer Science and Informatics (EECSI), pp. 1–6. IEEE (2017)
4. Blaauw, D., et al.: IoT design space challenges: circuits and systems. In: 2014 Symposium on VLSI Technology (VLSI-Technology): Digest of Technical Papers, pp. 1–2. IEEE (2014)
5. Bradley, A.P.: The use of the area under the roc curve in the evaluation of machine learning algorithms. Pattern Recogn. **30**(7), 1145–1159 (1997). https://doi.org/10.1016/S0031-3203(96)00142-2
6. Casale, P., Pujol, O., Radeva, P.: Approximate convex hulls family for one-class classification. In: Sansone, C., Kittler, J., Roli, F. (eds.) MCS 2011. LNCS, vol. 6713, pp. 106–115. Springer, Heidelberg (2011). https://doi.org/10.1007/978-3-642-21557-5_13
7. Casteleiro-Roca, J.L., Javier Barragan, A., Segura, F., Luis Calvo-Rolle, J., Manuel Andujar, J.: Intelligent hybrid system for the prediction of the voltage-current characteristic curve of a hydrogen-based fuel cell. Rev. Iberoamericana Autom. Inform. Ind. **16**(4), 492–501 (2019)
8. Casteleiro-Roca, J.L., Jove, E., Gonzalez-Cava, J.M., Pérez, J.A.M., Calvo-Rolle, J.L., Alvarez, F.B.: Hybrid model for the ANI index prediction using remifentanil drug and EMG signal. Neural Comput. Appl. **32**(5), 1249–1258 (2020). https://doi.org/10.1007/s00521-018-3605-z
9. Chen, F., Huo, Y., Zhu, J., Fan, D.: A review on the study on MQTT security challenge. Proceedings - 2020 IEEE International Conference on Smart Cloud, SmartCloud 2020, pp. 128–133 (2020). https://doi.org/10.1109/SmartCloud49737.2020.00032
10. Fawcett, T.: An introduction to ROC analysis. Pattern Recogn. Lett. **27**(8), 861–874 (2006)
11. Garcia-Teodoro, P., Diaz-Verdejo, J., Maciá-Fernández, G., Vázquez, E.: Anomaly-based network intrusion detection: techniques, systems and challenges. Comput. Secur. **28**(1), 18–28 (2009). https://doi.org/10.1016/j.cose.2008.08.003

12. Jove, E., et al.: Missing data imputation over academic records of electrical engineering students. Log. J. IGPL **28**(4), 487–501 (2020)
13. Jove, E., Casteleiro-Roca, J.L., Quintián, H., Méndez-Pérez, J.A., Calvo-Rolle, J.L.: A fault detection system based on unsupervised techniques for industrial control loops. Expert Syst. **36**(4), e12395 (2019)
14. Jove, E., Casteleiro-Roca, J.L., Quintián, H., Méndez-Pérez, J.A., Calvo-Rolle, J.L.: Virtual sensor for fault detection, isolation and data recovery for bicomponent mixing machine monitoring. Informatica **30**(4), 671–687 (2019)
15. Jove, E., Casteleiro-Roca, J.L., Quintián, H., Méndez-Pérez, J.A., Calvo-Rolle, J.L.: A new method for anomaly detection based on non-convex boundaries with random two-dimensional projections. Inform. Fusion **65**, 50–57 (2021)
16. Jove, E., Casteleiro-Roca, J.L., Quintián, H., Simić, D., Méndez-Pérez, J.A., Luis Calvo-Rolle, J.: Anomaly detection based on one-class intelligent techniques over a control level plant. Log. J. IGPL **28**, 502–518 (2020)
17. Jove, E., Casteleiro-Roca, J.L., Quintián, H., Zayas-Gato, F., Vercelli, G., Calvo-Rolle, J.L.: A one-class classifier based on a hybrid topology to detect faults in power cells. Log. J. IGPL (2021)
18. Juwita, P.S., Fadhil, R., Damayanti, T.N., Ramadan, D.N.: Smart parking management system using SSGA MQTT and real-time database. Telkomnika (Telecommun. Comput. Electron. Control) **18**(3), 1243–1251 (2020). https://doi.org/10.12928/TELKOMNIKA.v18i3.14869
19. Kolias, C., Kambourakis, G., Stavrou, A., Gritzalis, S.: Intrusion detection in 802.11 networks: empirical evaluation of threats and a public dataset. IEEE Commun. Surv. Tutor. **18**(1), 184–208 (2015)
20. Lee, J., Kao, A.: Industry 4.0 factory in big data environment. tec. News HARTING's Technol. Newsl. **26**(Industry 4.0 Factory in Big Data Environment), 8–9 (2014)
21. Liu, J., Kantarci, B., Adams, C.: Machine learning-driven intrusion detection for Contiki-NG-based IoT networks exposed to NSL-KDD dataset. In: WiseML 2020 - Proceedings of the 2nd ACM Workshop on Wireless Security and Machine Learning, pp. 25–30 (2020). https://doi.org/10.1145/3395352.3402621
22. Wollschlaeger, M., Sauter, T., Jasperneite, J.: The future of industrial communication. IEEE Ind. Electron. Mag. **11**(4), 17–27 (2017). https://doi.org/10.1109/MIE.2017.2649104
23. Sethi, P., Sarangi, S.R.: Internet of things: architectures, protocols, and applications. J. Electr. Comput. Eng. **2017** (2017)
24. Shalabi, L.A., Shaaban, Z.: Normalization as a preprocessing engine for data mining and the approach of preference matrix. In: 2006 International Conference on Dependability of Computer Systems, pp. 207–214, May 2006. https://doi.org/10.1109/DEPCOS-RELCOMEX.2006.38
25. Vega Vega, R., Quintián, H., Calvo-Rolle, J.L., Herrero, Á., Corchado, E.: Gaining deep knowledge of android malware families through dimensionality reduction techniques. Log. J. IGPL **27**(2), 160–176 (2019)
26. Wu, J., Zhang, X.: A PCA classifier and its application in vehicle detection. In: IJCNN 2001. International Joint Conference on Neural Networks. Proceedings (Cat. No. 01CH37222), vol. 1, pp. 600–604. IEEE (2001)
27. Zhang, X., Upton, O., Beebe, N.L., Choo, K.K.R.: IoT botnet forensics: a comprehensive digital forensic case study on Mirai botnet servers. Forensic Sci. Int.: Digit. Invest. **32**, 300926 (2020). https://doi.org/10.1016/j.fsidi.2020.300926

# Fuzzy Logic System for Risk and Energy Efficiency Estimation of Driving Maneuvers

F. Barreno[1](✉), M. Santos[2], and M. Romana[3]

[1] Computer Science Faculty, Complutense University of Madrid, 28040 Madrid, Spain
febarren@ucm.es

[2] Institute of Knowledge Technology, Complutense University of Madrid, 28040 Madrid, Spain
msantos@ucm.es

[3] Civil Engineering School, Technical University of Madrid, 28040 Madrid, Spain
manuel.romana@upm.es

**Abstract.** This paper presents a Soft Computing based system to identify risky driving maneuvers on conventional two-lane roads. Road design and vehicle dynamics are considered. Specifically, a fuzzy rule-based Mamdani-type inference system is applied. The vehicle dynamics features are measured by smartphone inertial sensors. The real data obtained from the GPS, accelerometer, and gyroscope are used to identify the driving maneuvers. A conventional two-lane road located in the Madrid Region, Spain is used for this research. The results obtained with the fuzzy system are promising and suggest that this intelligent system can be used to warn drivers of a risky maneuver in real time for a safer, more ecological and comfortable driving.

**Keywords:** Soft Computing · Fuzzy logic · Smartphone sensors · Risk · Ecological driving · Identification · Classification · Two-lane road

## 1 Introduction

Road driving user experience regarding both energy saving and driving safety, depends on different elements. Although the most influential and significant factor that impacts energy saving, safe and comfort is the way a driver behaves, the geometry of a road also influences safety driving [1, 2]. Moreover, road infrastructure deteriorates due to use, aging, etc., and may make risky a driving maneuver that otherwise could be a safe one.

The vehicle dynamics help define different types of maneuvers according to some measures. These features are, among others, inertial data such as accelerometers and gyroscope readings, and GPS measures, that have been shown useful in this field [3].

From another point of view, driving conditions may change, either because user perception can be different depending on the vehicle driven or because the road geometry may make the driving more complicated due to its worn out, sinuosity, etc. Hence, the driver subjective perceptions should be taken into account as the driver may carry out maneuvers that are incorrect, reckless, or risky [4].


J. J. Gude Prego et al. (Eds.): CISIS 2021/ICEUTE 2021, AISC 1400, pp. 94–104, 2022.
https://doi.org/10.1007/978-3-030-87872-6_10

In this work, an intelligent system includes driver´s perception of the acceleration while driving is designed. The final goal is to estimate the risk of a maneuver and energy consumption. To do so, a Soft Computing technique, particularly fuzzy logic, is applied to identify risky maneuvers based on inertial sensor measures. The results here obtained on conventional two-lane road of the Madrid Region, Spain, are encouraging and allow the determination of a more appropriate driving style for an ecological, comfortable, and safer driving.

This work also supports the interest of using the driving style recognition by inertial sensors and GPS data from smartphone devices. This low-cost system is able to provide way of driving information to detect reckless and risky maneuvers and hence, to prevent traffic accidents and obtain better vehicle energy regulation.

The identification of driving styles using artificial intelligence techniques has been previously addressed, supporting the suitability of these strategies. Indeed, a popular way of analysing driving behaviour is to move the focus to the manoeuvres as they give useful information about the driver who is performing them. So, in [5] two models that combine acceleration and deceleration and velocity data in an erratic driving detection algorithm are presented. In [6], the authors use vehicle inertial sensors to build a driver profile with the final aim of reducing the number of dangerous car maneuvers. In [7], a driving system is presented. It aims to educate the driver in energy-efficient and safe driving by using in-vehicle sensors to obtain information about the car, the driver, and the environment. An intelligent road detection system that includes pothole detection and roadway quality analysis to decrease the traffic accidents is presented in [8]. In [9], the problem of diagnosing people's driving skills under real driving conditions using GPS data and video records is considered. A comprehensive survey on the challenges associated to model and improve driving monitoring and assistance systems and their potential solutions is presented in [10].

The main difference of the work here proposed and those found in the literature is that the intelligent system designed combines, on the one hand, the geometrical features of the road as an input variable called lateral acceleration perceived by the driver. It includes the road geometry effect. On the other hand, the fuzzy system also takes into account the measures from the inertial sensors of a smartphone.

Thus, the main contribution of our paper can be summarized saying that the model proposed uses the knowledge about how drivers adjust their speed in curves so that the maximum lateral accelerations of the vehicle decrease at high speeds, together with the geometrical characteristic of the road.

This rest of the paper is structured as follows. Section 2 presents the two-lane conventional road design characteristics and the vehicle dynamics features, which constitute the basis for the identification system. In Sect. 3 the proposed fuzzy intelligent system that assigns the risk to a maneuver is described and applied to real data. Results are discussed in Sect. 4. Conclusions and future works end the paper.

## 2 Road and Vehicle Dynamics

While driving, the vehicle is subjected to forces that affect its dynamics [11]. If the brakes are activated, the vehicle decelerates. If the accelerator is pressed, the vehicle increases speed. If the vehicle turns by means of a steering maneuver, the higher the speed of the vehicle, the greater the centrifugal force on the vehicle.

Parameters influencing vehicle dynamics may change over time (tire pressure, vehicle weight, brake energy dissipation capacity, etc.) Thus, the energy that a vehicle spends changes and so does the energy-efficient ratio.

Roads are defined by their geometric characteristics, which determine whether a vehicle can travel at a certain speed with an adequate degree of comfort and safety. The design speed of a section ($V$), regardless of the maximum speed allowed by the administration, related to its geometrical features, is calculated as follows [11]:

$$V^2 = 127R\left(f_t + \frac{\rho}{100}\right) \tag{1}$$

where $V$ is design linear speed (km/h), $R$ is the curvature radius (m), $f_t$ is the maximum transversal friction mobilized coefficient, and $\rho$ is the inclination of the cross slope (%). Centripetal acceleration is then expressed as follows:

$$a_c = {V^2}/{R} \tag{2}$$

where $a_c$ is the centripetal acceleration (m/s$^2$).
A vehicle that does not surpass the appropriate speed will result in safe driving. If the maximum acceleration due to the minimum radius is not exceeded, the specific set speed will result in comfortable and energy-efficient driving.

The acceleration measured on the vehicle onward direction is the longitudinal acceleration, that is parallel to the linear velocity. The acceleration measured on the vehicle orthogonal direction is the lateral one. The acceleration due to the road is calculated, according to the AASHTO Green Book, as [12]:

$$\frac{V^2}{gR} = \rho + f_t \tag{3}$$

$$a_{road} = g(\rho + f_t) \tag{4}$$

where $a_{road}$ is the critical acceleration due to the road geometry effect (m/s$^2$), $g$ is the acceleration of gravity (m/s$^2$), $\rho$ is the road camber (m) and $f_t$ is the maximum transversal friction mobilized coefficient. This acceleration $a_{road}$ is the limit acceleration for a horizontal curve to maintain safety and comfort while driving.

Regardless of the road camber and the maximum mobilized coefficient of friction, that may be difficult to obtain, $a_{road}$ can be obtained from the measurements of inertial sensors and GPS. The acceleration due to road effect can be defined as [13]:

$$a_{road} = |\omega| \cdot v_l \tag{5}$$

where $\omega$ is the angular velocity (rad/s) and $v_l$ is the linear velocity of the vehicle (m/s).

The lateral acceleration perceived by the driver due to erratic driving is calculated by the following expression [12]:

$$a_p = |a_m| - a_{road} \tag{6}$$

Where $a_p$ is the driver's perceived acceleration (m/s$^2$), $a_m$ is the measured lateral acceleration (m/s$^2$). From the driver's point of view, this acceleration measures the "feeling" of safety and comfort of a drive.

In Fig. 1, some of these dynamic characteristics of a vehicle on the M-100 two-lane road (Madrid, Spain) are represented. That driving profile corresponds to a normal driving. In the upper image, the accelerometer measure (g, gravity) is shown; at the bottom, the gyroscope readings All of them for each coordinate, *x*, *y* and *z*.

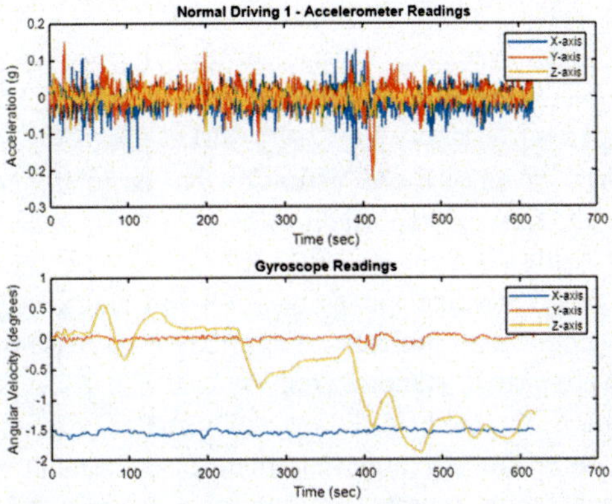


**Fig. 1.** Intertial measures of a driving on the M-100 two-lane conventional road.

## 3 Risky Maneuvers Identification by a Fuzzy Mamdani System

The data used belong to the public UAH-DriveSet dataset [14]. This dataset provides a large amount of information obtained from 6 different drivers with different vehicles (Table 1). It simulates 3 different behaviors (normal, drowsy, and aggressive), on two different road classes, a motorway (A-2, Spain) and a secondary road (M-100, Madrid, Spain). The information contains raw real-time data of measurements obtained using inertial sensors (accelerations and gyroscopes) and GPS located in a smartphone.
The raw smartphone data have a lot of noise from in-vehicle vibrations, thus data are cleaned using a Kalman filter. The filtered data have been proved to be well correlated with the vehicle movements. Drowsy driving represents slight sleepiness, which normally results in sporadic unawareness. Aggressive driving normally results in impatience and brusqueness while driving. A more detailed description of the variables and data of this dataset can be found in [14].

### 3.1 Variables of the Fuzzy Inference System

The variables we are working with are obtained at different sampling rates. The sample time of the inertial sensor is 10 Hz, and the GPS rate is 1 Hz. To synchronize those measures, GPS data have been assumed constant until the next sample, and the same for the accelerometers, that have been subsampled at 1 Hz.

**Table 1.** Driver and vehicles profiles collected at UAH-DriveSet [13].

| Driver | Genre | Age range | Vehicle | Fuel type |
|---|---|---|---|---|
| D1 | Male | 40–50 | Audi Q5 (2014) | Diesel |
| D2 | Male | 20–30 | Mercedes B180 (2013) | Diesel |
| D3 | Male | 20–30 | Citroën C4 (2015) | Diesel |
| D4 | Female | 30–40 | Kia Picanto (2004) | Gasoline |
| D5 | Male | 30–40 | Opel Astra (2007) | Gasoline |
| D6 | Male | 40–50 | Citröen C-Zero (2011) | Electric |

Linear velocity ($v_l$) is given by the GPS (km/h). The angular velocity ($\omega$) is obtained from the gyro sensor. Samples are reported in °/s.

Longitudinal acceleration is measured from the accelerometer ($z$-axis) in the vehicle onward direction. A positive value means acceleration and a negative one deceleration (braking). Large values may indicate an aggressive increase or decrease in speed, respectively. Samples are gravity standardized.

Lateral acceleration is obtained from the accelerometer ($y$-axis) and it evaluates the vehicle behavior in a curve since a large increment or decrement may be indicative of excessive turning speed. Again, samples are gravity standardized.
Therefore, the input variables of the fuzzy identification system that will be used to obtain the risky maneuvers and energy consumption are the following:

- Linear speed (from GPS).
- Longitudinal acceleration (from accelerometer, $z$-axis).
- Lateral acceleration perceived by the driver (Eq. 6).

### 3.2 Mamdani-Type Fuzzy Identification System

A Mamdani-type fuzzy [15] identification system is proposed for the classification of risky maneuvers (Fig. 2). The inputs are real measures obtained by the sensors incorporated in a smartphone on the vehicle while travelling.

The linguistic terms of these variables are represented by trapezoidal membership functions. The input variable longitudinal acceleration has five fuzzy sets assigned within the range [−0.5, 0.5]: {NH, negative high; NM, negative medium; Z, zero; PM, positive medium; PH, positive high} (Fig. 3, left). The input variable driver perceived acceleration is represented by four fuzzy sets in the range [0, 0.5]: {VL, very low; L, low; M, medium; H, high} (Fig. 3, right). The third input variable, linear velocity, has three fuzzy sets in the interval [0, 200] km/h: {L, low; M, medium; H, high} (Fig. 4).
The output are the risky driving degree and energy consumption. The first output can be normal or risky driving. The variable is in the range [0, 1] with two fuzzy sets: {Normal; Risky} (Fig. 5, left). The second output, in the interval [0, 10], can be: {L, low; M, medium; H, high} (Fig. 5, right).

The knowledge is represented by if-then rules that have three antecedents and two consequents, where $v_{ij}$ is the corresponding linguistic variable.

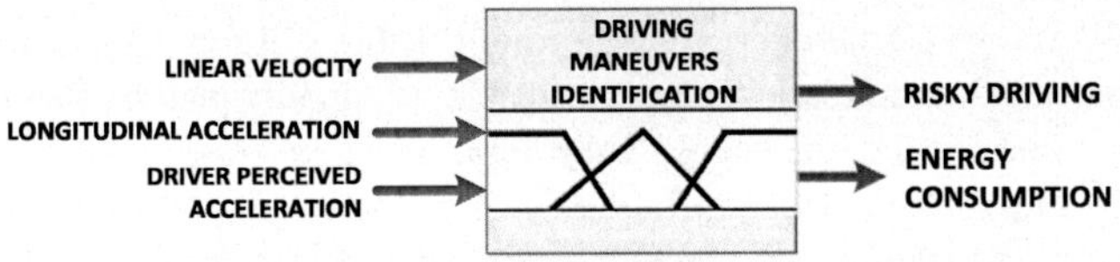


**Fig. 2.** Fuzzy rule-based system classifier of risky maneuvers.

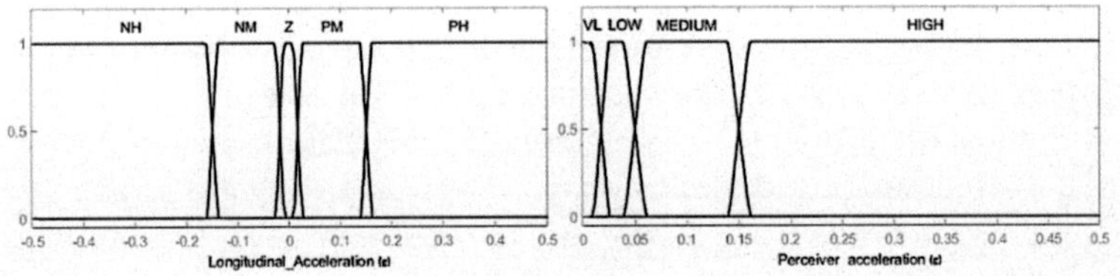


**Fig. 3.** Fuzzy input variables: longitudinal acceleration (left); lateral acceleration perceived by user (right).

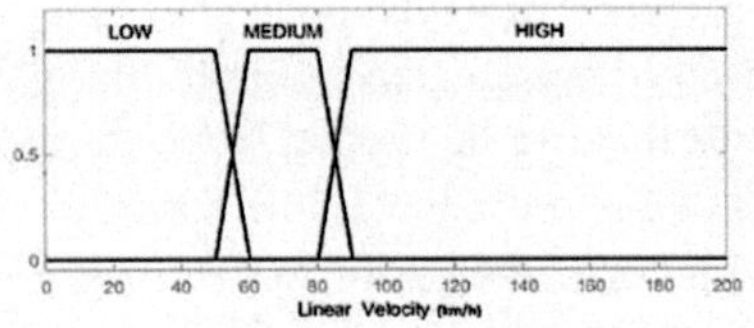


**Fig. 4.** Fuzzy input variable: linear velocity.

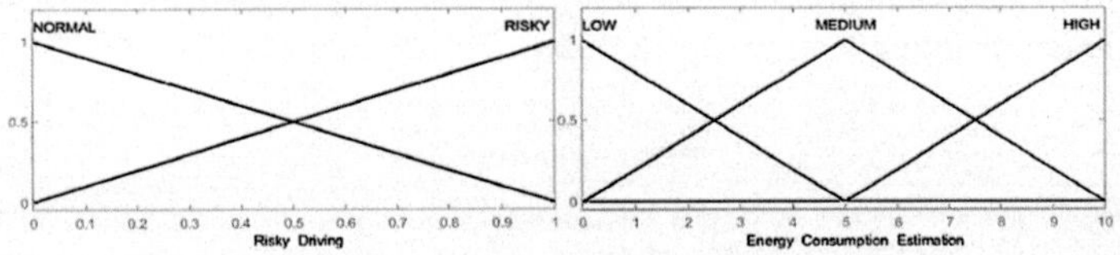


**Fig. 5.** Fuzzy soutput variables: risky driving (left) and energy consumption (right).

Ri: **IF** (*in_1 is* $v_{1j}$) **AND** (*in_2 is* $v_{2j}$) **AND** (*in_3 is* $v_{3j}$) **THEN** (*out_1 is Class_x*) and (*out_2 is Class_y*).

The approximate reasoning implemented considers the following knowledge-based criteria:

- Risky maneuvers: large accelerations and taking close curves (small radius) at relatively high speed increase the risk or recklessness of the maneuver.
- Energy consumption: large accelerations increase consumption.

The rules are the results of the combinations of these values. For example, if vehicle speed, longitudinal acceleration and acceleration perceived by the driver are all high, the maneuver is risky and the energy consumption is high; if the vehicle speed is high, the longitudinal acceleration is low, and the acceleration perceived by the driver is low, the maneuver is normal, and the energy consumption is low; if the vehicle speed is high or medium, the longitudinal acceleration is medium, and the acceleration perceived by the driver is low or medium, the maneuver is normal and the energy consumption is moderate. And so on.

In the same way, normal driving is associated with low perceived acceleration, since the curve entry is moderate, and the vehicle linear speed does not increase or decrease abruptly. If the driving is aggressive (unsafe driving behavior, deliberately carried out with disregard for safety) or drowsy (when driving while sleep deprived, the driver does not pay due attention to driving, which can lead to reckless maneuvers when losing control of the vehicle at a critical moment), perceived acceleration is greater.

## 4 Results and Discussion

The following figures show the results obtained on the M-100 two-lane road with real data with the driving profile D1 (first driver of Table 1). The upper graph shows the longitudinal acceleration (magenta) and driver's perceived acceleration (green), both in gravity units. The GPS speed is represented in km/h (blue). The second graph shows the maneuvers estimated risk that can be normal (cyan) or risky (red). The third graph represents the energy consumption rate: low (green), medium (cyan) and high (red).

Figure 6 shows the results obtained by the system with D1 normal driving profile. It can be observed that the estimated risk of a driving maneuver is larger when the driver's perceived acceleration is high (2nd graph). On the other hand, the longitudinal acceleration measured hardly varies, hence most of the maneuvers are not high energy consuming (3rd graph). In addition, the risky maneuvers coincide with the road sections that have greater changes.

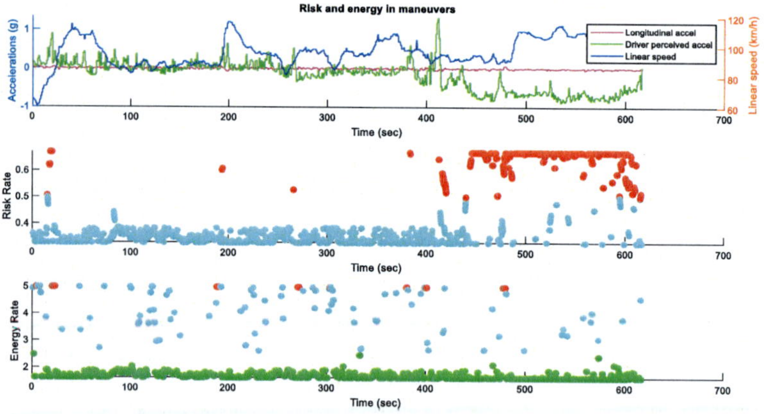


**Fig. 6.** Risky maneuvers and energy consumption for a D1 normal driving profile.

Similarly, in Fig. 7 the same variables are shown for a D1 drowsy driver profile. It can be observed that the estimated risk is greater when driver's perceived acceleration is high (2nd graph) and the energy consumption is low due to few longitudinal acceleration variations (3rd graph). Again the most influential variable is the driver's perception.

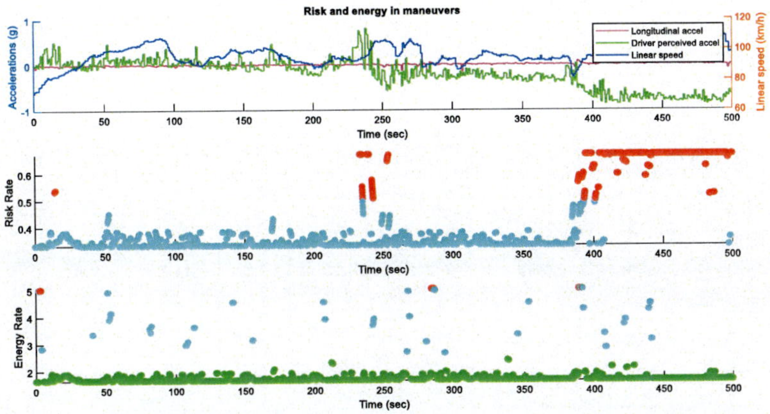


**Fig. 7.** Risky maneuvers and energy consumption of a D1 drowsy driver profile.

**Table 2.** Normal driving profile classification results.

| Driving profile | Risky driving (%) | | Energy consumption (%) | | |
|---|---|---|---|---|---|
| | Normal | Risky | Low | Medium | High |
| D1 | 52.38 | 47.62 | 83.63 | 16.37 | 0 |
| D2 | 28.01 | 71.98 | 82.77 | 17.23 | 0 |
| D3 | 31.01 | 68.99 | 86.11 | 13.89 | 0 |
| D4 | 5.01 | 94.99 | 80.29 | 19.71 | 0 |
| D5 | 33.51 | 66.49 | 92.38 | 7.62 | 0 |
| D6 | 1.31 | 98.69 | 90.49 | 9.51 | 0 |

The results of applying the Mamdani-type fuzzy system are shown in Tables 2, 3, and 4. They present the results for normal, drowsy and aggressive driving, respectively. Comparing them it is possible to say that aggressive and drowsy driving profiles are identified correctly because most of the data samples have high percentage. On the contrary, normal driving profiles D2, D3, D4, D5 and D6 are misclassified and only D1 is correctly identified. This may be explained because in normal driving there are more overtaking maneuvers, or the vehicle ahead slows down or stops, and in any case, because of the limited number of examples.

To summarize, the fuzzy rule-based system is able to identify most of the road style driving, and thus to classify risky maneuvers in normal driving. This tool can help achieve

**Table 3.** Drowsy driving profile classification results.

| Driving profile | Risky driving (%) | | Energy consumption (%) | | |
|---|---|---|---|---|---|
| | Normal | Risky | Low | Medium | High |
| D1 | 21.96 | 78.08 | 92.4 | 7.6 | 0 |
| D2 | 15.87 | 84.13 | 77.79 | 22.21 | 0 |
| D3 | 27.02 | 72.98 | 88.46 | 11.54 | 0 |
| D4 | 16.04 | 83.96 | 78.47 | 21.53 | 0 |
| D5 | 2.87 | 97.13 | 85.63 | 14.37 | 0 |
| D6 | 2.42 | 97.58 | 84.84 | 15.16 | 0 |

**Table 4.** Aggressive driving profile classification results (D6 profile not available).

| Driving profile | Risky driving (%) | | Energy consumption (%) | | |
|---|---|---|---|---|---|
| | Normal | Risky | Low | Medium | High |
| D1 | 28.42 | 71.58 | 87.18 | 12.82 | 0 |
| D2 | 19.14 | 80.86 | 77.79 | 22.21 | 0 |
| D3 | 12.36 | 87.64 | 73.57 | 26.43 | 0 |
| D4 | 16.04 | 83.96 | 73.2 | 26.8 | 0 |
| D5 | 39.74 | 60.26 | 73.28 | 26.72 | 0 |

a more safety and comfortable driving on conventional roads, using current measurements of smartphones and GPS. It can then lead to a more suitable and appropriate way of driving for these conventional two-lane roads, reducing risk and energy consumption.

## 5 Conclusions and Future Work

In this paper, a fuzzy logic system, Mamdani-type, has been designed and applied to identify risky driving maneuvers and energy consumption on conventional two-lane roads. It uses as inputs the real measures of accelerometers, gyroscope, and GPS. Based on those measurements, obtained with low-cost devices (smartphones) [16], it is possible to obtain the driver's perceived acceleration. This information, together with the geometric characteristic of the road, allows to identify risky or reckless maneuvers.

The results obtained with this intelligent system are interesting and useful. Some conclusions can be drawn. On the one hand, all driving profiles with risky maneuvers have been rightly identified. The database contains only 6 different driving profiles, each one with different characteristics. Besides, different driving styles (normal, drowsy, aggressive) have been considered. Indeed, depending on the driver, normal driving is understood differently. This may be the reason why this driving style has given the poorest results.

As expected, aggressive driving results in more energy consumption. Nevertheless, more information is necessary to accurately state the energy efficiency while driving, as some parts of the car may absorb the vehicle energy during maneuvers and this way they mask an inefficient, reckless or inattentive driving.

Different future works can be addressed. First, more data could be considered if available. This will directly impact the results and will allow to generalize some of the conclusions. In addition, following the idea presented in [17], drivers' subjective perception errors of the speed and accelerations of the preceding vehicle could be also taken into account. It would be also interesting to see the results if an in-car camera analyzing the driver's face/state of mind/eyes, corroborated with the car sensors analyzed, would help improve accuracy of the classification, potentially based on more expansive rules. Or new ways of identifying manoeuvres from vehicle telematics data, through motif detection in time-series, could be explored [18].

## References

1. Santos, M., López, V.: Fuzzy decision system for safety on roads. In: Lu, J., Jain, L.C., Zhang, G. (eds) Handbook on Decision Making, pp. 171–187. Springer, Berlin (2012). https://doi.org/10.1007/978-3-642-25755-1_9
2. Martín, S., Romana, M.G., Santos, M.: Fuzzy model of vehicle delay to determine the level of service of two-lane roads. Expert Syst. Appl. **54**, 48–60 (2016)
3. Arroyo, C., Bergasa, L.M., Romera, E.: Adaptive fuzzy classifier to detect driving events from the inertial sensors of a smartphone. In: 2016 IEEE 19th International Conference on Intelligent Transportation Systems ITSC, pp. 1896–1901. IEEE (2016)
4. Wu, C., Yu, D., Doherty, A., Zhang, T., Kust, L., Luo, G.: An investigation of perceived vehicle speed from a driver's perspective. PLoS ONE **12**(10), e0185347 (2017)
5. Zeeman, A.S., Booysen, M.J.: Combining speed and acceleration to detect reckless driving in the informal public transport industry. In: 16th International IEEE Conference on Intelligent Transportation Systems ITSC 2013, pp. 756–761. IEEE (2013)
6. Ly, M.V., Martin, S., Trivedi, M.M.: Driver classification and driving style recognition using inertial sensors. In: 2013 IEEE Intelligent Vehicles Symposium IV, pp. 1040–1045. IEEE (2013)
7. Yay, E., Madrid, N.M.: SEEDrive--an adaptive and rule based driving system. In: 2013 9th International Conference on Intelligent Environments, pp. 262–265. IEEE (2013)
8. Hsu, Y.W., Perng, J.W., Wu, Z.H.: Design and implementation of an intelligent road detection system with multisensor integration. In: 2016 International Conference on Machine Learning and Cybernetics ICMLC, vol. 1, pp. 219–225. IEEE (2016)
9. Pinilla, A.C.C., Quintero, M.C.G., Premachandra, C.: Intelligent driving diagnosis based on a fuzzy logic approach in a real environment implementation. In: 2014 IEEE Intelligent Vehicles Symposium Proceedings, pp. 102–107. IEEE (2014)
10. Khan, M.Q., Lee, S.: A comprehensive survey of driving monitoring and assistance systems. Sensors **19**(11), 2574 (2019)
11. Spanish Ministry of Development: Standard 3.1- IC. Road tracing. Order FOM/273/2016, of February 19 (2016). www.fomento.gob.es/recursos_mfom/norma_31ic_trazado_orden_fom_273_2016.pdf

12. AASHTO (American Association of State Highway and Transportation Officials): A Policy on Geometric Design of Highways and Streets, 6th edn. Washington, D.C. (2011)
13. Bergasa, L.M., Almería, D., Almazán, J., Yebes, J.J., Arroyo, R.: Drivesafe: an app for alerting inattentive drivers and scoring driving behaviors. In: 2014 IEEE Intelligent Vehicles Symposium Proceedings, pp. 240–245. IEEE (2014)
14. Romera, E., Bergasa, L.M., Arroyo, R.: Need data for driver behaviour analysis? Presenting the public UAH-DriveSet. In: 2016 IEEE 19th International Conference on Intelligent Transportation Systems ITSC, pp. 387–392. IEEE (2016)
15. Mamdani, E.H.: Application of fuzzy algorithms for control of simple dynamic plant. In: Proceedings of the Institution of Electrical Engineers, vol. 121(12), pp. 1585–1588. IET, (1974)
16. Khodairy, M.A., Abosamra, G.: Driving behavior classification based on oversampled signals of smartphone embedded sensors using an optimized stacked-LSTM neural networks. IEEE Access **9**, 4957–4972 (2021)
17. Zhang, J., Liu, M., Sun, Z.: A modified mixed car-following model considering that the connected and intelligent vehicle and non-connected vehicle. In: Journal of Physics: Conference Series, vol. 1910(1), p. 012019. IOP Publishing (2021)
18. Silva, M.I., Henriques, R.: Finding manoeuvre motifs in vehicle telematics. Accid. Anal. Prev. **138**, 105467 (2020)

# Machine Learning Ensemble Methods for Optical Network Traffic Prediction

Daniel Szostak(✉)

Department of Systems and Computer Networks, Wrocław University of Science and Technology, Wybrzeże Wyspiańskiego 27,
50-370 Wrocław, Poland
daniel.szostak@pwr.edu.pl

**Abstract.** This work focuses on finding efficient Machine Learning (ML) method for traffic prediction in optical network. Considering optical networks' characteristics, we predict fixed bitrate levels. For the considered problem, we propose two ML approaches, namely classification and regression, for which we compare performance of single ML algorithms and ensemble methods. Features for dataset instances are selected based on autocorrelation analysis. Prediction quality is analyzed using a new metric, which evaluates algorithms in terms of network resources usage and operational costs. Results obtained for datasets generated using real-world traffic characteristics show that for the examined prediction problem the regression approach yields better results than classification. Additionally, ensemble methods outperform single ML algorithms performance, i.e., the best committee returns 5,3% better results than the reference algorithm.

**Keywords:** Optical networks · Traffic forecasting · Machine Learning · Classification · Regression · Ensemble methods

## 1 Introduction

According to network reports provided by technology companies like CISCO [1] or Nokia [2], it can be seen that the Internet traffic constantly grows in the backbone optical networks. Moreover, situations like COVID-19 pandemic also cause a significant growth of traffic in the network. As a result, network operators have to constantly search for new solutions for optimization of network performance. To deliver operators tools for better network management, a concept of *cognitive optical network* [3] has been proposed. A cognitive optical network is equipped with cognitive process, which, often using Machine Learning (ML) techniques, monitor current network conditions, analyzes them and then adjust network operation to those conditions. More information about the application of ML techniques in optical networks can be found in [4].

Information about future network traffic can be highly beneficial for network operators. It allows them to optimize network resources usage, reduce operation costs, provide better quality of service, or detect anomalies in the traffic flow. It can also provide view for future expansion of physical network. There are few forecasting types in terms of


J. J. Gude Prego et al. (Eds.): CISIS 2021/ICEUTE 2021, AISC 1400, pp. 105–115, 2022.
https://doi.org/10.1007/978-3-030-87872-6_11

horizon time [5]: (i) real-time, in case of on-line traffic prediction, (ii) short-term, where traffic is forecasting up to several hours, (iii) middle-term, predicts a few days of future traffic and (iv) long-term, predict months or years of network traffic. Results of each strategy can be used as input to different network optimization tasks, as it was stated above. In this paper, we focus on a short-term traffic prediction, however our approach can be successfully applied to all forecasting types.

In this work, we consider an optical network in which traffic flows between nodes can be modeled as continuous and regular flows whose characteristics change over time. General trends, resulting from everyday user activities, can be noticed in the analyzed network traffic in a form of daily and weekly patterns. ML algorithms perceive such patterns, analyze them and base on obtained knowledge predict the future traffic in the network.

Considering fact that traffic characteristics change over the time, a large majority of works in the field, formulate problem of predicting network traffic as a *time series* problem. Main approaches used to solve such problem are Autoregressive Integrated Moving Average (ARIMA) and its numerous variations, as well as ML techniques [6]. Authors in [7] reduce required bandwidth by 18.9%, using ARIMA and SARMIA models for future traffic volumes prediction. Graph Convolutional Generative Adversarial Networks model is used in [8] to predict burst events in an optical network. The proposed algorithm outperformed the LSTM reference method. Paper [9] compares the allocation of data center traffic with and without traffic prediction process. All traffic allocation algorithms performance was improved by using ML techniques. Authors in [10] compared three algorithms during task of forecasting the amount of traffic in TCP/IP-based networks, i.e. ARIMA, Holt-Winters, and neural network. They obtained a 1–3% error for the 5 min forecasts, 11% to 17% error for 2 h forecast and 13–22% error for daily forecast. Although that many works have presented promising results, employment of ML methods for solving network problems is still in its early stage and there is a high demand for studding this topic.

This work is a continuation and extension of our recent, proof of concept paper [11], where we have introduced a ML procedure for short-term traffic volumes forecasting, using the Linear Discriminant Analysis (LDA) classifier. As the key novelty of this paper, comparing to the previous one, we propose a regression approach for the problem of predicting traffic in optical network. Additionally, we examine single ML algorithms and ensemble methods. Datasets instances for experiments are created based on features selected during autocorrelation analysis. For evaluation we use a new metric, which considers the usage of network resources and network operational costs. Motivation behind improving ML approaches and methods for traffic prediction is that even small increase of accuracy can benefit in significant increase of saved network resources and operational costs.

The rest of this paper is organized as follows. Section 2 presents dataset creation method. Next, in Sect. 3, network model and ML approach are depicted. Section 4 shows numerical results and a brief description of them. At the end, final conclusions are presented in Sect. 5.

## 2 Datasets

Datasets used for conducting experiments were generated with the use of the custom traffic generator, however as the input, the generator takes real-world traffic measurements made by Seattle Internet Exchange Point (SIX), which publishes its statistics online [12]. Therefore, the output of the traffic generator returns traffic flows which in general follows real traffic characteristics in terms of periodicity. The main motivation to use a special traffic generator instead of using directly the traffic observed at SIX is the fact that SIX aggregates traffic from many various networks up to almost 2 Tbps, while we want to examine traffic flows observed in an optical network between individual node pairs with more diversified fluctuations of the traffic. In consequence to generate the analyzed datasets, the whole generated traffic is divided into several web services with assigned share in whole expected traffic, as reported in [1]. Considered services and their share are the internet video (51%), IP VoD (22%), web data (18%), file sharing (8%) and gaming (1%). Each of such services has its own properties, which are taking part in final data generation. The traffic is divided and then assigned to nodes of network topology. This work considers the Euro28 backbone network [13] with 28 nodes. The distribution of traffic between nodes is inversely proportional to the distance between each pair of nodes. The output traffic is a series of tuples of 756 numbers, representing the volume of traffic for each pair of nodes in 5 min time slot. Traffic characteristic taken as a generator input were collected in SIX between 24th of October and 19th of December 2019. More about traffic generator can be found in [14].

As described above, traffic flow generated for each nodes pair can be characterized by its fluctuation. As fluctuation metric we take Mean Absolute Percentage Error (MAPE), which describes how the values of one function differ from values of the base function. To calculate MAPE of a single traffic flow, as the base function, we use the original SIX bitrate flows applied to generate the traffic datasets and normalized it to a common range of considered flow bitrate values. In other words, MAPE parameter indicates how the considered traffic flow differs from the SIX traffic. We have selected the following datasets:

- dataset_1 – traffic flow between a single nodes pair, MAPE equal to 3,4%,
- dataset_2 – traffic flow between a single nodes pair, MAPE equal to 8,2%,
- dataset_3 – traffic flow between a single nodes pair, MAPE equal to 13,4%,
- dataset_4 – traffic incoming to a single network node, MAPE equal to 1,3%,
- dataset_5 – original SIX traffic, MAPE equal to 0%.

Figure 1 shows visual representation of datasets used during experiments. The horizontal axis reflects time and the vertical axis presents normalized bitrate. Due to the fact that datasets bitrates ranges differ from each other, to present datasets visualization, datasets bitrates were normalized to the same range. It is clear that traffic flows differ in terms of their fluctuation.

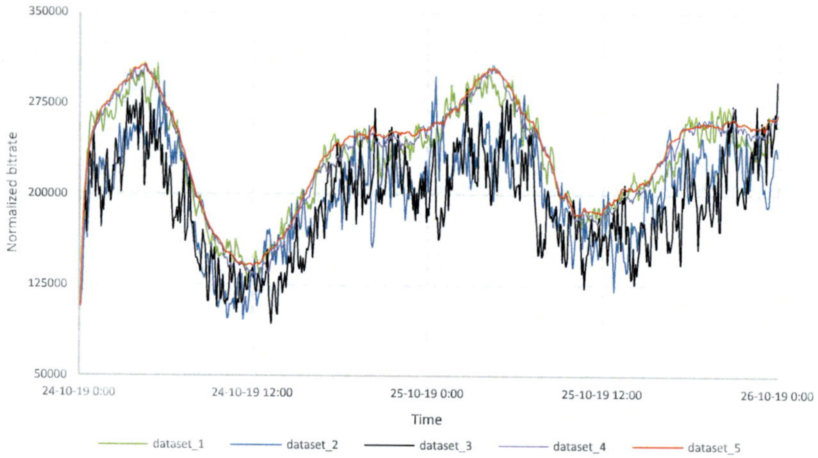


**Fig. 1.** Traffic flows

## 3 Network Model and Machine Learning Approach

The network works based on a time scale divided to *time intervals* (TIs), of the same 5 min size. For successive TIs, traffic volumes (bitrates) related to a single pair of nodes or whole traffic going through a single node create continuous and regular flows.

In optical technologies, such as Wavelength Division Multiplexing (WDM) or Elastic Optical Networks (EON) [15], each optical channel can carry a fixed amount of data (fixed bitrate) based on the characteristics of the used transceiver. For example, if a single optical channel can carry 100 Gbps, to serve traffic with bitrates of 110 Gbps or 190 Gbps, we need to establish two optical channels (two transceivers) for transmission, since we need 200 Gbps to provision each of the considered bitrates. Because of that fact, in the paper, traffic prediction task is realized by prediction future traffic levels (represent the number of optical channels that are needed to send data) instead of the exact traffic bitrate. Another motivation behind predicting traffic levels are Ethernet standards, which offer transmission of defined bitrate levels, e.g. 100 Gb Ethernet [16].

Figure 2 shows the process of defining traffic levels for traffic flows. Vertical axis shows bitrate and horizontal axis represents time. The red line illustrates real bitrate values and the orange line traffic levels that are assigned to them. Possible traffic levels are represented by grey horizontal lines. Traffic forecasting problem can be realized by predicting future orange line levels.

Two different supervised ML approaches for defining traffic levels are examined in this work: (i) classification when a forecasting problem is considered as a classification task, where classes are determined by possible traffic levels, (ii) regression for which first, the exact value of bitrate in TI is predicted. Next, the traffic levels in TIs are calculated by rounding up the predicted bitrate to the closest traffic level. To test each approach, different single ML algorithms and ensemble methods were applied. As a reference we use Linear Discriminant Analysis classifier presented in our previous work

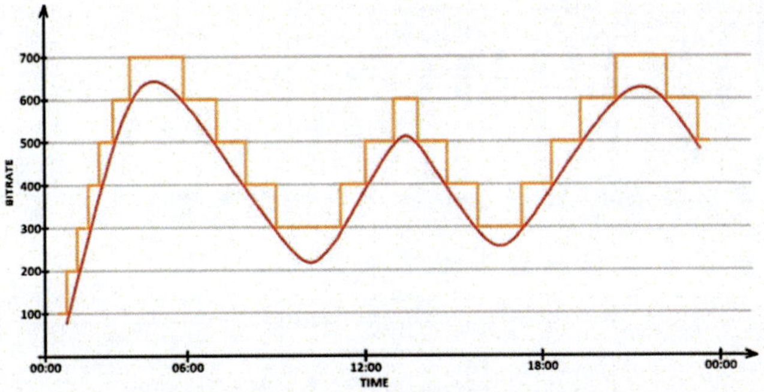


**Fig. 2.** Traffic levels

[11]. Additionally, to compare performance of single ML algorithms with ensemble methods, for each approach we use Decision Tree (DT) and different ensemble methods with DT as a base function. For classification, we examine following methods: Decision Tree Classifier (DTC), Bagging Classifier (BC), Random Forest Classifier (RF), Extra Trees Classifier (ETC), Gradient Boosting Classifier (GBC). In turn, for regression we use: Decision Tree Regressor (DTR), Random Forest Regressor (RFR), Extra Trees Regressor (ETR), Ada Boost Regressor (ABR), Gradient Boosting Regressor (GBR).

Shape of traffic flows is correlated with time, thus the data used in the experiment can be characterized as data streams [17]. Because of that, instead of classic k-fold cross validation, we train ML algorithms using datasets created based on consecutive TIs. In more detail, the whole dataset for the experiment consists of traffic flows information lasting about 58 days. We take first 28 days to create a training dataset and last 30 days to create a testing dataset.

Figure 3 shows the autocorrelation of dataset_1 traffic flow. In other words, it determines how much the value of a particular bitrate depends on bitrate values from previous TIs. Vertical axis presents autocorrelation values and horizontal axis contains of TIs. Visualization shows that the highest correlations occur for the closest preceding iterations. Additionally, every 12 h (every 144 TIs, since single time interval represents 5 min) occurs a meaningful correlation. Positive correlations (greater than zero) occur every 24 h, which refers to 288 time intervals. Based on autocorrelation characteristics, as well as fact that shape of traffic flows is correlated with time, 26 features have been chosen as model for creating datasets, in detail, ten previous TIs bitrates, information about number of the day during the week and minute during the day and traffic values (bitrates) from previous *24 h, 48 h, …, 336 h* (two weeks) TIs.

Figure 4 presents the model creation process for the classification task. Features considered in the example are information about day, minute and bitrate in previous TI. Possible traffic levels (classes) are multiples of 100. First TI is omitted during the model creation, because there is no information about its previous TIs.

Using accuracy as an evaluation metric, information about underpredictions (based on these, the network operator can't serve traffic) and overpredictions (based on these, the operator can serve traffic, however it requires additional network resources) is omitted. Such knowledge can result in better network resource usage. To provide evaluation

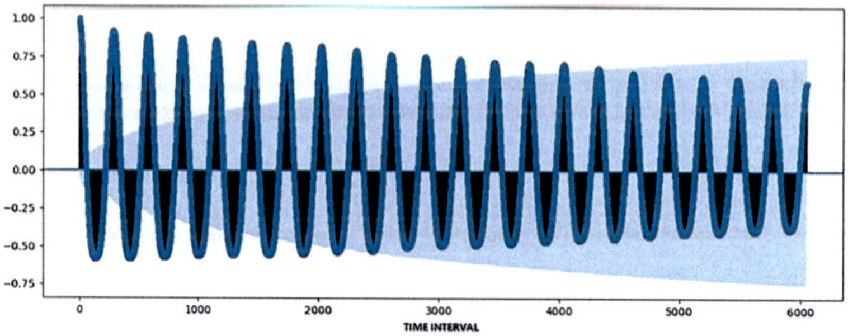


**Fig. 3.** Dataset_1 autocorrelation

| TI | Day | Minute | Real Traffic Volume | Traffic Level for time interval |
|---|---|---|---|---|
| 1 | 1 | 5 | 195 | 200 |
| 2 | 1 | 10 | 255 | 300 |
| 3 | 1 | 15 | 330 | 400 |
| 4 | 1 | 20 | 550 | 600 |
| 5 | 1 | 25 | 490 | 500 |
| 6 | 1 | 30 | 380 | 400 |
| 7 | 1 | 35 | 205 | 300 |
| 8 | 1 | 40 | 285 | 300 |
| 9 | 1 | 45 | 340 | 400 |
| 10 | 1 | 50 | 550 | 600 |

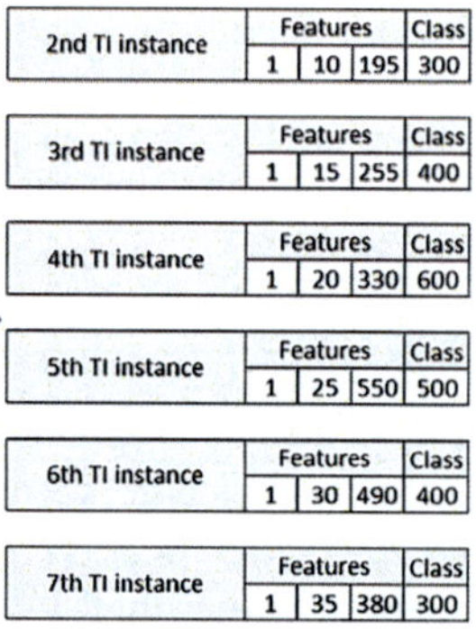


**Fig. 4.** ML model creation, features: day, minute, single previous TI bitrate, classification

giving a better insight, we propose new metric called *Traffic Level Prediction Quality* (TLPQ), which is calculated based on the following prediction variants:

a. predicted traffic level is correct,
b. underestimation of traffic level,
c. overestimation by one traffic level.

The equation below indicates the way to calculate the TLPQ ($\alpha$; $\beta$) metric:

$$TLPQ(\alpha;\beta) = a + \alpha * c - \beta * b \tag{1}$$

The values of $\alpha$ and $\beta$ tuning parameters can be freely determined by network operators, based on their expectations and needs. In this work, we consider three different network operator scenarios, differing from each other by values of $\alpha$ and $\beta$ parameters: (i) TLPQ (1;0), (ii) TLPQ (0,7;0,3), (iii) TLPQ (0,5;0,5).

## 4 Numerical Results

All experiments have been conducted based on five datasets depicted in Sect. 2, using ML algorithms presented in Sect. 3. Whole implementation has been made in in Python, using scikit-learn ML package [18]. All classifiers' and regressors' parameters were left as default. In each tested dataset, number of possible traffic levels is equal to seven.

Table 1 and Table 2 report accuracy, Root Mean Squared Percentage Error (RMSPE) and TLPQ values from each considered scenario obtained by tested ML algorithms. RMSPE indicates by how many percent on average considered algorithm traffic level prediction differs compared to real traffic level. Cells highlighted in green consist of the best results for the considered dataset, while those highlighted in red point the worst results for dataset, i.e., for each dataset (five adjacent rows) the best result is highlighted in green and the worst in red.

Based on results from tables, some general trends can be noticed. In both examined approaches (i.e., classification and regression), ensemble methods, with DT as a base model, outperform single DT results. Difference increases with increase of dataset MAPE value and reaches 14% for accuracy, 9 for TLPQ (1;0), 12 for TLPQ (0,7;0,3) and 15 for TLPQ (0,5;0,5) in case of dataset_3 and regression approach (Table 2). Committees in regression approach obtain better results than committees in classification

**Table 1.** Classification approach, accuracy, RMSPE and TLPQ

| | Classifier | Accuracy | RMSPE | TLPQ(1;0) | TLPQ(0,7;0,3) | TLPQ(0,5;0,5) |
|---|---|---|---|---|---|---|
| dataset_1 | DT | 76,8% | 7,93% | 88,5 | 81,5 | 76,8 |
| | BC | 83,3% | 6,77% | 90,7 | 85,7 | 82,3 |
| | RF | 82,4% | 6,90% | 89,6 | 84,3 | 80,8 |
| | ETC | 81,0% | 7,08% | 88,8 | 83,1 | 79,3 |
| | GBC | 83,8% | 6,70% | 92,0 | 87,2 | 83,9 |
| dataset_2 | DT | 56,8% | 13,76% | 78,2 | 65,6 | 57,2 |
| | BC | 65,1% | 11,74% | 80,7 | 70,3 | 63,3 |
| | RF | 63,5% | 12,08% | 80,1 | 69,3 | 62,0 |
| | ETC | 63,9% | 12,06% | 80,2 | 69,5 | 62,3 |
| | GBC | 66,4% | 11,85% | 83,9 | 73,9 | 67,2 |
| dataset_3 | DT | 50,1% | 19,58% | 73,2 | 58,9 | 49,4 |
| | BC | 58,0% | 16,64% | 75,8 | 63,3 | 55,1 |
| | RF | 58,8% | 16,52% | 76,0 | 63,8 | 55,7 |
| | ETC | 57,6% | 16,53% | 75,3 | 62,7 | 54,2 |
| | GBC | 61,3% | 15,93% | 79,5 | 68,0 | 60,4 |
| dataset_4 | DT | 85,6% | 5,05% | 92,8 | 88,5 | 85,6 |
| | BC | 88,8% | 4,32% | 92,9 | 89,5 | 87,3 |
| | RF | 89,6% | 4,19% | 93,4 | 90,2 | 88,2 |
| | ETC | 89,1% | 4,21% | 93,0 | 89,7 | 87,6 |
| | GBC | 90,5% | 3,91% | 94,7 | 91,8 | 90,0 |
| dataset_5 | DT | 91,4% | 4,84% | 96,0 | 93,4 | 91,7 |
| | BC | 92,2% | 4,59% | 95,8 | 93,4 | 91,9 |
| | RF | 92,3% | 4,58% | 95,9 | 93,6 | 92,0 |
| | ETC | 90,6% | 5,01% | 95,1 | 92,3 | 90,4 |
| | GBC | 93,1% | 4,48% | 96,9 | 94,8 | 93,4 |

**Table 2.** Regression approach, accuracy, RMSPE and TLPQ

| | Regressor | Accuracy | RMSPE | TLPQ(1;0) | TLPQ(0,7;0,3) | TLPQ(0,5;0,5) |
|---|---|---|---|---|---|---|
| dataset_1 | DTR | 78,0% | 7,72% | 88,8 | 82,2 | 77,8 |
| | RFR | 84,5% | 6,51% | 92,1 | 87,4 | 84,3 |
| | ETR | 84,5% | 6,50% | 92,0 | 87,3 | 84,2 |
| | ABR | 85,1% | 6,34% | 91,4 | 86,9 | 83,9 |
| | GBR | 84,9% | 6,46% | 92,4 | 87,8 | 84,8 |
| dataset_2 | DTR | 56,2% | 13,78% | 78,0 | 65,1 | 56,5 |
| | RFR | 67,4% | 11,63% | 84,1 | 74,4 | 67,9 |
| | ETR | 67,7% | 11,50% | 84,2 | 74,5 | 68,1 |
| | ABR | 67,5% | 12,07% | 87,1 | 77,4 | 70,9 |
| | GBR | 69,2% | 11,22% | 85,3 | 76,1 | 69,9 |
| dataset_3 | DTR | 50,1% | 19,33% | 73,3 | 59,0 | 49,5 |
| | RFR | 61,8% | 15,57% | 79,9 | 68,6 | 61,0 |
| | ETR | 61,8% | 15,53% | 79,7 | 68,3 | 60,7 |
| | ABR | 63,7% | 15,37% | 82,5 | 71,6 | 64,4 |
| | GBR | 64,0% | 15,09% | 80,9 | 70,2 | 63,0 |
| dataset_4 | DTR | 88,3% | 4,46% | 93,7 | 90,2 | 87,9 |
| | RFR | 91,2% | 3,88% | 95,3 | 92,7 | 91,0 |
| | ETR | 91,5% | 3,80% | 95,3 | 92,8 | 91,1 |
| | ABR | 90,0% | 4,00% | 93,8 | 90,8 | 88,8 |
| | GBR | 91,8% | 3,73% | 95,6 | 93,2 | 91,5 |
| dataset_5 | DTR | 93,5% | 4,48% | 97,5 | 95,5 | 94,2 |
| | RFR | 94,5% | 4,26% | 98,3 | 96,6 | 95,5 |
| | ETR | 93,7% | 4,39% | 97,4 | 95,5 | 94,2 |
| | ABR | 91,5% | 4,73% | 95,5 | 92,9 | 91,2 |
| | GBR | 94,2% | 4,28% | 97,5 | 95,7 | 94,6 |

**Table 3.** Difference between the best ensembles results and LDA reference classifier

| | Regression | | | Classification | | |
|---|---|---|---|---|---|---|
| | TLPQ(0;1) | TLPQ(0,7;0,3) | TLPQ(0,5;0,5) | TLPQ(1;0) | TLPQ(0,7;0,3) | TLPQ(0,5;0,5) |
| dataset_1 | 0,01% | 0,56% | 0,98% | -0,39% | -0,23% | -0,12% |
| dataset_2 | 2,42% | 2,71% | 2,96% | -1,33% | -1,94% | -2,44% |
| dataset_3 | 2,37% | 3,10% | 3,73% | -1,31% | -2,06% | -2,71% |
| dataset_4 | 2,30% | 3,23% | 3,89% | 1,30% | 1,76% | 2,09% |
| dataset_5 | 2,18% | 4,04% | 5,35% | 0,75% | 2,08% | 3,02% |

approach for all datasets. For a single ML algorithm, in case of dataset_2 and dataset_3, classification approach turned out to return better results (Table 1). For rest datasets regression approach outperform classification (Table 2). In case of classification approach, the best ensemble method was GBC. It achieved the highest values of all metrics for all datasets (Table 1). For regression approach, the best turned out to be ABR, GBR and RFR methods (depends on dataset number) (Table 2). It can also be seen that ML algorithms obtain better results for datasets with lower dataset MAPE value. With the increasing of dataset MAPE value, RMSPE of traffic level prediction also increases. This was equal around 4% for dataset_5 and reached almost 20% for dataset_3.

Figure 5 and Fig. 6 compare real bitrate (blue line) with traffic levels (red line) predicted based on results returned by GBC and GBR. In Fig. 5 green line presents bitrate values predicted by ensemble. Dotted line shows possible traffic levels. Charts present data from the first day of tested datasets. Figures visualize results presented in Tables 1 and 2. Both, regressor and classification ensembles predict correct traffic levels in case of most TIs. The algorithms are able to detect short (lasting only few TIs) changes of traffic level. Predicting traffic level directly as a result of classification (Fig. 6) causes more frequent wrong predictions than calculated traffic level based on prediction of bitrate in regression (Fig. 5).

As it was stated in Sect. 3, as reference we took classification algorithm from our previous work [11], namely LDA classifier. Table 3 presents percentage difference of TLPQ values for all considered scenarios and datasets between LDA classifier and the best ensembles results in case of classification and regression approaches. Green marks the biggest differences while red the smallest ones. It can be seen that in case of regression approach, ensembles achieved better results than reference ML algorithm. Different situation can be noticed in case of classification approach. For datasets related to traffic between single nodes pair (dataset_1–dataset_3) LDA classifier outperformed ensembles results, however for dataset_4 and dataset_5 committees obtained better results. For dataset_5 and TLPQ (0,5;0,5), ensemble method outperforms reference method by 5,35%.

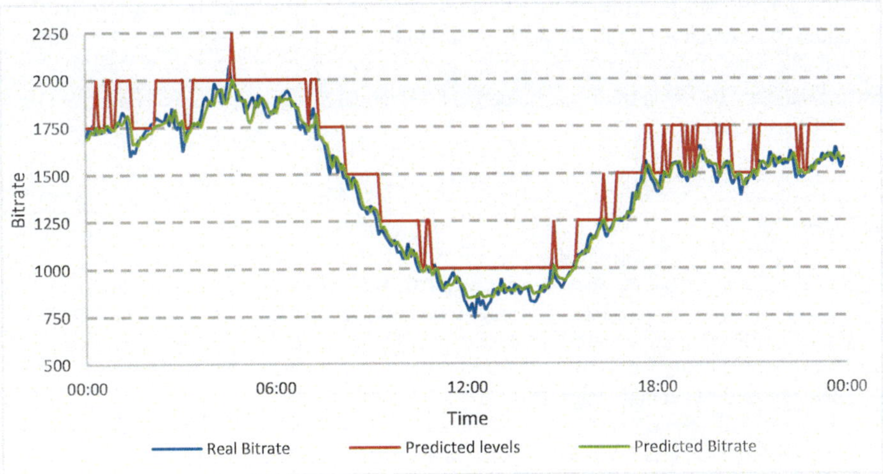


**Fig. 5.** Real traffic and predicted traffic levels, dataset_1, GBR

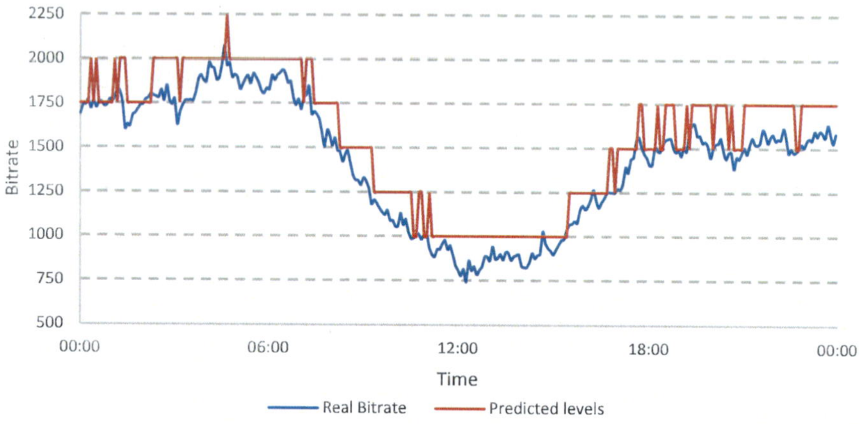


**Fig. 6.** Real traffic, predicted traffic and predicted traffic levels, dataset_1, GBC

## 5 Conclusions

This paper considers task of predicting traffic in optical network. To solve the problem, we have presented a method based on regression and have compared it with the classification approach. For each approach, we have used different single ML algorithms and ensemble methods. Our study showed that regression provides better results when compared to the results yielded by classification. Additionally, ensemble methods predict traffic more accurately than single ML algorithms. In case of regression approach, tested ensembles outperformed reference LDA classifier results by 5,3%. We have also reported that the result obtained by ML algorithm is strictly correlated with traffic flow characteristic expressed by the MAPE parameter.

**Acknowledgements.** This work was supported by National Science Centre, Poland under Grant 2017/27/B/ST7/00888.

## References

1. CISCO: Cisco Annual Internet Report (2018–2023). CISCO, San Jose, USA (2020)
2. Nokia Deepfield Network Intelligence Report Networks in 2020. Nokia, Finland (2020)
3. Chan, V.W.S.: Cognitive optical networks. In: ICC, Kansas City, USA, pp. 1–6 (2018)
4. Musumeci, F., et al.: An overview on application of machine learning techniques in optical networks. IEEE Commun. Surv. Tutor. **21**, 1383–1408 (2019)
5. Ding, X., et al.: Neural network based models for forecasting. In: Proceedings of Applied Decision Technologies Conference, San Antonio, USA, pp. 243–252 (1995)
6. Rzym, G., Boryło, P., Chołda, P.: Time-efficient shrinkage algorithm for fourier-based prediction enabling proactive optimization in software defined networks. Int. J. Commun. Syst. **33**, e4448 (2019)
7. Otoshi, T., Ohista, Y., et al.: Traffic prediction for dynamic traffic engineering. Comput. Netw. **85**, 36–50 (2015)

8. Vinchoff, C., et al.: Traffic prediction in optical networks using graph convolutional generative adversarial networks. In: ICTON, Bari, Italy, pp. 3–6 (2020)
9. Aibin, M.: Traffic prediction based on machine learning for elastic optical networks. Opt. Switch. Netw. **30**, 33–39 (2018)
10. Cortez, P., Rio, M., Rocha, M., Sousa, P.: Multi-scale internet traffic forecasting using neural networks and time series methods. Expert Syst. **29**, 143–155 (2012)
11. Szostak, D., Walkowiak, K., Włodarczyk, A.: Short-term traffic forecasting in optical network using linear discriminant analysis machine learning classifier. In: ICTON, Bari, Italy (2020)
12. Seattle Internet Exchange Point Homepage. https://www.seattleix.net/
13. Walkowiak, K., et al.: Dynamic routing in spectrally-spatially flexible optical networks with back-to-back regeneration. IEEE/OSA J. Opt. Commun. Netw. **10**(5), 523–534 (2018)
14. Włodarczyk, A., et al.: An algorithm for provisioning of time-varying traffic in translucent SDM elastic optical networks. In: ICTON, Bari, Italy (2020)
15. Walkowiak, K.: Modeling and Optimization of Cloud-Ready and Content-Oriented Networks. Studies in Systems, Decision and Control, vol. 56, Springer, Cham (2016).https://doi.org/10.1007/978-3-319-30309-3
16. Mukherjee, B., Tomkos, I., Tornatore, M., Winzer, P., Zhao, Y. (eds.): Springer handbook of optical networks. SH, Springer, Cham (2020). https://doi.org/10.1007/978-3-030-16250-4
17. Bifet, A., Morales, G., et al.: Big data stream learning with SAMOA. In: IEEE International Conference on Data Mining Work-shop, Shenzhen (2014)
18. Pedregosa, F., Varoquaux, G., et al.: Scikit-learn: machine learning in python. J. Mach. Learn. Res. **12**, 2825–2830 (2011)

# Assessing the Impact of Batch-Based Data Aggregation Techniques for Feature Engineering on Machine Learning-Based Network IDSs

Roberto Magán-Carrión[1(✉)], Daniel Urda[2], Ignacio Díaz-Cano[3], and Bernabé Dorronsoro[4]

[1] Network Engineering and Security Group, Department of Signal Theory, Communications and Telematics, University of Granada, Granada, Spain
rmagan@ugr.es
[2] Grupo de Inteligencia Computacional Aplicada (GICAP), Departamento de Ingeniería Informática, Escuela Politécnica Superior, Universidad de Burgos, Av. Cantabria s/n, 09006 Burgos, Spain
durda@ubu.es
[3] Applied Robotics Group, Department of Automatic, Electronic, Computer Architecture and Com. Net. Engineering, University of Cádiz, Cádiz, Spain
ignacio.diaz@uca.es
[4] Graphical Methods, Optimization and Learning (GOAL) Group, Department of Computer Engineering, University of Cádiz, Cádiz, Spain
bernabe.dorronsoro@uca.es

**Abstract.** Communication networks and systems are continuously threatened by a great variety of cybersecurity attacks coming from new malware that targets old and new systems' vulnerabilities. In this sense, Intrusion Detection Systems (IDSs) and, specifically, Network IDSs (NIDSs) are used to count on robust methods and techniques to detect and classify security attacks. One of the important parts in the assessment of NIDSs, is the Feature Engineering (FE) process, where raw datasets are transformed onto derived ones where both, features and observations are smartly transformed. In this work, the *ff4ml* framework, which includes the Feature as a Counter (FaaC) FE approach, is used to transform raw features into new ones that are counters of the originals. The FaaC approach aggregates raw observations by time intervals, thus limiting its use to network datasets containing timestamps. This work proposes a batch-based aggregation technique that allows applying FaaC in timestamp-less datasets and analyzes its impact on the performance of Machine Learning (ML)-based NIDSs in comparison to timestamp-based aggregation approaches.

**Keywords:** Machine learning · Feature engineering · NIDS · Cybersecurity · Information security


J. J. Gude Prego et al. (Eds.): CISIS 2021/ICEUTE 2021, AISC 1400, pp. 116–125, 2022.
https://doi.org/10.1007/978-3-030-87872-6_12

## 1 Introduction

Communication networks, systems and technological assets are getting more and more threatened by malicious actors that take advantage of known and unknown security vulnerabilities [7]. To face them, security solutions are needed, being the Intrusion Detection Systems (IDSs) one of the most relevant defense line nowadays. Therefore, robust and dynamic detection solutions are demanded for reducing false alarms and the mean time for attacks detection [12].

Artificial Intelligence (AI) techniques and algorithms have been applied in a large number of interesting fields as *healthcare*, *natural language processing*, etc., and particularly in cybersecurity for proposing Machine Learning (ML)-based IDSs [1]. The evaluation of this type of systems usually falls on the use of datasets previously gathered from the application scenario. The characteristics of the chosen dataset markedly influences the methods, algorithms or techniques upon which the IDS are based [2]. Moreover, the evaluation of the performance of IDS solutions involves not only the own ML model under evaluation, but other relevant pre- and post-processing tasks, as components of a working pipeline.

This work relies on the use of *ff4ml* framework [9] to build and evaluate supervised classification ML-based Network IDSs (NIDSs). The authors of *ff4ml* highlight the importance of the Feature Engineering (FE) stage in the performance of the solution. In fact, the authors employ the Feature as a Counter (FaaC) approach tested in this context in previous works [5,6], which transforms raw dataset features into new variables, which are counters of the originals. This procedure relies on the use of 1-minute time interval to aggregate raw dataset samples. However, the use of the FaaC approach is limited to network datasets whose samples have a timestamp. This is the case of the UGR'16 [8] dataset.

The requirement of timestamp information limits the applicability of *ff4ml* framework. Nevertheless, its use could be generalized if the FaaC algorithm would work on timestamp-less network datasets, e.g., the well-known and widely used NLS-KDD network dataset [11]. For that, the present work evaluates and compares the classification performance of two classical and widely used ML models, namely Linear Regression (LR) and Random Forest (RF), using the *ff4ml* framework with a modified version of FaaC process as an alternative strategy for this kind of datasets. Thus, the contribution of the present work is twofold:

- The proposal of an alternative raw observation aggregation technique for the FaaC FE process instead of using the original timestamp-based one. Two batch-based approaches are considered with different batch observation sizes.
- The evaluation and comparison of the performance of two ML-based NIDSs when using the timestamp- and batch-based aggregation methods to see their influence on the classification performance.

Preliminary results show that no significant differences can be found, no matter the use of the aggregation technique. Therefore, this study opens the door to the use of FaaC FE technique on datasets without timestamp and, more

**Table 1.** Number of flows per class in TEST partition of the UGR'16 dataset.

| Class | Background | Blacklist | Botnet | DoS | SSH scan | Scan | Spam | UDP scan |
|---|---|---|---|---|---|---|---|---|
| Flows | $\sim 4,000M$ | $\sim 18M$ | $\sim 2M$ | $\sim 9M$ | 64 | $\sim 6M$ | $\sim 78M$ | $\sim 1M$ |
| % | 97.14 | 0.46 | 0.04 | 0.23 | $\sim 0$ | 0.14 | 1.96 | 0.03 |

importantly, it makes possible the comparison of timestamp-based FE techniques against state-of-the-art NIDS methods proposed on those datasets.

The rest of the paper is organized as follows. Section 2 describes the dataset and ML models considered in this work. Then, Sect. 3 introduces the batch-based aggregation technique strategy proposed in order to deal with timestamp-less NIDS datasets, and describes the evaluation strategy used to assess the performance of ML models. Finally, results are presented in Sect. 4 and some conclusions and future works are outlined in Sect. 5.

## 2 Dataset and Methods

This section introduces both the network dataset used and the ML models selected in this work. Together with the methodology proposed in *ff4ml* framework [9], they will be used to analyze the impact on their predictive performance using different batch-based observation aggregation strategies, in comparison with a timestamp-based one.

### 2.1 UGR'16 Dataset

The UGR'16 [8] is a dataset containing four months of anonymized real network traffic flows (netflow), collected at the facilities of a Spanish Internet Service Provider (ISP). UGR'16 is split into two parts: CAL and TEST. The first one contains network communication flows captured from March to June 2016, while the second one was obtained from the last week of July to August of the same year. During this period, network attacks were synthetically built, orchestrated and launched together with the normal network traffic captured from the ISP facilities. For that, up-to-date attack generation tools were used, capable of covering difficult-to-detect and, in some cases, harmful attacks such as *low and high rate DoS* or *port scan*. Some other type of network attacks, as *Botnet*, were carefully added to the UGR'16 using network communication flow traces from other existing network datasets. Additionally, the authors tagged attacks that were suggested as network anomalies by state-of-the-art anomaly detectors, after conducting an in-depth analysis of traffic flows. Because of that, *UDP port scan*, *SSH scan* and *spam campaigns* were identified. Finally, certain flows were labeled as *blacklist* when their IP was found in well-known blacklist open sources.

Table 1 shows the distribution of the classes found in the TEST part of the UGR'16 dataset both in terms of the number of traffic flows and in percentage. As it could be expected in the context of NIDS problems, there are significant differences among the number of observations coming from attacks and those representing regular network traffic (*background*). For instance, the attack with the highest number of flows in the dataset is *spam*, and it represents only 1.96% of the total traffic flows. Although attacks samples can be found in both CAL and TEST partitions, the last one comprises all kind of network attacks samples. This fact motivates us to focus on TEST partition for the studies in this work.

### 2.2 Methods

Training supervised ML models to accurately identify attack patterns needs to employ techniques that allow transforming raw network datasets onto derived datasets, being the last ones suitable for the application in ML models. For this purpose, authors have used *ff4ml* framework [9], which preprocesses the raw dataset using the FaaC (Feature as a Counter) technique. Briefly, this Feature Engineering (FE) procedure transforms the input variables into new ones, which are just counters of the original ones, in a given interval of time that comprises an unknown number of raw dataset observations. Additionally, the use of *ff4ml* framework adds extra benefits in ML-based NIDSs such as the use of different components and stages into an organized execution pipeline. Indeed, it is one of the first attempts to provide a standard methodology in order to perform fair and honest comparisons among ML models in this domain.

Let us define a derived dataset ready to employ in supervised ML algorithms as $\mathcal{D} = \{\boldsymbol{x_i}, y_i\}_{i=1}^{n}$, where $n$ is the number of samples, $y_i \in [0, ..., k-1]$ is the dependent variable assuming a multi-classification problem of $k$ different labels or classes, and $\boldsymbol{x_i}$ is a vector of $p$ input features, commonly known as independent variables. In this work, $\mathcal{D}$ will be used to train and test two well-known ML methods based on different learning models. They are:

- Multinomial logistic regression (LR). It is a linear model [3] widely used to address classification problems in which the dependent variable is modelled as a linear combination of the independent variables. In particular, Eq. 1 shows this modelling for the binary classification setting (i.e., $y_i \in [0, 1]$),

$$y_i = f(\beta_0 + \beta_1 x_{i1} + \beta_2 x_{i2} + ... + \beta_p x_{ip}), \tag{1}$$

assuming that $f(\boldsymbol{\beta} X^T)$ is the logistic or sigmoid function. The parameters' vector $\boldsymbol{\beta}$ is learned by solving the minimization problem depicted in Eq. 2:

$$\boldsymbol{\beta}^* = \arg\min_{\boldsymbol{\beta}} \; ||\boldsymbol{y} - f(\boldsymbol{\beta} X^T)||_2^2 \tag{2}$$

In order to deal with multi-classification settings (i.e., $k > 2$ labels), a *One-vs-All* approach is used. Now, $k$ logistic regression models are trained, each one focusing on solving the corresponding binary classification problem. Therefore, the overall prediction of the LR model is computed using the *softmax* function depicted in Eq. 3.

$$Pr(y_i = k|X) = \frac{Pr(y_i = k|X)}{\sum_{z=1}^{k} Pr(y_i = z|X)} \tag{3}$$

- Random Forest (RF). It is a tree-based non-linear bagging model for which multiple decision trees are fitted to different views of the observed data [4]. Each view consists of a subset of observations randomly sampled as well as a random subset of input features. Overall RF predictions are computed by calculating the average (or weighted average according to the performance of each single decision tree on the out-of-bag samples) of the individual predictions provided by the multiple decision trees.

## 3 Experimental Design

As stated in the previous section, raw network datasets need to be transformed onto derived datasets suitable for the application of ML techniques. In this sense, different FE procedures will definitely lead to different derived datasets. For instance, the authors of the *ff4ml* framework uses the FaaC approach by aggregating 1-minute time interval of UGR'16 raw samples onto a only one derived observation. It is repeated every minute to built a complete derived dataset. On average, each derived observation comprises $\sim 85,000$ raw samples. Nevertheless, there are many other network datasets which do not include any timestamp information (e.g., the well-known and widely used NLS-KDD network dataset [11]) which limits the adoption of the FaaC approach and, in the end, the use of the *ff4ml* framework.

To address the previous issue, this work proposes to modify the FaaC process by using a batch-based strategy which do not need any timestamp to obtain the new aggregated counter variables. On the one hand, authors expect to overcome the previously mentioned limitation of *ff4ml* framework and increase its global adoption regardless the characteristics of the network dataset. On the other hand, authors aim at analyzing the impact of the batch-based alternative strategy on the performance of ML-based NIDS. For this purpose, three different settings will be analyzed using the UGR'16 dataset, described next:

- *Timestamp-based ($\sim 40,000$ observations).* In this approach, a derived dataset is built by aggregating 1-minute time intervals of the raw network dataset observations. It assumes the existence of a timestamp within the dataset to be analyzed (as it is the case of UGR'16) and, in this work, it will be used as a baseline for comparison purposes against batch-based approaches.
- *Batch-based ($10,000$ observations).* This approach fixes the desired number of observation in the derived dataset to be equal to 10,000 and, consequently, generates fixed-size batches of raw samples. In particular, the size of any batch within this approach consists of $395,082$ raw samples.
- *Batch-based ($20,000$ observations).* This approach is identical to the previous one but varying the desired number of observations in the derived dataset to $20,000$, thus resulting on batches consisting of $197,541$ raw samples. This modification will allow to analyze the impact (i.e., positive, negative or neutral) of the size of the resulting derived dataset in the performance of ML-based NIDSs.

**Table 2.** Class distributions (%) per derived dataset corresponding to the application of the FaaC procedure using the three described settings: timestamp- and batch-based ones.

| Class | Timestamp-based (~ 40,000 obs.) | Batch-based (10,000 obs.) | Batch-based (20,000 obs.) |
|---|---|---|---|
| Background | 63.65 | 55.97 | 58.23 |
| Spam | 33.76 | 39.86 | 38.23 |
| Botnet | 1.26 | 1.66 | 1.34 |
| DoS | 0.88 | 1.41 | 1.3 |
| Scan | 0.37 | 0.79 | 0.62 |
| SSH scan | 0.06 | 0.24 | 0.12 |
| UDP scan | 0.02 | 0.08 | 0.05 |

Table 2 shows the class distribution on each derived dataset obtained through the application of the FaaC procedure by using the three settings previously proposed. Although the final size, in terms of the number of observations, is significantly different across datasets (which implies the use of different batch sizes to compute the aggregated counter variables), minor differences can be finally appreciated on the class distribution, thus resulting on similar problems to be addressed using ML techniques. Nevertheless, lower-size datasets will be more challenging since the application of ML models typically benefits from the existence of more (and higher quality) samples. Therefore, one could expect a bit worse performance of ML models trained in lower-size datasets (i.e., batch-based with $10,000$ observations) than in bigger-size ones (i.e., timestamp-based with $\sim$ 40,000 observations).

All the experimentation carried out in this analysis was based on the *ff4ml* framework with the FaaC variations previously described. A 5-fold cross-validation (5-CV) evaluation strategy is employed in such a way that each derived dataset is splitted into five non-overlapping folds of equal size and keeping the same class distribution within each fold. On each iteration of the 5-CV, four folds (known as train set) are used to train the ML models described in Sect. 2.2 and the remaining fold (known as test set) is used to evaluate their performance. Furthermore, this partitioning procedure is repeated 20 times in order to guarantee the randomness of the partitioning process and the absence of any bias that could be be introduced during the train/test partition procedure. Several well-known and widely used performance metrics were adopted to evaluate the performance of the ML models considered in this work and for the multi-classification problem analyzed: Precision (*Pr*), Recall (*Re*), F1-score (*F1*), Area Under the Receiving Operation Characteristic Curve ($AUC$) and Accuracy ($ACC$). Additionally, a *Weighted avg.* metric is calculated which considers the relative presence of each class in a weighted average. Interested readers are kindly referred to [9] for a more detailed explanation.

## 4 Results and Discussion

In order to corroborate the viability of the batch-based aggregation approaches in comparison to the timestamp-based one, we carried out several experiments with the two selected ML models (Sect. 2.2), following the methodology proposed in [9].

Table 3 shows a complete view of the classification performance obtained by each ML model for each class, considering five different metrics, and for the three derived datasets under study. We emphasize in **bold font** the best obtained result for every class and performance metric (overall on the three datasets).

As it was suggested in Sect. 3, Table 3 reveals how using the batch-based derived dataset with fewer observations (i.e., 10,000 samples) penalizes the prediction quality of the models, for all metrics. This result is specially relevant for *Pr*, *Re* and *F1* metrics on those classes with less presence in the dataset (e.g., DoS and Botnet attacks) and the LR model. Instead, the RF model performs similarly no matter the derived dataset used even outperforming the timestamp-based approach for the batch-based one of 20,000 observations. In general, non linear models outperforms the linear one due to the nature of the problem addressed here where attack classes are not easily differentiated from the background network traffic as can be expected in real scenarios indeed.

$AUC$ and $ACC$ metrics are less sensitive than the others to the dataset selection. At this point, it is worth noting that $ACC$ metric has been included here for completeness but, though it is still being used to evaluate ML-based solutions, accuracy is not recommended when addressing imbalanced classification problems [10]. On the contrary, $AUC$ is considered a robust metric in this kind of problems, as it mitigates the impact of using imbalanced datasets. Because of that, $AUC$ results are specifically visualized in Fig. 1. This figure presents a comparison between $AUC$ performances for each class and derived dataset, and no significant differences are found in the obtained values, no matter the approach used.

Focusing on the results obtained for the *Weighted avg.* metric (shaded in grey color in Table 3), which summarizes the overall performance of the methods on all classes, it can be seen that the best results are clearly obtained by RF on the batch-based dataset composed by 20,000 observations. Additionally, LR also obtains the best results for this metric on the same dataset. However, differences in the results for this metric are not significant: the performance of RF in the batch-based dataset with 20,000 observations is 0.55% better, in average, than for the timestamp-based one.

All in all, the analysis presented in this work concerning the use of different strategies to transform raw network datasets onto derived ones suitable for ML-based NIDS through the FaaC procedure in the *ff4ml* framework, concludes the viability of the batch-based aggregation approach in comparison to the timestamp-based one. Indeed, no significant differences can be found through the use of one or another approach.

**Table 3.** Average performance results obtained by the different ML models considered on the three derived datasets analyzed.

| Dataset | Model | Class | Performance Metrics | | | | |
|---|---|---|---|---|---|---|---|
| | | | **Pr** | **Re** | **F1** | **AUC** | **ACC** |
| timestamp-based (~ 40,000 obs.) | LR | Background | 0.814 | 0.919 | 0.863 | 0.776 | 0.815 |
| | | DoS | 0.933 | **0.916** | 0.924 | **0.957** | **0.999** |
| | | Botnet | 0.965 | 0.891 | 0.926 | 0.945 | 0.998 |
| | | Scan | 0.80 | 0.916 | 0.851 | 0.958 | **0.999** |
| | | Spam | 0.798 | 0.607 | 0.689 | 0.764 | 0.815 |
| | | *Weighted avg.* | *0.811* | *0.813* | *0.805* | *0.776* | *0.820* |
| | RF | Background | **0.912** | **0.958** | **0.934** | 0.898 | 0.915 |
| | | DoS | 0.973 | 0.886 | **0.926** | 0.943 | **0.999** |
| | | Botnet | **0.977** | **0.924** | **0.95** | **0.962** | **0.999** |
| | | Scan | 0.935 | **0.926** | 0.93 | **0.963** | **0.999** |
| | | Spam | 0.916 | 0.824 | 0.868 | 0.893 | 0.915 |
| | | *Weighted avg.* | *0.914* | *0.911* | *0.911* | *0.898* | *0.917* |
| batch-based (10,000 obs.) | LR | Background | 0.808 | 0.873 | 0.839 | 0.805 | 0.813 |
| | | DoS | 0.692 | 0.855 | 0.758 | 0.924 | 0.99 |
| | | Botnet | 0.685 | 0.898 | 0.771 | 0.946 | 0.992 |
| | | Scan | 0.825 | 0.888 | 0.850 | 0.943 | 0.997 |
| | | Spam | 0.814 | 0.721 | 0.765 | 0.806 | 0.823 |
| | | *Weighted avg.* | *0.805* | *0.81* | *0.805* | *0.809* | *0.824* |
| | RF | Background | 0.896 | 0.945 | 0.92 | 0.903 | 0.908 |
| | | DoS | 0.988 | 0.769 | 0.862 | 0.884 | 0.996 |
| | | Botnet | **0.977** | 0.882 | 0.926 | 0.941 | 0.998 |
| | | Scan | 0.948 | 0.922 | 0.933 | 0.961 | **0.999** |
| | | Spam | 0.926 | 0.845 | 0.884 | 0.9 | 0.911 |
| | | *Weighted avg.* | *0.909* | *0.898* | *0.902* | *0.901* | *0.913* |
| batch-based (20,000 obs.) | LR | Background | 0.812 | 0.89 | 0.849 | 0.801 | 0.816 |
| | | DoS | 0.873 | 0.864 | 0.867 | 0.931 | 0.997 |
| | | Botnet | 0.911 | 0.877 | 0.892 | 0.938 | 0.997 |
| | | Scan | 0.682 | 0.904 | 0.772 | 0.951 | 0.997 |
| | | Spam | 0.813 | 0.695 | 0.75 | 0.798 | 0.822 |
| | | *Weighted avg.* | *0.813* | *0.814* | *0.81* | *0.804* | *0.824* |
| | RF | Background | 0.91 | 0.953 | 0.931 | **0.911** | **0.918** |
| | | DoS | **0.993** | 0.824 | 0.899 | 0.912 | 0.998 |
| | | Botnet | 0.974 | 0.915 | 0.943 | 0.958 | **0.999** |
| | | Scan | **0.963** | 0.916 | **0.937** | 0.958 | **0.999** |
| | | Spam | **0.93** | **0.856** | **0.891** | **0.908** | **0.92** |
| | | *Weighted avg.* | ***0.919*** | ***0.912*** | ***0.914*** | ***0.91*** | ***0.921*** |

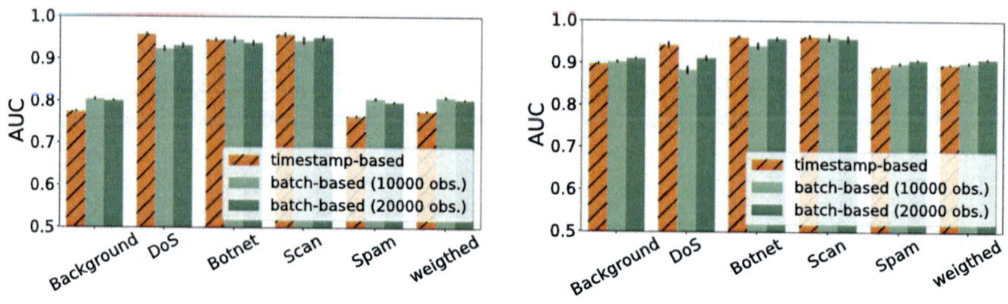


(a) LR model (b) RF model

**Fig. 1.** AUC performance results comparison for each class and dataset including a weighted average of the corresponding performance metric.

## 5 Conclusions and Future Work

In this work, a batch-based data aggregation strategy was presented as the counterpart to the timestamp-based one for those network datasets with no timestamp information. This technique mitigates the limitation of the FaaC FE approach included in the *ff4ml* framework for NIDS evaluation. Independently of the chosen strategy, their goal is to transform raw network datasets onto derived datasets suitable for the application of ML-based techniques. Particularly, two different settings varying the batch size were analyzed as a modified version of the FaaC process to assess its impact on the performance of ML-based NIDS. Some preliminary results obtained allows to conclude the viability of the batch-based approach in comparison to the timestamp-based one since no significant differences were found in the performance of the ML models no matter the strategy used. Consequently, the batch-based approach arises as an alternative strategy to be considered within the FaaC FE approach when dealing with timestamp-less network datasets. Therefore, the main result of this work is that the proposed technique allows the application of timestamp-based ML-based NIDS solutions to datasets without timestamp, merging this way these two main trends in the field, and making the comparison of the state-of-the-art techniques in both kinds of approaches possible.

As future work, further experimentation should be performed to firmly conclude the relevance of using different dataset aggregation approaches and its impact on the classification performance. This includes extending the presented study with the evaluation of the performance of other well-known ML methods. Besides, all the proposed techniques should be evaluated with different network datasets too.

**Acknowledgments.** This work was supported by the Spanish Ministerio de Ciencia, Innovación y Universidades and the ERDF under contracts RTI2018-100754-B-I00 (iSUN) and RTI2018-098160-B-I00 (DEEPAPFORE), ERDF under project FEDER-UCA18-108393 (OPTIMALE), and Junta de Andalucía and ERDF (GENIUS – P18-2399).

## References

1. Ali, R., Ali, A., Iqbal, F., Khattak, A.M., Aleem, S.: A systematic review of artificial intelligence and machine learning techniques for cyber security. In: Tian, Y., Ma, T., Khan, M.K. (eds.) ICBDS 2019. CCIS, vol. 1210, pp. 584–593. Springer, Singapore (2020). https://doi.org/10.1007/978-981-15-7530-3_44
2. Bhuyan, M.H., Bhattacharyya, D.K., Kalita, J.K.: Network anomaly detection: methods, systems and tools. IEEE Commun. Surv. Tutor. **16**(1), 303–336 (2014)
3. Bishop, C.: Pattern Recognition and Machine Learning. Springer, New York Inc., Information Science and Statistics, Berlin (2006)
4. Breiman, L.: Random forests. Mach. Learn. **45**(1), 5–32 (2001)
5. Camacho, J., García-Giménez, J.M., Fuentes-García, N.M., Maciá-Fernández, G.: Multivariate Big Data Analysis for intrusion detection: 5 steps from the haystack to the needle. Comput. Secur. **87**, 1–11 (2019)
6. Camacho, J., Pérez-Villegas, A., García-Teodoro, P., Maciá-Fernández, G.: PCA-based multivariate statistical network monitoring for anomaly detection. Comput. Secur. **59**, 118–137 (2016)
7. ENISA: ENISA Threat Landscape Report (2020). https://bit.ly/3gdsB1O. Accessed 9 June 2020
8. Maciá-Fernández, G., Camacho, J., Magán-Carrión, R., García-Teodoro, P., Therón, R.: UGR'16: a new dataset for the evaluation of cyclostationarity-based network IDSs. Comput. Secur. **73**, 411–424 (2018)
9. Magán-Carrión, R., Urda, D., Diaz-Cano, I., Dorronsoro, B.: Towards a reliable comparison and evaluation of network intrusion detection systems based on machine learning approaches. Appl. Sci. **10**(5), 1775 (2020)
10. Stapor, K., Ksieniewicz, P., García, S., Woźniak, M.: How to design the fair experimental classifier evaluation. Appl. Soft Comput. **104**, 107219 (2021)
11. Tavallaee, M., Bagheri, E., Lu, W., Ghorbani, A.A.: A detailed analysis of the KDD CUP 99 data set. In: 2009 IEEE Symposium on Computational Intelligence for Security and Defense Applications, pp. 1–6 (2009)
12. Wiafe, I., Koranteng, F.N., Obeng, E.N., Assyne, N., Wiafe, A., Gulliver, S.R.: Artificial intelligence for cybersecurity: a systematic mapping of literature. IEEE Access **8**, 146598–146612 (2020)

# Gait-Based Authentication Using a RGB Camera

Verónica Toral-Álvarez, Claudia Álvarez-Aparicio(✉), Ángel Manuel Guerrero-Higueras, and Camino Fernández-Llamas

Robotics Group, University of León, Campus de Vegazana S/N, 24071 León, Spain
vtoraa00@estudiantes.unileon.es
{calvaa,am.guerrero,camino.fernandez}@unileon.es
http://robotica.unileon.es

**Abstract.** This paper focuses on the effectiveness of the gait-based authentication of users by using a camera. The authentication is possible as walking movement is considered a behavioural biometric technology, so it is inherent and individual to each human being. People can be successfully identified by walking 4 m in a corridor in the most natural way. Different machine learning models were trained and tested achieving good accuracy rates, particularly good results were achieved with Linear Discriminant Analysis (LDA) with a 97.33% of accuracy. This biometric system allows authentication without contact and using masks which make it safe to users in the current COVID-19 pandemic situation.

**Keywords:** Behavioural biometrics · Biometry authentication · COVID-19 · Gait-based authentication · Machine learning

## 1 Introduction

The development of authentication methods is known to be an important challenge in scientific research so the introduction of biometric techniques is a growing concern. These techniques refer to the identification of a person based on an aspect of its biology as well as their behaviour. Over the last few years, due to the introduction of Artificial Intelligence, behavioural authentication methods are booming because machine learning techniques are usually used. The implementation of these new ways of authentication results in an increase in cybersecurity. Allowing a greater robustness in the face of credential theft and the theft of physical devices of authentication such as cards. This form of biometry not only improves the security but also the speed of authentication as it is an inherent characteristic of a human being.

It is well-known that most of the behavioural methods imply contact with a surface, such as the use of a mouse or keystroke dynamics, which means typing on a keyboard. However, due to the current worldwide situation of COVID-19, methods that avoid contact or allow identification with masks have achieved


J. J. Gude Prego et al. (Eds.): CISIS 2021/ICEUTE 2021, AISC 1400, pp. 126–135, 2022.
https://doi.org/10.1007/978-3-030-87872-6_13

greater significance. Some of these methods may be voice authentication, gait-based methods using inertial sensors, or the identification of hand gestures.

Biometric authentication systems have been increased in the last years, some works have been focused on behavioural biometrics using several techniques and achieving good results.

Keystroke and mouse dynamics enable human authentication even when they submit the same password due to the application of machine learning concepts on personal patterns and different features on information gathering while using computing devices. These techniques provide low error rates. However, future work suggests improving architectures and reaching to a wider population [4,5,12,18].

Even better results are achieved when the work focuses on hand and wrist gestures, especially when they are recorded over a long period of time. Authors propose continuing the investigation by improving datasets and by trying to authenticate people in movement since the existing projects do it with costumers seated. A possible disadvantage could be the high cost of individual devices for authentication [17,24].

An interesting field of investigation is facial expression identification, which is not widely explored and presents good results and potential applications. According to the sanitary situation this method cannot be applied because of the masks [19].

There is no agreement on results as regards about hand-written signatures. There are different results depending on the dataset used and the language of the sample. All works related to signatures propose improving datasets to avoid artificially enlarging them and having overfitting. It would be interesting to discern between false and real signatures [14,15,20].

Finally, gait-based authentication proposed an interesting future because it posits the creation of larger datasets. Using new sensors and devices would improve the current results which are different depending on the sample [11,23]. Last year, some works were created by considering authentication during COVID-19 pandemic as the one that investigates periocular person identification [16] and the issues fingerprint authentication raised [21].

This paper focuses on developing a method of authentication based on human gait due to its sanitary security as no contact is needed and it is compatible with the use of masks, making usability in the current pandemic situation easier. Moreover, its low cost and ease of implementation are another strength of the method. We used OpenPose [3] software to obtain the coordinates of different parts of the body, so a new application is found for this tool. This method would be suitable for implementation in a company for employees' authentication as they walk along a corridor. This work proves the feasibility of the method and offers a wide dataset of people walking which is insufficient in many other works.

## 2 Materials and Methods

A set of experiments has been carried out to evaluate the accuracy of the method. In this section, the main elements of the experiment are described in depth, as well as the methodology used to evaluate the accuracy of the new approach.

## 2.1 OpenPose

OpenPose is the first open-source realtime system for multi-person 2D pose detection, including body, foot, hand, and facial keypoints. This system may reach almost 135 keypoints in total on single images. This output can be used by many computer vision or machine learning systems to understand and interpret humans and interactions [3].

In this project, we only use 25 points to identify their position in each of the frames of a video. The keypoints obtained, composed of their coordinates, were: nose, neck, right shoulder, right elbow, right wrist, left shoulder, left elbow, left wrist, right hip, right knee, right ankle, left hip, left knee, left ankle, right eye, left eye, right ear, left ear, left big toe, left small toe, left heel, right big toe, right small toe and right heel. These keypoints are shown in Fig. 1.

(a) Keypoints obtained from each frame

(b) OpenPose output for a frame

**Fig. 1.** OpenPose outputs

## 2.2 Dataset

The data for this dataset (available online, DOI: 10.5281/zenodo.4923011) has been gathered in a corridor with a RGB (Red, Blue and Green) camera at 30 FPS (Frames Per Second).

The dataset is composed by 10 523 JSON files created from videos of 14 people walking a distance of 4 m in the most natural way as shown in Fig. 2. This process was repeated 10 times for each person to obtain a large sample. The entire dataset is composed by 140 videos (an average of 751.64 frames for each person) which were split in train (70%) and test (30%) data. After that, videos were processed with OpenPose to obtain the keypoints in JSON format needed to identify a human.

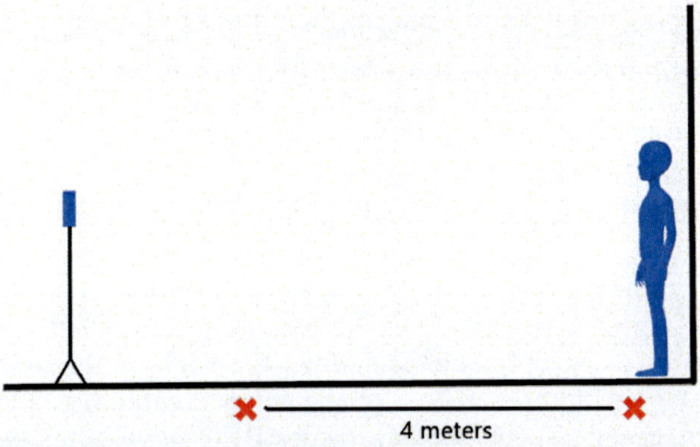


**Fig. 2.** Recording environment

### 2.3 Data Labeling

The keypoints obtained from OpenPose in JSON format were processed with Python scripts, combined to create 7 different configurations in which frames were correlated to provide the sensation of movement and dynamism. Each configuration is composed by a vector of (x, y) coordinates for each point of the body provided by OpenPose and the latest value is the label associated to this person. This data is formatted in a CSV file needed for fitting later the proposed models.

An explanation of each configuration created is described hereafter:

- Configuration 1 ($C_1$): Each frame was labelled individually. This one does not offer the movement sensation, however is a good point to compare the results obtained from an static image to the following data in movement. For other works focused on physical constitution, this dataset would be useful.
- Configuration 2 ($C_2$): 2 frames were concatenated to show a slight sense of movement.
- Configuration 3 ($C_3$): 2 frames were concatenated with a leap to avoid the close similarity between consecutive frames.
- Configuration 4 ($C_4$): 4 frames were concatenated to show a sense of movement.
- Configuration 5 ($C_5$): 4 frames were concatenated with a leap to avoid the close similarity between consecutive frames.
- Configuration 6 ($C_6$): 6 frames were concatenated to show a great sense of movement.
- Configuration 7 ($C_7$): 6 frames were concatenated with a leap to avoid the close similarity between consecutive frames.

### 2.4 Model Fitting

All configurations described in Subsect. 2.3 have been used to train a set of machine learning models using MoEv tool. MoEv [7] is an open-source tool developed in Python [22] for building machine learning models from datasets. Our dataset was input as a CSV file, evaluating with it several machine learning algorithms, getting as output accuracy rate, confusion matrix, recall, precision, and F1-score. MoEv also allows validate the models to ensure the optimal generalization of them. Once a model, or a set of them, has been fitted, it is validated by using a different dataset. This tool has been successfully used in different research areas such as attack detection on real-time location systems [8]; malicious network traffic detection [2], or prediction of students' academic success [9,10].

Training has been executed on Calendula, the High performance computing (HPC) cluster located in SCAYLE, a public entity created by Junta de Castilla y León and University of León which provides HPC services to research centers and companies.

### 2.5 Methods of Evaluation

Our configurations have been evaluated through MoEv tool. As shown in the Sect. 2.4, the configurations have been trained over several machine learning algorithms. Three of them were selected for an in-depth consideration, due to their good accuracy rate. MoEv allows evaluating models fitted through another dataset. As explained in Sect. 2.2 the 70% of the data gathered have been used to train the models. The remaining 30% have been used to test these models.

The classification models with a higher accuracy rate for people authentication were Linear Discriminant Analysis (LDA) [1], Logistic Regression Classifier (LR) [13] and Quadratic Discriminant Analysis (QDA) [6]. Accuracy was computed as shown in Eq. 1, where $T_P$ is the true-positive rate, $T_N$ is the true-negative rate, $F_P$ is the false-positive rate, and $F_N$ is the false-negative rate. Other KPIs were obtained through the confusion matrix: precision ($P$), recall ($R$) and F1-score ($F_1$). These metrics were computed as shown in Eqs. 2, 3, 4.

LDA is a commonly used technique for data classification and dimensional reduction, which easily handles the case where the within-class frequencies are unequal and their performances have been examined on randomly generated test data. LR is one of the most important statistical and data mining techniques which can be extended to multi-class classification. QDA is also well-known as a common tool for classification.

$$accuracy = \frac{T_P + T_N}{T_P + F_P + T_N + F_N} \tag{1}$$

$$\mathcal{P} = \frac{T_P}{T_P + F_P} \quad (2) \qquad \mathcal{R} = \frac{T_P}{T_P + F_N} \quad (3) \qquad \mathcal{F}_1 = 2\frac{\mathcal{P} \times \mathcal{R}}{\mathcal{P} + \mathcal{R}} \quad (4)$$

## 3 Results

Table 1 shows the results obtained in the different experiments by the 3 models which were in-depth analysed. Accuracy is shown in Table 1 as well as additional KPIs like P, R, and F1-score for such models and experiments.

The different trend of each model over each configuration is shown on Fig. 3 according to the accuracy rate, both in train and test.

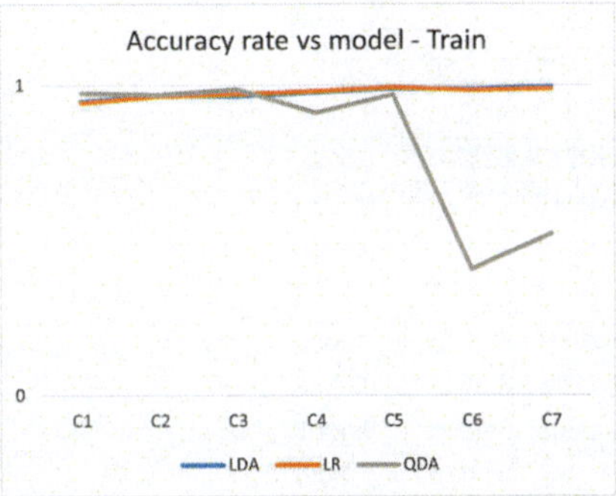


(a) Train

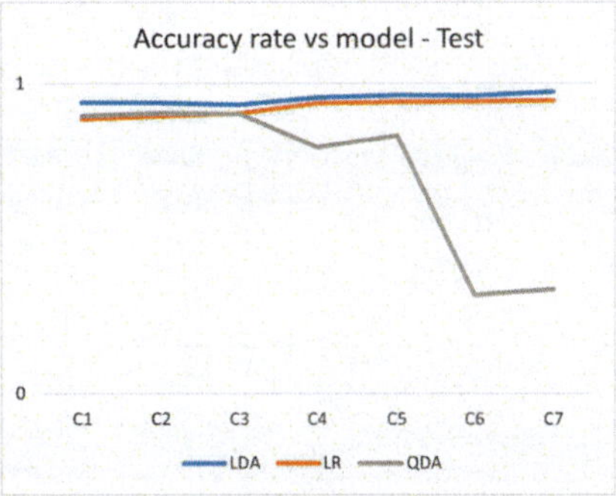


(b) Test

**Fig. 3.** Accuracy rate variation

## 4 Discussion

As it is shown in Table 1 we obtained good results in all the experiments, even when movement was not considered. However, when we concatenated frames to show a movement sensation, the accuracy raises to 97.33%.

QDA classifier earns good results in the first configuration, however as we added new frames concatenated accuracy decreased to an unacceptable model as shown in Fig. 3. Accuracy rate of QDA model decreases from 0.9867 in the best configuration to 0.4064 in worst one in training and from 0.9051 to 0.3188 in testing. We can conclude that QDA is a good classifier for $C_1$-$C_3$, meaning 1 frame or 2 frames concatenated.

LDA and LR models have similar results, being in most configurations over 90% which makes them available as classifiers in our method. LDA gets its best results, both in train and test, in $C_7$ with 0.9958 and 0.9733 respectively. However, LR obtains its best accuracy rate in $C_5$ (0.9929) and $C_7$ (0.9457).

Furthermore, KPI's $P$, $R$ and $F_1$ has the higher value in LDA model using the $C_5$ and $C_7$ configurations. Respectively, $C_5$ configuration has the following KPI's: for training $P = 0.9901, R = 0.9899, F_1 = 0.9899$ and for testing $P = 0.9650, R = 0.9622, F_1 = 0.9622$. On the other hand, $C_7$ configuration obtains for training $P = 0.9958, R = 0.9958 F_1 = 0.9958$ and for testing $P = 0.9746, R =$

**Table 1.** Results obtained in different experiments

| | Classifier | Accuracy | $P$ | $R$ | $F_1$ |
|---|---|---|---|---|---|
| Train $C_1$ | LDA | 0.9491 | 0.9514 | 0.9491 | 0.9494 |
| | LR | 0.9449 | 0.9452 | 0.9449 | 0.9446 |
| | QDA | 0.9756 | 0.9768 | 0.9756 | 0.9757 |
| Test $C_1$ | LDA | 0.9395 | 0.9435 | 0.9395 | 0.9400 |
| | LR | 0.8856 | 0.8901 | 0.885633 | 0.8845 |
| | QDA | 0.8970 | 0.9113 | 0.8970 | 0.8945 |
| Train $C_2$ | LDA | 0.9689 | 0.9697 | 0.9689 | 0.9690 |
| | LR | 0.9656 | 0.9659 | 0.9656 | 0.9655 |
| | QDA | 0.9705 | 0.9713 | 0.9705 | 0.9706 |
| Test $C_2$ | LDA | 0.9383 | 0.9416 | 0.9383 | 0.9382 |
| | LR | 0.8947 | 0.9012 | 0.8947 | 0.8930 |
| | QDA | 0.9051 | 0.9162 | 0.9051 | 0.9045 |
| Train $C_3$ | LDA | 0.9655 | 0.9661 | 0.9655 | 0.9656 |
| | LR | 0.9726 | 0.9728 | 0.9726 | 0.9726 |
| | QDA | 0.9867 | 0.9869 | 0.9867 | 0.9867 |
| Test $C_3$ | LDA | 0.9314 | 0.9383 | 0.9314 | 0.9309 |
| | LR | 0.9037 | 0.9102 | 0.9037 | 0.9006 |
| | QDA | 0.9028 | 0.9183 | 0.9028 | 0.9008 |
| Train $C_4$ | LDA | 0.9783 | 0.9788 | 0.9783 | 0.9783 |
| | LR | 0.9808 | 0.9811 | 0.9808 | 0.9808 |
| | QDA | 0.9125 | 0.9275 | 0.9125 | 0.9135 |
| Test $C_4$ | LDA | 0.9550 | 0.9581 | 0.9550 | 0.9543 |
| | LR | 0.9368 | 0.9409 | 0.9368 | 0.9356 |
| | QDA | 0.7933 | 0.8602 | 0.7933 | 0.7972 |
| Train $C_5$ | LDA | 0.9899 | 0.9901 | 0.9899 | 0.9899 |
| | LR | 0.9929 | 0.9929 | 0.9929 | 0.9929 |
| | QDA | 0.9698 | 0.9714 | 0.9698 | 0.9699 |
| Test $C_5$ | LDA | 0.9622 | 0.9650 | 0.9622 | 0.9622 |
| | LR | 0.9418 | 0.9466 | 0.9418 | 0.9404 |
| | QDA | 0.8322 | 0.8952 | 0.8322 | 0.8409 |
| Train $C_6$ | LDA | 0.9866 | 0.9868 | 0.9866 | 0.9867 |
| | LR | 0.9837 | 0.9838 | 0.9837 | 0.9837 |
| | QDA | 0.4064 | 0.5343 | 0.4064 | 0.3823 |
| Test $C_6$ | LDA | 0.9614 | 0.9643 | 0.9614 | 0.9612 |
| | LR | 0.9430 | 0.9497 | 0.9430 | 0.9413 |
| | QDA | 0.3188 | 0.4999 | 0.3188 | 0.2887 |
| Train $C_7$ | LDA | 0.9958 | 0.9958 | 0.9958 | 0.9958 |
| | LR | 0.9882 | 0.9882 | 0.9882 | 0.9881 |
| | QDA | 0.5205 | 0.7576 | 0.5205 | 0.4941 |
| Test $C_7$ | LDA | 0.9733 | 0.9746 | 0.9733 | 0.9732 |
| | LR | 0.9457 | 0.9532 | 0.9457 | 0.9446 |
| | QDA | 0.3360 | 0.5474 | 0.3360 | 0.2900 |

0.9733, $F_1 = 0.9732$. Results, as shown in Fig. 3, are regular over the different configurations with a positive tendency. Leaping a frame while concatenation is a good practise for getting better results.

## 5 Conclusions

As conclusion, our system has a good accuracy rate and it is a good method of authentication due to its limited cost, as the processing may be done in a multi-purpose computer and any camera may be used. Moreover, no personal devices are needed which eases scalability and it is not intrusive.

It has been demonstrated that it is safe from a health point of view because every person during data gathering was wearing his mask and did not touch any surface. Therefore, our authentication method can be applied during pandemic situations, in the chemical industry or in healthcare environments.

Although our method has not been tested as a real-time authentication system, this can be used as a proof of concept to implement it. The implementation is going to undergo a slight delay of 1 s in the worst case, as 12 frames are needed to identify a person. This delay could be overcome with a better processor for subsequent processing in OpenPose. This work can be considered a new application for OpenPose open-source pose system detection.

Most of the related works found datasets a disadvantage and we tried to overcome this lack of data with a contribution of keypoints coordinates corresponding to people in movement. A sample of 14 people were recorded walking ten times so that results in 10 523. Keypoints data which could contribute to future related works. These may include real-time implementation in a real environment and a proof of concept with a 3D camera.

Even when we have good results with many models and configurations, the best one is the $C_7$ configuration, in which we concatenated 6 frames with a leap to avoid too many similarities between frames. The best machine learning model is LDA with an accuracy of 97.33% and KPI's $P = 0.9746$, $R = 0.9733$ and $F_1 = 0.9732$.

**Acknowledgements.** The research described in this article has been partially funded by Instituto Nacional de Ciberseguridad de España (INCIBE), under the grant "ADENDA 4: detección de nuevas amenazas y patrones desconocidos (red Regional de Ciencia y Tecnología)", addendum to the framework agreement INCIBE-Universidad de León, 2019–2021; the Spanish Ministry of Science, Innovation, and Universities RTI2018-100683-B-I00 grant; and the regional Government of Castilla y León under the grant BDNS (487971).

Authors would like to thank Lidia Castro, Carlos Ujados, Marina Valdajos, Azucena Morala, Iris Fernández, Javier Toral, Carolina Junco, Ana Aroca, Alejandra Pérez, Martín Morala, Emilio José Luque, Igor Penín, Belén Miguel their support during the data gathering.

## References


1. Balakrishnama, S., Ganapathiraju, A.: Linear discriminant analysis - a brief tutorial (1995)
2. Campazas-Vega, A., Crespo-Martínez, I.S., Guerrero-Higueras, Á.M., Fernández-Llamas, C.: Flow-data gathering using netflow sensors for fitting malicious-traffic detection models. Sensors **20**(24), 7294 (2020)
3. Cao, Z., Hidalgo Martinez, G., Simon, T., Wei, S., Sheikh, Y.A.: Openpose: real-time multi-person 2d pose estimation using part affinity fields. IEEE Trans. Pattern Anal. Mach. Intell. **43**(1), 172–186 (2019)
4. Chong, P., Elovici, Y., Binder, A.: User authentication based on mouse dynamics using deep neural networks: a comprehensive study. IEEE Trans. Inf. Forensics Secur. **15**, 1086–1101 (2020). https://doi.org/10.1109/TIFS.2019.2930429
5. Darabseh, A., Pal, D.: Performance analysis of keystroke dynamics using classification algorithms. In: 2020 3rd International Conference on Information and Computer Technologies (ICICT), pp. 124–130 (2020). https://doi.org/10.1109/ICICT50521.2020.00027
6. Friedman, J.H.: Regularized discriminant analysis. J. Am. Stat. Assoc. **84**(405), 165–175 (1989). http://www.jstor.org/stable/2289860
7. Guerrero-Higueras, Á.M., Campazas-Vega, A., Crespo-Martínez, I.S.: Module evaluator (moev). Technical report, Robotics group, Universidad de León (2020). https://doi.org/10.5281/zenodo.4114127
8. Guerrero-Higueras, Á.M., DeCastro-Garcia, N., Matellan, V.: Detection of cyber-attacks to indoor real time localization systems for autonomous robots. Robot. Auton. Syst. **99**, 75–83 (2018)
9. Guerrero-Higueras, Á.M., DeCastro-García, N., Rodriguez-Lera, F.J., Matellán, V., Conde, M.Á.: Predicting academic success through students' interaction with version control systems. Open Comput. Sci. **9**(1), 243–251 (2019)
10. Guerrero-Higueras, Á.M., Fernández Llamas, C., Sánchez González, L., Gutierrez Fernández, A., Esteban Costales, G., González, M.Á.C.: Academic success assessment through version control systems. Appl. Sci. **10**(4), 1492 (2020)
11. Guo, C., Ito, S.I., Ito, M., Fukumi, M.: Personal authentication by walking motion using kinect. In: 2019 International Symposium on Intelligent Signal Processing and Communication Systems (ISPACS), pp. 1–2 (2019). https://doi.org/10.1109/ISPACS48206.2019.8986388
12. Handa, J., Singh, S., Saraswat, S.: A comparative study of mouse and keystroke based authentication. In: 2019 9th International Conference on Cloud Computing, Data Science Engineering (Confluence), pp. 670–674 (2019). https://doi.org/10.1109/CONFLUENCE.2019.8776953
13. Hosmer Jr, D.W., Lemeshow, S., Sturdivant, R.X.: Applied Logistic Regression, vol. 398. John Wiley & Sons, Hoboken (2013)
14. Jain, A., Singh, S., Singh, K.: Handwritten signature verification using shallow convolutional neural network. Multimedia Tools Appl. **79**(27), 19993–20018 (2020). https://doi.org/10.1007/s11042-020-08728-6
15. Kancharla, K., Kamble, V., Kapoor, M.: Handwritten signature recognition: a convolutional neural network approach. In: 2018 International Conference on Advanced Computation and Telecommunication (ICACAT), pp. 1–5 (2018). https://doi.org/10.1109/ICACAT.2018.8933575
16. Kumari, P., Seeja, K.R.: A novel periocular biometrics solution for authentication during COVID-19 pandemic situation. J. Ambient Intell. Humanized Comput. 1–17 (2021). https://doi.org/10.1007/s12652-020-02814-1

17. Li, B., Wang, W., Gao, Y., Phoha, V.V., Jin, Z.: Hand in motion: Enhanced authentication through wrist and mouse movement. In: 2018 IEEE 9th International Conference on Biometrics Theory, Applications and Systems (BTAS), pp. 1–9 (2018). https://doi.org/10.1109/BTAS.2018.8698577
18. Manabe, T., Yamana, H.: Two-factor authentication using leap motion and numeric keypad. In: Moallem, A. (ed.) HCII 2019. LNCS, vol. 11594, pp. 38–51. Springer, Cham (2019). https://doi.org/10.1007/978-3-030-22351-9_3
19. Mei Yin, D.B., Mukhlas, A.A., Wan Chik, R.Z., Talib Othman, A., Omar, S.: A proposed approach for biometric-based authentication using of face and facial expression recognition. In: 2018 IEEE 3rd International Conference on Communication and Information Systems (ICCIS), pp. 28–33 (2018). https://doi.org/10.1109/ICOMIS.2018.8644974
20. Mohapatra, R.K., Shaswat, K., Kedia, S.: Offline handwritten signature verification using CNN inspired by inception v1 architecture. In: 2019 Fifth International Conference on Image Information Processing (ICIIP), pp. 263–267 (2019). https://doi.org/10.1109/ICIIP47207.2019.8985925
21. Okereafor, K., Ekong, I., Markson, I., Enwere, K.: Fingerprint biometric system hygiene and the risk of COVID-19 transmission. JMIR Biomed. Eng. **5**, 1–15 (2020). https://doi.org/10.2196/19623
22. Python: About python (2021). https://www.python.org/about/. Accessed 5 June 2021
23. Tao, S., Zhang, X., Cai, H., Lv, Z., Hu, C., Xie, H.: Gait based biometric personal authentication by using mems inertial sensors. J. Ambient Intell. Humanized Comput. **9**(5), 1705–1712 (2018). https://doi.org/10.1007/s12652-018-0880-6
24. Wong, A.M.H., Furukawa, M., Ando, H., Maeda, T.: Dynamic hand gesture authentication using electromyography (EMG). In: 2020 IEEE/SICE International Symposium on System Integration (SII), pp. 300–304 (2020). https://doi.org/10.1109/SII46433.2020.9026294

# The Neverending Story: Memory Corruption 30 Years Later

Oscar Llorente-Vazquez[1](✉), Igor Santos[2], Iker Pastor-Lopez[1], and Pablo Garcia Bringas[1]

[1] University of Deusto, Bilbao, Spain
{oscar.llorente,iker.pastor,pablo.garcia.bringas}@deusto.es
[2] Mondragon Unibertsitatea, Arrasate-Mondragon, Spain
isantos@mondragon.edu

**Abstract.** Memory errors have been present in software for more than three decades now, leading to numerous security issues. Low-level languages like C and C++ that are prone to this class of errors are in widespread use, meaning that a large number of current systems are susceptible to attacks that target memory corruption vulnerabilities. Therefore, hardening programs and protecting against this type of attacks is of the utmost importance to build secure systems. This paper explains the basis of memory errors and their characteristics, analyzes the huge research effort that has been carried out in relation to memory error detection, exploitation and attack prevention, and depicts the most relevant approaches in this area.

## 1 Introduction

Among all the software errors that lead to security vulnerabilities, memory corruption is one of the most dangerous and oldest problems in which the systems security community has focused on, and despite the numerous approaches proposed to mitigate them, they are still found in generalized and specialized software [10,48]. Memory corruption occupies the top positions in the CWE most dangerous software weaknesses list [11], appearing separately as out-of-bounds writes, out-of-bounds reads, Use-after-Frees (UAF), and so on.

Several factors influence the prevalence of memory errors, but one of the main reasons is that a large amount of proposed detection and protection solutions are not deployed in practice due to performance overhead, compatibility problems, lack of completeness in protection, or dependence conflicts [44,47].

The primary intention of this paper is to provide a general understanding about memory errors, pointing out the most recent research in detection, exploitation and convenient defense methods.

The remainder of this paper is organized as follows. Section 2 depicts the basis of memory errors and classifies them in accordance with their properties. Section 3 systematizes the main approaches in the context of memory corruption detection. Then, Sect. 4 describes software attacks that exploit memory corruption focusing on code-reuse attacks and data-only attacks, the two most actively


J. J. Gude Prego et al. (Eds.): CISIS 2021/ICEUTE 2021, AISC 1400, pp. 136–145, 2022.
https://doi.org/10.1007/978-3-030-87872-6_14

researched classes. Section 5 analyzes attack countermeasures. Section 6 reviews the research related to our work. Finally, Sect. 7 concludes the paper.

## 2 Background

Memory corruption becomes apparent as a consequence of the lack of memory and type safety in low-level languages such as C and C++. Since these languages provide low level control over implementation details for example through manual memory management, they open up the possibility for the program to access memory in unintended ways.

It is commonly said that the execution of a program is memory safe as long as a set of memory errors never occur.

Memory errors include every access to undefined memory, i.e., memory that the program has not allocated or not initialized, but also accesses to not designated memory regions (i.e., a pointer dereference that does not point to its assigned memory area). According to their characteristics, they are commonly classified into spatial and temporal memory errors.

- **Spatial Memory Errors:** On the one hand, spatial errors comprise every memory access outside the established bounds of its referent. The most prominent example are buffer overflows, where data is written beyond the limit of a buffer, potentially overwriting other values. Therefore, we can observe out-of-bounds reads and writes as part of this class.
- **Temporal Memory Errors:** On the other hand, temporal errors are those caused by the usage of pointers whose referents are not valid at the time of dereference, because they have not been initialized or they have been previously freed, for instance. In that context, we can observe Use-After-Free (UAF) errors, uses of uninitialized memory, null pointer dereferences and illegal frees (i.e., calling free to a non-allocated or an already-freed pointer).

## 3 Memory Error Detection

Over the years, many detection systems that use different strategies and program analysis techniques have been proposed, usually putting together diverse analysis methods [42,44].

In the same way as user-space programs, operating system kernels such as Linux suffer from memory corruption. Detection in OS kernels entails its own challenges since their characteristics (e.g., generalized use of machine-level code, large and complex code bases) differ from userland programs. Among the proposed approaches, we can find static-analysis-based solutions [22,29] and dynamic analysis frameworks [41].

In this section we refer to general detection methods, not specific to any domain.

### 3.1 Static Analysis

Given the source or binary code of a program, several approaches make use of static analysis methods to find memory errors without running the program. The most common include: pointer analysis [50], model checking [19], abstract interpretation [9] and pattern matching [29].

In general, pure static analysis implementations have the advantage of examining the whole program rather than only executed paths, but are more likely to report false positives.

### 3.2 Instrumentation and Runtime Monitoring

Program instrumentation is a remarkably versatile method that enables a great variety of analyses. The so-called sanitizers and many other vulnerability detection tools leverage instrumentation capabilities from e.g., compiler toolchains, to insert reference monitors into the program to then enforce a security policy at runtime [42].

To detect memory errors, a diverse range of strategies have been implemented and numerous sanitizers have been proposed [43]. To illustrate, AddressSanitizer [36] introduces red zones around memory objects to detect out-of-bounds accesses and delays the reuse of freed memory regions to discover temporal safety issues. Moreover, SoftBound+CETS [24,25] track pointers with bounds and allocation status metadata and perform checks on dereference operations to provide full memory safety at the cost of higher overhead and compatibility issues.

Likewise, when program source code is not available, Dynamic Binary Instrumentation (DBI) frameworks like Valgrind [26] or DynamoRIO [3] provide the means to build memory corruption detection tools. Excellent examples are Memcheck [37] and Dr. Memory [4]. However, DBI usually introduces excessive performance overhead to the execution.

### 3.3 Fuzzing

Fuzz testing is a dynamic approach that has become one of the most used and effective techniques in recent years, while developing several new strategies and algorithms [18]. The main idea behind fuzzing resides in randomly generating inputs to test target programs, sometimes with the aim of discovering memory errors [48].

Since one of the main problems of sanitizers and other instrumentation-based detection tools is that they can only detect vulnerabilities when they are getting triggered, the combination with fuzzing turned out natural [13,28]. Fuzzing has also been shown to be effective in finding memory corruption in embedded systems, even though it exhibits different challenges to those of commodity systems [23].

### 3.4 Symbolic Execution

Rather than regularly executing a target program, input values and variables are represented as symbolic values instead of concrete values in a symbolic execution. These values generate path conditions that represent the state of the program and transformations between states, opening up the possibility to analyze all possible states [1].

These techniques have been successfully used to detect memory errors [31], and combined with other techniques such as Natural Language Processing (NLP) [49].

In practice, symbolic execution approaches are infrequently adopted because of its limitations in terms of scalability due to the path explosion problem, and complex constraint solving [1].

## 4 Memory Corruption Exploitation

Over the last years, there has been a constant arms race between memory corruption attacks and defenses. Triggering a memory error is usually the first step to carry out several attacks. By making a pointer go out of bounds or making it become dangling and then dereferencing it, a potential attacker could modify code pointers and variables, for instance [44].

The most usual objective is to divert program execution away from the intended control flow. In that context, code corruption attacks try to modify program code in memory while code injection schemes introduce new code, in any case to make the program execute the attacker-provided code. However, they are easily prevented using current processor features. Memory errors can be further exploited to disclose memory contents [27], usually to leak secrets in order to bypass several defenses [35].

Amid existing attacks, the security community has lately paid more attention to two classes: code-reuse and data-only attacks.

### 4.1 Code-Reuse Attacks

Since general attack mitigations became ubiquitous and code corruption and injection were no longer viable, new attacks emerged that reuse program code already present in memory [32].

Return-into-libc is the simplest form of these kind of attacks. By redirecting the control-flow to libc functions, an attacker is able to execute sequences of arbitrary function calls chained together [12].

These attacks evolved into Return-Oriented Programming (ROP), that instead of targeting whole functions, chains the execution of short instruction sequences (i.e., gadgets) that end with a return instruction [38,46]. While maintaining the basis of the technique, further ROP attacks have been proposed over the years to overcome emerging defenses [33], showing its effectiveness.

### 4.2 Data-Only Attacks

Unlike previous approaches, data-only attacks (also called non-control-data attacks) do not require hijacking the control-flow of the target program. Instead, this class of attacks focus on modifying security-sensitive data such as user input, configuration data, user identity, or decision-making data to bypass authentication checks, escalate privileges and so on [7].

Moreover, Hu et al. [15] proposed FlowStitch, a tool that successfully automates the process of generating data-oriented attacks from common memory errors by systematically joining data flows in the program. Afterwards, a follow-up work [16] introduced the notion of Data-Oriented Programming (DOP), a systematic technique to construct Turing-complete data-only exploits by finding data-oriented gadgets, in a similar fashion to ROP. Likewise, Ispoglou et al. [17] proposed Block-Oriented Programming (BOP), which uses entire basic blocks as gadgets and produces payloads that follow the valid Control-Flow Graph of the program but not its Data-Flow Graph.

## 5 Defense Strategies

As a consequence of the increasing number of developed attacks over the years, several countermeasures have been equally proposed. However, many of them are not deployed in practice, or a less strict version is used, since they usually incur high performance overhead.

The most basic and widely used defenses include: Address Space Layout Randomization (ASLR), the Write XOR Execute policy for memory pages, and stack canaries. ASLR randomizes the arrangement of the process memory layout so that attackers do not know the location of interesting targets. Write XOR Execute ensures that memory pages can be writable or executable but not both to thwart code injection. While stack canaries introduce values (i.e., canaries) into the stack that are checked when a function terminates to prevent control-flow hijacking. Unfortunately, they have been shown to be ineffective against more sophisticated attacks [32,35] and consequently more advanced defenses have been developed.

### 5.1 Enhanced Software Diversity

Since the main problem of secret-based defenses like ASLR is that they are susceptible to information disclosure attacks [27], a lot of research effort has been put in developing leakage-resistant software diversity techniques [20]. As an example, ASLRGuard [21] uses a secure memory region to store most code locators and encryption for the rest of the code pointers with the aim of preventing its disclosure. Another approach is live re-randomization, which prevents the attacker from exploiting the obtained knowledge about the program by systematically re-randomizing the address space to invalidate current code pointers [2,8].

### 5.2 Control-Flow Integrity

An interesting alternative are integrity-based defenses, being Control-Flow Integrity (CFI) [5] the most extended approach. CFI builds the Control-Flow Graph (CFG) of the program to obtain possible execution paths and then ensures that execution follows one of those paths at runtime. Nevertheless, precise CFI enforcement entails high performance overhead and, consequently, several implementations trade security for performance. Therefore we discern two broad classes of CFI: a more relaxed integrity checking called coarse-grained CFI [45,51] and a more precise fine-grained CFI [30].

In order to improve overall performance, researchers have explored the application of hardware features, successfully enforcing CFI using Intel's processor trace [14] for example.

### 5.3 Data-Flow Integrity

In a similar fashion, Data-Flow Integrity (DFI) aims to mitigate non-control data attacks by preventing the program from deviating from the intended data-flow.

With the aid of static analysis (i.e., reaching definitions analysis) a Data-Flow Graph (DFG) is built that, in short, maps instructions to definition identifiers. Then, read and write instructions are instrumented to ensure at runtime that all data flows are within the DFG [6,40]. Any deviation from the DFG terminates the execution.

### 5.4 Software Fault Isolation

To deal with untrusted code, Software Fault Isolation (SFI) isolates untrusted modules in a sandbox in the host's address space so that it cannot access memory outside the sandbox. As representative examples, Google's NaCl [34] loads the untrusted code into a predefined area and instruments it to confine its memory and instruction references to the sandbox, while ARMlock [52] uses the ARM hardware support for memory domains to create sandboxes that constrain every memory access inside them.

## 6 Related Work

In the context of this research, several systematization of knowledge works have been already proposed.

Van der Veen et al. [47] provide a historical overview and analysis of memory errors. They found that memory errors are unlikely to lose significance in the near future and that state-of-the-art detection and containment techniques fail to protect against motivated attackers.

Szekeres et al. [44] organize the knowledge about diverse protection methods by setting up a general model for memory corruption attacks to find that stronger protection policies are needed.

Shoshitaishvili et al. [39] propose angr, a unified framework for the automated identification and exploitation of memory corruption vulnerabilities at binary-level.

In relation to defense approaches, Larsen et al. [20] systematize the understanding of software diversity and highlight fundamental trade-offs between fully automated approaches.

With regard to program reasoning and vulnerability detection, Baldoni et al. [1] survey the main aspects of symbolic execution and the most notable techniques used in testing and security applications, while Klees et al. [18] perform an extensive evaluation of the most recent fuzzers and analyze their experimental methodologies.

Lastly, Song et al. [42] bring together the knowledge about sanitizers. They propose a new taxonomy of available tools and the security vulnerabilities they cover, describe their performance and compatibility properties, and emphasize different trade-offs.

## 7 Conclusion

As observed, memory errors have been extensively studied from different perspectives but they still remain as one of the principal difficulties to build secure systems, which suggests that further research is needed.

By using techniques like runtime monitoring or fuzzing, researchers have effectively mitigated memory corruption. While, software diversity and integrity-based defenses have been found to be capable of preventing complex attacks. However, in order to keep exploiting memory vulnerabilities, attacks have evolved from basic code-injection to more advanced code-reuse and data-only attacks.

We hope that this review provides security researchers with solid background knowledge to contribute to the field.

**Acknowledgments.** This work is partially supported by the Basque Government under a pre-doctoral grant given to Oscar Llorente-Vazquez.

## References

1. Baldoni, R., Coppa, E., D'elia, D.C., Demetrescu, C., Finocchi, I.: A survey of symbolic execution techniques. ACM Comput. Surv. (CSUR) **51**(3), 1–39 (2018)
2. Bigelow, D., Hobson, T., Rudd, R., Streilein, W., Okhravi, H.: Timely rerandomization for mitigating memory disclosures. In: Proceedings of the 22nd ACM SIGSAC Conference on Computer and Communications Security (CCS), pp. 268–279 (2015)
3. Bruening, D., Amarasinghe, S.: Efficient, transparent, and comprehensive runtime code manipulation. Ph.D. thesis, Massachusetts Institute of Technology, Department of Electrical Engineering (2004)
4. Bruening, D., Zhao, Q.: Practical memory checking with dr. memory. In: International Symposium on Code Generation and Optimization (CGO 2011), pp. 213–223. IEEE (2011)

5. Burow, N., et al.: Control-flow integrity: precision, security, and performance. ACM Comput. Surv. (CSUR) **50**(1), 1–33 (2017)
6. Castro, M., Costa, M., Harris, T.: Securing software by enforcing data-flow integrity. In: Proceedings of the 7th Symposium on Operating Systems Design and Implementation (OSDI), pp. 147–160 (2006)
7. Chen, S., Xu, J., Sezer, E.C., Gauriar, P., Iyer, R.K.: Non-control-data attacks are realistic threats. In: 14th USENIX Security Symposium (USENIX Security 2005), vol. 5 (2005)
8. Chen, X., Bos, H., Giuffrida, C.: Codearmor: virtualizing the code space to counter disclosure attacks. In: 2017 IEEE European Symposium on Security and Privacy (EuroS&P), pp. 514–529. IEEE (2017)
9. Clang static analyzer. https://clang-analyzer.llvm.org/
10. Cloosters, T., Rodler, M., Davi, L.: Teerex: discovery and exploitation of memory corruption vulnerabilities in sgx enclaves. In: 29th USENIX Security Symposium (USENIX Security 2020), pp. 841–858 (2020)
11. CWE - 2020 CWE top 25 most dangerous software weaknesses. https://cwe.mitre.org/top25/archive/2020/2020_cwe_top25.html
12. Designer, S.: Return-to-libc attack. Bugtraq (1997)
13. Dinesh, S., Burow, N., Xu, D., Payer, M.: Retrowrite: statically instrumenting cots binaries for fuzzing and sanitization. In: 2020 IEEE Symposium on Security and Privacy (SP), pp. 1497–1511. IEEE (2020)
14. Ge, X., Cui, W., Jaeger, T.: Griffin: guarding control flows using intel processor trace. In: 22nd International Conference on Architectural Support for Programming Languages and Operating Systems, ASPLOS 2017, pp. 585–598. Association for Computing Machinery (2017)
15. Hu, H., Chua, Z.L., Adrian, S., Saxena, P., Liang, Z.: Automatic generation of data-oriented exploits. In: 24th USENIX Security Symposium (USENIX Security 2015), pp. 177–192 (2015)
16. Hu, H., Shinde, S., Adrian, S., Chua, Z.L., Saxena, P., Liang, Z.: Data-oriented programming: On the expressiveness of non-control data attacks. In: 2016 IEEE Symposium on Security and Privacy (SP), pp. 969–986. IEEE (2016)
17. Ispoglou, K.K., AlBassam, B., Jaeger, T., Payer, M.: Block oriented programming: Automating data-only attacks. In: Proceedings of the 2018 ACM SIGSAC Conference on Computer and Communications Security (CCS), pp. 1868–1882 (2018)
18. Klees, G., Ruef, A., Cooper, B., Wei, S., Hicks, M.: Evaluating fuzz testing. In: Proceedings of the 2018 ACM SIGSAC Conference on Computer and Communications Security (CCS), pp. 2123–2138 (2018)
19. Kroening, D., Tautschnig, M.: CBMC–C bounded model checker. In: International Conference on Tools and Algorithms for the Construction and Analysis of Systems, pp. 389–391. Springer, Heidelberg (2014)
20. Larsen, P., Homescu, A., Brunthaler, S., Franz, M.: Sok: automated software diversity. In: 2014 IEEE Symposium on Security and Privacy (SP), pp. 276–291. IEEE (2014)
21. Lu, K., Song, C., Lee, B., Chung, S.P., Kim, T., Lee, W.: ASLR-guard: stopping address space leakage for code reuse attacks. In: Proceedings of the 22nd ACM SIGSAC Conference on Computer and Communications Security (CCS), pp. 280–291 (2015)
22. Machiry, A., Spensky, C., Corina, J., Stephens, N., Kruegel, C., Vigna, G.: DR.CHECKER: a soundy analysis for linux kernel drivers. In: 26th USENIX Security Symposium (USENIX Security 2017), pp. 1007–1024 (2017)

23. Muench, M., Stijohann, J., Kargl, F., Francillon, A., Balzarotti, D.: What you corrupt is not what you crash: challenges in fuzzing embedded devices. In: Proceedings of the 2018 Annual Network and Distributed System Security Symposium (NDSS) (2018)
24. Nagarakatte, S., Zhao, J., Martin, M.M., Zdancewic, S.: Softbound: highly compatible and complete spatial memory safety for c. In: Proceedings of the 30th ACM SIGPLAN Conference on Programming Language Design and Implementation (PLDI), pp. 245–258 (2009)
25. Nagarakatte, S., Zhao, J., Martin, M.M., Zdancewic, S.: Cets: compiler enforced temporal safety for C. In: Proceedings of the 2010 International Symposium on Memory Management, pp. 31–40 (2010)
26. Nethercote, N., Seward, J.: Valgrind: a framework for heavyweight dynamic binary instrumentation. ACM SIGPLAN Not. **42**(6), 89–100 (2007)
27. Oikonomopoulos, A., Athanasopoulos, E., Bos, H., Giuffrida, C.: Poking holes in information hiding. In: 25th USENIX Security Symposium (USENIX Security 2016), pp. 121–138 (2016)
28. Österlund, S., Razavi, K., Bos, H., Giuffrida, C.: Parmesan: sanitizer-guided greybox fuzzing. In: 29th USENIX Security Symposium (USENIX Security 20), pp. 2289–2306 (2020)
29. Padioleau, Y., Lawall, J., Hansen, R.R., Muller, G.: Documenting and automating collateral evolutions in linux device drivers. ACM SIGOPS Oper. Syst. Rev. **42**(4), 247–260 (2008)
30. Payer, M., Barresi, A., Gross, T.R.: Fine-grained control-flow integrity through binary hardening. In: International Conference on Detection of Intrusions and Malware, and Vulnerability Assessment (DIMVA), pp. 144–164. Springer (2015)
31. Ramos, D.A., Engler, D.: Under-constrained symbolic execution: correctness checking for real code. In: 24th USENIX Security Symposium (USENIX Security 2015), pp. 49–64 (2015)
32. Roemer, R., Buchanan, E., Shacham, H., Savage, S.: Return-oriented programming: systems, languages, and applications. ACM Trans. Inf. Syst. Secur. (TISSEC) **15**(1), 1–34 (2012)
33. Schuster, F., Tendyck, T., Liebchen, C., Davi, L., Sadeghi, A.R., Holz, T.: Counterfeit object-oriented programming: on the difficulty of preventing code reuse attacks in C++ applications. In: 2015 IEEE Symposium on Security and Privacy (SP), pp. 745–762. IEEE (2015)
34. Sehr, D., et al.: Adapting software fault isolation to contemporary cpu architectures. In: 19th USENIX Security Symposium (USENIX Security 2010) (2010)
35. Seibert, J., Okhravi, H., Söderström, E.: Information leaks without memory disclosures: remote side channel attacks on diversified code. In: Proceedings of the 2014 ACM SIGSAC Conference on Computer and Communications Security (CCS), pp. 54–65 (2014)
36. Serebryany, K., Bruening, D., Potapenko, A., Vyukov, D.: Addresssanitizer: a fast address sanity checker. In: 2012 USENIX Annual Technical Conference (USENIX ATC 2012), pp. 309–318 (2012)
37. Seward, J., Nethercote, N.: Using valgrind to detect undefined value errors with bit-precision. In: USENIX Annual Technical Conference (USENIX ATC 2005), pp. 17–30 (2005)
38. Shacham, H.: The geometry of innocent flesh on the bone: return-into-libc without function calls (on the x86). In: Proceedings of the 14th ACM conference on Computer and Communications Security (CCS), pp. 552–561 (2007)

39. Shoshitaishvili, Y., et al.: Sok:(state of) the art of war: offensive techniques in binary analysis. In: 2016 IEEE Symposium on Security and Privacy (SP), pp. 138–157. IEEE (2016)
40. Song, C., Lee, B., Lu, K., Harris, W., Kim, T., Lee, W.: Enforcing kernel security invariants with data flow integrity. In: Proceedings of the 2016 Annual Network and Distributed System Security Symposium (NDSS) (2016)
41. Song, D., et al.: An effective probing and fuzzing framework for the hardware-OS boundary. In: Proceedings of the 2019 Annual Network and Distributed System Security Symposium (NDSS) (2019)
42. Song, D., et al.: Sok: sanitizing for security. In: 2019 IEEE Symposium on Security and Privacy (SP), pp. 1275–1295. IEEE (2019)
43. Stepanov, E., Serebryany, K.: Memorysanitizer: fast detector of uninitialized memory use in C++. In: 2015 IEEE/ACM International Symposium on Code Generation and Optimization (CGO), pp. 46–55. IEEE (2015)
44. Szekeres, L., Payer, M., Wei, T., Song, D.: SoK: eternal war in memory. In: 2013 IEEE Symposium on Security and Privacy (SP), pp. 48–62. IEEE (2013)
45. Van Der Veen, V., et al.: A tough call: mitigating advanced code-reuse attacks at the binary level. In: 2016 IEEE Symposium on Security and Privacy (SP), pp. 934–953. IEEE (2016)
46. van der Veen, V., Andriesse, D., Stamatogiannakis, M., Chen, X., Bos, H., Giuffrdia, C.: The dynamics of innocent flesh on the bone: Code reuse ten years later. In: Proceedings of the 2017 ACM SIGSAC Conference on Computer and Communications Security (CCS), pp. 1675–1689 (2017)
47. Van der Veen, V., Cavallaro, L., Bos, H., et al.: Memory errors: the past, the present, and the future. In: International Conference on Recent Advances in Intrusion Detection (RAID), pp. 86–106. Springer (2012)
48. Wang, H., et al.: Typestate-guided fuzzer for discovering use-after-free vulnerabilities. In: 2020 IEEE/ACM 42nd International Conference on Software Engineering (ICSE), pp. 999–1010. IEEE (2020)
49. Wang, J., et al.: NLP-EYE: detecting memory corruptions via semantic-aware memory operation function identification. In: 22nd International Symposium on Research in Attacks, Intrusions and Defenses (RAID 2019), pp. 309–321 (2019)
50. Yan, H., Sui, Y., Chen, S., Xue, J.: Spatio-temporal context reduction: a pointer-analysis-based static approach for detecting use-after-free vulnerabilities. In: 2018 IEEE/ACM 40th International Conference on Software Engineering (ICSE), pp. 327–337. IEEE (2018)
51. Zhang, C., et al.: Practical control flow integrity and randomization for binary executables. In: 2013 IEEE Symposium on Security and Privacy (SP), pp. 559–573. IEEE (2013)
52. Zhou, Y., Wang, X., Chen, Y., Wang, Z.: Armlock: hardware-based fault isolation for arm. In: Proceedings of the 2014 ACM SIGSAC conference on Computer and Communications Security (CCS), pp. 558–569 (2014)

# StegoPass – Utilization of Steganography to Produce a Novel Unbreakable Biometric Based Password Authentication Scheme

Konstantinos Karampidis(✉), Euangelos Linardos, and Ergina Kavallieratou

University of the Aegean, Samos, Greece
{karampidis,kavallieratou}@aegean.gr,
euangelos.linardos@icsd.aegean.gr

**Abstract.** In the digital era we live, trustworthy verification schemes are required to ensure security and to authenticate the identity of an individual. Traditional passwords were proved to be highly vulnerable to attacks and the need of adopting new verification schemes is compulsory. Biometric factors have gained a lot of interest during the last years due to their uniqueness, ease of use, user convenience, and ease of deployment. However, recent research showed that even this unique authentication factors are not inviolable techniques. Thus, it is necessary to employ new verification schemes that cannot be replicated or stolen. In this paper we propose the utilization of steganography as a tool to provide unbreakable passwords. More specifically, we obtain a biometric feature of a user and embed it as a hidden message in an image. This image is then utilized as a password, the so-called StegoPass. Reversely, when a legit user or an attacker tries to unlock a device or an application, the same biometric feature is captured and embedded with the same steganography algorithm into the picture. The hash key of the resulted stego image in both cases is produced and if there is a complete match, user is considered as authenticated. To ensure that the proposed StegoPass cannot be replicated, we have conducted experiments with state-of-the-art deep learning algorithms. Moreover, it was examined whether Generative Adversarial Networks could produce exact copies of the StegoPass to fool the suggested method and the results showed that the proposed verification scheme is extremely secure.

**Keywords:** Steganography · Biometrics · Neural networks · Generative adversarial networks

## 1 Introduction

As the utilization of digital devices and applications keeps growing exponentially, the need of authenticating legit users is compulsory. It is estimated that by 2030 around 50 billion of Internet of Things (IoT) devices will be in use around the world [1], creating a massive web of interconnected devices spanning everything from smartphones to kitchen appliances. Moreover, with over 200 billion downloads of applications only in 2020 and the utilization of a huge number of systems like access controls, it becomes obvious that


J. J. Gude Prego et al. (Eds.): CISIS 2021/ICEUTE 2021, AISC 1400, pp. 146–155, 2022.
https://doi.org/10.1007/978-3-030-87872-6_15

the need of employment of complex passwords in order to prevent attacks and spoofing is getting compulsory.

Traditional password schemes are vulnerable to attacks. Traditional passwords can be lost, forgotten, stolen or even intercepted. Under this context, new means i.e., biometric passwords were adopted to authenticate a legit user. Biometrics are biological measurements — or physical characteristics — that can be used to uniquely identify individuals. Furthermore, the so-called morphological biometrics i.e., physical traits like fingerprint mapping, facial recognition, and iris scans are widely used as authentication systems and are getting higher deployment. Each person's unique biometric identity is used to replace or supplement traditional password systems for digital devices e.g., computers, smartphones etc., and/or restrict access to rooms and buildings. Biometrics cannot be lost, stolen, or forgotten.

However, although morphological biometrics seems to be more secure than the traditional password schemes, recent research showed that these methods also have security flaws, and these flaws concern the presentation attack system [2–5]. An attacker could replicate a fingerprint artifact, either by printing it and then presenting it to a fingerprint sensor as a legit one or by constructing a fake fingerprint from materials like gelatin, silicone or latex. Presentation attack is the ability of a system to detect whether the captured biometric feature -e.g., facial points- is real (from a live person) or fake (from a spoof artifact) [6]. Regarding face recognition systems, there are three ways to be spoofed: a) by presenting a photograph of a valid user, b) by presenting a video of a valid user and c) by presenting a 3D model of a valid user. Therefore, in order to overcome these security flaws and to ensure that the captured facial points come from a living person, specialized equipment must be supplementally installed. This, in many cases is difficult, makes the model more complex and adds more time to the identification procedure.

Thus, it is crucial to propose an authentication scheme which will a) be unique for each individual, b) provide the maximum security, c) can be applied to the majority of devices and applications, d) cannot be replicated, e) cannot be forgotten f) be easy to implement. In this paper, we propose the utilization of steganography in conjunction with biometrics to enhance security of authentication schemes and to overcome the security flaws that the aforementioned authentication techniques have. Steganography is a widespread ancient Greek technique which hides messages in a cover medium in a way that the hidden message becomes unnoticed. It is very easy to be deployed and since cryptography is forbidden in many countries [7] it became very popular among criminals who try to find a way to communicate with each other without being noticed. However, in this paper we utilize this technique in conjunction with biometrics, as a security mean. More specifically, the 68 facial points that uniquely characterize an individual, are extracted and embedded in an image which is the cover medium for steganography. The facial points are embedded to the image in a way that it could be almost inevitable someone to reproduce the method. To the best of our knowledge, the most related work are the watermarking techniques. These techniques utilize steganography to embed hidden information to images, in order to help professionals -e.g., photographers- to track their work and prove ownership when their work is found to be illegally used. However, watermarking is not our case. The novelty of the presented work relies on the fact that we propose an authentication method that is unique, fast and simple, it cannot be replicated, it provides maximum security, and it can be easily expanded. These requirements cannot be met in watermarking techniques, since it was proved that they have security flaws [8].

The resulted image is the so-called StegoPass i.e., a password that contains each person unique biometric identity embedded in an image.

The major contributions of our work are summarized as follows: a) a novel authentication method is proposed b) the proposed method is fast and simple c) the proposed method cannot be replicated and thus it provides maximum security d) the produced password i.e., the StegoPass is unique for each user e) it can be deployed to a vast number of devices and applications like IoT devices, computers, smartphones etc. f) the proposed method can be easily expanded.

The rest of the paper is structured as follows. In Sect. 2 we describe the proposed method along with terms and definitions. The dataset and the experimental results are described in Sect. 3. Finally, in Sect. 4 the conclusions and future work are given.

## 2 Methodology

Steganography is the art of hiding information in a cover medium in a way that nobody can understand its presence while it is in plain sight [9] (Fig. 1). It is an ancient technique, and the etymology of the word comes from the ancient Greek words: steganos (cover) + grapho (write). This technique of hiding messages is very common nowadays since the hidden messages become unnoticed, while in cryptography the message cannot be read but everybody acknowledges its existence.

**Fig. 1.** (left) An "innocent" stego image (right) the hidden image.

The cover medium could be an image, a text file, a sound file, a video file, or a network packet. However, images are the most utilized cover mediums as they can be found everywhere. In this paper, images were selected as the cover medium to hide the embedded message, but the proposed method could be deployed to any of the possible aforementioned cover mediums.

The most utilized method to embed the hidden message is the Least Significant Bit (LSB) method. In this method hidden message is embedded in the LSB plane of the cover medium (Fig. 2). It is a very simple and effective technique, but the major limitation is that since the attacker knows where to look for the hidden message, it is very easy to retrieve it. For this reason, different steganography techniques were developed such as LSB with sets where specific sets of pixels are used in order to select where the message will be hidden.

There are a lot of different sets that can be employed to perform steganography e.g., the Sieve of Eratosthenes, a Fibonacci generator, the n-th Fermat number, the sequence of Syracuse etc. In this paper we have utilized the Sieve the Eratosthenes [10], but every other available set could be used to define the pixels where the message would be embedded. The sieve of Eratosthenes is an ancient algorithm that finds all prime numbers up to a given limit. Since the message is not embedded in consecutive LSB bits, it is much harder for an attacker to identify where the embedded information is hidden and eventually retrieve the hidden message.

**Fig. 2.** a) A clean image b) a stego image c) LSB of clean image d) LSB of stego image.

The biometric factor selected as the embedded message, is the 68 facial marks that uniquely characterize a human being and distinguishes one face from another. For example, one person may have wide set of eyes or a small nose. Identifying the location of these landmarks can help us compare faces. There are different estimation models based on the number of face landmark points. The 5 facial points model is the simplest one and it only detects the edges of each eye and the bottom of the nose. Other models -which show better accuracy- include 68 different points on the face including eyes, nose, lips and face shape (Fig. 3). The process of detecting facial landmarks consists of two steps: a) to localize the face on the image, b) to detect the facial landmarks.

A python script was written and recursively processed each one of the images of the dataset and extracted the 68 facial points. These points were reshaped into a vector form and embedded into the cover medium with the LSB method described earlier. Thus, the proposed method can be shown on Fig. 4.

The security of the proposed method is extremely high, and it can be further strengthened if more biometric factors like fingerprint minutiae moments were also embedded

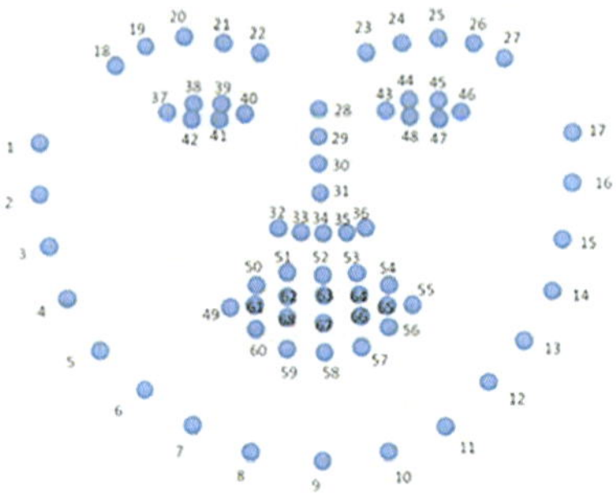

**Fig. 3.** The 68 facial landmarks

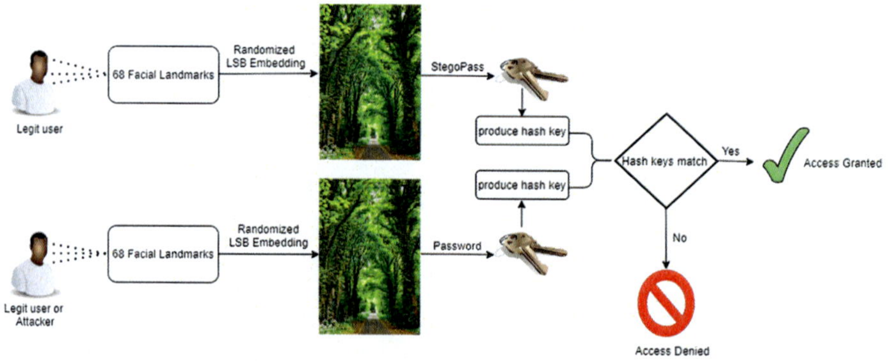


**Fig. 4.** The proposed method

to the cover medium (along with the 68 facial landmarks). Therefore, we could produce the so-called 2 Factor Stego Password (2-FSP) or by deploying -and merging- numerous biometric factors as the hidden message, we could propose a n-FSP, where n denotes the number of the embedded biometric factors. It is obvious that as the number of the embedded biometric factors increases, the security of the proposed password also increases. To test the security of the proposed method (1-FSP) and examine whether the proposed method could be fooled, we trained two different Generative Adversarial Networks (GAN) [11] to produce synthetic images and afterwards these produced images were compared against the original ones.

Generative Adversarial Networks are generative models and therefore they are able to generate new content. GANs consists of two networks i.e., a generator and a discriminator. The generator produces new content -starting from pure noise-, while the discriminator decides whether the produced content is real or fake. The quality output of GANs outperform other deep generative models [12]. A GAN is flexible since it can train any type of generator network while other generative algorithms have certain limitations for the generator [13], and furthermore there is no restriction on the latent variable's size. The aforementioned benefits made GANs to be considered as the models with the top performance in generating artificial data and more specifically synthetic images [14].

In this paper the GAN proposed by Goodfellow et al. [11] and the Wasserstein GAN (WGAN) [15], were utilized and trained with the training data of the produced dataset. When the GANs were trained, a test set was presented to the trained models and the generators produced synthetic images which were then compared against the original ones i.e., the StegoPass.

There are many proposed qualitative and quantitative measures that are utilized to evaluate GANs performance [16, 17]. Qualitative measures are not numerical and frequently require human evaluation, while the quantitative measures calculate specific numerical scores used to review the quality of the generated images. The most utilized quantitative measures are the Inception Score (IS) and the Frechet Inception Distance (FID).

However, these measures are not applicable to our case. The aforementioned measures -and all the others proposed to the literature- refer to the evaluation of the quality of the generated images. In our case, in addition to the quality of the generated fake images we also want them to be exact copies of the originals. For example, even if only one bit is changed to the generated image in comparison with the original one, it will not affect the quality. Nevertheless, since there is an embedded message to the LSB plane of the original image, this small failure will result to a totally different meaning, i.e., the embedded message will be completely different.

Thus, we need to deploy a measure that besides quality will also compare -bit by bit- the original and the generated fake images. For this reason, the hash value of the images was adopted as the chosen evaluation metric. The hash values (MD5 algorithm) of the original and the generated fake images were calculated and compared, and the results showed that the proposed technique offers extremely high security. Hash-based verification ensures that if the two images (the original and the fake) are identical the embedded message -the 68 facial points- is identical with the one of the legit user and therefore the user is considered as authenticated.

## 3 Experimental Results

The original dataset that was utilized to extract the 68 facial points, was the Flickr-Faces-HQ dataset [18] (Fig. 5). This dataset was originally created as a benchmark for Generative Adversarial Networks (GAN) and comprises of png images sized 1024x1024 pixels. It contains considerable variation in terms of age, ethnicity and image background. It also has good coverage of accessories such as eyeglasses, sunglasses, hats, etc.

The images were crawled from Flickr and only images under permissive licenses were collected. A limited modified version of this dataset was utilized, and the inclusion criteria were a) only adults' images were allowed b) the subject did not wear sunglasses, c) only one subject could be in every image and d) only frontal view was allowed. After the deployment of these filters, 2000 images resized to $256 \times 256$ pixels -due to computational resources limitations- were selected to form the dataset that will train the GANs.

Afterwards, for each one of the images of the produced dataset, the 68 facial points were extracted (Fig. 6) and the extracted facial point vector was embedded to the cover medium. To perform the embedding of the message, the Stegano Python library [19]

**Fig. 5.** A sample image of the dataset

was utilized and more specifically the LSB with sets algorithm was chosen (the Sieve of Eratosthenes). This resulted to the produce of the so-called StegoPass or 1-FSP. Therefore, 2000 StegoPass images were produced and utilized to train the aforementioned GANs. Furthermore, one hundred (100) images were utilized as a test set.

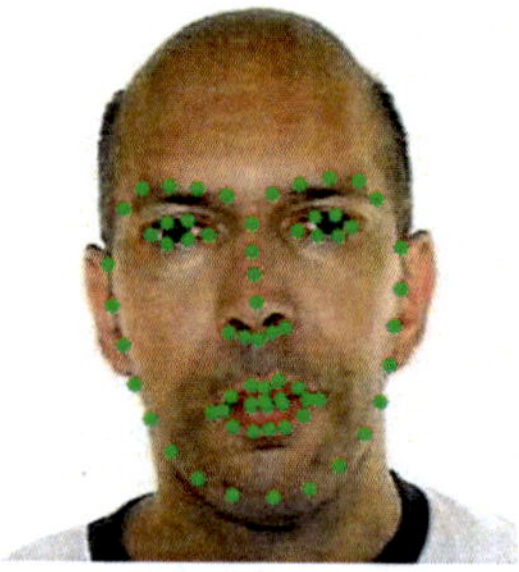

**Fig. 6.** The extracted facial points

Both GANs were trained with the Adam optimizer and a learning rate of 0.00002. The batch size was 1024 images for the GAN [11], while for the WGAN [15] it was 512. When the GANs were eventually trained, the 100 images of the test set were presented to the trained models, and it generated 100 images for each GAN. It must be stated that both GANs have no convolutional layers, either to the generator or the discriminator. Figure 7 shows the training evolution -for the two GANs- from epoch 1 (Fig. 7f - pure noise) to epoch 5000, for a sample image (Fig. 7a) of the test set. Afterwards, the hash values of the 100 images of the test set and the 100 images that each GAN produced, were calculated and compared. As it can been seen from Table 1 there are no produced images that are identical to the images of the test set since their hash values are different. If these values match, the file is presumed to be unmodified. Therefore, it was proved that the proposed method offers extremely high security even when state-of-the-art algorithms i.e., GANs try to attack to our model.

It is quite promising that none of the synthetic images was identical to the original ones. This was expected since the message -i.e., the facial landmarks- was hidden in a way that an intruder i.e., the GANs in this case, could not know. In the unlikely scenario that

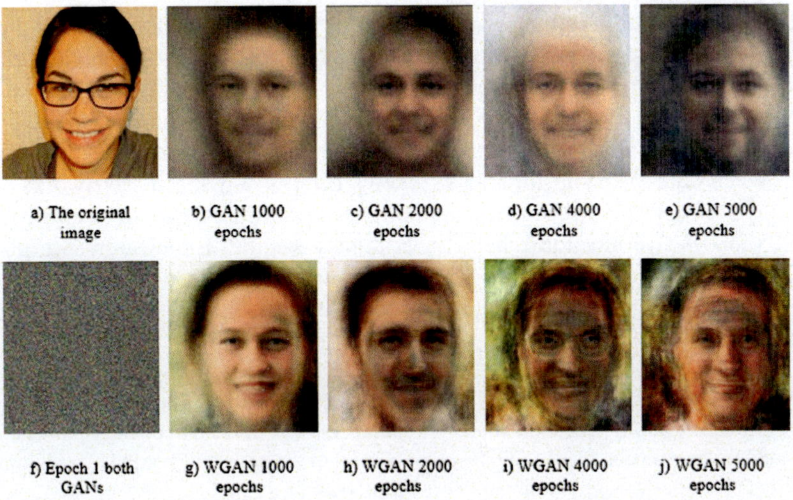


**Fig. 7.** a) The original StegoPass b-e) Training evolution of GAN [11] g-j) Training evolution of WGAN [15] f) The fake image after 1 epoch for both GANS.

**Table 1.** Confusion matrix for both GANs

| | Hash value (MD5) match | |
|---|---|---|
| | Original images | Fake images |
| Original images | 100 | 0 |
| Fake images | 0 | 100 |

in the future, new deep learning algorithms could eventually generate synthetic images which were identical with the originals, the proposed method could be easily expanded. This could be possible in many ways. For example, more biometric factors like iris and/or fingerprint minutiae moments could be embedded to the image as well, resulting to a multi-modal hidden biometric message. This multi-modal biometric message could be embedded to an image in a different -more random- way than the one in the proposed method, making almost impossible for an intruder to identify a) the bits that hide the message and b) the message itself. Even if we assume that an intruder managed to overcome all the aforementioned difficulties, we could further enhance the security by encrypting the multi-modal biometric message.

## 4 Conclusion and Future Work

In this paper a novel verification scheme that cannot be replicated or stolen, was proposed. More specifically, steganography was utilized as a tool to provide unbreakable passwords. A biometric feature of a user i.e., the facial landmarks was acquired and embedded as a hidden message in an image. This image was then utilized as a password. When an attacker or a legit user tries to unlock a device or an application, the same biometric feature is captured and embedded with the same stego algorithm into the image. The hash key of the resulted stego image in both cases is produced and if there is a complete match, the user is considered as authenticated. Moreover, it was examined whether state-of-the-art algorithms like Generative Adversarial Networks could produce exact copies of the stego images utilized as passwords, and the results showed that the proposed verification scheme is extremely secure. In the future, we plan to make experiments with at least two biometric factors as embedded message and to test the security of the proposed authentication scheme against more state-of-the-art deep learning algorithms.

**Acknowledgments.** This research is co-financed by Greece and the European Union (European Social Fund- ESF) through the Operational Programme «Human Resources Development, Education and Lifelong Learning 2014–2020» in the context of the project "Creation of a Multimodal Biometric Password by using Steganography" (MIS 5050338)."

## References

1. Number of connected devices worldwide (2030). https://www.statista.com/statistics/802690/worldwide-connected-devices-by-access-technology/. Accessed 13 June 2021
2. Ramachandra, R., Busch, C.: Presentation attack detection methods for face recognition systems: a comprehensive survey. ACM Comput. Surv. **50**(1), 1–37 (2017). https://doi.org/10.1145/3038924
3. Chen, Y., Ma, B., Ma, H.C.: Biometric Authentication Under Threat: Liveness Detection Hacking. Black hat, Las Vegas (2019)
4. Czajka, A., Bowyer, K.W.: Presentation attack detection for iris recognition: An assessment of the state-of-the-art. ACM Comput. Surv. **51**, 1–35 (2018). https://doi.org/10.1145/3232849
5. Sousedik, C., Busch, C.: Presentation attack detection methods for fingerprint recognition systems: a survey. IET Biometr. **3**, 219–233 (2014). https://doi.org/10.1049/iet-bmt.2013.0020
6. ISO - ISO/IEC 30107–1: Information technology — Biometric presentation attack detection — Part 1: Framework (2016). https://www.iso.org/standard/53227.html. Accessed 13 Jan 2021
7. Koops, B.-J.: Crypto Law Survey. http://www.cryptolaw.org/cls2.htm#busigov. Accessed 7 Dec 2017
8. Lebcir, M., Awang, S.: A review on type of attacks on fingerprint image and watermarking techniques. IOP Conf. Ser. Mater. Sci. Eng. **551**, 012071 (2019). https://doi.org/10.1088/1757-899X/551/1/012071
9. Karampidis, K., Kavallieratou, E., Papadourakis, G.: A review of image steganalysis techniques for digital forensics. J. Inf. Secur. Appl. **40**, 217–235 (2018). https://doi.org/10.1016/J.JISA.2018.04.005
10. O'Neill, M.E.: The genuine sieve of eratosthenes. J. Funct. Program. **19**, 95–106 (2009). https://doi.org/10.1017/S0956796808007004
11. Goodfellow, I.J.,et al.: Generative adversarial nets. In: International Conference on Neural Information Processing Systems (NIPS 2014). pp. 2672–2680, Montreal (2014)
12. Goodfellow, I.: NIPS 2016 Tutorial: Generative Adversarial Networks. (2016)

13. Kodali, N., Abernethy, J., Hays, J., Kira, Z.: On Convergence and Stability of GANs (2017)
14. Wang, Z., She, Q., Ward, T.E.: Generative adversarial networks in computer vision: a survey and taxonomy. ACM Comput. Surv. **54**, 1–38 (2021). https://doi.org/10.1145/3439723
15. Gulrajani, I., Ahmed, F., Arjovsky, M., Dumoulin, V., Courville, A.: Improved training of Wasserstein GANs. Adv. Neural. Inf. Process. Syst. **2017**, 5768–5778 (2017)
16. Lucic, M., Kurach, K., Michalski, M., Gelly, S., Bousquet, O.: Are GANs created equal? A large-scale study. Adv. Neural. Inf. Process. Syst. **2018**, 700–709 (2017)
17. Borji, A.: Pros and cons of GAN evaluation measures. Comput. Vis. Image Underst. **179**, 41–65 (2018)
18. Karras, T., Laine, S., Aila, T.: A style-based generator architecture for generative adversarial networks. In: Proceedings of the IEEE Computer Society Conference on Computer Vision and Pattern Recognition, pp. 4396–4405. IEEE Computer Society (2019). https://doi.org/10.1109/CVPR.2019.00453
19. stegano PyPI. https://pypi.org/project/stegano/. Accessed 13 June 2021

# A Repeated Mistake is a Choice: Considering Security Issues and Risks in Quantum Computing from Scratch

Danel Arias(✉), Borja Sanz, José Gaviria de la Puerta, Iker Pastor, and Pablo García Bringas

University of Deusto, 48007 Bilbao, Spain
danel.arias@opendeusto.es, {borja.sanz,jgaviria,iker.pastor, pablo.garcia.bringas}@deusto.es

**Abstract.** Quantum computing is set to revolutionize the world of software engineering. Currently, quantum paradigms and languages are being developed that are bringing the physical principles of quantum information closer to solving problems in multiple application domains. These paradigms and languages will continue to evolve and bring quantum computing closer to software engineering, and to an increasingly massive use of it. Unfortunately, as quantum software engineering advances in bringing these new capabilities closer to widespread industry exploitation, attackers will turn the spotlight on quantum software vulnerabilities, on what will be the new battlefront in the world of cybersecurity. In this paper, we approach this security from scratch, starting with an introduction to the current state of quantum programming languages. Then, some current hardware limitations are presented and how these affect quantum software engineering. Followed, we make an overview on the analysis of the quantum languages in terms of security together with some of the research made up to date. As a conclusion, we express the necessity to avoid the many times made mistake of not including security considerations from the first steps in the design of computing paradigms. An error that is repeated a second time is not an error; it is, indeed, a choice.

**Keywords:** Quantum computing · Quantum software engineering · Quantum software security · Quantum program security · Quantum code analysis · Quantum programming language security

## 1 Introduction

In recent decades, more and more it is heard about how quantum computing is going to revolutionize the computing paradigm, in a qualitative leap analogous to that produced by the transistor, but there are still many loose ends to solve before we have quantum hardware working in our personal computers. A more than significant advance has been made in hardware, existing quantum computers that we could use (IBMQ, IonQ, D-Wave, Microsoft Quantum, Google Quantum, Honeywell, etc.), and even take advantage of their computational capacity. To achieve the latter, there are quantum algorithms that


J. J. Gude Prego et al. (Eds.): CISIS 2021/ICEUTE 2021, AISC 1400, pp. 156–166, 2022.
https://doi.org/10.1007/978-3-030-87872-6_16

take advantage of the quantum characteristics of these computers. However, we must, still, look further, because there is no use in having hardware in which you cannot program these quantum algorithms. For this, the so-called quantum languages (QPL) are proposed.

In recent years there have been many proposals for quantum languages, software development kits (Forest, Qiskit, QDK, Orquestra, Cirq) and platforms (Quantum Inspire, IQ Experience, Quantum Playground, Forge, LIQ*Ui*|⟩) but quantum computing is still under development.

Thus, seeing the hardware and development environments more and more established, we must begin to consider how programmers will develop this quantum software. In order to deploy any type of product or service regarding software, security is a main issue to consider. The code is executed for a purpose and if the code is well-written, it should reach that purpose. Robin Miller provided the following slogan: a well-typed programs cannot "go wrong" [1]. The appropriate formalization of this slogan depends on the style of formal semantics used for a particular language. In the case of quantum programming languages is not different. That is why is important to narrow down on a type of QPL, that way a deep and specific research can be done.

## 2 Quantum Programming Languages (QPL)

These quantum languages, as the classical ones, can be categorized into 2 groups, imperative and functional languages. Well, the latter is not strictly true, there are language proposals that do not fall within this classification, such as the work of Freedman, Kitaev, and Wang [2] that propose a radically different direction in the semantics of quantum computation exploiting connections between quantum computation and topological quantum field theories (TQFTs).

But, sticking to the 2 main groups, although there has been progress in the development of functional quantum languages (QML [3], LIQ*Ui*|⟩ [4], Quipper [5]) since the first proposal of one (QFC[6]) and the future improvements by the linear lambda calculus for quantum computation described by van Tonder [7, 8], the most advanced and recognized are the imperative languages for their involvement in the hardware model.

### 2.1 Imperative Languages

To begin to speak of imperative languages it is necessary to mention that Deutsch (1985) [9] defined the Quantum Turing Machine (QTM). This was the first hardware model for general quantum computers. From this model several modifications and improvements were developed until Bernstein and Vazirani (1997) [10] give a construction for an efficient universal QTM and discuss the programming primitives. The imperative language development thread was started by Knill (1996) [11] who proposed a set of conventions for quantum pseudo-code that would be implemented in a QRAM (quantum random access machine). Knill's proposal significantly influenced future imperative language developments, although the proposed pseudo-code is not precise enough to be implemented in the form of a language directly.

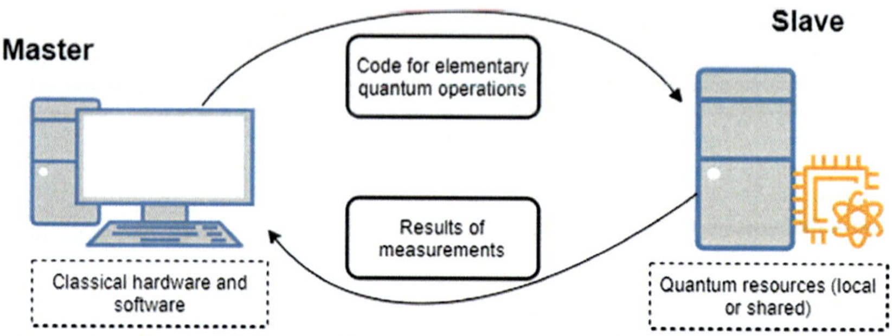


**Fig. 1.** Simplified scheme of a QRAM machine. The classical hardware drives the quantum resources in a master-slave configuration.

**QRAM Model.** Since this model is the basis for later languages, it is necessary to mention how this works (Fig. 1).

The QRAM is made up of a quantum memory and a quantum ALU (qALU) which in turn is made up of quantum logic gates. This model is controlled by classical instructions, which gives rise to a hybrid system in which the classical machine provides the code for quantum operations to the quantum subsystem and this, as a black box, returns the results after the operations, acting with a slave role. The role of the quantum subsystem is to decode the classical instructions and carry out the pertinent operations through quantum gates at the indicated qubits.

The development of quantum languages is not easy because although phenomena such as superposition or entanglement, typical of quantum particles, is what gives quantum computing its power, they present a problem when they want to reach a level of abstraction necessary to a high-level programming language. For this reason, the most developed QPLs present a way of defining quantum circuits for a quantum algorithm to be described by deconstructing it into instructions of the QASM language [12] that then will be sent to the QRAM.

The definition of these circuits is done through a succession of quantum gates. Instruction by instruction is indicated, sequentially, which gate would apply to each qubit or qubits indicated. The measurements are also indicated like the rest of the gates. In the end, the circuit is defined and ready to be executed in the hardware. Due to the need for such low-level programming, the programmer in question must be knowledgeable about the logic gates used and how they work, just as they should understand the basic quantum behaviors that provide that essential functioning for these computers. This knowledge is essential for the programmer since a higher level of abstraction is currently not available.

**QPLs Currently.** During the years following Knill's proposal, Ömer (1998, 2002 and 2003) [13–15] developed what is considered the first quantum programming language, QCL, which has a syntax based on C. Bettelli et al. (2003) [16] proposed a high-level language based on C+ + that was implemented as a library for the latter. Subsequently, Sanders & Zuliani (2000) [17] explored non-deterministic and mixed-state programming with the qGCL language. All these languages were the basis of the languages that have proliferated today and the precursors in their development. Some examples are Q# [18], Q|SI⟩ [19], Qiskit [20] and Scaffold [21].

It is important to mention that although the designers of some of these languages intend to use these languages as descriptions of application-specific quantum circuits to later be used by future software tools to generate quantum circuit designs, the programs of all these "high level" languages maybe compiled into low-level QASM programs to be executed by quantum processors.

Today two of the most advanced and used QPLs based on the QRAM model are Q# developed by Microsoft and Qiskit by IBM. However, these languages are still in an early phase of development, in fact Qiskit is in version 0.27 and Microsoft Quantum Development Kit is in version 0.15. Both languages can be considered extensions of classical languages, since they are not only based on these, but Qiskit, for example, works as a Python library [20]. On the other hand, Q# is based on F#, C# and Python [18]. This base in classical languages makes sense because they will serve to define the behaviour of our quantum circuit, because the quantum behaviour of the hardware is not applied directly to it since the instructions will only describe, sequentially, which operations to carry out to the qubits according to the definition of the quantum circuit.

## 3 Limitations of the Hardware

As mentioned in the introduction, several physical hardware is being developed. However, the current state of the art in terms of quantum hardware still suffers some core shortcomings, which include a small number of qubits, restrictive connectivity, or limited native instructions. Moreover, compared to classical bits, quantum bits (qubits) are much more fragile, unstable, error-prone and deciduous. Quantum circuits are, of course, affected by these errors, which keeps the quality of the generated outputs in low levels, and with high requirements of expert supervision. These devices are termed as noisy intermediate-scale quantum (NISQ) computers [22, 23].

These errors and instabilities are extremely attractive for attackers, which are already thinking in quantum computing as the next battlefield of cybersecurity. Due to this background, it is key to identify the errors and its causes:

- Quantum bits suffer from a short lifetime and unstable gate and measure operations. On the one hand, the relaxation problem implies that a qubit in state $|1>$ spontaneously can lose energy and end with state $|0>$. Besides, the dephasing phenomena can produce a similar effect. On the other hand, measurements and gate dependencies can suffer alterations when read out, even accumulating errors in parallel operations.
- Limited native set of (small) gates, so every quantum program needs to be decomposed in a relatively high number of gates, which increases the complexity of the circuit, the need for a long life of the qubits, and -in general- the probability of error.
- Coupling constraint, and limited connectivity of the qubits, that introduces the need of adding additional connection instructions (swap operations), of course increasing the run-time and the number of gates of the quantum program, and -again- the feasibility of error.

- Quantum computing is far from becoming a personal commodity due to cost and needs of the hardware (ultra-cold temperature [24, 25], shielded environment, and complex wiring for control). Therefore, cloud-based access to quantum computers is the logical path forward where the hardware is hosted in a remote location. In addition to hardware, high-performance quantum simulators also need cloud offerings as personal computers can hit memory limit while simulating a reasonably large circuit. And likewise, cloud-based access is already providing quantum capabilities from vendors to the scientific society that are even giving free access to their quantum processors and simulators. This access through the cloud has become the only option, but it also introduces security problems such as malicious assignment of hardware or reporting of outputs and errors, or the feasibility of suffering man-in-the-middle situations.
- The specific way of the encoding of a quantum circuit can provide enough clues to infer sensitive information of the problem.
- Given the essential need of optimization that quantum circuits have, it is also mandatory to consider third party compilers, able to perform this improvement of the circuits. These compilers should be taken into consideration in a trusted way; deficiencies or shortcomings in the compilers would open the door of the quantum program to tamper and exploit attacks, and also to reverse engineering.
- Application oriented vulnerabilities, in particular for one of the most promising applications of quantum computing, as is Quantum Machine Learning. On the one hand, QML applications can suffer in particular Creep Attacks, subtle deviations driven to guide the ML reasoning model to a subverted output. On the other hand, quantum circuits have already been seen producing vanishing gradients that become an obstacle to several (important) ML optimization methods.

Therefore, noise-resilient methods are of paramount importance. Due to this, besides of engineering better qubits, several other independent approaches like parameter optimization in variational algorithms, noise-aware mapping, and scheduling are pursued to make quantum programs resilient to noise. The theory of Quantum Error Correction (QEC) [26–31] enables fault tolerant computation [27, 32, 33] using redundant qubits, but full fault tolerance is still too much expensive for current noisy devices. However, lot of effort is being put in applying feasible countermeasures (listed in [34]) directly into the hardware or to the compiler that might reduce error occurrence.

Hence, writing quantum programs does not seem as trivial anymore, at least nowadays with the hardware limitations. Even a well-typed program might be affected by these errors making it appear wrong. Some of these problems, at the stage QPLs are today, might be handled by the programmer or at least be considered in an analysis of the program. This enforces the idea of either a prior analysis of the quantum code or, if possible, a dynamic analysis in which, the unintended behaviour of the software because of a noise error can be spotted.

## 4 Analysis for Security in Quantum Languages

Putting the focus on the use of these languages, a code analysis is necessary in all senses to ensure the intended operation of the program. Hence, a security analysis must be done to identify the weak points and prevent them. That way, the security of the program is ensured. This analysis can be categorized as follows:

### 4.1 Analysis of the Base Language

Most quantum languages are built on top of a classical language. Imperative languages as Q# [18] or Qiskit [20] for example, have in this case its base language on Python. In the case of Qiskit, is implemented as a Python library. Another example in functional languages is QML [3] and Quipper [5] which are based on Haskell. This makes the code verification process more challenging as more sections must be considered. But, on the other hand, the current extensive knowledge of classical languages becomes an indispensable tool for that analysis.

### 4.2 Analysis of Errors in the Definition of Circuits

It must be considered that, like all programming languages, QPLs are subject to a syntax and semantics that the programmer must know and that errors arise after incorrect use. Not only that, but several errors can come from ignorance or confusion in the definition of these quantum circuits. As with the Array Index Out of Bounds, the programmer must consider certain characteristics of the circuit he defines to avoid these errors.

Many of these are usual and constantly repeated as an incorrect definition of the number of classical bits in which to map the measurement results of the qubits, the correct deallocation of qubits etc. We can, therefore, separate these into 2 main groups, those that arise from a lack of knowledge of the language itself and those that arise from not considering certain quantum characteristics.

**Errors caused by the language.** These are independent of the QPL you are working on. Each language has certain characteristics and allows certain behaviours that others do not. This can lead to programmer confusion just like in classical languages. Moreover, certain problems may not even be present in certain languages. For example, when defining circuits in Qiskit, it is necessary to indicate how many classical bits are going to be used, while in Q# the same number of classical bits as qubits are always defined. Another section to consider is the use of libraries or external code that usually implies incompatibility with the quantum language, this, for example, in the case of Qiskit, is very notable as it works as a Python library and not as a language in itself.

**Errors caused by quantum characteristics.** The ignorance of the existence of these characteristics or of how they are reflected in the QPL in which it is being programmed can cause a variety of bugs. There are many quantum phenomena which indirectly or directly affect the circuit defined in this program such as superposition of states, entanglement, etc. In addition to certain crucial concepts such as measurement or the need to deallocate the qubits. Each language deals with these concepts differently which can confuse the programmer. What is more, as stated in Sect. 3, there might be some quantum hardware inconveniences that might be handled by the programmer.

In the realm of Qiskit, the work of Zhao et al. (2021) [35] shows an excellent research in this section through identifying bug patterns in the Qiskit case. Although this research is focused on Qiskit, many of these problems can be generalized to other languages based on gates because, although the syntax is different, it is similar in most aspects.

As an early conclusion, most of the problems mentioned in this section can be managed thanks to a static analysis of the code. An analysis of the semantics, syntax and behaviour of the language would provide certain patterns that could be analysed to find a way to avoid them. On the other hand, it is also presented the need to define certain standards or conventions in the syntax of these languages to mainly reduce the errors caused by the disparity between languages.

### 4.3 Dynamic Analysis of the Program

Due to the quantum characteristics of the qubits, measuring their actual values during the execution for debugging following the classic trail, is impossible. The measured qubit will collapse to either one of the 2 binary states losing the information of the superposition it might have had and other quantum consequences like the determination of the state of other entangled qubits. Having into account that both, the superposition, and the entanglement are present in most quantum algorithms, if not all, making measurements would affect to the program's execution.

Even if the dynamic analysis presents this type of difficulty, recently there have been research in this section.

The proposal of Liu et al. (2020) [36] is to implement certain extra qubits that could be measured without interrupting the execution by indirectly verifying the desired condition regarding the qubits under test. Thus, they propose certain quantum circuits that could perform dynamic assertion not only for classical values but also for superposition and entanglement states. Subsequently, after the measurement of what they called ancilla qubits, an algorithm called Quantum Phase Estimation (QPE) [28, 37–39] is carried out. This is proposed in the same paper. This algorithm, based on the inverse Quantum Fourier Transformation (QFT) [40, 41], estimates the value of the qubit.

This approach depends on the fact that these quantum circuits do not suffer from errors since these errors could not only affect the measurement but also spread to the qubits under test. This branch of research is barely developed and although it is currently not totally reliable, it is very promising.

Another approach is the proposal of Huang and Martonosi (2019) [42] that focuses on carrying out a statistical study of the quantum program. First, they define 3 types of assertions: classical state, superposition, and entanglement, with their respective operating hypotheses. These statistical tests use multiple measures to decide whether to reject the hypothesis. With enough measurements, a statistical test can determine if an assertion is not true, indicating an error in the program. The set of quantum states considered in this proposal, classical states, superposition, and entanglement, are a subset of possible quantum states. In the general case, quantum phase estimation, quantum state tomography and quantum process tomography are necessary to be able to examine general quantum states. However, as is mentioned in the paper, these processes are extremely expensive and cannot be used as efficient assertion checks with this method.

As an early conclusion, both proposals present an approach to the dynamic analysis, that seem like a reasonable approach for the future since they are still very recent and have big drawbacks to overcome.

### 4.4 Work of Mention

A proposal to consider is that of Abhari et al. (2014) [43]. They propose "ScaffCC", a compiler for the Scaffold language [21] that would integrate code analysis tools. This compiler receives the code in Scaffold as input and returns the representation in quantum assembly (QASM) [44] as output. In the paper they also expose methods of analysis not only of the quantum program itself but also of the classical-quantum duality to which these languages are subjected, in this case Scaffold is based on the C language.

Besides this, Microsoft has already taken the first step in ensuring the execution of Q# programs. For this, it provides programmers with tools for testing and debugging programs in this quantum language [45]. This approach is focused on the testing and debugging of quantum programs executed in a simulator. This tool allows simulating the behaviour of the program and correcting errors before being executed on the hardware itself. This, while useful, is subject to the simulation's own limitations and how realistic the simulation is.

## 5 Conclusions

Quantum computers are great tools that will help in solving problems thanks to their great computing power and they are no longer as utopian as many believed, but to work with them, a safe way to program these quantum algorithms is necessary. That way takes advantage of the quantum characteristics of these computers. This security is a difficult task to achieve because:

- Firstly, even if the hardware has advanced, current quantum computers suffer from various noises or errors that could be (are going -for sure- to be) used by attackers in order to subvert the correct functioning of quantum programs, and converted into security breaches, once quantum computing became massively used in industry.
- And secondly, there is the barrier on how the languages themselves are defined. This, together with the current impossibility of the generalization of the classical methodology for an analysis of the code, encourages the development of new techniques.

Thus, in light of the background presented, it is reasonable to state a new research hypothesis in the sense that the study and experimentation about the state of the art in stability and security approaches from conventional to quantum can produce a Quantum Software Security Model useful to prevent the malicious impact of malware-related knowledge on Quantum Programs; model that can be substantiated into a Tool to measure the security level of these quantum programs.

In particular, the static analysis of the code would be essential in ensuring security during the development of quantum software, and also as an analytical base for structuring of further dynamic analysis that could eventually be proposed when debugging of quantum program execution became more stable. Therefore, applying current knowledge from the binary world to the development and implementation of a static analysis model for quantum programming would be crucial in order not to repeat three decades of a generalized mistake of considering security in conventional software after the design, not from the scratch, and allow quantum software a full development of its potential.

**Acknowledgements.** The authors would like to thank the Basque Government through its partial funding support through the TRUSTIND ELKARTEK R&D project (ref. KK-2020/00054).

## References

1. Milner, R.: A theory of type polymorphism in programming. J. Comput. Syst. Sci. **17**, 348–375 (1978). https://doi.org/10.1016/0022-0000(78)90014-4
2. Freedman, M.H., Kitaev, A., Wang, Z.: Simulation of topological field theories by quantum computers. Commun. Math. Phys. **227**, 587–603 (2002). https://doi.org/10.1007/s002200200635
3. Altenkirch, T., Grattage, J.: QML: Quantum data and control (2005)
4. LIQ$Ui|\rangle$Github. https://stationq.github.io/Liquid/. Accessed 24 June 2021
5. Quipper Homepage. https://www.mathstat.dal.ca/~selinger/quipper/. Accessed 24 June 2021
6. Selinger, P.: Towards a quantum programming language. Math. Struct. Comput. Sci. **14**, 527–586 (2004). https://doi.org/10.1017/S0960129504004256
7. van Tonder, A.: A lambda calculus for quantum computation. SIAM J. Comput. **33**, 1109–1135 (2004). https://doi.org/10.1137/S0097539703432165
8. van Tonder, A., Dorca, M.: Quantum Computation, Categorical Semantics and Linear Logic. ArXivquant-Ph0312174 (2011)
9. Deutsch, D., Penrose, R.: Quantum theory, the Church-Turing principle and the universal quantum computer. Proc. R. Soc. Lond. Math. Phys. Sci. **400**, 97–117 (1985). https://doi.org/10.1098/rspa.1985.0070
10. Bernstein, E., Vazirani, U.: Quantum complexity theory. SIAM J. Comput. **26**, 1411–1473 (1997). https://doi.org/10.1137/S0097539796300921
11. Knill, E.: Conventions for Quantum Pseudocode. Los Alamos National Lab (LANL), Los Alamos, NM (United States) (1996)
12. Svore, K.M., Aho, A.V., Cross, A.W., Chuang, I., Markov, I.L.: A layered software architecture for quantum computing design tools. Computer **39**, 74–83 (2006). https://doi.org/10.1109/MC.2006.4
13. Ömer, B.: A procedural formalism for quantum computing. (1998)
14. Ömer, B.: Procedural quantum programming. AIP Conf. Proc. **627**, 276–285 (2002). https://doi.org/10.1063/1.1503695
15. Ömer, B.: Structured Quantum Programming (2003)
16. Bettelli, S., Calarco, T., Serafini, L.: Toward an architecture for quantum programming. Eur. Phys. J. D-Atom. Mol. Optical Phys. **25**(2), 181–200 (2003). https://doi.org/10.1140/epjd/e2003-00242-2
17. Sanders, J.W., Zuliani, P.: Quantum programming. In: Backhouse, R., Oliveira, J.N. (eds.) MPC 2000. LNCS, vol. 1837, pp. 80–99. Springer, Heidelberg (2000). https://doi.org/10.1007/10722010_6
18. Azure Quantum Documentation: QDK & Q# API Reference. https://docs.microsoft.com/en-us/azure/quantum/. Accessed 24 June 2021
19. Liu, S., et al.: Q$|SI\rangle$: A Quantum Programming Environment. ArXiv171009500 Quant-Ph (2017)
20. Qiskit documentation. https://qiskit.org/documentation/. Accessed 24 June 2021

21. Abhari, A.J., et al.: Scaffold: Quantum Programming Language. Princeton University NJ, Department of Computer Science, Princeton (2012)
22. Preskill, J.: Quantum computing in the NISQ era and beyond. Quantum **2**, 79 (2018). https://doi.org/10.22331/q-2018-08-06-79
23. Almudever, C.G., et al.: The engineering challenges in quantum computing. In: Design, Automation Test in Europe Conference Exhibition (DATE), 2017, pp. 836–845 (2017). https://doi.org/10.23919/DATE.2017.7927104
24. Krantz, P., Kjaergaard, M., Yan, F., Orlando, T.P., Gustavsson, S., Oliver, W.D.: A quantum engineer's guide to superconducting Qubits. Appl. Phys. Rev. **6**, 021318 (2019). https://doi.org/10.1063/1.5089550
25. Bruzewicz, C.D., Chiaverini, J., McConnell, R., Sage, J.M.: Trapped-ion quantum computing: progress and challenges. Appl. Phys. Rev. **6**, 021314 (2019). https://doi.org/10.1063/1.5088164
26. Devitt, S.J., Munro, W.J., Nemoto, K.: Quantum error correction for beginners. Rep. Prog. Phys. **76**, 076001 (2013). https://doi.org/10.1088/0034-4885/76/7/076001
27. Gottesman, D.: An introduction to quantum error correction and fault-tolerant quantum computation. In: Quantum Information Science and Its Contributions to Mathematics, Proceedings of Symposia in Applied Mathematics (2010)
28. Nielsen, M.A., Chuang, I.L.: Quantum Computation and Quantum Information: 10th, Anniversary Cambridge University Press, USA (2011)
29. Preskill, J.: Lecture notes for physics 229: Quantum information and computation. Calif. Inst. Technol. **16**, 1–8 (1998)
30. Preskill, J.: Reliable quantum computers. Proc. R. Soc. Lond. Ser. Math. Phys. Eng. Sci. **454**, 385–410 (1998). https://doi.org/10.1098/rspa.1998.0167
31. Roffe, J.: Quantum error correction: an introductory guide. Contemp. Phys. **60**, 226–245 (2019). https://doi.org/10.1080/00107514.2019.1667078
32. Campbell, E.T., Terhal, B.M., Vuillot, C.: Roads towards fault-tolerant universal quantum computation. Nature **549**, 172–179 (2017). https://doi.org/10.1038/nature23460
33. Preskill, J.: Fault-tolerant quantum computation. In: Introduction to quantum computation and information. pp. 213–269. World Scientific (1998)
34. Saki, A.A., Alam, M., Phalak, K., Suresh, A., Topaloglu, R.O., Ghosh, S.: A Survey and Tutorial on Security and Resilience of Quantum Computing. ArXiv210606081 Quant-Ph. (2021)
35. Zhao, P., Zhao, J., Ma, L.: Identifying Bug Patterns in Quantum Programs. ArXiv210309069 Quant-Ph (2021)
36. Liu, J., Byrd, G.T., Zhou, H.: Quantum circuits for dynamic runtime assertions in quantum computation. In: Proceedings of the Twenty-Fifth International Conference on Architectural Support for Programming Languages and Operating Systems, pp. 1017–1030. Association for Computing Machinery, New York, NY, USA (2020). https://doi.org/10.1145/3373376.3378488.
37. Kitaev, A.Y.: Quantum measurements and the Abelian Stabilizer Problem. ArXivquant-Ph. 9511026. (1995)
38. IBM Quantum phase estimation. https://quantum-computing.ibm.com/composer/docs/iqx/guide/quantum-phase-estimation. Accessed 08 July 2021
39. Coppersmith, D.: An approximate Fourier transform useful in quantum factoring. ArXivquant-Ph. 0201067 (2002)
40. Qiskit Quantum Fourier Transform. https://community.qiskit.org/textbook/ch-algorithms/quantum-fourier-transform.html. Accessed 08 July 2021

41. Huang, Y., Martonosi, M.: Statistical assertions for validating patterns and finding bugs in quantum programs. In: Proceedings of the 46th International Symposium on Computer Architecture. pp. 541–553. Association for Computing Machinery, New York, NY, USA (2019). https://doi.org/10.1145/3307650.3322213
42. Abhari, A.J., Patil, S., Kudrow, D., Heckey, J., Lvov, A., Chong, F.T., Martonosi, M.: ScaffCC: a framework for compilation and analysis of quantum computing programs. In: Proceedings of the 11th ACM Conference on Computing Frontiers, pp. 1–10. Association for Computing Machinery, New York, NY, USA (2014). https://doi.org/10.1145/2597917.2597939
43. Cross, A.W., Bishop, L.S., Smolin, J.A., Gambetta, J.M.: Open Quantum Assembly Language. ArXiv170703429 Quant-Ph. (2017)
44. Test and debug quantum programs - Azure Quantum, https://docs.microsoft.com/en-us/azure/quantum/user-guide/testing-debugging. Accessed 24 June 2021
45. Selinger, P.: A brief survey of quantum programming languages. In: Kameyama, Y., Stuckey, P.J. (eds.) FLOPS 2004. LNCS, vol. 2998, pp. 1–6. Springer, Heidelberg (2004). https://doi.org/10.1007/978-3-540-24754-8_1
46. Menon, P.S., Ritwik, M.: A comprehensive but not complicated survey on quantum computing. IERI Procedia. **10**, 144–152 (2014). https://doi.org/10.1016/j.ieri.2014.09.069
47. Elhoushi, M., El-Kharashi, M.W., Elrefaei, H.: Modeling a quantum processor using the QRAM model. In: Proceedings of 2011 IEEE Pacific Rim Conference on Communications, Computers and Signal Processing, pp. 409–415 (2011). https://doi.org/10.1109/PACRIM.2011.6032928
48. Rüdiger, R.: Quantum programming languages: an introductory overview. Comput. J. **50**, 134–150 (2007). https://doi.org/10.1093/comjnl/bxl057
49. Gay, S.J.: Quantum programming languages: survey and bibliography. Math. Struct. Comput. Sci. **16**, 581–600 (2006). https://doi.org/10.1017/S0960129506005378
50. Tao, R., Shi, Y., Yao, J., Hui, J., Chong, F.T., Gu, R.: Gleipnir: toward practical error analysis for quantum programs (Extended Version). ArXiv210406349 Quant-Ph. (2021)

# Advanced 3D Visualization of Android Malware Families

Nuño Basurto[1](✉), Héctor Quintián[2], Daniel Urda[1], José Luis Calvo-Rolle[2], Álvaro Herrero[1], and Emilio Corchado[3]

[1] Grupo de Inteligencia Computacional Aplicada (GICAP), Departamento de Ingeniería Informática, Escuela Politécnica Superior, Universidad de Burgos, Av. Cantabria s/n, 09006 Burgos, Spain
{nbasurto,durda,ahcosio}@ubu.es

[2] Department of Industrial Engineering, University of A Coruña, CTC, CITIC, Avda. 19 de febrero s/n, 15405 Ferrol, A Coruña, Spain
{hector.quintian,jlcalvog}@udc.es

[3] Departamento de Informática y Automática, Universidad de Salamanca, Plaza de la Merced, s/n, 37008 Salamanca, Spain
escorchado@usal.es

**Abstract.** The number of attacks aimed at compromising smartphones in general, and Android devices in particular, is acknowledged as one of the main security concerns of these devices. Accordingly, a great effort has been devoted in recent years to deal with such incidents. However, scant attention has been paid to study the application of different visualization techniques for the analysis of malware. To bridge this gap, the present paper proposes the application of a novel technique called Hybrid Unsupervised Exploratory Plots (HUEPs) for the visualization of an Android malware dataset. Thanks to the advanced 3D visualization that is obtained, the proposed solution provides with an overview of the structure of the malware families, supporting the analysis of their internal organization. Experimentation has been carried out with the popular Android Malware Genome (Malgenome) dataset, obtaining promising results.

**Keywords:** Android malware · Malware families · Exploratory projection pursuit · Clustering · 3D visualization

## 1 Introduction

Since the first smartphones came onto the market in the late 90s, sales on that sector have increased constantly until present days. Among all the available operating systems, Google's Android is the most popular mobile platform [1]. The number of sold Android-run units, as well as the number of apps (those available at Google Android's official store) remain at very high rates [2]. Concerning security issues, Android has been the top mobile malware platform as well [3] and it is forecast that the volume of Android malware will continue growing [4]. This operating system is an appealing target for


J. J. Gude Prego et al. (Eds.): CISIS 2021/ICEUTE 2021, AISC 1400, pp. 167–177, 2022.
https://doi.org/10.1007/978-3-030-87872-6_17

bad-intentioned apps, mainly because of its open mentality, in contrast to iOS or some other operating systems.

Smartphone security and privacy are nowadays major concerns. In order to address these issues, it is required to understand the malware and its nature. Otherwise, it will not be possible to practically develop an effective solution [5]. According to this idea of gaining deeper knowledge about malware nature, the present study is focused on the analysis of Android malware families. To do so, Malgenome (a real-life publicly-available) dataset [6] has been analyzed by means of several neural visualization models. From the samples contained in such dataset, several alarming statistics were found [5], that motivate further research on Android malware. That is the case of the 36.7% of the collected samples that leverage root-level exploits to fully compromise the security of the whole system or the fact that more than 90% of the samples turn the compromised phones into a botnet controlled through network or short messages.

To characterize malware families, this study proposes the use of neural networks-based and clustering models which allow visualizing a high-dimensionality dataset and that are further described in Sect. 2. Each individual sample from the dataset (a malware app) encodes the subset of selected features using a binary representation (details on Sect. 3.1). These individuals are grouped by families and then visualized trying to identify patterns that exist across dimensional boundaries in the high dimensional dataset by changing the spatial coordinates of family data. The idea is to obtain an intuitive visualization of the malware families to draw conclusions about the structure of the dataset.

Neural visualization techniques have been previously applied to massive security datasets, such as those generated by network traffic [7, 8], SQL code [9], and honeynets [10], among others. In the present paper, such methods are combined with clustering ones and applied in a novel way to the analysis of malware.

Up to now, a growing effort has been devoted to detect Android malware [11]. Machine learning [12, 13] has been applied to differentiate between legitimate and malicious Android apps, as well as knowledge discovery [14], and weighted similarity matching of logs [15] among others. Although some visualization techniques have been applied to cybersecurity [16, 17] in general, and to the detection of malware in particular [18], few visualization-based proposals for Android malware detection are available at present time. In [19] Pythagoras tree fractal is used to visualize the malware data, being all apps scattered, as leaves in the tree. Authors of [20] proposed graphs for deciding about malware by depicting lists malicious methods, needless permissions and malicious strings. In [21], visualization obtained from biclustering on permission information is described. Behavior-related dendrograms are generated out of malware traces in [22], comprising nodes related to the package name of the application, the Android components that has called the API call and the names of functions and methods invoked by the application. Graphs were proposed in [23] as a way of visualizing and detecting Android malware. More recently, in [24] it is proposed a hybrid neural model combining Convolutional Neural Networks and Long Short Term Memories. Such deep learning model is applied to malware family identification by converting malware binaries into gray scale images. That is, a 2D "footprint" of each app is generated and then, classified in a supervised way.

As an initial proposal, [25] validated the application of Exploratory Projection Pursuit (EPP) methods for the visualization of the Android Malware Genome dataset. Going one step further, the present work combines such methods together with clustering techniques in a new application of the recently proposed Hybrid Unsupervised Exploratory Plots (HUEP) technique [26]. HUEPs have been previously applied to other problems such as enterprise management [26] and predictive maintenance [27], but they are applied for the first time to security concerns.

The rest of this paper is organized as follows: HUEPs are briefly introduced in Sect. 2, the Android Malware Genome dataset and the obtained results are presented in Sect. 3, while the conclusions of this study and the proposals for future work are stated in Sect. 4.

## 2 HUEPs

The complete results of EPP and clustering techniques are combined, together with the glyph metaphor, in a novel and different way, thanks to the Hybrid Unsupervised Exploratory Plots (HUEPs) [26]. HUEPs are a general-purpose visualization tool, in which any EPP and clustering technique could be combined to generate 3D visualizations from high-dimensional data. These representations are informative and intuitive, not only for ML experts, but also for security staff (without requiring any knowledge about the applied techniques).

A HUEP is a scatter plot where each data is considered as a 3D vector. These three-dimensional vectors are obtained from (raw) original data by means of an EPP method and a clustering one, according to what is shown in Fig. 1.

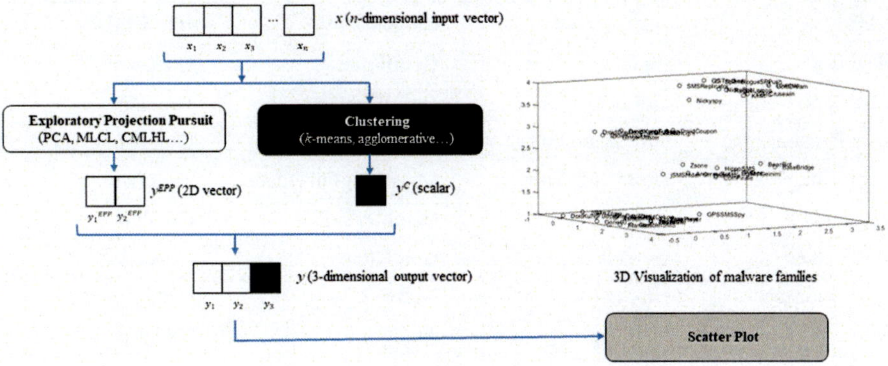


**Fig. 1.** HUEP generation process.

HUEPs are hybrid as they combine both exploratory (dimensionality reduction) techniques with clustering ones. Additionally, all these techniques employ unsupervised learning (no target class or value is provided to be reproduced for new data instances).

In the original formulation of HUEPs [26], PCA [28], MLHL [29] and CMLHL [30] were applied as EPP techniques. Complementarily, k-means [31] and Hierarchical Methods were applied as clustering techniques. In the present paper, the same techniques have been applied and their combinations are combined in many different HUEPs.

## 3 Experimental Study

As previously mentioned, HUEPs (see Sect. 2) have been applied to analyze Android malware. The dataset under study, together with the analysis of the obtained results are presented in this section.

### 3.1 Android Malware Genome Dataset

The Malgenome dataset [5], generated in the Android Malware Genome Project [6], is studied in the present study. It is the first large collection of Android malware (1,260 samples) that was split in malware families (49 different ones). It covered the majority of existing Android malware, collected from the beginning of the above-mentioned project.

Data related to many different apps from a variety of Android app repositories were accumulated over more than one year. Additionally, malware apps were thoroughly characterized based on their detailed behavior breakdown, including the installation, activation, and payloads.

Collected malware was split into families, that were obtained by "carefully examining the related security announcements, threat reports, and blog contents from existing mobile antivirus companies and active researchers as exhaustively as possible and diligently requesting malware samples from them or actively crawling from existing official and alternative Android Markets" [5]. The defined families are in Table 1.

**Table 1.** Malware famlies

| ADRD | AnserverBot | Asroot | BaseBridge |
|---|---|---|---|
| BeanBot | BgServ | CoinPirate | Crusewin |
| DogWars | DroidCoupon | DroidDeluxe | DroidDream |
| DroidDreamLight | DroidKungFu1 | DroidKungFu2 | DroidKungFu3 |
| DroidKungFu4 | DroidKungFuSapp | DoidKungFuUpdate | Endofday |
| FakeNetflix | FakePlayer | GamblerSMS | Geinimi |
| GGTracker | Gin-gerMaster | GoldDream | Gone60 |
| GPSSMSSpy | HippoSMS | Jifake | jSMSHider |
| Kmin | Lovetrap | NickyBot | Nickyspy |
| Pjapps | Plank-ton | RogueLemon | RogueSPPush |
| SMSRep-licator | SndApps | Spitmo | TapSnake |
| Walkinwat | YZHC | zHash | Zitmo |
| Zsone | | | |

Samples of 14 of the malware families were obtained from the official Android market, while samples of 44 of the families came from unofficial markets.

The dataset to be analyzed consists of 49 samples (one for each family) and each sample is described by 26 different features derived from a study of each one of the apps. The features are divided into the following six categories:

- Category #1: Installation. (1) Repackaging, (2) Update, (3) Drive-by download, and (4) Standalone
- Category #2: Activation. (5) BOOT, (6) SMS, (7) NET, (8) CALL, (9) USB, (10) PKG, (11) BATT, (12) SYS, and (13) MAIN.
- Category #3: Privilege escalation. (14) exploid, (15) RATC/zimperlich, (16) ginger break, (17) asroot, and (18) encrypted.
- Category #4: remote control. (19) NET, and (20) SMS.
- Category #5: financial charges. (21) phone call, (22) SMS, and (23) block SMS.
- Category #6: personal information stealing. (24) SMS, (25) phone number, and (26) user account.

The features describing each family take the values of 0 (if that feature is not present in that family) or 1 (if the feature is present).

## 3.2 Obtained HUEP Visualizations

As stated in Sect. 3, several EPP and clustering methods are combined to obtain different visualizations under the frame of HUEPs. Furthermore, clustering methods have been applied with varying values for the *k* parameter (number of clusters): 2, 3, and 4. Due to length limitations, only the most interesting depictions of the dataset are presented in this section. For comparison purposes HUEPs obtained by varying some of its features are also shown.

**HUEPs Incorporating the PCA Projection**

The best HUEP visualization obtained from the PCA projection is shown in Fig. 2. To generate it, the two principal components have been combined with the output of the hierarchical clustering technique. As the value for the *k* parameter has been set up to 4, it results in a sparse representation of the android families. Obviously, it is sparser than the visualization obtained with a *k* value of 2.

Differentiating from what was obtained in the 2D PCA projection [25], most families are not gathered in one main group but split into different ones. It is worth mentioning the case of cluster number 2 (according to the Z axis). It contains many families; while many of them are grouped together, two of them (BaseBridge and AnserverBot) are visualized far away from the other ones. These are the only two families in the dataset that combine repackaging and update installation, clearly differentiating from the other ones. In a similar way, cluster 1 includes four families (DroidKungFu1, DroidKungFu2, DroidKungFu3 and DroidKungFuSapp) that are separated from the other families in the cluster as these are the only ones in the dataset presenting the encrypted privilege escalation.

As it happened in the 2D PCA projection, some families are projected at the very same place as they take the very same values for all the features. This is the case of DroidKungFu1, DroidKungFu2, DroidKungFu3 and DroidKungFuSapp on the one hand

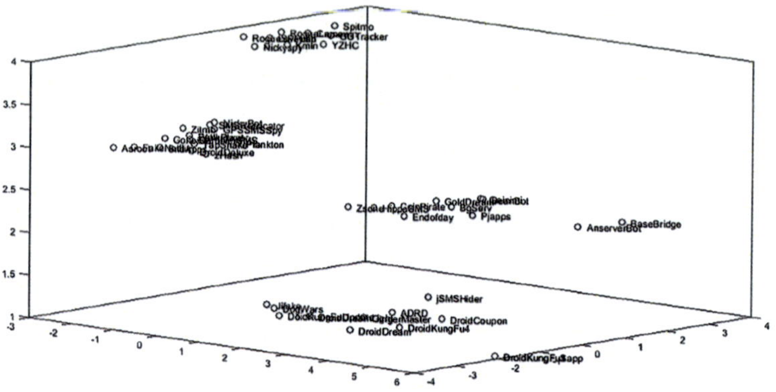


**Fig. 2.** HUEP visualization of Malgenome families: PCA + hierarchical clustering ($k = 4$).

and Walkinwat and FakePlayer on the other hand. It means that, by taking into account the features in the analysed dataset, it will not be possible to distinguish Walkinwat from FakePlayer malware or any of the 4 mentioned variants of DroidKungFu malware.

**HUEPs Incorporating the MLHL Projection**

The best HUEP visualization obtained from the MLHL projection is shown in Fig. 3.

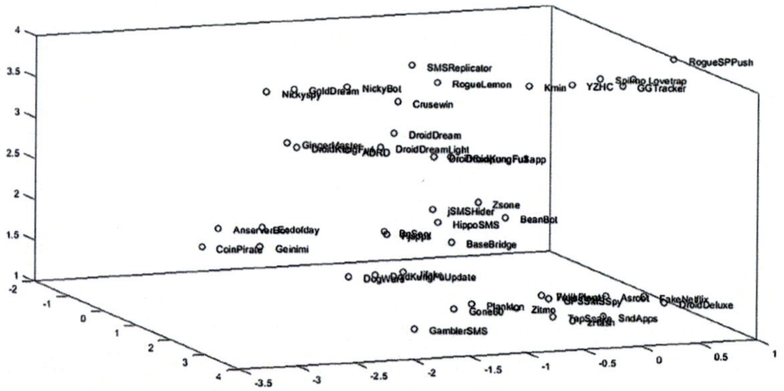


**Fig. 3.** HUEP visualization of Malgenome families: MLHL + $k$-means clustering ($k = 4$).

In a similar way to what was revealed by the 2D MLHL projection [25], MLHL-based HUEP shows the structure of the malware dataset in a way that a kind of ordering can be seen identified. However, as it is more clearly shown in the CMLHL-based HUEP (Fig. 4), the MLHL-based one is no further described and shown only for comparison purposes.

**HUEPs Incorporating the CMLHL Projection**

Similar visualizations from the CMLHL projection are obtained by combining it with the two clustering methods, as can be seen in the following figures. The one incorporating the output of hierarchical clustering (*k* equal to 4) is shown in Fig. 4.

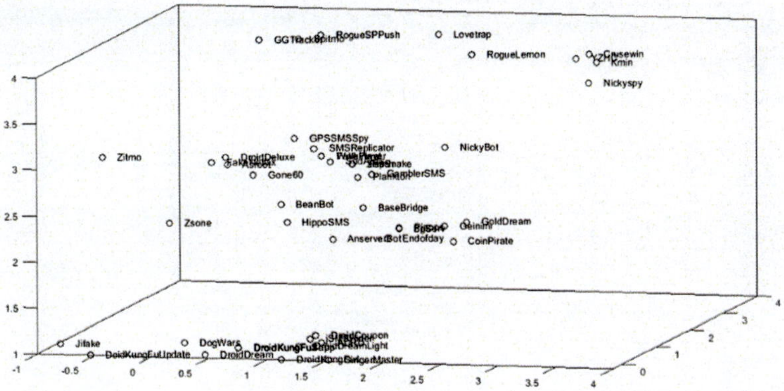


**Fig. 4.** HUEP visualization of Malgenome families: CMLHL + hierarchical clustering ($k = 4$).

It can be said about this figure that this HUEP depicts the data with a certain ordering. Due to the chosen perspective, clusters 2 and 3 seem to overlap, although they are clearly separated.

For comparison purposes, some alternative HUEPs incorporating the CMLH projection are shown in Fig. 5. All of them have been generated with the output of the *k*-means method and varying values for the *k* parameter.

As it can be seen in Fig. 5 and has been previously mentioned, the higher the number of clusters, the more sparse visualization of the data. Regarding this issue, it is worth highlighting that the analyzed dataset contains a reduced number of instances (49). Further research is required to know whether this trend can be also observed for larger malware datasets. Additionally, it must be mentioned that if the number of clusters is too high, it might result in a confusing visualization due to the slight margin between clusters in the visualization itself and the reduced number of instances within each cluster.

For the thorough analysis of the results, the distribution of data in the different clusters (generated by *k*-means) has been compared with their distribution in the subgroups identified in the 2D CMLHL projection [25]. This comparison is shown in Table 2.

According to the data in Table 2, it can be seen that not all the CMLHL subgroups are located in the same cluster, for any of the *k* values. The most relevant case is the CMLHL subgroup 2.4 that is split in the 4 different clusters when the *k* parameter takes the value of 4.

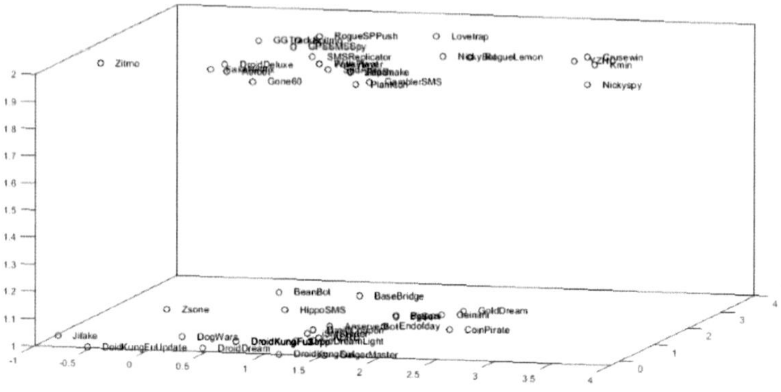


a) $k = 2$

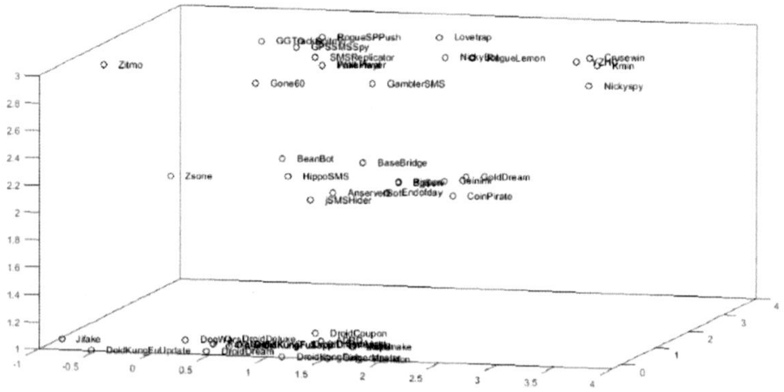


b) $k = 3$

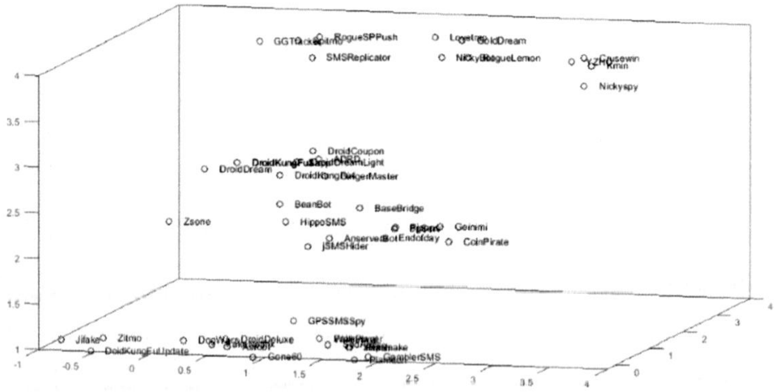


c) $k = 4$

**Fig. 5.** HUEP visualization of Malgenome families: CMLHL + $k$-means clustering.

**Table 2.** Families distribution in CMLHL subgroups and clusters.

| CMLHL subgroup | Families | $k = 2$ | $k = 3$ | $k = 4$ |
|---|---|---|---|---|
| 1.1 | BaseBridge, BeanBot | 1 | 2 | 2 |
| 1.2 | Zsone | 1 | 2 | 2 |
| 1.3 | GGTracker, GPSSMSSpy, HippoSMS, RogueSPPush, Spitmo | 1, 2 | 3 | 1, 2, 4 |
| 1.4 | BgServ, Geinimi, GoldDream, Lovetrap, Pjapps | 1, 2 | 2, 3 | 2, 4 |
| 2.1 | Jifake, Zitmo | 1, 2 | 1, 3 | 1 |
| 2.2 | DroidKungFuUpdate | 1 | 1 | 1 |
| 2.3 | Asroot, DogWars, DroidDeluxe, DroidDream, DroidKungFu1, DroidKungFu2, DroidKungFu3, DroidKungFuSapp, FakeNetflix | 1, 2 | 1 | 1, 3 |
| 2.4 | ADRD, AnserverBot, DroidCoupon, DroidDreamLight, Endofday, FakePlayer, jSMSHider, SMSReplicator, SndApps, TapSnake, Walkinwat, zHash | 1, 2 | 1, 2, 3 | 1, 2, 3, 4 |
| 2.5 | DroidKungFu4, GamblerSMS, GingerMaster, Gone60, Plankton | 1, 2 | 1.2, 3 | 1, 3 |
| 2.6 | CoinPirate, NickyBot, RogueLemon | 1, 2 | 2, 3 | 2, 4 |
| 2.7 | Crusewin, Kmin, YZHC | 2 | 3 | 4 |
| 2.8 | Nickyspy | 2 | 3 | 4 |

This difference between the outputs of the CMLHL and *k*-means methods is understood as a positive issue. It is caused by the different characteristics of the data that each method considers. Thanks to the combination of these two outputs in the same visualization, it is more informative than each one of the outputs on its own, as it is generated by considering the different characteristics.

## 4 Conclusions and Future Work

Based on the obtained results, it can be concluded that the EPP and clustering techniques, combined under the frame of HUEPs, are a valuable technique for visualizing Android malware families. More specifically, when studying such malware items, HUEP-generated visualizations let security staff gain deep knowledge about the nature of the malware and its structure. Similarities and differences of the studied families are identified thanks to the obtained visualizations in an intuitive way that does not require any knowledge about the applied ML techniques.

After the comparison of the obtained HUEPs, it is worth mentioning that the HUEP incorporating the CMLHL projection is the most informative one, when combined with any of the clustering techniques and a high number of clusters (4). The revealed structure of the dataset is consistent with the seminal characterization of the Malgenome dataset [6]. Finally, it can be said that thanks to the combination of the clustering output together with the projection, the structure of the dataset is revealed more clearly than using one of these techniques in isolation.

As further steps, it will be interesting to combine new clustering and projection methods, under the frame of HUEPs, to the same dataset to obtain even better visualizations. Thanks to them, more advanced knowledge about the nature of Android malware could be generated.

## References


1. Statista - The Statistics Portal. http://www.statista.com/statistics/266219/global-smartphone-sales-since-1st-quarter-2009-by-operating-system/. Accessed 21 May 2021
2. AppBrain Stats. http://www.appbrain.com/stats/stats-index. Accessed 20 May 2021
3. Micro, T.: The Fine Line: 2016 Trend Micro Security Predictions (2015)
4. Android Security: Click Fraud Apps Drove 100% Malware Increase in Google Play for 2018. https://www.trendmicro.com/vinfo/us/security/news/mobile-safety/android-security-click-fraud-apps-drove-100-malware-increase-in-google-play-for-2018. Accessed 20 May 2021
5. Yajin, Z., Xuxian, J.: Dissecting android malware: characterization and evolution. In: 2012 IEEE Symposium on Security and Privacy, pp. 95–109 (2012)
6. Malgenome Project. http://www.malgenomeproject.org/. Accessed 20 May 2021
7. Corchado, E., Herrero, Á.: Neural visualization of network traffic data for intrusion detection. Appl. Soft Comput. **11**, 2042–2056 (2011)
8. Sánchez, R., Herrero, Á., Corchado, E.: Visualization and clustering for SNMP intrusion detection. Cybern. Syst. Int. J. **44**, 505–532 (2013)
9. Pinzón, C.I., De Paz, J.F., Herrero, Á., Corchado, E., Bajo, J., Corchado, J.M.: idMAS-SQL: intrusion detection based on MAS to detect and block SQL injection through data mining. Inf. Sci. **231**, 15–31 (2013)
10. Zurutuza, U., Ezpeleta, E., Herrero, Á., Corchado, E.: Visualization of misuse-based intrusion detection: application to honeynet data. In: Corchado, E., Snášel, V., Sedano, J., Hassanien, A.E., Calvo, J.L., Ślęzak, D. (eds.) Soft Computing Models in Industrial and Environmental Applications, 6th International Conference SOCO 2011, pp. 561–570. Springer, Berlin, Heidelberg (2011). https://doi.org/10.1007/978-3-642-19644-7_59
11. Razgallah, A., Khoury, R., Hallé, S., Khanmohammadi, K.: A survey of malware detection in android apps: recommendations and perspectives for future research. Comput. Sci. Rev. **39**, 100358 (2021)
12. Cen, L., Gates, C.S., Si, L., Li, N.: A probabilistic discriminative model for android malware detection with decompiled source code. IEEE Trans. Depend. Secur. Comput. **12**, 400–412 (2015)
13. Sanz, B., et al.: MAMA: manifest analysis for malware detection in android. Cybern. Syst. **44**, 469–488 (2013)
14. Teufl, P., Ferk, M., Fitzek, A., Hein, D., Kraxberger, S., Orthacker, C.: Malware detection by applying knowledge discovery processes to application metadata on the android market (Google Play). Secur. Commun. Netw. **9**, 389–419 (2016)
15. Jang, J.-W., Yun, J., Mohaisen, A., Woo, J., Kim, H.K.: Detecting and classifying method based on similarity matching of android malware behavior with profile. Springerplus **5**, 1–23 (2016)

16. Zhao, J., Masood, R., Seneviratne, S.: A Review of Computer Vision Methods in Network Security. IEEE Communications Surveys & Tutorials, pp. 1–1 (2021)
17. Herrero, Á., Corchado, E., Sáiz, J.M.: MOVICAB-IDS: visual analysis of network traffic data streams for intrusion detection. In: Corchado, E., Yin, H., Botti, V., Fyfe, C. (eds.) IDEAL 2006. LNCS, vol. 4224, pp. 1424–1433. Springer, Heidelberg (2006). https://doi.org/10.1007/11875581_169
18. Wagner, M., et al.: A survey of visualization systems for malware analysis. In: EG Conference on Visualization (EuroVis)-STARs, pp. 105–125 (2015)
19. Paturi, A., Cherukuri, M., Donahue, J., Mukkamala, S.: Mobile malware visual analytics and similarities of attack toolkits (malware gene analysis). In: Collaboration Technologies and Systems (CTS), 2013 International Conference on, pp. 149–154 (2013)
20. Park, W., Lee, K.H., Cho, K.S., Ryu, W.: Analyzing and detecting method of android malware via disassembling and visualization. In: 2014 International Conference on Information and Communication Technology Convergence (ICTC), pp. 817–818 (2014)
21. Moonsamy, V., Rong, J., Liu, S.: Mining permission patterns for contrasting clean and malicious android applications. Futur. Gener. Comput. Syst. **36**, 122–132 (2014)
22. Somarriba, O., Zurutuza, U., Uribeetxeberria, R., Delosières, L., Nadjm-Tehrani, S.: Detection and visualization of android malware behavior. J. Electr. Comput. Eng. **2016**, 1–17 (2016)
23. Zhang, Y., et al.: Visual analysis of android malware behavior profile based on $$PMCG_{droid}$$ : a pruned lightweight APP call graph. In: Lin, X., Ghorbani, A., Ren, K., Zhu, S. (eds.) Security and Privacy in Communication Networks: 13th International Conference, SecureComm 2017, Niagara Falls, ON, Canada, October 22–25, 2017, Proceedings, pp. 449–468. Springer International Publishing, Cham (2018). https://doi.org/10.1007/978-3-319-78813-5_23
24. Akarsh, S., Poornachandran, P., Menon, V.K., Soman, K.P.: A detailed investigation and analysis of deep learning architectures and visualization techniques for malware family identification. In: Hassanien, A.E., Elhoseny, M. (eds.) Cybersecurity and Secure Information Systems: Challenges and Solutions in Smart Environments, pp. 241–286. Springer International Publishing, Cham (2019)
25. González, A., Herrero, Á., Corchado, E.: Neural visualization of android malware families. In: Graña, M., López-Guede, J.M., Etxaniz, O., Herrero, Á., Quintián, H., Corchado, E. (eds.) SOCO/CISIS/ICEUTE -2016. AISC, vol. 527, pp. 574–583. Springer, Cham (2017). https://doi.org/10.1007/978-3-319-47364-2_56
26. Herrero, Á., Jiménez, A., Bayraktar, S.: Hybrid unsupervised exploratory plots: a case study of analysing foreign direct investment. Complexity **2019**, 6271017 (2019)
27. Redondo, R., Herrero, Á., Corchado, E., Sedano, J.: A decision-making tool based on exploratory visualization for the automotive industry. Appl. Sci. **10**, 4355 (2020)
28. Hotelling, H.: Analysis of a complex of statistical variables into principal components. J. Educ. Psychol. **24**, 417–444 (1933)
29. Corchado, E., MacDonald, D., Fyfe, C.: Maximum and minimum likelihood Hebbian learning for exploratory projection pursuit. Data Min. Knowl. Disc. **8**, 203–225 (2004)
30. Corchado, E., Fyfe, C.: Connectionist techniques for the identification and suppression of interfering underlying factors. Int. J. Pattern Recognit. Artif. Intell. **17**, 1447–1466 (2003)
31. Macqueen, J.: Some methods for classification and analysis of multivariate observations. In: Fifth Berkeley Symposium on Mathematical Statistics and Probability, pp. 281–297 (1967)

# SPECIAL SESSION: Building Trust in Ecosystems and Ecosystem Components

# BIECO Runtime Auditing Framework

Antonello Calabrò[1], Emilia Cioroaica[2], Said Daoudagh[1(✉)], and Eda Marchetti[1]

[1] ISTI-CNR, Pisa, Italy
{antonello.calabro,said.daoudagh,eda.marchetti}@isti.cnr.it
[2] Fraunhofer IESE, Fraunhofer-Platz 1, Kaiserslautern, Germany
emilia.cioroaica@iese.fraunhofer.de

**Abstract.** *Context:* Within digital ecosystems avoiding the propagation of security and trust violations among interconnected parties is a mandatory requirement, especially when a new device, a software component, or a system component is integrated within the ecosystem. *Objective:* The aim is to define an auditing framework able to assess and evaluate the specific functional and non-functional properties of the ecosystems and their components. *Method:* In this paper, we present the concept of predictive simulation and runtime monitoring for detecting malicious behavior of ecosystem components. *Results and Conclusion:* We defined a reference architecture allowing the automation of the auditing process for the runtime behavior verification of ecosystems and their components. Validation of the proposal with real use-cases is part of the future BIECO's activities.

**Keywords:** Auditing · Predictive simulation · Runtime monitoring · BIECO

## 1 Introduction

Digital Ecosystems, which are the emerging extension of Systems of Systems (SoSs), are increasingly influencing our societies on multiple levels. Around the typical interconnected and collaborating system, ecosystems contain various actors (such as organizations, developers or, users) that may have different collaborative and competitive goals, which significantly influences the dynamics within these ecosystems. The emerging dynamism requires more attention in the integration and collaboration of the Information and Communication Technology (ICT) components and devices which exercise various functionalities at the operational level to fulfill tactical decisions and satisfy higher-level strategic goals of a business within an ecosystem. For example, in a vehicles platoon scenario, multiple software smart agents executed on a vehicle platform implement diverse *operational goals* (OG) such as: sharing of context information or


J. J. Gude Prego et al. (Eds.): CISIS 2021/ICEUTE 2021, AISC 1400, pp. 181–191, 2022.
https://doi.org/10.1007/978-3-030-87872-6_18

activating speed limits. The correct implementation of OG further up satisfy the fulfilment of *tactical goals* (TG) such as: creation of vehicles platoon that consist of vehicles driving together in the same direction. By satisfying the TG of driving together in close proximity, due to reduced air friction, fuel consumption is reduced as well, and this contributed to the fulfilment of *strategic business goals* within automotive digital ecosystems.

When a dynamic hierarchical analysis of goals within an ecosystem is performed, the end operational goals represent the core asset for virtual technical evaluation and testing directed towards prompt discovering of security and safety issues. A general unpleasant situation is that systems within an ecosystem could be affected by intended hidden faults inserted into software components, leading to the expression of malicious behavior that propagates to other interconnected parties within an ecosystem. Particularly, when a new device, a software component, or system component is integrated into an ecosystem, techniques for efficiently and effectively assess and prevent anomalies and dangerous situations are requires. This is especially important in context of safety critical digital ecosystems, formed around safety critical system, as in the above example from the autonomous domain. Among these techniques, a commonly adopted one is the audit of components through a monitoring system [12]. This technique provides a dynamic mechanism for the online analysis of functional and non-functional properties of an entity against well-stated conditions, such as contractual conditions for trust. In this context, an auditing framework can collect events at different levels of dynamic goal evaluation (strategic-tactical-operational) and from various systems and system components (including sensors). It uses the collected data to infer complex patterns that indicate specific functional and non-functional properties. The derived complex patterns represent the observed normal or abnormal behavior of the monitored system and its components. For this, the auditing framework should be able to: i) collect and analyze data coming from the different SoS sources (e.g., sensors, software and hardware components or devices); ii) assess the run time behavior of these SoS (components or devices); iii) promptly rise up alarms in case of violations; and moreover, an intelligent, complex monitoring can iv) put in place countermeasures if necessary.

Within the BIECO (Building Trust in Ecosystems and Ecosystem Component) project, we are working for advancing the state of the art on runtime monitoring by addressing the emerging needs for auditing techniques specific for interconnected ICT systems (including Cyber-Physical Systems) within digital ecosystems. In what follows, Sect. 2 presents related work. Section 3 presents the conceptual idea that provides the description of the behavioral process of the auditing framework. Section 4 presents in details the auditing process with its main components: predictive simulation and monitoring, and Sect. 6 concludes the paper and reports future work.

## 2 Related Work

Various techniques for qualitative and quantitative monitoring of CPS exist in the literature, as it is reported in [2] which presents a review of techniques and tools. In particular, the technique reported in of [3] describes offline monitors for identifying violations against policies. Another approach is the monitoring for validity presented in [4]. Similar to our approach, it utilizes internal simulation models of the robot, interacting actors and the environment. Our approach, different than other approaches, does not rely on selection of a set of predefined actions, but dynamically generates the actions exploiting the results of decision events. In particular, our validity monitoring activities involve selection of most probable environmental changes and iterations through multiple internal simulation models of the environment.

## 3 Conceptual Idea

In this section, we describe the concept of auditing for malicious detection, considering the situation in which a new component or device needs to be included into an ecosystems as an integral part of a SoS. In Fig. 1, we depict the basic idea of the proposed auditing framework. As in the figure, for assessing the behavior of the new device during runtime, we consider two different parallel executions: On the right side, the execution of the device or component in a Controlled Environment (CE), i.e., in a real or realistic environment that can be exploited for validation and testing; on the left side: the execution of abstractions of the software components within a Simulation Environment (SE) fed with real-time data. For this, we rely on the Digital Twin (DT) representation of the component as presented in [7] independently derived from the component specification. DTs are abstract, trusted representations of components that can be executed in a simulation environment.

Following the prediction phase, the runtime monitor will collect both the predicted events computed by the simulated environment as well as real events coming from the CE and will compare them according to predefined rules with the scope of assessing the correctness of the behavior of the component itself against the behavior of the predictive simulation. If the behavior of the component executed in the real world is considered not trustworthy, then monitor can immediately rise up alarms so in order to trigger an operational fail-over behavior. For this, the behavior of the DT becomes the specification against which the behavior of the real world system is validated and therefore our solution represents a first attempt of using monitoring data coming from combined simulated and CE/real world software executions to promptly detect malicious behaviors and provide a timely reaction. Because the trustworthiness of the component is judged by executing the behavior of its DT in a simulated environment, the conformity between the DT's behavior and the specification needs to be assured pre-deployment. It is out of the scope of this paper going into detail of the conformity checking, and we refer to [5] for more information.

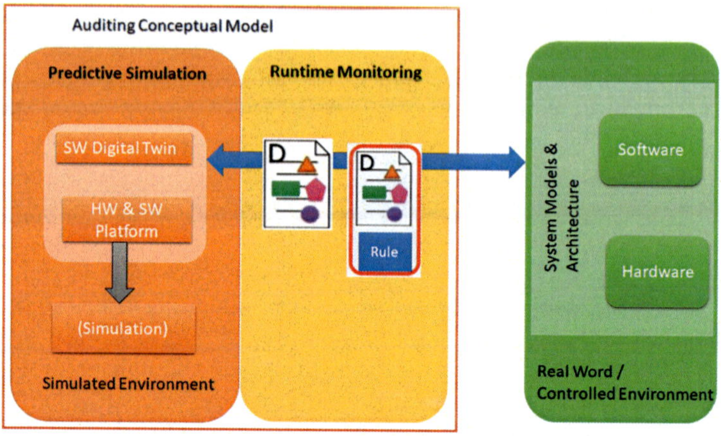


**Fig. 1.** Runtime auditing conceptual model.

## 4 BIECO Auditing Process

In this section, we illustrate in more details the idea presented in the previous section, by explicitly referring to the auditing process implemented within the BIECO project.

As reported in Fig. 2, two main participants are considered: the BIECO Framework and the Auditing component. In this section, however, we focus only on the description of the latter, which is in turn composed of two different roles: Predictive Simulation and Runtime Monitoring. Looking at the diagram, the user of Bieco framework can set up both the Predictive simulation and the runtime Monitoring. It could also set up the Controlled Environment, however because it is part of the BIECO Framework currently under development, we do not include it in this figure. Considering the Runtime monitoring, the procedural activities are: (1) `Monitor Set up`: during this phase the monitoring engine will establish all the communication needed from and to other components of the BIECO framework to which the information encapsulated in events will flow. In case of detected deviation, these events can be alarms or warnings that trigger a fail-over behavior. An event-based communication pattern guarantees a loosely coupled architecture with possibility of exchanging different components. (2) `Get Blueprints`: the information required for the monitoring activity are retrieved from the `Data Store`. The information includes the expected boundaries, i.e. admissible values and constraints for each event to be observed together with previously evaluated and trusted behavior of different ecosystem components. This information is used in deriving the initial rules to be monitored. (3) `Generate Monitoring Rules`: The blueprints are used for deriving a set of initial monitoring rules. In this activity the *Collection Data Object* generated from the task `Generate and Instrument Simulated environment and Digital Twin` contains possible names of the events that will be fired by the DT. This data

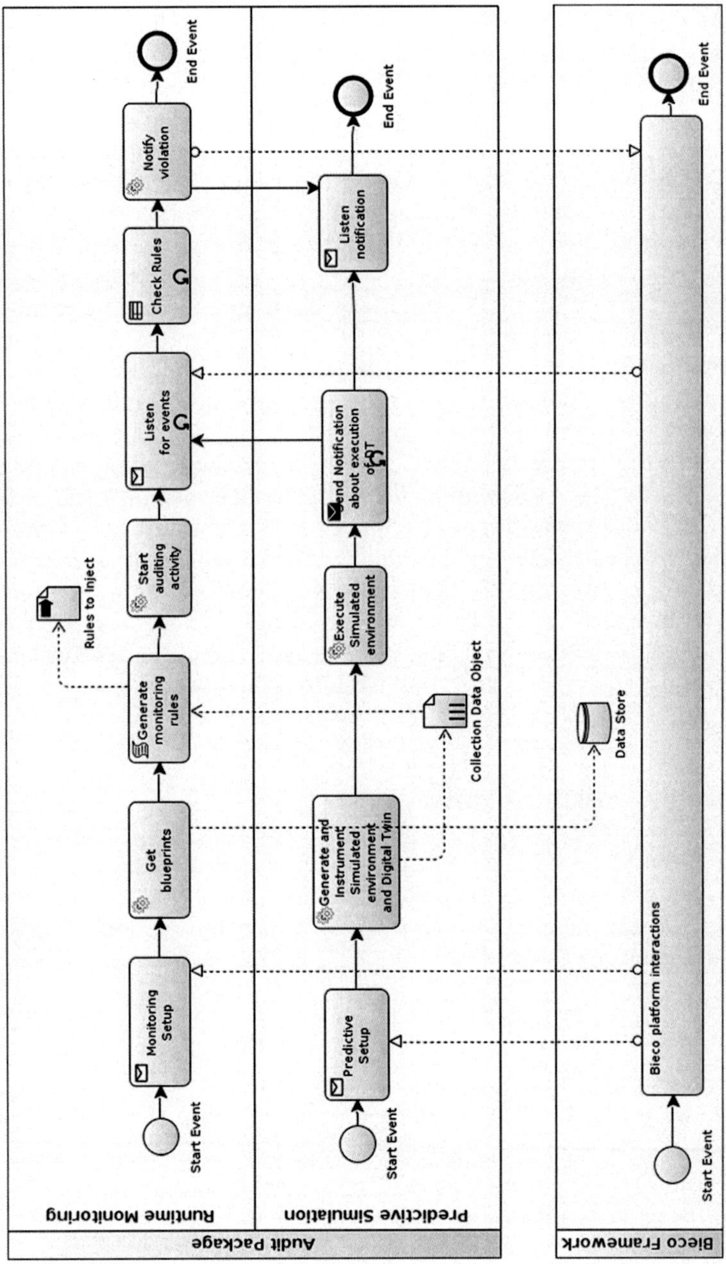


Fig. 2. Auditing process.

together with blueprints are input for the `Generate monitoring rules` which generate the final rules needed for monitoring the runtime execution. The output of this phase is a set of rules against which the predicted behaviour of the DT will be evaluated together with the parameters provided within the blueprints. The blueprints contain design time evidence of trust that speeds up the runtime rule derivation process. (4) `Start Auditing Activity`: When the CE and the SE have been set up, the auditing activity can start. (5) `Listen for Events`: this activity collects and synchronizes the event coming from the CE and the SE (6) `Check Rules`: the event of the CE and SE are compared and the assessment of the Component performed (7) `Notify violation`: in case a rule violation an alarm message is generated and sent to both the Controlled Environment (i.e.,) and the Predictive Simulation.

The conceived activities for the predictive simulation are: (1) `Set up the simulation scenario`: During this phase a virtual representation of a concrete technical situations in which behavior needs evaluation is created. The concrete technical situation is a composition of external environmental information and internal system configuration. During this phase, DTs are executed in collaboration with stubs that represent abstractions of the hardware components and/or platform with which the software under evaluation interacts. (2) `Instrument the execution of Digital Twins (DT)`: DTs, which are abstract models fed with real time data are executed by a simulation environment at a faster speed than the wall clock. As a results, the predictive simulation provides an event signature that consists of the type, number and order of events who have been virtually validated. (3) `Send Notification about execution of DT`: during this phase an event signature is encapsulated into probes that the runtime monitoring can read.

## 5 Runtime Auditing Framework

In this section, we describe the auditing framework used within the BIECO project. This framework allows supporting the execution of the auditing process described in Sect. 4, and it is composed of two main modules: Predictive Simulation and Runtime Monitoring.

### 5.1 Predictive Simulation

The predictive simulation component is in charge of running a virtual evaluation environment through execution of Digital Twins (DTs). DTs are abstract models representing expressed in executable abstractions of ecosystem components under evaluation (ICT systems, ICT system components such as software components). Because a software smart expressed can hide a malicious behavior, it is necessary to execute abstractions of these components instead. These abstractions are directed towards the scope of the evaluation, and for example: if functional interaction between components needs to be evaluated, DTs that represent the functional behavior of the components are executed. If on the other hand, scheduling behavior needs runtime evaluation, DTs representing time behavior of interacting components are evaluated instead. By restricting the logic of expressing behavior while keeping the overall actions and event decisions that cross the architectural boundaries, the components under virtual evaluation cannot detect whether they interact with real world or virtual world entities. Consequently, malicious behavior will be visible in the virtual world without affecting the real world.

In the linked predictive simulation in particular, the current state of the system is used to predict behavior in the future. For enabling detection of malicious behavior hidden within software components, a Domain Specific Language (DSL) that enables definition of control functions is currently under development. The predictive simulation works in collaboration with the Runtime Monitoring component.

The predictive simulation module is based on FERAL simulator [10] and comes as an extension of the platform presented in [6], which enables testing of automotive smart ecosystems in scenarios composed by virtual and real-world entities collaborating with each other. Typically used for the rapid development of architecture prototypes through coupling of simulators, simulation models, and high-level design models, FERAL enables the coupling of abstract simulation models with very detailed ones. In this way, the predictive simulation framework can integrate multiple abstraction versions of virtual agents and execute their behavior at different frequencies. Information from the predictive simulation is passed to the monitoring engine in the form of ActiveMQ messages.

Figure 3 depicts the main components of the predictive simulation environment. The instantiation of a technical setting for evaluation is performed through a configuration engine which sets scene models. The scene models contain simulation models that are executed by the FERAL simulator in a predictive manner. The results of the prediction are events that are monitored and validated w.r.t. goals. Internally, the prediction performs a validity monitoring of the raised events, and only based on evidence of trustworthy behavior, the signature events are sent to the monitoring engine described in the following section.

### 5.2 Runtime Monitoring

The component is in charge of setting up and managing monitoring component. The Runtime Monitoring is based on event messages. In particular, it enables the collection of specific events that flows during controlled environment, real execution and predictive simulation [12] among the different virtual and real entities (e.g., DT, sensors and ecosystem components) and it infers one or more complex events about the runtime execution (Complex Event Processing (CEP)). Complex events inference is based on a set of derived rules, i.e., "if-then-else" grammar expressions that define sequences of attended or un-attended events patterns. Thus, Runtime Monitoring includes a set of generic rules templates (meta-rules) that can be instantiated at runtime in concrete technical scenarios in which behavior is evaluated. The technical scenarios are a combination of both

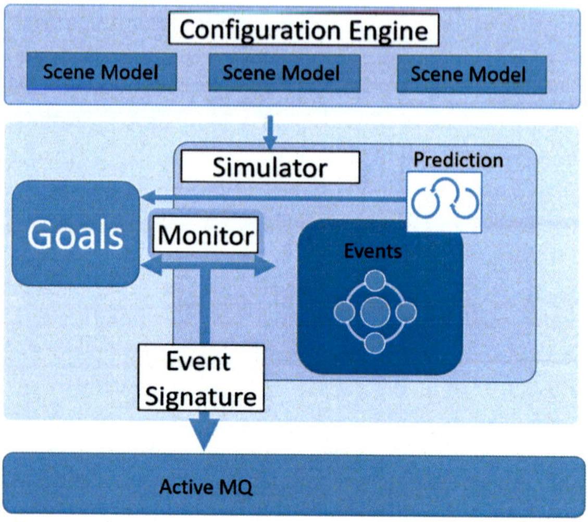


**Fig. 3.** Predictive simulation.

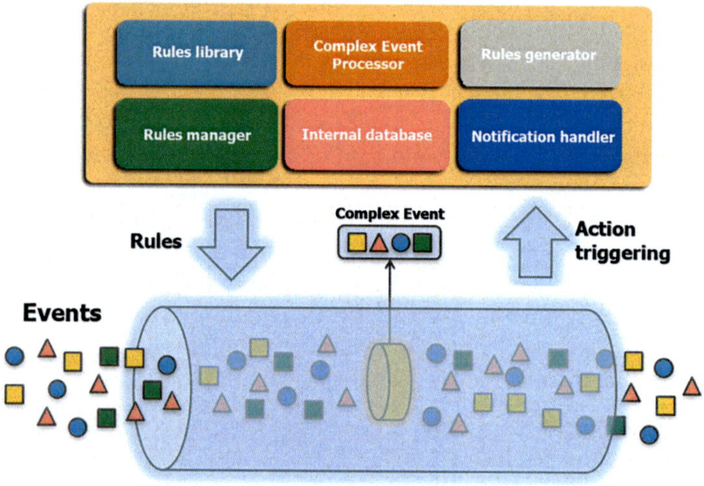


**Fig. 4.** Monitoring infrastructure.

external environment and internal configuration of a system capable to accommodate different components over time. The events that trigger the execution of a rule are generated by a probe, i.e., a piece of code injected in the entities to be observed during the runtime execution able to notify the occurrence of the events to the monitor engine. At the technical level, the Runtime Monitoring is based on ActiveMQ[1] messaging protocol, and it also exposes a REST interface.

[1] https://activemq.apache.org/.

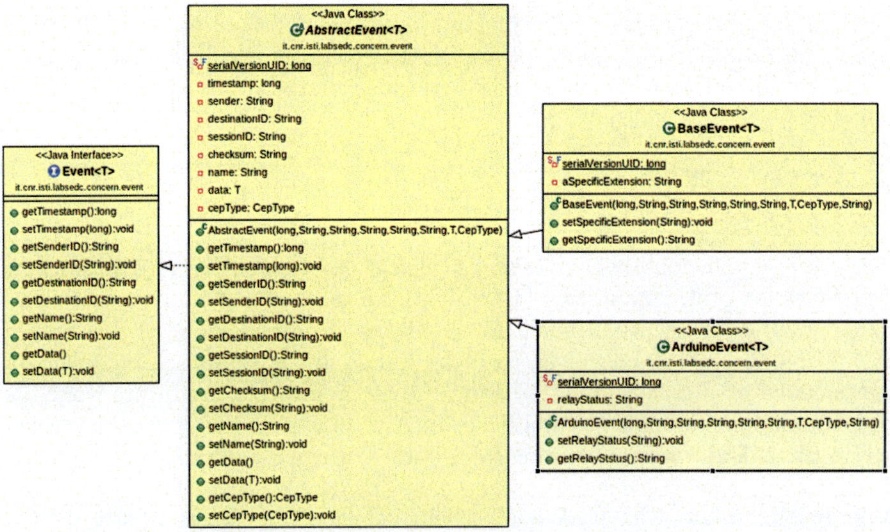


**Fig. 5.** UML class diagram of the monitoring event.

Figure 4 shows our reference monitoring architecture, whose main components are:

1) *Complex Event Processing (CEP):* i.e., a rule engine which analyzes the events generated by probes and correlates them to infer more complex events [1]. If the event triggers no rule, the event is just collected into the Event Stream of the CEP. Figure 5 depicts the structure of the events considered inside the monitoring component.
2) *Notification Handler:* i.e., a registry that keeps track of the requests for monitoring sent to the monitoring infrastructure. Once it receives the advice of a rule firing, pattern completion, or property violation from the CEP, the component sends the evaluation to the requester.
3) *Rules Generator:* i.e., the component in charge to generate the rules using the templates stored into the Rules Library. These rules are generated according to the specific properties to be monitored. A generic rule consists of two main parts: i) the events to be matched and the constraints to be verified are specified; and ii) the events/actions to be notified after the rule evaluation.
4) *Rules Library:* i.e., an archive of predetermined rules templates that will be instantiated by the Rules Generator when needed. A rule template is a rule skeleton, that is specified by the instantiation of a set of template-dependent placeholders. For the sake of completeness, the Rules Template Repository can also include sets of static rules.
5) *Rules Manager:* i.e., a component in charge to instruct the Complex Event Processor, to load and unload a set of rules. The Rule Generator will instantiate the template with appropriate values inferred from the specific properties

to be monitored. Once a rule is instantiated this is loaded by the Rule Manager into the Complex Event Processor. The complex event detection process depends directly on the operations performed by the Rules Manager component.

## 6 Conclusions and Future Work

Detection of malicious behavior within digital ecosystems [9,11] requires new techniques for runtime auditing [8,13]. In this paper, we have introduced a reference auditing conceptual model for monitoring the behaviour of SoS within BIECO platform. The conceived auditing framework is composed of two main components: predictive simulation and runtime monitoring. As future work, we are planning to thoroughly validate our approach by considering three use cases within the BIECO project, coming from both academic and industrial contexts.

**Acknowledgement.** This work was partially supported by the EU H2020 BIECO project GA No. 952702 (www.bieco.org).

## References

1. de Almeida, V.P., Bhowmik, S., Lima, G., Endler, M., Rothermel, K.: DSCEP: an infrastructure for decentralized semantic complex event processing. In: 2020 IEEE International Conference on Big Data (Big Data), pp. 391–398. IEEE (2020)
2. Bartocci, E., et al.: Specification-based monitoring of cyber-physical systems: a survey on theory, tools and applications. In: Bartocci, E., Falcone, Y. (eds.) Lectures on Runtime Verification. LNCS, vol. 10457, pp. 135–175. Springer, Cham (2018). https://doi.org/10.1007/978-3-319-75632-5_5
3. Basin, D., Caronni, G., Ereth, S., Harvan, M., Klaedtke, F., Mantel, H.: Scalable offline monitoring. In: Bonakdarpour, B., Smolka, S.A. (eds.) RV 2014. LNCS, vol. 8734, pp. 31–47. Springer, Cham (2014). https://doi.org/10.1007/978-3-319-11164-3_4
4. Blum, C., Winfield, A.F.T., Hafner, V.V.: Simulation-based internal models for safer robots. Front. Robot. AI **4**, 74 (2017)
5. Brilliant, S.S., Knight, J.C., Leveson, N.G.: Analysis of faults in an N-version software experiment. IEEE Trans. Softw. Eng. **16**(2), 238–247 (1990)
6. Cioroaica, E., Kuhn, T., Bauer, T.: Prototyping automotive smart ecosystems. In: 2018 48th Annual IEEE/IFIP International Conference on Dependable Systems and Networks Workshops (DSN-W). IEEE (2018)
7. Cioroaica, E., Kuhn, T., Buhnova, B.: (Do not) trust in ecosystems. In: 2019 IEEE/ACM 41st International Conference on Software Engineering: New Ideas and Emerging Results (ICSE-NIER), pp. 9–12. IEEE (2019)
8. Fotescu, R.P., Constantinescu, R., Alexandrescu, B., Burciu, L.M.: System for monitoring the parameters of vehicle. In: Advanced Topics in Optoelectronics, Microelectronics and Nanotechnologies X, vol. 11718, p. 117180A (2020)
9. Hidayanti, F.: Design and application of monitoring system for electrical energy based-on internet of things. Helix **10**(01), 18–26 (2020)

10. Kuhn, T., Forster, T., Braun, T., Gotzhein, R.: Feral—framework for simulator coupling on requirements and architecture level. In: 2013 Eleventh IEEE/ACM International Conference on Formal Methods and Models for Codesign (MEMOCODE), pp. 11–22. IEEE (2013)
11. Santos, M.A., et al.: Online heart monitoring systems on the internet of health things environments: a survey, a reference model and an outlook. Inform. Fusion **53**, 222–239 (2020)
12. Ullo, S.L., Sinha, G.R.: Advances in smart environment monitoring systems using IoT and sensors. Sensors **20**(11), 3113 (2020)
13. Won, M.: Intelligent traffic monitoring systems for vehicle classification: a survey. IEEE Access **8**, 73340–73358 (2020)

# Basic Aspects in Redundancy-Based Intrusion Tolerance

Felicita Di Giandomenico and Giulio Masetti(✉)

Institute of Science and Technology "A. Faedo", Consiglio Nazionale delle Ricerche, 56124 Pisa, Italy
giulio.masetti@isti.cnr.it

**Abstract.** This paper addresses intrusion tolerance in ICT systems (that is, how to guarantee the correct operation even when some of its parts are compromised), by adopting a redundancy approach. As an initial step to the process of incorporating redundancy features to enhance robustness of a system/subsystem to cyber attacks that end up with an intrusion, a conceptual framework encompassing the major aspects to be taken into account, regarding both system and redundancy characteristics, is proposed. This is the basis for reasoning on effective redundancy solutions, so a useful support for defining systems that need to satisfy resilience properties in hostile environments where threats include both intentional attacks and accidental faults.

**Keywords:** Intrusion tolerance · Redundancy · Diversity

## 1 Introduction and Related Work

The security of ICT components is increasingly challenged by a variety of attacks, and countermeasures are needed to cope with them. A primary defense against cyber attacks has been traditionally based on preventing intrusions, through methods such as authentication and access control, or on intrusion detection but without systematic forms of processing the intrusion symptoms. Despite the utmost contribution provided by preventive solutions, unfortunately reality shows that their effectiveness is limited and intrusions are permanently happening [14].

Therefore, intrusion tolerance emerged, with the aim to guarantee that a system works correctly even when some of its parts are compromised. Initial intrusion tolerance approaches are dated early ninetees, although they appear as isolated works, mainly focusing on protocols (see [21] for an overview of initial works on this subject). Many other papers addressed this topic in more recent years, some with emphasis on analysis and evaluation of security oriented strategies to manage system resources in order to reach desired reliability/availability levels (e.g., [9,17,23]), others on architectural solutions for specific application domains (e.g., [3,22]) or on proposing variations/extensions to some classical fault tolerance architectures towards secure service (e.g., [16]), or investigating


J. J. Gude Prego et al. (Eds.): CISIS 2021/ICEUTE 2021, AISC 1400, pp. 192–202, 2022.
https://doi.org/10.1007/978-3-030-87872-6_19

trade-offs between competing requirements (e.g., [8]). Intrusion tolerance in certain contexts can benefit from achievements obtained in other areas, such as multi-party computations and threshold cryptography; however, the downside is that new challenges have to be addressed [5].

The objective of this paper is to start a research activity aiming at a more general and deeper investigation of aspects and practices around redundancy-based intrusion tolerance, so to promote easier access and exploitation of this approach in a variety of system contexts. In particular, with the purpose to provide guidance in the selection and customization of appropriate intrusion tolerance approaches relying on the usage of redundancy, the offered contribution consists in the identification of the various aspects that need to be considered, regarding both system and redundancy characteristics. This is a preliminary, but essential step, to the correct design and deployment of the intrusion tolerance solution, guaranteeing its effectiveness. This work builds upon several, but fragmented solutions existing in the literature, and advances by developing a high level, general reference framework where the several dimensions, to be taken into account when resorting to redundancy-based intrusion tolerance techniques, are indicated and easily exploitable.

The rest of the paper is organized as follows. Section 2 is an overview of redundancy-based intrusion tolerance, and ends with the identification of the four major dimensions composing the proposed conceptual framework for intrusion tolerance, each one then addressed in the subsequent four sections. Specifically, Sect. 3 describes how cyber attacks are characterized in the proposed framework; Sect. 4 discusses the system components that are considered the subject of potential attacks; Sect. 5 characterizes the system under analysis; Sect. 6 focuses on the types of redundancy that are expected to be more relevant than others for intrusion tolerance. Section 7 draws final considerations and outlines the next steps.

## 2 Overview of Redundancy-Based Intrusion Tolerance

Prompt detection of attacked components is primarily pursued in security, to prevent the successful completion of an attack path, resulting in a high number of (potentially critical) compromised components. This is absolutely correct. However, detection as the only measure in place may be insufficient to reach desired levels of security, so forms of intrusion management are needed [21].

The dependability community has proposed since several decades fault tolerance techniques, with the aim of avoiding service failure in the presence of faults. As described in [2], fault tolerance includes techniques for *error detection* and *system recovery*, these last consisting of *error handling* and *fault handling* mechanisms. Although the kind of faults foreseen in [2] encompasses both accidental faults and deliberate attacks, so targeting dependability and security within a common framework, fault tolerance has been primarily thought as a means for dealing with accidental faults. However, as discussed in [21], the essential concepts are applicable to malicious faults, thus leading to intrusion tolerance.

Indeed, the fault model in security resembles similar dynamics as for accidental faults: the presence of a *vulnerability*, which can be exploited with malicious intention by an attacker to launch an *attack* that, if successful, leads to an *intrusion*. The parallel with an example of accidental fault could be the following: a hardware component that is not properly shielded (pseudo-vulnerability) may be penetrated by particles from an electromagnetic field (pseudo-attacker), resulting in an operational fault (pseudo-intrusion) that generates errors in the performed computation. Of course, there are specificities of an intentional attack that need to be carefully analyzed, in order to take the most effective measure to tolerate it. In particular, the *logic* behind the attack requires more attentive reasoning than in the case of pure *accidental* events.

For terminology clarity, and in line with the AVI (Attack Vulnerability Intrusion) fault model in [21], we consider the fault chain *attack* → *vulnerability* → *intrusion* be the source of the generated error(s), and intrusion tolerance to be inserted as a barrier to prevent the generated error to result into a failure.

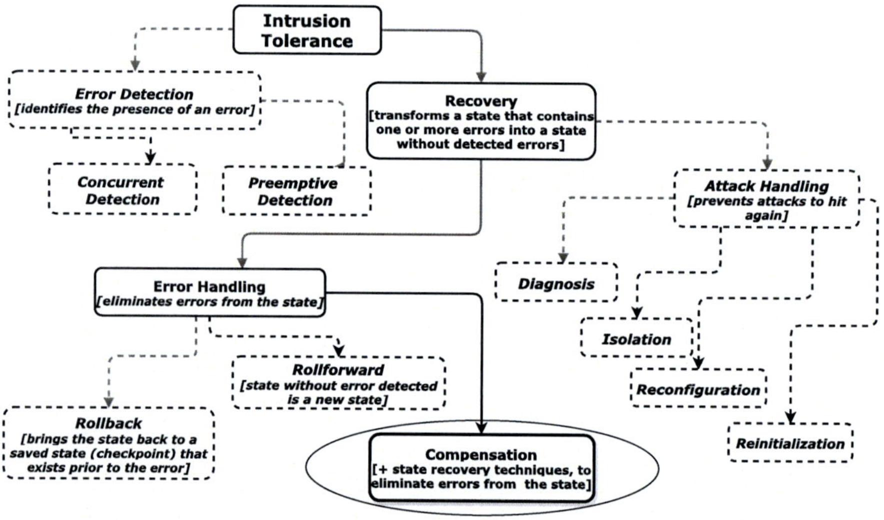


**Fig. 1.** Intrusion tolerance, focus on compensation.

In Fig. 1, the organization of fault tolerance mechanisms from [2], but declined in terms of attack tolerance, is proposed, with solid arrows identifying the path of interest in this paper. Specifically, among the wide spectrum of research directions under the hat of *Intrusion Tolerance* as depicted in Fig. 1, we concentrate on techniques for handling the errors generated through the intrusion, and in particular on those solutions that adopt redundancy to mask the presence of error(s). They correspond to the category of solutions labelled *Compensation*. Redundancy means the exploitation of more components, to count on enough good results/values that mask, through a properly defined adjudication function that operates the selection among the multiple outcomes of the redundant structure, the presence of compromised computations/stored values/transmitted messages. Well known redundancy-based fault tolerant structures are *Recovery Block* [15] and *N-Version Programming* [1], and we are interested to borrow the concepts underlying them (and other schema as well) in the intrusion tolerance domain. Redundancy in time (i.e., retry of a computation or transmission), typically effective with transient accidental faults, appears inappropriate in the context of attacks (an attack remains solid, or increases along time).

Exploiting redundancy to mask the presence of errors makes error detection not immediately needed, although it remains a fundamental step of the chain, to identify and help confining the intrusion from propagating within the system. Indeed, masking the effect of intrusions can result as the only choice to satisfy other system requirements, for example when sufficiently accurate attack detection implies too large latency for the timing requirements of the application at hand. Moreover, when dealing with stateful computation (that is, the outcome of a computation depends on the input and the internal state of the computing unit), it is advisable to pair masking with state recovery techniques, to eliminate errors from the sate of the failed element(s) of the redundant structure.

Based on the long-dated experience with fault tolerance, and considering the peculiarities of an intentional attack, the proposed general conceptual framework for intrusion tolerance develops along four major dimensions. These last can be considered the *ingredients* to work with in the design of redundancy-based intrusion tolerance mechanisms when facing a specific application context:

- Attack model, that is how the cyber attack is characterized (in Sect. 3);
- Categories of system components targeted by attacks, that is the identification of which part(s) of the ICT system at hand can be the target of a cyber attack (in Sect. 4);
- System model and failure assumptions, namely whether the system at hand is a single system component, or a structured, distributed system (in Sect. 5);
- Type of redundancy, that is the characterization of the forms of redundancy that can be put in place (in Sect. 6).

## 3 Cyber Attack Model

A large variety of cyber attacks have been discovered and studied in the last decades. Several threat repositories exist, such as MITREs ATT&CK [20], CAPEC [18] and CWE [19], where a description on how an adversary can accomplish each individual attack, as well as the consequences of the attack itself, are generally provided. They are certainly helpful support to developers of countermeasures to cope with them, especially how to prevent their occurrence.

From the intrusion tolerance perspective, essentially what is relevant is to know the consequences of the attack, more than the dynamics on which vulnerabilities it exploits and the path to the successful intrusion. These last are vital information to accomplish the attack detection and treatment, but less relevant to carry out masking of the generated intrusion. Attack consequences are then connected with the security attribute that is primarily impacted, namely the well known triad *confidentiality*, *integrity* and *availability*. As a brief recall from [2]: i) confidentiality is preserved in absence of unauthorized disclosure of information; ii) integrity is preserved in absence of improper system alteration; and iii) availability is preserved with readiness for correct service.

Similar to [16], the manifestation of a successful attack on the compromised components can be summarized as follows:

- *Functionality Change*, that is the delivered results are incorrect. This means that the compromised components experience a failure in the value domain. The impact is mainly on the integrity property;
- *Performance Degradation*, that is the results are delivered late or, in the extreme case, they are omitted. This means that the compromised components experience a failure in the time domain. The impact is mainly on the availability property;
- *Information Leakage*, that is sensitive information are revealed. The impact is mainly on the confidentiality property.

Intrusion tolerance techniques by masking the presence of compromised components are mainly directed to preserve integrity and availability. Regarding confidentiality, such techniques are not effective; attack prevention is the reference approach for this property. Therefore, confidentiality is left out from the conceptual intrusion tolerance framework under development in this paper.

## 4 Categories of System Components Targeted by Attacks

Cyber attacks can be launched to all the components of an ICT system, and typically an attack develops through several of them to be successful and lead to an intrusion. From the attacker perspective, the ICT components can be considered as belonging to three major categories:

- *Computing element*, that is a component that is devoted to perform some kind of functionality, to provide a service to the requesting entity (a user or another component). Operating systems primitives, software applications and enterprise software are typical examples of this category;
- *Communication element*, that is the means through which information is delivered to/from computing elements, users and storage. The internet and the several wireless networks technologies are typical examples of this category;
- *Data storage element*, which includes different storage technologies used to retain digital data within a computer system architecture. The term storage may refer both to a user's data generally and, more specifically, to the integrated hardware and software systems used to capture, manage and prioritize the data. This includes information in applications, databases, data warehouses, archiving, backup appliances and cloud storage.

## 5 System Model and Failure Assumptions

Existing ICT systems (seen in isolation, or as part of a larger system, or as composed by a number of smaller systems) include a variety of components belonging to categories recalled in Sect. 4 and are organized according to different models, following methodologies and practices of the reference application domain. Without going in the details, such system models range from monolithic structure to distributed interacting components of different granularities, including SOA (Service Oriented Architecture) [7], microservice and SoS (Systems-of-Systems) paradigm [6]. There are pros and cons with each system model, but this discussion is out of the scope of this paper. The architectural solution is typically chosen on the basis of the functional and non-functional requirements, as well as cost implications. Since flexibility and scalability are among the most relevant requirements to drive the selection of the system architecture, the current trend is to increasingly evolve from the monolithic structure to forms of distributed computation. However, in addition to other considerations, it needs to be mentioned that there are long-lived systems, originally developed as a monolithic architecture, which cannot undergo significant redesign, but need to be enhanced from the resilience perspective. Therefore, the interest in monolithic-based solution appears to be still significant.

Another important aspect associated to the structure and operation of an ICT system is the assumed *failure model* for the system components (due to accidental faults and/or intentional attacks). In line with the discussion on effects of attacks in Sect. 3, experienced failures can be in the *value domain* (an incorrect value is delivered/transmitted/stored) or in the *time domain* (a value that violates the time constraints is delivered/transmitted).

## 6 Type of Redundancy

Redundancy for fault tolerance purposes was initially exploited in the form of replicas, that is identical copies of the same system component, both hardware or software, to cope with accidental faults. The problem of common mode failures, mainly originated by design faults, raised attention on the dependencies among replicas such that a problem in one replica is actually present also in all the other replicas. Research on how to obtain diverse components, in order to promote sufficiently high probability of failure independence, was then triggered, leading to a shift of the paradigm of redundancy from replicas to variants, which are functionally equivalent components developed with some form of *diversity*.

In the security context, redundancy in the form of replicas has been recognized as not effective: the attacker can easily exploit the same attack to compromise all the replicas. Therefore, forms of diversity are needed.

As presented in [14], *diversity* in an ICT system can affect several aspects (the axes of diversity), ranging from software components (application level functionality, as well as supporting middleware features, libraries and lower level OS primitives) to hardware components, but also location, administrative and security methods. Indeed, diversity to hardware and software components has been practised since long time, through a variety of solutions. A thorough discussion of practices for enhancing diversity in software designs is in [13], including methods such as employment of diverse development teams, enforcing functional diversity and adopting different development environments (e.g., different programming languages, compilers, run-time supports, etc.). Data diversity, mainly consisting in using random perturbations of inputs or algorithm specific re-expression of inputs, is another form of diversity that can be applied alone, or in conjunction with other methods to develop diverse fault intrusion systems.

Administrative diversity plays a more important role when security is targeted rather than in case of accidental faults only. In fact, it is well known that many security compromises are carried out through social engineering [24]. Therefore, distributing the components of an intrusion tolerant system across different administrative domains, applying different security management policies, helps to mitigate the exploitation of social engineering by an intruder.

Concerning location diversity, consisting in placing several hw/sw components in different sites, it is another important defense against both physical and cyber threats (e.g., [3]).

Note that security methods are a category of system components that take great advantage of the diversity principle. As for application functions, using several security methods to enforce a given security attribute (e.g., encryption techniques), increases the chance that the attribute is not violated if a subset of these methods is compromised. Since security methods play a critical role, increasing their security is a priority when designing and deploying them.

A few considerations about diversity in relation with intrusion tolerance are in the following. It has been already recalled that degrees of diversity can be obtained in a variety of practices. However, when attacks are to be counteracted, practices such as just the adoption of different OS or programming languages may result too light. Functional diversity looks certainly more effective, although more expensive in general. Moreover, diversity at one layer leads to states of the underlying layers which are different, even if they are not diversified. This suggests that the higher is the level where diversity is applied and the greater is the expected benefit in terms of protection against an intruder. Moreover, since general categories of attacks can be identified and the efficacy of defenses varies between them, the decision on where and how to apply diversity methods for security should take into consideration such knowledge rather than simple enforcement of randomly picked techniques, to improve in efficacy.

Another aspect to consider cautiously is the granularity of the units that are selected as the entities to replicate for intrusion tolerance. Keeping the size of such units small has benefits in terms of error propagation and diagnosis, but reduces the possibility of diversity. In addition, small size means a higher number of units that are made redundant, thus increasing the attack surface.

Of course, diversity introduces also new problems. The major ones are:

- inexact voting: correctly working diverse variants will often produce different, but correct results. So, a generalized decision function is required, typically a distance function over the space of possible results, and rules to group results into consensus groups. Such definition may be application specific, making it more difficult to create a standard (sw or hw) voter component.

- Consistency among correct variants: decision points (e.g., expressions like if $x > a$ in a program) in the versions may be source of inconsistent actions taken by correct variants. Bypassing this problem by having versions decide on a consensus value of the variable before the decision point has the drawback of potential reduction in terms of diversity and performance degradation.

## 7 Final Considerations and Next Steps

The system structure and the failure model, together with the resilience properties the system has to satisfy (e.g., a specified level of reliability and/or availability) drive the identification of which system components are primarily critical and deserve to be made more robust to undesirable events. When redundancy-based intrusion tolerance solutions are selected for such purpose, the same information is helpful to decide on the degree of redundancy (failure assumptions play a primary role), on the adjudication function for the redundant structure (able to cope with diverse redundancy), as well as on how to organize the redundancy from the operational perspective (sequential versus parallel execution, also depending on the presence of real-time constraints).

In this paper, we focused on redundancy-based intrusion tolerance with the objective of identifying the basic aspects around an ICT system that need to be considered in order to make an effective choice of the redundancy structure to adopt. The discussed conceptual framework, graphically depicted in Fig. 2, encompasses four identified dimensions that are primarily relevant when building redundancy-based intrusion tolerance.

**Fig. 2.** Proposed conceptual framework for redundancy-based Intrusion Tolerance.

Moving from this initial, but important identification of major aspects that have to be considered in combination when addressing the employment of redundancy-based intrusion tolerance, the final goal of this research line is to provide guidelines to practitioners on which would be suitable solutions to adopt for the system at hand. To reach this goal, in the next steps we will:

i) revisit from the security perspective the classical redundancy-based fault tolerant approaches, namely the two extremes N-Version Programming [1] and Recovery-Block [15], together with the several variants in between (such as NSCP [12] and SCOP [4]). Literature review will be the starting point for this investigation, such as [16];

ii) discuss the relationship between the aspects overviewed in this paper and the characteristics of the intrusion tolerance solutions identified at the previous point. For example, in a microservice environment a somewhat natural way to deal with variants of a computational component is through N-version programming, with the adjudicator cautiously designed[1]. If the diversity degree of the variants constituting the redundant structure is such that the adjudication function would not be sufficiently accurate, while the application domain is more prone to define effective acceptance tests on the results provided by a variant, then sequential execution according to the Recovery Block paradigm would be preferable. However, assumptions on communication attacks are to be taken into account when deciding the redundancy structure: sequential execution of the available variants can be defeated by the communication attacks, while parallel execution is a more robust alternative. To continue with exemplifications, the assumption on the attack model leads to deciding on the degree of redundancy: if the impact of the attack is omission, lower redundancy is required than in the case of failure in the value domain. Relationships as those just presented will be carried on in a systematic way.

We believe that this research activity will be practically useful to developers of nowadays complex systems employed in critical domains. Among others, the BIECO project aims at developing a holistic framework, populated by a variety of methods and techniques to understand and manage the cybersecurity risks and threats in complex interconnected ICT systems. The BIECO's vision is to target a variety of domain sectors, thus calling for as much as possible general solutions, suitable to be customized to fit specific contexts. The proposed approach to redundancy-based intrusion tolerance constitutes an initial contribution in BIECO, as well as to other similar initiatives.

**Acknowledgment.** This work was supported by the project BIECO (https://www.bieco.org/) that received funding from the European Unionâs Horizon 2020 research and innovation programme under grant agreement No. 952702.

## References

1. Avizienis, A.: The N-version approach to fault-tolerant software. IEEE Trans. Softw. Eng. **11**(12), 1491–1501 (1985)
2. Avizienis, A., Laprie, J.C., Randell, B., Landwehr, C.: Basic concepts and taxonomy of dependable and secure computing. IEEE Trans. Dependable Secure Comput. **1**(1), 11–33 (2004)

[1] As a separate voter, possibly enclosed within the root of trust [16], or a voting logic distributed onto the variants [10], maybe introducing also camouflage units [11].

3. Babay, A., Tantillo, T., Aron, T., Platania, M., Amir, Y.: Network-attack-resilient intrusion-tolerant SCADA for the power grid. In: 48th Annual IEEE/IFIP International Conference on Dependable Systems and Networks (DSN), pp. 255–266 (2018)
4. Bondavalli, A., Di Giandomenico, F., Xu, J.: A cost-effective and flexible scheme for software fault tolerance. J. Comput. Syst. Sci. Eng. **8**, 234–244 (1993)
5. Brandão, L.T.A.N., Mouha, N., Vassilev, A.: Threshold schemes for cryptographic primitives. Technical report NISTIR8214, National Institute of Standards and Technology (2019)
6. Ceccarelli, A., Bondavalli, A., Froemel, B., Hoeftberger, O., Kopetz, H.: Basic Concepts on Systems of Systems, pp. 1–39 (2016)
7. Erl, T.: Service-Oriented Architecture: Concepts, Technology, and Design (2005)
8. Gashi, I., Povyakalo, A., Strigini, L.: Diversity, safety and security in embedded systems: modelling adversary effort and supply chain risks. In: 12th European Dependable Computing Conference (EDCC), pp. 13–24 (2016)
9. Haphuriwat, N., Bier, V.M.: Trade-offs between target hardening and overarching protection. Eur. J. Oper. Res. **213**(1), 320–328 (2011)
10. Hardekopf, B., Kwiat, K., Upadhyaya, S.: Secure and fault-tolerant voting in distributed systems. In: IEEE Aerospace Conference Proceedings, pp. 1117–1126 (2001)
11. Kwiat, K., Taylor, A., Zwicker, W., Hill, D., Wetzonis, S., Ren, S.: Analysis of binary voting algorithms for use in fault-tolerant and secure computing. In: 5th International Conference on Computer Engineering Systems, pp. 269–273 (2010)
12. Laprie, J.C., Arlat, J., Beounes, C., Kanoun, K.: Definition and analysis of hardware- and software-fault-tolerant architectures. Computer **23**(7), 39–51 (1990)
13. Littlewood, B., Strigini, L.: A discussion of practices for enhancing diversity in software designs. Technical report DISPO, Centre for Software Reliability, City University (2000)
14. Obelheiro, R.R., Bessani, A.N., Lung, L.C., Correia, M.: How practical are intrusion-tolerant distributed systems? Technical report TR-06-15, Dept. of Informatics, Univ. of Lisbon (2006)
15. Randell, B.: System structure for software fault tolerance. IEEE Trans. Softw. Eng. **1**(2), 220–232 (1975)
16. Rodriguez, M., Kwiat, K.A., Kamhoua, C.A.: Modeling fault tolerant architectures with design diversity for secure systems. In: IEEE Military Communications Conference (MILCOM), pp. 1254–1263 (2015)
17. Tarraf, D.C., Kamhoua, C.A., Kwiat, K.A., Njilla, L.: Majority is not always supreme: less can be more when voting with compromised nodes. In: IEEE 18th International Symposium on High Assurance Systems Engineering (HASE), pp. 9–12 (2017)
18. The MITRE Corporation: Common Attack Pattern Enumeration and Classification (CAPEC). https://capec.mitre.org/data/definitions/3000.html
19. The MITRE Corporation: Common Weakness Enumeration (CWE). https://cwe.mitre.org
20. The MITRE Corporation: Mitre attack. https://attack.mitre.org
21. Veríssimo, P.E., Neves, N.F., Correia, M.P.: Intrusion-tolerant architectures: concepts and design. In: Architecting Dependable Systems, pp. 3–36 (2003)
22. Vöelp, M., Verissimo, P.: Intrusion-tolerant autonomous driving. In: IEEE 21st International Symposium on Real-Time Distributed Computing (ISORC), pp. 130–133 (2018)

23. Wang, L., Ren, S., Korel, B., Kwiat, K.A., Salerno, E.: Improving system reliability against rational attacks under given resources. IEEE Trans. Syst. Man Cybern. Syst. **44**(4), 446–456 (2014)
24. Winkler, I.S., Dealy, B.: Information security technology? Don't rely on it - a case study in social engineering. In: 5th USENIX UNIX Security Symposium (USENIX Security), pp. 1–6 (1995)

# A Survey on the State of the Art of Vulnerability Assessment Techniques

Eva Sotos Martínez(✉), Nora M. Villanueva, and Lilian Adkinson Orellana

Gradiant. Carretera Vilar, no. 56-58, 36214 Vigo, Spain
{esotos,nmvillanueva,ladkinson}@gradiant.org
https://www.gradiant.org/en/

**Abstract.** A suitable vulnerability assessment process improves the overall security of an organization, and therefore, reduces the chances that an attacker could breach the IT system. The detection of vulnerabilities is one of the first steps in software security lifecycle, but in some cases, it is not enough. Understanding how a vulnerability can affect the rest of the system as well as forecasting its exploitability are important issues when assessing its real impact. Under this premise, this paper presents an overview of the state of the art, including different approaches related to the detection, forecasting and propagation of vulnerabilities in software, focusing mainly on Machine Learning and Data Mining techniques.

**Keywords:** Vulnerability assessment · Vulnerability detection · Vulnerability forecasting · Vulnerability propagation · Machine learning · Data mining

## 1 Introduction

One of the main purposes of cybersecurity is to guarantee the properties of the CIA triad: Confidentiality, Integrity, and Availability [1]. These properties are the security requirements a computer system should accomplish. A balance between them guarantees high levels of security and trust.

One of the main issues we can face when it comes to the fulfilment of such properties is the existence of security vulnerabilities. A security vulnerability is a weakness that can be exploited by an attacker in order to compromise the CIA triad of a system. The number of disclosed vulnerabilities is increasing every year, and those that are present in widely-used systems can cause severe economic, reputational and even societal harms. Therefore, it is essential to identify these vulnerabilities at an early stage of the systems' development life cycle, and improve the assessment processes and tools that allow to detect, classify, evaluate and mitigate them in an accurate manner.

Substantial research has been devoted to techniques that analyse source code in order to detect and characterize security vulnerabilities, but also to evaluate how a vulnerability could propagate to other elements of a software supply chain.


J. J. Gude Prego et al. (Eds.): CISIS 2021/ICEUTE 2021, AISC 1400, pp. 203–213, 2022.
https://doi.org/10.1007/978-3-030-87872-6_20

For this reason, in this paper the vulnerability identification and characterization process will focus mainly on three topics: vulnerability detection, vulnerability forecasting, and vulnerability propagation.

The rest of the paper is structured as follows. Section 2 presents an overview of vulnerability detection methods, mainly based on Machine Learning (ML) and Data Mining (DM) techniques. Section 3 includes a description of vulnerability forecasting techniques, focusing on the period in which a certain vulnerability could be exploited. Section 4 presents an overview of techniques that could be of use in order to determine how a certain vulnerability could propagate across a software system. Finally, Sect. 5 summarizes the conclusions.

## 2 Vulnerability Detection Techniques

In general, vulnerabilities require a different identification process than software defects, and are considered as a subset of faults which occur, and are discovered, much less frequently. Vulnerabilities are often not realized during the development or normal operation of the system, while defects are more easily noticed.

To date, several tools and techniques can be used for detecting vulnerabilities in software applications. Particularly, we will focus on three main topics [2]: anomaly detection-based techniques, which model the normal behaviour of the software, considering vulnerabilities as anomalies; code pattern recognition, that is focused on defining models and patterns of vulnerable code segments; and vulnerability prediction models, which predict the presence of vulnerabilities based on classical software metrics (e.g., code churn, code complexity...). Anomaly detection-based techniques and code pattern recognition include studies that analyse program syntax and semantics, unlike vulnerability prediction model approaches, in which the majority of works do not analyse them.

### 2.1 Anomaly Detection-Based Techniques

Anomaly detection addresses the problem of finding unusual events that differ significantly from the majority of the data, usually by applying ML and DM algorithms. In the context of software security, vulnerabilities can be considered as an "anomaly", as they are undesired events that can appear during software development, and which occurrence should be minimized as far as possible. In this section, several works based mainly on DM techniques are reviewed, such as association rules or frequent itemset mining.

Engler at el. [3] pointed out the need for extracting rules that could be used in bug-finding tools. Their approach demonstrated a static analysis technique to automatically extract rules from source code without prior knowledge of the system. Years later, Li and Zhou [4] presented the PR-Miner technique to extract implicit undocumented programming rules and detect violations on large software code. The tool used frequent itemset mining to find sets of functions, variables and data types that tend to appear together. Wasylkowski et al. [5] extended this work taking also ordering into account, identifying additional defects. Furthermore, they introduced a defect indicator, implemented by the tool JADET (Java Anomaly Detector), which allowed to rank the anomalies.

Gruska et al. [6] improved the mechanism of JADET, adding cross-project anomaly detection. They introduced a method which allowed enabling to analyse only selected parts of a source code. This new approach offered a set of frequent temporal properties more expressive, and improved the ranking system.

The works [4,6] have some limitations when multiple APIs are involved across different procedures. For this reason, Acharya et al. [7] presented an approach to automatically extract frequent partial-orders from an API client code. However, their approach is limited to mining function call ordering rules. In contrast, Chang et al. [8] created an approach general enough to discover function call ordering rules, as well as their preconditions and postconditions ordering rules. Thus, the user indicates just a few restrictions on the context of the rules that have to be searched, rather than specific rule templates.

A related approach was Alattin [9], which tried to reduce false positives produced by the frequent sub-graph mining approaches. To this end, authors introduced a new mining algorithm, ImMiner, that extracted alternative patterns classifying them as balanced, where all patterns are frequent, and imbalanced, where some of them are infrequent. Although [8,9] approaches target neglected conditions, ImMiner is much more scalable due to its imbalance frequent itemset mining algorithm.

Another form of analysis complementary to [8] approach is DynaMine [10], an automatic way to discover usage patterns combining revision history information with dynamic analysis. Although the results obtained are more precise, it requires extensive history of software revisions in repositories and, as with PR-Miner, it may suffer from issues of a high number of false positives.

Even though some of the aforementioned studies reveal good results in terms of vulnerability detection, they are still subject to high false positive rates since they do not take into account rare checks or the use of improper APIs. In addition, these techniques do not provide the type of the vulnerability or its location.

### 2.2 Vulnerable Code Pattern Recognition

Vulnerable code pattern recognition methods use ML techniques to automatically identify patterns of vulnerable code segments, analysing and mining features from the source code. In order to reduce manual effort, different approaches have been proposed to statistically detect these vulnerable code patterns.

Yamaguchi et al. [11] proposed a method to assist a security analyst during source code audit. The method extracts an Abstract Syntax Tree (AST) from the source code and determines its structural patterns, which allows to break down a known vulnerability and suggests code with similar properties. Nevertheless, AST is not suitable for advanced analysis. For this reason in [12], authors introduced a new graph representation named Code Property Graph. Using this representation, in [13] they presented a method for automatically inferring search patterns for taint-style vulnerabilities from the source code, combining unsupervised ML (i.e., clustering) and static program analysis. However, the approach still required a considerable amount of manual auditing and analysis work.

Input validation and sanitization are also essential secure coding techniques used to protect programs from common vulnerabilities. Shar and Tan [14] proposed a set of input sanitization code attributes that can be statically collected. However, their predictive capability is limited as it depends on the classification of the input validation and sanitization code patterns. To address these limitations, Shar et al. [15] proposed a hybrid analysis combining static and dynamic analysis, and using a semi-unsupervised approach, unexplored in this domain till the date, along the supervised approach to predict vulnerabilities.

The aforementioned works, with the exception of [15], are based on static analysis techniques focused on source code analysis. Grieco et al. [16] presented the first large scale study on vulnerability discovery for binary code, applying a scalable ML approach to predict if a binary program was likely to contain an easily exploitable memory corruption vulnerability. They also developed VDiscover, a tool to predict vulnerabilities in test cases. This study increased the possibility to find vulnerabilities at operating system scale.

In the last years, deep learning-based (DL) techniques have had a big success in many domains, including software vulnerability detection. The first systematic framework for using DL to detect vulnerabilities was presented by Zhen Li et al. [17]. The framework, called SySeVR, extracted program representations that can accommodate syntax and semantic information relevant for the vulnerabilities such as function call and pointer usage. Zhen Li et al. [18] also propose the use of DL to detect vulnerabilities at the slice level. In this approach, authors vectorized multiple semantically related lines of code as an input to the model. To this end, they implemented VulDeePecker, a DL-based vulnerability detection system.

Although vulnerable code pattern recognition is a promising approach, the studies presented do not identify the type of the vulnerability and, while it is possible to detect the location of the vulnerability within the source code, it is still not fine-grained.

### 2.3 Vulnerability Prediction Models

Vulnerability Prediction Models (VPM) [19] rely on ML techniques to determine which software components are most likely to contain a certain vulnerability. As mentioned earlier, vulnerabilities occur less frequently than defects, and therefore, VPMs have to deal typically with highly unbalanced datasets.

The first work related to VPM was published in 2007 by Neuhaus et al. [19]. The authors proposed to fit a ML model using the imports and the function calls as independent variables. This approach is based on the intuition that vulnerable files are likely to share similar sets of imports and function calls. The idea is in accordance with the work of Schroter et al. [20], which aims at predicting defects by the usage relationships between components.

There is extensive literature for VPM approaches that use software engineering metrics computed from the source code to build their models. Their starting point is the study carried out by Shin et al. [21]. Some years later, in addition to complexity metrics, code churn and develop activity metrics are evaluated as indicators of software vulnerabilities [22]. Other studies use code complexity, code churn, and other static alerts to predict attack-prone or vulnerable components, as in the case of [23], among others.

Additionally, Morrison et al. [24] replicated the study of Zimmermann et al. [25], who studied typical code metrics as well as dependency metrics and evaluated their correlation with post-release vulnerabilities. Besides, Morrison et al. investigated the performance of the models using the same metrics on finer levels of granularity (source file level). They used a total of 29 metrics classified as churn metrics, complexity metrics, dependency metrics, legacy metrics and size metrics. More recently, Younis et al. [26] described the attributes of code that contain vulnerabilities that are more likely to be exploitable.

Newly, Bilgin et al. [27] examined how to predict software vulnerabilities from source code by employing ML techniques. To this end, they developed a source code representation that enables to perform intelligent analysis on the Abstract Syntax Tree (AST) form of source code and then investigate whether ML can distinguish vulnerable and non-vulnerable code fragments.

In general terms, VPM are able to identify which software component is more likely to be vulnerable, which could be a complementary way to find vulnerability locations and to direct further inspections and testing. However, there are still open issues in the construction of an effective VPM, as the choice of granularity, the statistical learner choice [28] or the classification performance [29].

## 3 Vulnerability Exploitability Forecasting Techniques

The term forecasting refers to the process of making predictions of the future as accurately as possible, based on past and present data available or even knowledge of any future events. In the context of vulnerability assessment, forecasting refers to the estimation of certain information such as the number, type or time of occurrence of an event, which is related to a vulnerability. To date, substantial research has been dedicated to techniques that detect security vulnerabilities in the source code. However, limited research has focused on forecasting aspects of the security vulnerabilities that are reported after a software release.

One interesting and widely studied approach is estimating the likelihood of a vulnerability to be exploited. Some contributions have been described in the literature such as Jacob et al. [30] and Bhatt et al. [31]. Unlike these research works which study if a Common Vulnerability and Exposure (CVE) is used in an exploit, an important question growing between researchers is forecasting when an exploit might appear. This issue is especially important for system administrators, as they need to devote scarce resources to take corrective actions when a new vulnerability appears [32]. When vulnerability information is disclosed, there is an attacker's tendency to exploit vulnerabilities, and consequently, vendors' tendency to release patches. Since patching is expensive, many vulnerabilities go unpatched because of the limited resources to tackle with large patching tasks

[33]. Thus, it is important to prioritize patching based on the results of prediction models about when a vulnerability will be exploited.

It should be noted that there are two types of exploits: proof of concepts (PoC), generated in controlled environments, and real world exploits. In particular, [34,35] focus on predicting public PoC exploits. Sabottke et al. [36] developed methods to predict both PoC and real-world exploits, demonstrating that prediction accuracy could be improved by using Twitter data. Recently, more efforts have been made in order to study real world exploits [35] where, apart from National Vulnerability Database (NVD) and Exploit Database Archive (EDB), Zero Day Initiative (ZDI) data, and Dark and Deep Web posts are used.

On the other hand, even though Common Vulnerability Scoring System (CVSS) has become an industry standard for assessing vulnerabilities, some limitations have been identified, such as the absence of an authoritative entity to update the metric values and lack of data to inform the score. Following this idea, Haipeng et al. [32] described a method to predict when a vulnerability will be exploited based on CVE ID and Twitter discussion data, without the need for CVSS scores. In addition, Jacob et al. [30] proposed a threat scoring system called Exploit Prediction Scoring System (EPSS) for predicting the probability that a vulnerability will be exploited within 12 months after public disclosure.

## 4 Vulnerability Propagation Models

Software systems can be composed by several entities or components that are related to each other, conforming a software supply chain. However, this interdependency can impact the system's security. A vulnerability from a specific area of a source code could be propagated and escalated through the software supply chain via multiple paths, making a fragment of the code with no vulnerabilities to be affected by the propagation, leading to a different system security risk. In order to deal with these issues, several works have been presented, trying to model how a vulnerability can be propagated across the software supply chain.

Program dependency relations have been the goal of many researchers to evaluate how a change in one source code entity propagates to other entities [37]. In the last decades, many studies tried to identify the effects of those changes using Change Impact Analysis [38], a technique to predict the impact of a change in software. These dependencies can lead to the introduction of security vulnerabilities in software systems by third-party components, as open source libraries with vulnerable code. Thus, several tools have been developed to detect if a third-party component with a known vulnerability has been used [39,40].

Another approach to consider is the impact of unvalidated inputs, or taint data. Taintedness is propagated in the obvious way, as strings derived from tainted strings are also considered tainted. Perl's taint mode [41] prevents the use of taint data in sensitive functions that can affect the local system creating files, sending data over the network or running local commands among others. Inspired by this idea, Haldar et al. [42] introduced a technique for tagging and tracking the taintedness of untrusted inputs throughout the lifetime of the application.

Many researchers have proposed the use of graph methods, such as Bayesian Networks and attack graphs, to provide an estimation of risk factors and thus analyse the complexity and uncertainty of the propagation of a vulnerability. To make these methods more effective, it is important to use vulnerability propagation analysis tools that are capable of proactively assessing propagation. This can be obtained by using graph optimization methods, such as Ant Colony Optimization (ACO) [43]. Combining ACO and Bayesian Networks, Feng et al. [44] proposed a Security Risk Analysis Model (SRAM) that allowed to determine which of the propagation paths presented the highest probability and risk. Other authors like Hu et al. [45] proposed a search algorithm based on lazy strategy in terms of the macro component level, and on propagation difficulty when it comes to the microcode class level, reducing the search space complexity.

Most of the works in attack graph analysis have focused on prioritizing vulnerabilities instead of their attack paths. Agrawal and Khan [46] used the idea of Breadth First Search (BFS) (a graph and tree search algorithm) to propose an algorithm for computing an attribute's vulnerability propagation, and introduced a measure called Attribute Vulnerability Ratio (AVR). More recently, Garg et al. [47] developed a hybrid methodology to estimate attack path score by first calculating the risk score of individual vulnerabilities using CVSS metrics, and Page rank and Markov models to calculate the attack path risk score.

## 5 Conclusions

This paper presents a review of the state of the art on the detection, forecasting and propagation of software vulnerabilities. Vulnerability detection techniques were divided in three categories (anomaly detection, code pattern recognition and vulnerability prediction models). However, it is important to note that the choice of the technique to use depends on the goal we are trying to achieve. To detect blocks of source code that are prone to be vulnerable, anomaly detection based on the use of frequent pattern mining techniques or VPM are suitable approaches. Anomaly detection techniques can be used also in order to find bugs and possible malformed APIs. To reduce false positive rates, the detection of missing checks can also be included, as in the ImMiner approach. In regard to the detection of vulnerabilities in a more fine grained way, or the detection of a particular type of vulnerability [14], approaches based on vulnerable code pattern recognition seem promising. Recently, the use of DL algorithms is gaining increasing attention in this field, offering acceptable results. Additionally, an interesting line of research could be to combine vulnerability detection techniques, such as pattern recognition methods and VPM, in order to narrow down the parts of the code under analysis and improve the accuracy of the results.

It should also be noted that, although detection is a crucial step, identifying which vulnerabilities are more likely to be exploitable could be even more essential in order to prioritize their management. It is important to highlight that some of the VPM, such as Younis et al. [26], consider this issue, unlike the other two approaches. Moreover, the estimation of when a vulnerability is more likely to be exploited can be also an strategic aspect of the assessment process. The use of public sources as additional features can increase the probability of success in determining when a vulnerability can be exploited, as shown in [35]. Finally, vulnerability propagation is also an important issue to take under consideration. Modeling how the vulnerabilities can impact the rest of the system can help to understand their severity and prioritise their patching. To this goal, the use of graph methods combined with optimization methods (such as in SRAM) provide a good approximation. For a more complete approach, it could be useful to estimate an attack path score using CVSS metrics [47], among others.

**Acknowledgment.** This work was supported by the project BIECO (www.bieco.org) that received funding from the European Union's Horizon 2020 research and innovation programme under grant agreement No. 952702, and by the Ayudas Cervera para Centros Tecnológicos grant of the Spanish Centre for the Development of Industrial Technology (CDTI) under the project ÉGIDA (CER-20191012).

## References

1. Samonas, S., Coss, D.: The cia strikes back: redefining confidentiality, integrity and availability in security. J. Inf. Syst. Securi. **10**(3), 21–45 (2014)
2. Ghaffarian, S.M., Shahriari, H.R.: Software vulnerability analysis and discovery using machine-learning and data-mining techniques: a survey. ACM Comput. Surv. (CSUR) **50**(4), 1–36 (2017)
3. Engler, D., Chen, D.Y., Hallem, S., Chou, A., Chelf, B.: Bugs as deviant behavior: a general approach to inferring errors in systems code. ACM SIGOPS Oper. Syst. Rev. **35**(5), 57–72 (2001)
4. Li, Z., Zhou, Y.: PR-Miner: automatically extracting implicit programming rules and detecting violations in large software code. ACM SIGSOFT Softw. Eng. Not. **30**(5), 306–315 (2005)
5. Wasylkowski, A., Zeller, A., Lindig, C.: Detecting object usage anomalies. In: Proceedings of the 6th Joint Meeting of the European Software Engineering Conference and the ACM SIGSOFT Symposium on the Foundations of Software Engineering (2007)
6. Gruska, N., Wasylkowski, A., Zeller, A.: Learning from 6,000 projects: lightweight cross-project anomaly detection. In: Proceedings of the 19th International Symposium on Software Testing and Analysis, pp. 119–130 (2010)
7. Acharya, M., Xie, T., Pei, J., Xu, J.: Mining API patterns as partial orders from source code: from usage scenarios to specifications. In: Proceedings of the 6th Joint Meeting of the European Software Engineering Conference and the ACM SIGSOFT Symposium on the Foundations of Software Engineering (2007)
8. Chang, R.Y., Podgurski, A., Yang, J.: Discovering neglected conditions in software by mining dependence graphs. IEEE Trans. Soft. Eng. **34**(5), 579–596 (2008)
9. Thummalapenta, S., Xie, T.: Alattin: mining alternative patterns for detecting neglected conditions. In: 2009 IEEE/ACM International Conference on Automated Software Engineering, pp. 283–294. IEEE (2009)

10. Livshits, B., Zimmermann, T.: Dynamine: finding common error patterns by mining software revision histories. ACM SIGSOFT Softw. Eng. Not. **30**(5), 296–305 (2005)
11. Yamaguchi, F., Lottmann, M., Rieck, K.: Generalized vulnerability extrapolation using abstract syntax trees. In: Proceedings of the 28th Annual Computer Security Applications Conference, pp. 359–368 (2012)
12. Yamaguchi, F., Golde, N., Arp, D., Rieck, K.: Modeling and discovering vulnerabilities with code property graphs. In: 2014 IEEE Symposium on Security and Privacy, pp. 590–604. IEEE (2014)
13. Yamaguchi, F., Maier, A., Gascon, H., Rieck, K.: Automatic inference of search patterns for taint-style vulnerabilities. In: 2015 IEEE Symposium on Security and Privacy, pp. 797–812. IEEE (2015)
14. Shar, L.K., Tan, H.B.K.: Predicting SQL injection and cross site scripting vulnerabilities through mining input sanitization patterns. Inf. Softw. Technol. **55**(10), 1767–1780 (2013)
15. Shar, L.K., Briand, L.C., Tan, H.B.K.: Web application vulnerability prediction using hybrid program analysis and machine learning. IEEE Trans. Depend. Secure Comput. **12**(6), 688–707 (2014)
16. Grieco, G., Grinblat, G.L., Uzal, L., Rawat, S., Feist, J., Mounier, L.: Toward large-scale vulnerability discovery using machine learning. In: Proceedings of the Sixth ACM Conference on Data and Application Security and Privacy, pp. 85–96 (2016)
17. Li, Z., Zou, D., Xu, S., Jin, H., Zhu, Y., Chen, Z.: SySeVR: a framework for using deep learning to detect software vulnerabilities. Trans. Depen. Secure Comput. (2021)
18. Li, Z., et al.: Vuldeepecker: a deep learning-based system for vulnerability detection. *arXiv preprint* arXiv:1801.01681 (2018)
19. Neuhaus, S., Zimmermann, T., Holler, C., Zeller, A.: Predicting vulnerable software components. In: Proceedings of the 14th ACM Conference on Computer and Communications Security, pp. 529–540 (2007)
20. Schröter, A., Zimmermann, T., Zeller, A.: Predicting component failures at design time. In: Proceedings of the 2006 ACM/IEEE International Symposium on Empirical Software Engineering, pp. 18–27 (2006)
21. Shin, Y., Williams, L.: An empirical model to predict security vulnerabilities using code complexity metrics. In: Proceedings of the Second ACM-IEEE International Symposium on Empirical Software Engineering and Measurement, pp. 315–317 (2008)
22. Shin, Y., Meneely, A., Williams, L., Osborne, J.A.: Evaluating complexity, code churn, and developer activity metrics as indicators of software vulnerabilities. IEEE Trans. Softw. Eng. **37**(6), 772–787 (2010)
23. Gegick, M., Williams, L., Osborne, J., Vouk, M.: Prioritizing software security fortification throughcode-level metrics. In: Proceedings of the 4th ACM Workshop on Quality of Protection, pp. 31–38 (2008)
24. Morrison, P., Herzig, K., Murphy, B., Williams, L.: Challenges with applying vulnerability prediction models. In: Proceedings of the 2015 Symposium and Bootcamp on the Science of Security, pp. 1–9 (2015)
25. Zimmermann, T., Nagappan, N., Williams, L.: Searching for a needle in a haystack: predicting security vulnerabilities for windows vista. In: 2010 3rd International Conference on Software Testing, Verification and Validation. IEEE (2010)

26. Younis, A., Malaiya, Y., Anderson, C., Ray, I.: To fear or not to fear that is the question: code characteristics of a vulnerable function with an existing exploit. In: Proceedings of the 6th ACM Conference on Data and Application Security and Privacy (2016)
27. Bilgin, Z., Ersoy, M.A., Soykan, E.U., Tomur, E., Çomak, P., Karaçay, L.: Vulnerability prediction from source code using machine learning. IEEE Access **8**, 150672–150684 (2020)
28. Hastie, T., Tibshirani, R., Friedman, J.: The elements of statistical learning: data mining, inference, and prediction. Springer Science & Business Media (2009)
29. Shin, Y., Williams, L.: Can traditional fault prediction models be used for vulnerability prediction? Empirical Softw. Eng. **18**(1), 25–59 (2013)
30. Jacobs, J., Romanosky, S., Edwards, B., Roytman, M., Adjerid, I.: Exploit prediction scoring system (epss). *arXiv preprint* arXiv:1908.04856 (2019)
31. Bhatt, N., Anand, A., Yadavalli, V.S.S.: Exploitability prediction of software vulnerabilities. Qual. Ability Eng. Int. **37**(2), 648–663 (2021)
32. Chen, H., Liu, R., Park, N., Subrahmanian, V.S.: Using twitter to predict when vulnerabilities will be exploited. In: Proceedings of the 25th ACM SIGKDD Internacional Conference on Knowledge Discovery & Data Mining, pp. 3143–3152 (2019)
33. Farris, K.A., Shah, A., Cybenko, G., Ganesan, R., Jajodia, S.: Vulcon: a system for vulnerability prioritization, mitigation, and management. ACM Trans. Priv. Secur. (TOPS) **21**(4), 1–28 (2018)
34. Edkrantz, M., Said, A.: Predicting cyber vulnerability exploits with machine learning. In: SCAI, pp. 48–57 (2015)
35. Almukaynizi, M., Nunes, E., Dharaiya, K., Senguttuvan, M., Shakarian, J., Shakarian, P.: Proactive identification of exploits in the wild through vulnerability mentions online. In: 2017 International Conference on Cyber Conflict (CyCon US), pp. 82–88. IEEE (2017)
36. Sabottke, C., Suciu, O., Dumitraş, T.: Vulnerability disclosure in the age of social media: exploiting twitter for predicting real-world exploits. In: 24th {USENIX} Security Symposium ({USENIX} Security 2015), pp. 1041–1056 (2015)
37. Hassan, A.E., Holt, R.C.: Predicting change propagation in software systems. In: 20th IEEE International Conference on Software Maintenance. Proceedings. IEEE (2004)
38. Li, B., Sun, X., Leung, H., Zhang, S.: A survey of code-based change impact analysis techniques. Softw. Test. Verif. Reliab. **23**(8) (2013)
39. Cadariu, M., Bouwers, E., Visser, J., van Deursen, A.: Tracking known security vulnerabilities in proprietary software systems. In: IEEE 22nd International Conference on Software Analysis, Evolution, and Reengineering (SANER). IEEE (2015)
40. Plate, H., Ponta, S.E., Sabetta, A.: Impact assessment for vulnerabilities in open-source software libraries. In: 2015 IEEE International Conference on Software Maintenance and Evolution (ICSME), pp. 411–420. IEEE(2015)
41. Christiansen, T., Wall, L., Orwant, J., et al.: Programming Perl: Unmatched Power for Text Processing and Scripting. O'Reilly Media, Inc. (2012)
42. Haldar, V., Chandra, D., Franz, M.: Dynamic taint propagation for Java. In: 21st Annual Computer Security Applications Conference (ACSAC 2005). IEEE (2005)
43. Abadi, M., Jalili, S.: An ant colony optimization algorithm for network vulnerability analysis. Iran. J. Electr. Electron. Eng. **2**(3) (2006)
44. Feng, N., Wang, H.J., Li, M.: A security risk analysis model for information systems: causal relationships of risk factors and vulnerability propagation analysis. Inf. Sci. **256**, 57–73 (2014)

45. Hu, W., Wang, Y., Liu, X., Sun, J., Gao, Q., Huang, Y.: Open source software vulnerability propagation analysis algorithm based on knowledge graph. In: IEEE International Conference on Smart Cloud (SmartCloud), pp. 121–127. IEEE (2019)
46. Agrawal, A., Khan, R.A.: Impact of inheritance on vulnerability propagation at design phase. ACM SIGSOFT Soft. Eng. Notes **34**(4), 1–5 (2009)
47. Garg, U., Sikka, G., Awasthi, L.K.: Empirical analysis of attack graphs for mitigating critical paths and vulnerabilities. Comput. Secur. **77** (2018)

# Data Based Message Validation as a Security Cornerstone in Loose Coupling Software Architecture

Oliviu Matei[1,2], Rudolf Erdei[2](✉), Daniela Delinschi[2], and Laura Andreica[2]

[1] Technical University of Cluj Napoca, North University Centre of Baia Mare, Baia Mare, Romania
oliviu.matei@cunbm.utcluj.ro
[2] HOLISUN, Baia Mare, Romania
{rudolf.erdei,daniela.delinschi,laura.andreica}@holisun.com

**Abstract.** Far from being a hype, cybersecurity is one of the most important aspects of our daily lives. Gaining trust in a cyber system is not an easy task, as problems and security hazards can be willingly or unwillingly introduced in it. This paper describes a security-centric architecture of the communication bus in loosely coupled heterogeneous systems. The concept is modular and designed having in mind the fulfillment of the principles of Service Oriented Architecture as well as the common functional and non-functional requirements regarding such communications.

**Keywords:** Service oriented architecture · Cybersecurity · Loose coupling · Communication bus

## 1 Introduction

`Loosely Coupled Architecture` (LCA) [12] has gained a lot of traction lately, due to the emergence of improvements within implementations like `Service Oriented Architecture` (SOA) [28] and `Microservices` [20], but also out of need to reuse parts of the system. This `Divide and Conquer` approach has enabled developers to build flexible, optimized and highly performing systems, but, at the same time, are less secure, due to the need for message exchange between the system components (the `Services`), messages that can be intercepted and decoded, even altered, making the system unstable or even maleficent [21].

Thus, embedding security in the design of a software system becomes a primary objective. Communication security means designing the messaging channel in a resilient and secure way, building mechanisms that will close the channel to all unauthorized access, while also providing multiple layers of security [32]. These mechanisms are highly important in the context of the increase in malicious attacks that can overpower and bring down entire software systems.


J. J. Gude Prego et al. (Eds.): CISIS 2021/ICEUTE 2021, AISC 1400, pp. 214–223, 2022.
https://doi.org/10.1007/978-3-030-87872-6_21

An important concept of SOA, although not essential, is the `Enterprise Service Bus` (ESB) [25]. Regardless, the lack of an ESB severely limits the usefulness as well as the communication between the services. The main role of the ESB is to provide an abstraction layer between the services themselves as well as between services and clients. This means that clients can benefit from composed services that work seamlessly together. However, unlike the classic SOA, where the ESB facilitates communication but does not abstract or simplify it, in the approach proposed in this article, the message exchange needs to be taken to a higher level of abstraction [33].

SOA Services usually know beforehand the recipient for their messages, but, in order to upgrade the implementation of loose coupling, we propose an architecture apriori agnostic to the intimate implementation and characteristics of the services. The platform itself will act as a facilitator for communication (an orchestrator).

### 1.1 Related Work

A significant amount of research has been invested in conceptualizing, formalizing and implementing Loosely Coupled Systems.

**SOA and Loose Coupling.** Loose coupling in the context of web applications, is thoroughly discussed by Pautasso and Wilde [22]. They also propose some metrics for quantifying loose coupling. They argue over the degree of coupling as viewed from different perspectives (e.g. an API can be loose coupled as it is completely described, but tight coupled as any change to it can brake client applications).

But messaging is not the only option in the case of SOA. There are other options that can and need to be reviewed in order to design a fully working solutions, as suggested by Wei [30].

Other work that incorporates multi-layered approaches for loose-coupling architecture implementations are discussed in works by Matei et al. [16,17]

**Messaging System as a Communication Option in Loose Coupling.** Mosshammer et al. [19] propose a lightweight Simulation Message Bus (SMB) that is able to increase the speed of message exchange in a heterogeneous system (continuous and time-triggered subsystems). A large scale prototype for a smart distribution grid was also demonstrated.

Smart software agents are used in a Industry 4.0 concept by Vogel-Heuser et al. [29], that use an approach based on Cyber-Physical Production Systems (CPPS) in order to design a loosely coupled smart system for yogurt production. One of the components enables external users to input orders, so the system can respond as quickly as possible to unforeseeable dynamic changes. Many types of platforms are used, along with many programming languages (e.g., C++, C#), so this approach also demonstrates heterogeneous abilities. Some of these ideas are also discussed in detail by Hasan [11]. The way that a Cyber-Physical System interferes with automation architectures is discussed by Riedl et al. [23], while recovery from faults in Real-Time Service Oriented Architecture is handled by Alho and Mattila [2].

Fiege et al. [9] proposed the notion of scopes that both abstract and encapsulate the high level notion of events and messages between bundles of system components. The proposed system has been integrated in a prototype for a stock trading platform, with good results.

Rodrigues et al. [24] present an interesting method to incorporate Message Oriented Middleware (MOM) into the design of loosely coupled third party systems, in this case API's that work with Virtual Reality (VR). They design a whole system dedicated to designing VR, by using API based VR tools. The concept has been validated with a demo application.

**Security in Loose Coupling and SOA.** An overview of security threats in a SOA based system and possible ways to mitigate them, is presented by Tan et al. [26], while Epstein et al. [7] presents some 13 possible threats in a badly designed SOA based platforms. Mitigation techniques can also be implemented by analyzing the book by Buecker et al. [5].

Security approaches for loose coupling architecture is a subject discussed rather rarely. Available propositions involve a layered approach, in which all elements of the system are secured in their own way, always considering that the outer landscape of the component is already compromised (also a Divide and Conquer approach). In this regard, Yarygina and Bagge [31] propose a simple security framework for a loose coupling architecture that involves an 11% extra performance overhead.

## 2 Context of the Work

The resulting platform will serve a collection of tools that need near-realtime access to each-other, in order to do assessments and collaborate for realtime predictions.

Within our platform, events, states, data and other information are sent as `messages` throughout the framework and are in need of resilience as well as mechanisms that cope with the framework's complexity and security.

Framework components can be accessed synchronously or asynchronously [18] and, depending on their functionalities, some may need real-time or quasi-real-time data exchange with it.

Location and access constraints are enforced by the service providers, meaning that these services as well as their locations can frequently change (i.e., their IP address) or be restricted at various levels and in many ways. This is why, hard-coding these locations would be not only a bad design choice, but rather an impossible approach of building the platform. Service providers must be able to register their services, setup configuration options for them and, in some cases, even remove them from the platform. In this case, the platform itself becomes a middle layer, an orchestrator, that is in charge of managing communications, integrating results and presenting them to the user.

The process of registering and removing services from the platform can be a tedious and problematic task, as these operations should inform the other services about their availability or unavailability. In this case, communication would be very verbose and the platform itself would not have complete control over what it is possible with it. The user would not know what the platform is capable at a specific moment. These problems lead to the need of a centralized knowledge base, where all information about the services, their needs and statuses are kept up-to-date at all times. A `Heartbeat` [13] mechanism assures that all services are online and ready.

The centralized registration system also has an added advantage. Services do not have to locally store details of the platform and availability of other services, as the platform takes care of that. This limits the verbosity of the communication, while also adhering to the `DRY` principle of engineering (Don't Repeat Yourself) [3,6].

In this case, the orchestrator itself is responsible with the following tasks:

- keeping an updated database with the services and their settings;
- keeping an updated situation of the statuses of services (online/offline/disabled);
- keeping accurate records on the direction and flow of data between the services;

In order to properly route the messages between services, a `Publish-Subscribe` pattern [8] of communication is designed, that links services together only when they are needed, depending on the job being processed. The sequence of services that is necessary for a job is called a `Workflow`. Workflows define how the services communicate to each other, in order to properly exchange information and also contain the needed inputs as well as provided outputs for each service.

The publish-subscribe methodology defines the notion of `Communication Channel`, that is used to link the services. A single instance of a communication channel is used to forward data from one or more services, called producers or publishers, to one or more destinations, called consumers or subscribers. As the services can be, at the same time, both publishers and subscribers, more instances of communication channels are used, so that the flow will be correct.

It is imperative that this communication is done at the highest `throughput` possible, as any delays would translate into possible data loss or limited functionality.

These principles presented above, abide to the key principles of SOA [14]:

- **Standardized Service Contract** - the available services are registered in the Yellow Pages, where a complete description of them is available. Services can be executed as single or as part of a `Workflow`;
- **Loose Coupling** - services don't communicate directly and don't depend on each other in order to be able to produce output. As long as the needed information is provided at its input, an output can be produced;

- **Service Abstraction** - services are autonomous pieces of software, treated like black-boxes. They run either in Docker containers, either at the developer's location, completely transparent;
- **Service Reusability** - the resulting platform logic is split into services that can be mixed and composed in order to generate a Workflow. These Workflows offer the necessary logic for a use-case or scenario, hence maximising the re-usage of code;
- **Service Autonomy** - services have exclusive control over their execution. They only receive the needed inputs and job ID and provide their outputs, based on the information registered in the Yellow Pages.
- **Service Statelessness** - although services might keep their own history, in order to improve functionalities, they don't maintain a general state. Along with the provided inputs, they will also be provided with a job ID, that will be used as a means to differentiate jobs.
- **Service Discoverability** - all the available services can be discovered by a simple query to the Yellow Pages.
- **Service Composability** - complex platform services can be designed with the use of Workflows, that define the sequences as well as inputs and expected outputs.
- **Service Interoperability** - this principle requires the use of communication standards. The platform defines industry standard formats, like JSON/XML, as well as compression standards (like ZIP and TAR/GZ) that are to be used in order to properly communicate.

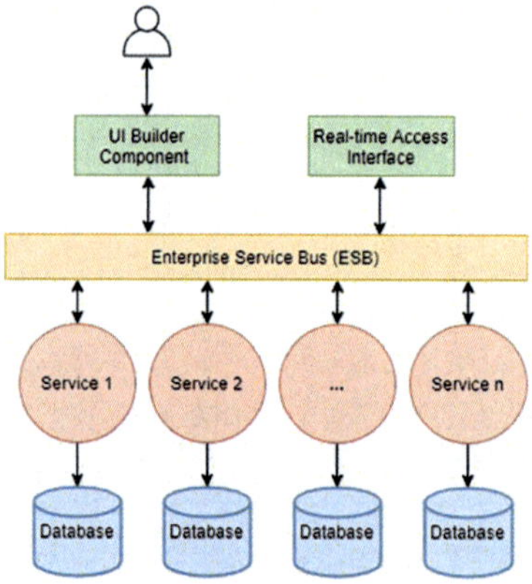


**Fig. 1.** Software architecture for SOA based systems

Implementation of these principles has led to a customized architecture implementation of SOA, that is depicted in Fig. 1. The architecture clearly states that an intermediate layer is needed. This layer is implemented by using a `Publish-Subscribe` approach and by paying extra attention to the security part, as stated before. The architecture of the communication channel, that is part of the intermediate layer, is discussed in Sect. 3.

The functional (FR) and non-functional requirements (NFR) are partially imposed by well-established requisites, as those defined by Akhunzada et al. [1], and partially elicited using the Walt-Disney Method, which is based on the interplay of three roles: the dreamer (visionary, idea supplier), the realist (maker) and the critic (quality manager) [27].

### 2.1 Functional Requirements for the Communication Channel

In order to be properly integrated in our approach, the Communication Channel must comply with the following functional requirements:

- **FR1. Real-time message transfer**: the channel must not introduce lags in the system;
- **FR2. Security certification**: the system must be able to certify that a service and message are valid and able to be forwarded by the channel;
- **FR3. Integration Flexibility**: the system must be able to adapt to different types of services and integration means;
- **FR4. Message Integrity**: integrity of the messages must be ensured by the system;
- **FR5. Channel Security**: the channel must be immune to outside threats.

### 2.2 Non-functional Requirements for the Communication Channel

- **NFR1. Confidentiality**: using different channels for targeted communication, as well as using password encrypted message bodies;
- **NFR2. Traceability**: the system must be able to trace the route of messages, on demand, and confirm the correct delivery to a specific service;
- **NFR3. Encapsulation**: messages must not depend on each other;
- **NFR4. Exchangeability/Interchangeability**: the communication channel should not be specific, it must be abstracted so channels are interchangeable;
- **NFR5. Canonicity**: messages and services must abide to the formats registered in the Yellow Pages [10];
- **NFR6. Atomicity**: each message is processed as a single unit of information;
- **NFR7. Throughput**: the usage of compression in order to speed up communication.

## 3 Architecture of the Communication Channel

The proposed architecture of the communication channel is depicted in Fig. 2.

The communication channel has been designed with both the functional requirements as well the non-functional ones in mind, as defined in Sects. 2.1 and 2.2. The channel consists of the following modules:

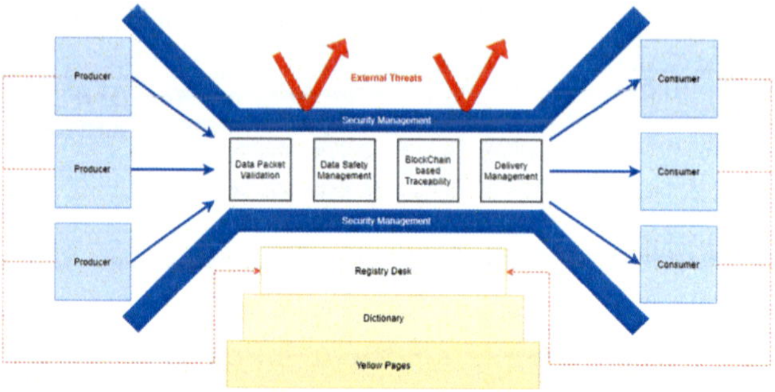


**Fig. 2.** Architecture of the communication channel

- `Security Management`: provides a security layer for preventing unauthorized access and other outside threats;
- `Data Packet Validation`: provides a message validation mechanism to ensure that all forwarded messages are not corrupted;
- `Data Safety Management`: provides a level of safety to ensure that only the intended messages pass and they do not contain harmful information and/or malicious code;
- `BlockChain Based Traceability`: provides a persistence layer that enables system logging and post-identification of potential problems;
- `Delivery Management`: provides a mechanism that ensures safe delivery of the messages to all the registered consumers, prior to removing them from the queue.

For identifying the producers and consumers, a `Yellow Pages` mechanism [4,10] is used, which needs to keep an updated list of the connected services, with all their options and parameters. On top of the `Yellow Pages` module, a `Dictionary` is built for quicker access to the available options and parameters as well as the exact message format that the specific channel forwards (extracted from the `Yellow Pages`). The `Registry Desk` is in charge with extracting only the connected producers and consumers from the `Yellow Pages`, also for quicker access.

## 4 Architecture Validation

The project is now in its infancy, so for validating the architecture of the communication channel, the Software Architecture Analysis Method (SAAM) [15] has been used. SAAM is a methodology which can determine how specific application quality attributes were achieved and how possible changes in the future will affect quality attributes based on hypothetical cases studies. Common quality attributes that can be utilized by this methodology include modifiability, robustness, portability, and extensibility.

An extra validation has been performed against functional and non-functional requirements, described in Sects. 2.1 and 2.2. This analysis shows in which way each requirement was met, and yielded the information discussed in Table 1.

**Table 1.** Addressing the Use-cases in the design of the communication channel

| Use-case | Successfully addressed |
|---|---|
| Communication Channel (CC) provides a secure mode of transferring information between services | Yes |
| CC provides secure access to each of the services | Yes |
| CC is resilient to outside threats, by blocking unauthorized access as well as malefic behaving services (i.e., services that try to perform unauthorized actions) | Yes |
| CC is resilient to DDOS attacks | Yes |
| CC provides persistence mechanisms, enabling the system to be analyzed | Yes |

## 5 Conclusions

The article presents an innovative, flexible and resilient architecture of a communication channel, that can be used for any communication approach, especially in loosely coupled and heterogeneous software architectures. It fulfills the principles of SOA formalized by Huhns and Singh [14] as well as the functional and non-functional requirements stated in Sect. 2.1, which apply to any loose heterogeneous software system.

The security of the communication, a cornerstone requirement, is assured on at least two levels: implementation of the communication channels, as well as at the message level.

The communication channels are secured by the resilience mechanism, that blocks outside threats and the continuous monitoring of already approved services, in order to `Just-In-Time` detect and ban maleficent behaviour. The message is secured by the use of security layers and message encryption. Also, a persistence layer will assure that all communication will be able to be analyzed in order to further improve the security of the resulting Platform.

**Acknowledgment.** This work was supported by the project BIECO (https://www.bieco.org/) that received funding from the European Union's Horizon 2020 research and innovation programme under grant agreement No. 952702.

## References

1. Akhunzada, A., Ahmed, E., Gani, A., Khan, M.K., Imran, M., Guizani, S.: Securing software defined networks: taxonomy, requirements, and open issues. IEEE Commun. Mag. **53**(4), 36–44 (2015)
2. Alho, P., Mattila, J.: Software fault detection and recovery in critical real-time systems: an approach based on loose coupling. Fusion Eng. Des. **89**(9–10), 2272–2277 (2014)
3. Baggen, R., Correia, J.P., Schill, K., Visser, J.: Standardized code quality benchmarking for improving software maintainability. Softw. Qual. J. **20**(2), 287–307 (2012)
4. Bettoni, M., Bernhard, W., Borter, F., Dönnges, G.: The yellow tool–making yellow pages more social and visible. In: Proceedings of the 8th European Conference on Knowledge Management (ECKM 2007). Academic Publishing, Reading Ma, pp. 118–124 (2007)
5. Buecker, A., et al.: Understanding SOA Security Design and Implementation. IBM Redbooks (2008)
6. Cunningham, H.C.: Multiparadigm programming with python 3 (2018)
7. Epstein, J., Matsumoto, S., McGraw, G.: Software security and soa: danger, will robinson! IEEE Secur. Priv. **4**(1), 80–83 (2006)
8. Fernandez, E.B., Yoshioka, N.: Two patterns for distributed systems: enterprise service bus (ESB) and distributed publish/subscribe. In: Proceedings of the 18th Conference on Pattern Languages of Programs, pp. 1–10 (2011)
9. Fiege, L., Mezini, M., Mühl, G., Buchmann, A.P.: Engineering event-based systems with scopes. In: Magnusson, B. (ed.) ECOOP 2002. LNCS, vol. 2374, pp. 309–333. Springer, Heidelberg (2002). https://doi.org/10.1007/3-540-47993-7_14
10. Gretsch, S., Mandl, H., Schätz, R.: Implementation process of a knowledge management initiative: yellow pages. New Res. Knowl. Manag. Models Methods **14**(1), 311–332 (2012)
11. Hasan, S.: Loose Coupling in Heterogeneous Event-Based Systems via Approximate Semantic Matching and Dynamic Enrichment. National University of Ireland, Galway (2016)
12. Hofmeister, C., Nord, R., Soni, D.: Applied Software Architecture. Addison-Wesley Professional, Boston (2000)
13. Hou, Z., Huang, Y., Zheng, S., Dong, X., Wang, B.: Design and implementation of heartbeat in multi-machine environment. In: 17th International Conference on Advanced Information Networking and Applications, 2003. AINA 2003, pp. 583–586. IEEE (2003)
14. Huhns, M.N., Singh, M.P.: Service-oriented computing: key concepts and principles. IEEE Internet Comput. **9**(1), 75–81 (2005)
15. Kazman, R., Bass, L., Abowd, G., Webb, M.: Saam: a method for analyzing the properties of software architectures. In: Proceedings of 16th International Conference on Software Engineering, pp. 81–90. IEEE (1994)
16. Matei, O., Anton, C., Bozga, A., Pop, P.: Multi-layered architecture for soil moisture prediction in agriculture 4.0. In: Proceedings of the International Conference on Computers and Industrial Engineering, CIE, vol. 2, pp. 39–48 (2017)
17. Matei, O., Anton, C., Scholze, S., Cenedese, C.: Multi-layered data mining architecture in the context of internet of things. In: 2017 IEEE 15th International Conference on Industrial Informatics (INDIN), pp. 1193–1198. IEEE (2017)

18. Molli, P., Skaf-Molli, H., Oster, G., Jourdain, S.: Sams: synchronous, asynchronous, multi-synchronous environments. In: The 7th International Conference on Computer Supported Cooperative Work in Design, pp. 80–84. IEEE (2002)
19. Mosshammer, R., Kupzog, F., Faschang, M., Stifter, M.: Loose coupling architecture for co-simulation of heterogeneous components. In: IECON 2013-39th Annual Conference of the IEEE Industrial Electronics Society, pp. 7570–7575. IEEE (2013)
20. Namiot, D., Sneps-Sneppe, M.: On micro-services architecture. Int. J. Open Inf. Technol. **2**(9), 24–27 (2014)
21. Papazoglou, M.P., Traverso, P., Dustdar, S., Leymann, F.: Service-oriented computing: state of the art and research challenges. Computer **40**(11), 38–45 (2007)
22. Pautasso, C., Wilde, E.: Why is the web loosely coupled? A multi-faceted metric for service design. In: Proceedings of the 18th International Conference on World Wide Web, pp. 911–920 (2009)
23. Riedl, M., Zipper, H., Meier, M., Diedrich, C.: Cyber-physical systems alter automation architectures. Annu. Rev. Control. **38**(1), 123–133 (2014)
24. Rodrigues, F., et al.: Coupling virtual reality open source software using message oriented middleware. In: SEARIS Worksphop 2009 (2009)
25. Schmidt, M.T., Hutchison, B., Lambros, P., Phippen, R.: The enterprise service bus: making service-oriented architecture real. IBM Syst. J. **44**(4), 781–797 (2005)
26. Tan, V., et al.: Security issues in a SOA-based provenance system. In: Moreau, L., Foster, I. (eds.) IPAW 2006. LNCS, vol. 4145, pp. 203–211. Springer, Heidelberg (2006). https://doi.org/10.1007/11890850_21
27. Tausch, S., Nußberger, F., Hußmann, H.: Supporting the disney method with an interactive feedback system. In: Proceedings of the 33rd Annual ACM Conference Extended Abstracts on Human Factors in Computing Systems, pp. 1013–1018 (2015)
28. Valipour, M.H., AmirZafari, B., Maleki, K.N., Daneshpour, N.: A brief survey of software architecture concepts and service oriented architecture. In: 2009 2nd IEEE International Conference on Computer Science and Information Technology, pp. 34–38. IEEE (2009)
29. Vogel-Heuser, B., Diedrich, C., Pantförder, D., Göhner, P.: Coupling heterogeneous production systems by a multi-agent based cyber-physical production system. In: 2014 12th IEEE International Conference on Industrial Informatics (INDIN), pp. 713–719. IEEE (2014)
30. Wei, X.: Communications solutions for heterogeneous systems based on SOA. Phys. Procedia **25**, 1738–1742 (2012)
31. Yarygina, T., Bagge, A.H.: Overcoming security challenges in microservice architectures. In: 2018 IEEE Symposium on Service-Oriented System Engineering (SOSE), pp. 11–20. IEEE (2018)
32. Yildiz, M., Abawajy, J., Ercan, T., Bernoth, A.: A layered security approach for cloud computing infrastructure. In: 2009 10th International Symposium on Pervasive Systems, Algorithms, and Networks, pp. 763–767. IEEE (2009)
33. Zhang, H., Kishore, R., Sharman, R., Ramesh, R.: Agile integration modeling language (AIML): a conceptual modeling grammar for agile integrative business information systems. Decis. Support Syst. **44**(1), 266–284 (2007)

# A Comparative Study of the Most Important Methods for Forecasting the ICT Systems Vulnerabilities

O. Cosma, M. Macelaru, P. C. Pop(✉), C. Sabo, and I. Zelina

Technical University of Cluj-Napoca, North University Center of Baia Mare, Dr. V. Babes 62A, 430083 Cluj-Napoca, Romania
{ovidiu.cosma,mara.hajdu,petrica.pop,cosmin.sabo, ioana.zelina}@mi.utcluj.ro

**Abstract.** Nowadays, companies are facing plenty of IT secure attacks and to guarantee safe, untroubled, and continuous functioning of their business, they should detect and forecast the volume of IT security vulnerabilities and be prepared for future threats. The aim of this paper is to present a comparative study of the most important and promising methods for forecasting the ICT systems vulnerabilities.

**Keywords:** Security vulnerabilities · Forecasting · Time series forecasting · Neural networks

## 1 Introduction

Due to the exponential growing trend in ICT system vulnerabilities and cyber threats, security became more and more important. Therefore, the companies have become preoccupated with the prediction, forecasting and propagation of the vulnerabilities.

A computer system vulnerability can be defined as a weakness within the system or network that might be taken advantage of, to generate damage or to permit attackers to exploit the network in some way. The vulnerabilities may appear because of unexpected intercommunications between distinct software programs, network components or weakness of an individual program. Vulnerabilities exist in every network, and it is impossible to find and to focus on all of them due to the highly complex structure of modern network architecture.

Even though there are different approaches to cover the process of software vulnerability analysis and discovery, most of them are approximate solutions with lack of soundness or completeness, or even both [27]. In this sense the previous research lines focus on providing an improvement on some specific aspects of the process, as the precision, efficiency, or vulnerability coverage.

In this paper we will focus on the forecasting the ITC systems vulnerabilities. The existing methods from the literature for forecasting ITC systems vulnerabilities can be classified in three categories:


J. J. Gude Prego et al. (Eds.): CISIS 2021/ICEUTE 2021, AISC 1400, pp. 224–233, 2022.
https://doi.org/10.1007/978-3-030-87872-6_22

1. Time series analysis-based models: Autoregressive integrating moving average (ARIMA), Exponential Smoothing, etc.
2. Artificial intelligence-based models: Neural Networks, Support Vector Machine (SVM), Bayesian Network, k-Nearest Neighbor, etc.
3. Statistical based models: Regression techniques, Linear Regression, Logistic Regression, Random Forest, Naïve Bayes, Decision Tree, Least Mean Square, Reliability Growth models, Statistical Code analysis, etc.

For a comprehensive survey on forecasting ITC systems vulnerabilities, we refer to Yasasin et al. [6], Roumani et al. [12].

The aim of our paper is to present a comparative study of the most important and promising methods for forecasting ITC systems vulnerabilities. We describe six methods belonging to the first two categories of models.

The remaining of our paper is organized as follows: in Sect. 2 we describe in detail the considered methods for forecasting ITC systems vulnerabilities and in Sect. 3 we present our conclusions and as well some future research directions.

## 2 Methods of Forecasting the ICT Systems Vulnerabilities

### 2.1 Time Series Forecasting Based Models

A *time series* is a series of data points indexed in time order and are usually plotted using run charts. Time series have several applications such as: economic forecasting, stock market analysis, inventory studies, weather forecasting, earthquake predictions, etc. *Time series analysis* contains techniques for examining time series data in order to obtain relevant statistics and other features from the data. *Time series forecasting* makes use of a model to forecast future values based on previously noticed values. For more information on this topic, we refer to Hyndman and Athanosoupulos [13].

**Autoregressive Integrating Moving Average (ARIMA) Models**

ARIMA models are advanced statistical models and are considered to be the most general models for forecasting time series that can be made stationary. ARIMA models give descriptions of the autocorrelations in the data [13] and not description of trend and seasonality as exponential smoothing methodologies do.

A stationary time series is a time series whose statistical properties do not rely on the time at which the series is noticed, they remain constant over time. Time series with trends or seasonality are not stationary. If nonstationary data is used, the estimators don't have the asymptotic normality and consistency properties, for the time series models. When an ARIMA model is constructed, the first step is to decide if the time series is stationary. If the series to be studied is not stationary it is transformed in order to become stationary.

The ARIMA models include autoregressive terms (AR), moving average terms (MA) and differencing operations (Integrated). Autoregression means regression of the variable against itself. In an autoregression model, the value of the variable is forecast using a linear combination of its past values. An autoregressive model of order $p$, AR($p$), for a time series $y$, where $y_t$ is the value of the variable at moment $t$, is described as

$y_t = c + \varphi_1 y_{t-1} + \varphi_2 y_{t-2} + \cdots + \varphi_p y_{t-p} + \varepsilon_t$, where $\varepsilon_t$ is white noise. The parameters $\varphi_1, \varphi_2, \ldots, \varphi_p$ are the autocorrelation coefficients and they define the time series patterns. For $p = 1$ the constraint for parameter $\varphi_1$ is $-1 < \varphi_1 < 1$ and for $p = 2$ the constraints for the parameters are $1 < \varphi_2 < 1, \varphi_1 + \varphi_2 < 1, \varphi_2 - \varphi_1 < 1$.

The moving average model MA of order $q$, MA($q$), uses past forecast errors instead of using previously observed values of the forecast variable, $y_t = c + \varepsilon_t + \theta_1 \varepsilon_{t-1} + \theta_2 \varepsilon_{t-2} + \cdots + \theta_q \varepsilon_{t-q}$.The parameters $\theta_1, \theta_2, \ldots, \theta_q$ define the time series patterns. The constraints for these parameters are $-1 < \theta_1 < 1$ for $p = 1$ and $-1 < \theta_2 < 1, \theta_1 + \theta_2 > -1, \theta_1 - \theta_2 < 1$ for $p = 2$.

If the time series is non-stationary, differencing is a way to make it stationary. Differencing means computing the differences between consecutive observations in the non-stationary series and the differenced series terms $y_t^{'} = y_t - y_{t-1}$ describe the change between two consecutive terms in the original series.

If the differenced data is not stationary, then we can repeat the differentiation. The data can be differenced again to obtain a stationary series: $y_t'' = y_t' - y_t' - 1 = y_t - 2t_{t-1} + y_{t-2}$ The process of differentiation can be repeated until the series become stationary. Usually, first or second differentiation is enough to build good models.

A general ARIMA ($p$, $d$, $q$) model is obtained combining differencing with autoregression and moving average model, and $p$ is the number of autoregressive terms, $d$ is the number of differences necessary for stationarity and $q$ is the number of lagged forecast errors [13]: $y'_t = c + \varphi_1 y'_{t-1} + \cdots + \varphi_p y'_{t-p} + \theta_1 \varepsilon_{t-1} + \theta_2 \varepsilon_{t-2} + \cdots + \theta_q \varepsilon_{t-q} + \varepsilon_t$.

**Exponential Smoothing**

Exponential smoothing is a popular forecasting method that aims to produce a smooth Time Series. The method was proposed by Brown [14], Holt [15] and Winter [16]. Its main feature is that it assigns higher weight in forecasting to recent observations than to older observations. Unlike the ARIMA models, exponential smoothing model do not require the time series to be stationary.

Single exponential smoothing (SES) is the simplest of the exponential smoothing methods and is applied to data series with no trend or seasonality. The equation in this case is $y_t = \alpha y_{t-1} + (1 - \alpha) y_t - 2$, where $0 \le \alpha \le 1$ is the smoothing parameter.

Double exponential smoothing (DES) proposed by Holt [15] is an extension of SES and is used to forecast data series with trend. The smoothing equations in this case are $L_t = \alpha y_t + (1 - \alpha)(L_{t-1} + b_{t-1})$, $b_t = \gamma(L_t - L_{t-1}) + (1 - \gamma) b_{t-1}$, where $\alpha$, $\gamma$ are the smoothing parameters for the level and for the trend, $0 \le \alpha, \gamma \le 1$, $L_t$ is the level of the series at time $t$ and $b_t$ is the trend of the series at time $t$. The forecast equation is $y_{t+1} = L_t + b_t$, meaning that the 1 step ahead forecast is the sum between the estimated level at time $t$ and the trend value at time $t$.

To also capture the seasonality, Holt and Winters [1] proposed a model with smoothing equations for level, for trend and for seasonality. We present next the Holt-Winters additive exponential smoothing model that was proposed by Roumani et al. [12] for vulnerability analysis and forecasting. The equations that describe the Holt-Winters additive exponential smoothing model are:

$$L_t = \alpha(Y_t - S_{t-s}) + (1 - \alpha)(L_{t-1} - b_{t-1}) \tag{1}$$

$$b_t = \gamma(L_t - L_{t-1}) + (1 - \gamma)b_{t-1} \tag{2}$$

$$S_t = \delta(Y_t - L_t) + (1 - \delta)S_{t-s} \tag{3}$$

$$F_{t+m} = L_t + mb_t + S_{t+m-s} \tag{4}$$

where $\alpha$, $\gamma$ and $\delta$ are the smoothing parameters, $Y_t$ represents the number of vulnerabilities at time $t$ calculated in months, $m$ represents the number of future periods to predict (12 months), $s$ represents the length of the seasonality (12 months), $L_t$ represents the level of the series at time $t$, $b_t$ represents the tendency of the series at time $t$ and $S_t$ is the seasonal component at the time $t$.

The first equation (level) describes the relative magnitude of the number of vulnerabilities, the second one (trend) describes the gradual upward or downward long-term movement of the number of vulnerabilities, the third one (seasonality) describes the short-term regular variations of the number of vulnerabilities at regular intervals and finally the last equation describes the vulnerability for a given period $m$.

## 2.2 Artificial Intelligence-Based Models

The foundations of artificial neural networks (ANNs) were established in 1943 by Warren McCulloch and Walter Pitts. They described in [1] how neurons might work. The back propagation algorithm (BP) for multilayer ANNs was proposed by Rummelhart et al. in 1986 [2].

ANNs are inspired by the operation of the human brain. They are made up of a set of interconnected components in several layers, called neurons or perceptrons. The inputs of each neuron are connected to the outputs of neurons in the previous layer. The neurons in the first layer get the inputs of the model, and the outputs of the neurons in the final layer provide the result. The structure of a neuron is presented in Fig. 1.a, where $x_i, i = 1, n$ are the inputs, $w_i, i = 1, n$ are the weights associated to the inputs, $b$ is the bias, $f$ is the activation function and $y$ is the output. The neuron operation is defined by relation (5). It calculates a weighted sum of the inputs and bias, and then it passes it through the activation function to determine the output. The most widely used activation functions are *Identity* (6), *Sigmoid* (7), *ReLu* (8), *Tanh* (9) and *Softplus* (10).

$$y = f\left(b + \sum_{i=1}^{n} w_i x_i\right) \tag{5}$$

$$f_1(x) = x \tag{6}$$

$$f_2(x) = \frac{1}{1 + e^{-x}} \tag{7}$$

$$f_3(x) = \max(0, x) \tag{8}$$

$$f_4(x) = \frac{e^x - e^{-x}}{e^x + e^{-x}} \tag{9}$$

$$f_5(x) = \ln(1 + e^x) \tag{10}$$

**Multilayer Perceptron Models**

A Multilayer Perceptron (MLP) model with an input layer, a hidden layer, and an output layer is shown in Fig. 1.b. The outputs of the model are given by relations (11) and (12), where $f$ is the activation function of the output layer neurons and $g$ is the activation function of the hidden layer neurons. A more compact representation of this model is given in Fig. 2.a, where $x$ and $y$ represent the input and the output vectors, $W$ and $V$ are the hidden and the output layers weights matrices, $b^h$ and $b^o$ are the hidden and the output layers bias vectors, $h$ is the hidden layer output vector, and $g$ and $f$ are the activation functions of the hidden and the output layers. The output is given by relations (13) and (14), that are compact representations of (11) and (12). The model can be extended by adding more hidden layers. Training such a model involves adjusting the weights and biases in such a way as to provide an output as close as possible to the correct value for each input. The most widely used methods of training ANNs are based on the Gradient Descent – Back Propagation (GD-BP) algorithms [3] or genetic algorithms.

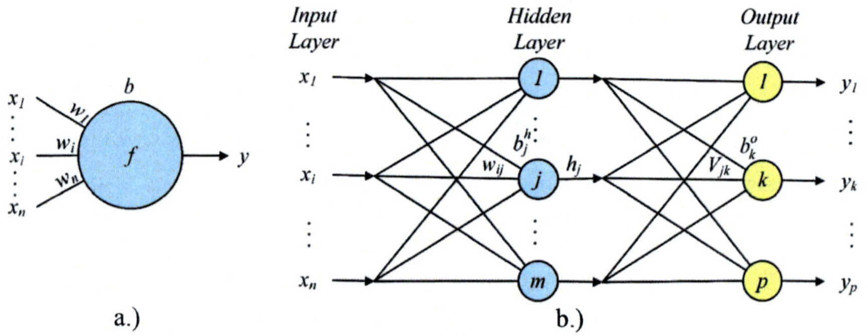


**Fig. 1.** a) Perceptron structure, b) MLP model with one hidden layer.

$$h_j = g\left(b_j^h + \sum\nolimits_{i=1}^{n} W_{ij}x_i\right), j = 1, \ldots, m \tag{11}$$

$$y_k = f\left(b_k^o + \sum\nolimits_{j=1}^{m} V_{jk}h_j\right), k = 1, \ldots p \tag{12}$$

$$h = g(b^h + xW) \tag{13}$$

$$y = f(b^o + hV) \tag{14}$$

**Recurrent Neural Networks**

Recurrent Neural Nets RNNs were designed to process sequential data. They can memorize previous states and use them to determine the current state. The structure of a RNN is shown in Fig. 2.b. Its state is composed by the hidden layer output vector $h$. The state vector $h$ calculated at step $t-1$ is processed as an entry at step $t$. The total number of trainable parameters does not depend on the number of steps, but only on the dimensions of the layers. Thus, for the network structure in Fig. 2.b, the total number of parameters is $(n+m+1)m+(m+1)p$, where $n$ is the size of the input vector $x$, $m$ is the size of the status vector $h$, and $p$ is the size of the output vector $y$. The outputs at step $t$ are given by relations (15) and (16). They depend not only on the entries at step $t$, but on their entire evolution, starting from the initial step.

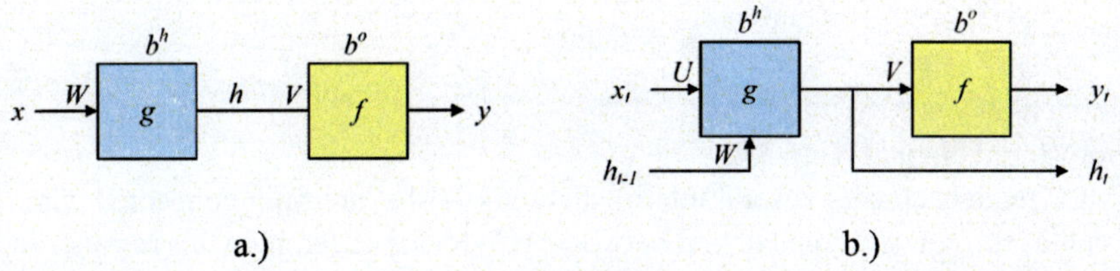


**Fig. 2.** a) Compact representation of the MLP model with one hidden layer, b) RNN structure.

$$h_t = g(b^h + x_t\mathrm{U} + h_{t-1}\mathrm{W}) \tag{15}$$

$$y_t = f(b^o + h_t\mathrm{V}) \tag{16}$$

**Long Short-Term Memory Models**

The main disadvantage of RNNs lies in the fact that they cannot learn long sequences, because of the vanishing gradient problem [3]. This problem occurs in deep networks with many hidden layers. RNNs do not have long-term memory. The Long Short-Term Memory (LSTM) models have been projected to solve this problem. The structure of a LSTM cell is presented in Fig. 3, where $f$ is the forget gate, $i$ is the input gate, $g$ is the input updater, $O$ is the output gate, $C_t$ is the current cell state and $h$ is the hidden state. Its operation is defined by relations (17)–(22).

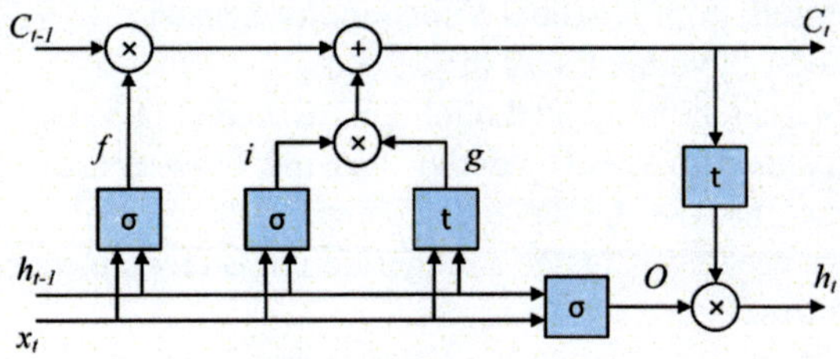


**Fig. 3.** LSTM cell structure.

$$f = \sigma(x_t U^f + h_{t-1} W^f) \tag{17}$$

$$i = \sigma(x_t U^i + h_{t-1} W^i) \tag{18}$$

$$g = \tanh(x_t U^g + h_{t-1} W^g) \tag{19}$$

$$C_t = C_{t-1} \circ f + g \circ i \tag{20}$$

$$O = \sigma(x_t U^o + h_{t-1} W^o) \tag{21}$$

$$h_t = \tanh(C_t) \circ O \tag{22}$$

**Convolutional Neural Networks**

The MLP models contain several intermediate layers that are fully connected with the adjacent layers. convolutional neural networks (CNN) are different in this respect. After the input layer follow several layers of different types [3]. The first type of layer is the convolution layer, that applies convolution operations, using a set of small filters, that contain weights trained to capture different features. The output of each filter is an activation map. The output of a convolution layer is composed of all the activation maps created by its filters. A convolution layer is sparsely connected with the previous layer.

Another type of layer in a CNN is the pooling layer, also sparsely connected with the previous layer. It has the purpose to downsample data, based on different strategies, such as: max poling designed to catch the peaks, or average pooling.

The third type of layer in a CNN is the dense layer, that is fully connected with the previous layers. This type of layer is similar with the hidden layers of the MLP model. The dense layers are usually placed at the end of the CNN model, before the output layer.

The common way of training a CNN is the GD-BP algorithm.

## 2.3 ICT Vulnerabilities Forecasting

Roumani et al. [12] used time series analysis to build predictive models for five well-known browsers: Chrome, Firefox, Internet Explorer, Safari and Opera, on a collection of vulnerability datasets from National Vulnerability Database (NVD) and concluded that ARIMA is the best fit vulnerability model.

Yasasin et al. [6] used different methodologies to predict IT vulnerabilities of different system and software packages (operating systems, browsers, and office solutions). The conclusion is that the ARIMA method achieved low forecasting errors for all types of software that have been investigated and therefore it is recommended for forecasting software vulnerabilities.

An ANN-based time series forecasting model uses observations at previous times as inputs, and the outputs of the model represent the forecast values. Neural networks have multiple advantages over classic models. They can produce quality forecasts, even if they work with the original data. There is no need to eliminate trend and seasonality. The model can be easily generalized to perform multivariate forecasts, by increasing the number of inputs. ANN can implicitly detect complex nonlinear relationships between dependent and independent variables.

The main disadvantage of ANN lies in the fact that their training is a complex process that requires a lot of processing power, and they can be overfitted.

Feed Forward-ANNs have been successfully used both to forecast software systems vulnerabilities and in other forecasting applications. The experimental data needed to train the vulnerability forecasting models are taken from the National Vulnerability Database (NVD) [11] or other public sources.

Most of the proposed models contain a single hidden layer of neurons [4–7], but there are also variants with several intermediate layers [8]. There are several proposals that present optimization strategies to find the best dimension of the input and intermediary layers [4, 5].

The majority of the proposed models are trained by the Gradient Descent and the Back Propagation (GD-BP) algorithms [5, 7, 9], but there are also some proposals of evolutionary computation training [8, 9] and hybrid training algorithms that use an adaptive differential evolution algorithm [10] or a particle swarm optimization algorithm [9] in combination with GD-BP. These proposals speculate that evolutionary computational techniques are making significant progress in the first part of the optimization process and manage better to avoid local minima than GD-BP. When evolution begins to stagnate, they switch to GD-BP, which reaches the optimal solution faster.

Most of the proposed models use the sigmoid activation function [6, 7] or the tanh activation function [5] in the hidden layer, but there are other proposals, such as the sinusoidal function [8]. Almost all proposed models use the identity activation function in the output layer, but there are also solutions based on the modified sigmoid activation function [7].

The proposed forecasting models are usually compared with other known techniques (Exponential smoothing, Croston's methodology, ARIMA, Support Vector Machines, Vulnerability Discovery Models) in terms of forecast accuracy and forecast bias. Most of the models trained by GD-BP are outperformed by the models they are compared with [4, 6], but there are also exceptions [5]. The models trained with evolutionary computation or hybrid algorithms outperform the other models [8–10].

The best forecast models proposed recently have LSTM cells in their composition and are trained by BP-GD. A comparative analysis of several types of deep neural networks is presented in [23], and it is concluded that the Convolutional Neural Networks (CNN), MLP, RNN and LSTM models perform well for one step forecasting and less satisfactory for multiple steps forecasting. Another comparative analysis of different forecasting models is presented in [25]. The bottom line is that LSTM models give the best forecasts, but CNN models are the most robust to changes in configuration parameters. Various forecast models are compared with LSTM models in [18, 21 and 22]. In general LSTM offers the best performance, except for the [18] work, in which CNN and shallow ANN models are found to be better. Different hybrid CNN – LSTM models are proposed in [17, 19, 20, 24, 26]. They outperform the other models they are compared with (MLP, CNN, RNN, LSTM, ARIMA) in all works, even though some authors report longer training times.

## 3 Conclusions

In this article we have presented some of the forecasting techniques that have been proven to be effective in the case of software vulnerabilities. In order to have an overview of the latest results in the field, we have not strictly limited our study to software vulnerabilities, but we have also considered other applications for forecasting natural phenomena. It is difficult to draw a conclusion because the analyzed works used different data sets. An objective conclusion could be drawn only if all models were tested under the same conditions, on the same data sets. Most authors have implemented several different models to compare their effectiveness, or to compare their proposed methods with the existing ones. However, it is known that the performance of a model depends on the parameters of the model, how it is optimized, how the experimental data is preprocessed, how the training is done, etc.

Overcoming all these uncertainties, based on the results analyzed we can conclude that the latest CNN – LTSM hybrid forecasting models seem to be the most accurate, but they require more complex training. However, the performance of these models must also be proven in the case of software vulnerabilities forecasting. There are several studies that recommend the classic Croston and ARIMA models, as they consistently obtain quality predictions. Several studies show that superior performance can be achieved by changing the GD-BP with other algorithms, such as evolutionary, hybrid, or other types, for training the ANN models. Another conclusion is that CNN, MLP, RNN and LSTM models perform well for one step forecasting and less satisfactory for multiple steps forecasting.

Starting from this study, we aim to test the efficiency of hybrid ANN models for forecasting software vulnerabilities, and to use other training techniques, different from GD – BP, that have proven their effectiveness for the FF-ANN models.

**Acknowledgment.** This work was supported by the project BIECO (www.bieco.org) that received funding from the European Union's Horizon 2020 research and innovation programme under grant agreement No. 952702.

## References

1. McCulloch, W.S., Pitts, W.: A logical calculus of the ideas immanent in nervous activity. Bull. Math. Biophys. **5**, 115–133 (1943)
2. Rumelhart, D., Hinton, G., Williams, R.: Learning representations by back-propagating errors. Nature **323**, 533–536 (1986)
3. Konasani, V.R., Kadre, S.: Machine Learning and Deep Learning Using Python and Tensor Flow. McGraw Hill, New York (2021)
4. Pokhrel, N.R., Rodrigo, H., Tsokos, C.P.: Cybersecurity: time series predictive modeling of vulnerabilities of desktop operating system using linear and non-linear approach. J. Inf. Secur. **8**, 362–382 (2017)
5. Movahedi, Y., Cukier, M., Gashi, I.: Vulnerability prediction capability: A comparison between vulnerability discovery models and neural network models. Comput. Secur. **87**, 101596 (2019)

6. Yasasin, E., Prester, J., Wagner, G., Schryen, G.: Forecasting IT security vulnerabilities –an empirical analysis. Comput. Secur. **88**, 101610 (2020)
7. Bennett, C., Stewart, R.A., Beal, C.D.: ANN-based residential water end-use demand forecasting model. Expert Syst. Appl. **40**, 1014–1023 (2013)
8. Chang, P.-C., Wang, D., Zhou, C.: A novel model by evolving partially connected neural for stock price trend forecasting. Expert Syst. App. **39**, 611–620 (2012)
9. Zhang, J.-R., Zhang, J., Lok, T.-M., Lyu, M.R.: A hybrid particle swarm optimization–back-propagation algorithm for feedforward neural network training. Appl. Math. Comput. **185**, 1026–1037 (2007)
10. Wang, L., Zeng, Y., Chen, T.: Back propagation neural network with adaptive differential evolution algorithm for time series forecasting. Expert Syst. Appl. **42**, 855–863 (2015)
11. National Vulnerability Database. https://nvd.nist.gov/. Accessed 24 Apr 2021
12. Roumani, Y., Nwankpa, J.K., Roumani, Y.F.: Time series modeling of vulnerabilities. Comput. Secur. **51**, 32–40 (2015)
13. Hyndman, R., Athanosoupulos, G.: Forecasting: Principles & Practice, 2nd edn. OTexts, Melbourne, Australia (2021)
14. Brown, R.G.: Statistical Forecasting for Inventory Control. McGraw/Hill, New York (1959)
15. Holt, C.E.: Forecasting Seasonal and Trends by Exponentially Weighted Averages (O.N.R. Memorandum No. 52). Carnegie Institute of Technology, Pittsburgh (1957)
16. Winters, P.R.: Forecasting sales by exponentially weighted moving averages. Manage. Sci. **6**(3), 324–342 (1960)
17. Lu, W., Li, J., Li, Y., Sun, A., Wang, J.: A CNN-LSTM-based model to forecast stock prices. Complexity **2020**, 1–10 (2020)
18. Wunsch, A., Liesch, T., Broda, S.: Groundwater level forecasting with artificial neural networks: a comparison of long short-term memory (LSTM), convolutional neural networks (CNNs), and non-linear autoregressive networks with exogenous input (NARX). Hydrol. Earth Syst. Sci. **25**(3), 1671–1687 (2021)
19. Shipman, R., et al.: We got the power: Predicting available capacity for vehicle-to-grid services using a deep recurrent neural network. Energy **221**, 119813 (2021)
20. Ma, M., Mao, Z.: Deep-convolution-based LSTM network for remaining useful life prediction. IEEE Trans. Industr. Inf. **17**(3), 1658–1667 (2021)
21. Hitimana, E., Bajpai, G., Musabe, R., Sibomana, L., Kayalvizhi, J.: Implementation of IoT framework with data analysis using deep learning methods for occupancy prediction in a building. Future Internet **13**, 67 (2021)
22. Budiharto, W.: Data science approach to stock prices forecasting in Indonesia during Covid-19 using Long Short-Term Memory (LSTM). J. Big Data **8**(1), 1–9 (2021). https://doi.org/10.1186/s40537-021-00430-0
23. Kaushik, R., Jain, S., Jain, S., Dash, T.: Performance evaluation of deep neural networks for forecasting time-series with multiple structural breaks and high volatility. CAAI Trans. Intell. Technol. **6**, 1–16 (2021)
24. Somu, N., Raman, G.M.R., Ramamritham, K.: A deep learning framework for building energy consumption forecast. Renew. Sustain. Energy. Rev. **137**, 110591 (2021)
25. Lara-Benitezy, P., Carranza-Garcia, M., Riquelme, J.C.: An experimental review on deep learning architectures for time series forecasting. Int. J. Neural Syst. **31**(3), 2130001 (2021)
26. Ioannis, E., Kiriakidou, N., Stavroyiannis, S., Pintelas, P.: An advanced CNN-LSTM model for cryptocurrency forecasting. Electronics **10**, 287 ( 2021)
27. Jhala, R., Majumdar, R.: Software model checking. ACM Comput. Surv. **21**, 1–57 (2009)

# Comparison of Safety and Security Analysis Techniques

Emilia Cioroaica[1](✉), Smruti Ranjan Kar[2], and Ioannis Sorokos[1]

[1] Fraunhofer IESE, Kaiserslautern, Germany
{emilia.cioroaica,ioannis.sorokos}@iese.fraunhofer.de
[2] Magna Electronics Europe GmbH & Co. GmbH, Kurfuerst-Eppstein-Ring 9, Sailauf, Germany
SmrutiRanjan.Kar@magna.com

**Abstract.** Growing adoption of new technological advancements within the automotive domain is highlighting multiple safety and security concerns. Aiming at reducing the errors humans make on the roads, deployment of novel and intelligent technological solutions are likely to introduce multiple vulnerabilities that can be exploited by attackers. In automated driving scenarios, in particular, vehicles communicating with each other using ad-hoc networks are becoming vulnerable to specialized cyber-physical attacks. A single compromised vehicle will provide an attack entry point for all linked vehicles, putting lives of passengers at risk. In this paper, we present an overview and evaluation of safety and security analysis methods applied on a use case from the automotive domain.

**Keywords:** Safety · Security · Automotive · Analysis · Techniques

## 1 Introduction

Vehicles continue to transition into real cyber-physical systems. We can already connect vehicles to the internet and access their data with smartphones. In the future, vehicles will likely have even more options to connect: neighbouring vehicles, roadside infrastructure, and even smart homes could be potentially supported. In this way, automotive systems become part of a larger ecosystem which aggregates not only other systems, but human actors as well. Actors within automotive digital ecosystems, such as human drivers, OEMs (Original Equipment Manufacturers) and Tier-1/Tier-n companies are becoming increasingly concerned with the security vulnerabilities likely to threaten the safe operation of vehicles, resulting in general distrust within an ecosystem. Distrust in technology suppliers can affect the health of an ecosystem and negatively impact the business within. The ecosystem health is a measure of how well the business within an ecosystem is performing [19].

For addressing the emerging transition in the automotive domain, AUTOSAR [1] made an evaluation which showed that emerging requirements


J. J. Gude Prego et al. (Eds.): CISIS 2021/ICEUTE 2021, AISC 1400, pp. 234–242, 2022.
https://doi.org/10.1007/978-3-030-87872-6_23

cannot be realized by typical software architectures. Instead, they proposed an architecture for vehicles that is more flexible, highly available and capable to adapt over time to specific application requirements. In this way, the POSIX-based AUTOSAR platform enables a segregated delivery and execution of control functions on ECUs (Embedded Control Units) in order to facilitate the rapid development of safe autonomous driving. The flexibility of this emerging vehicle infrastructure supports the business growth within automotive smart ecosystems, by enabling vehicles to communicate and collaborate dynamically during operation. At the same time, this evolution brings additional safety and security concerns due to the communication channels being introduced by the comparably more open environment.

In this paper we are summarizing evaluation results of the state-of-the-art on safety and security analysis methods within automotive domains, focusing on collaborative vehicles. Modern vehicles are complex systems, considered to be a network of ECUs interacting with each other, and expected to accommodate runtime updates or adaptation of their functionality. ECUs within a vehicle may be connected to multiple internal networks where they perform different functions. For instance, some ECUs may serve as gateways that transform and share information between different types of networks. Architecture flaws can expose intra-vehicle information to inter-vehicle communication without an information barrier, resulting in vulnerabilities. Therefore, from a security attack perspective, such ECUs represent an effective target. Moreover, the safety-critical nature of a vehicle enables us to reason of the safety implications of malicious attacks. Last but not least, even though addressing safety considerations in the automotive domain is well-established, security considerations have also grown more prominent over the last decade.

In what follows, Sect. 2 provides a background on relevant regulatory approaches for developing automotive systems, Sect. 3 presents an overview of safety and security analysis from the current state-of-the-art, Sect. 4 provides a summary of comparison between the methods and Sect. 5 presents conclusions and future work.

## 2 Background

Following the 2011 ISO 26262 recommendation of "reasonably foreseeable misuse" as a factor risk analysis [12], further elaborations of misuse with malicious attacks have emerged, with recommendations for security practices being published 9 years later [5]. Due to the fact that (a) at the moment when security practices started to emerge, safety was well established in the automotive domain and (b) security incidents can compromise safety, first security approaches have proposed a joint addressing of safety and security at all stages of system development, initially in a component-oriented fashion [18]. As reported in [17], **safety** is concerned with *preventing accidents* through identification of potential weaknesses, events, internal hazards and potentially hazardous states, followed by identification and implementation of appropriate mitigation mechanisms to reduce the risk to a tolerable level. On the other hand, **security** is concerned with *protecting assets* against internal or external threats and vulnerabilities that can compromise them. A mechanism for protecting assets is through the use of control strategies that reduce the risk of compromising the functionality to an acceptable level.

Driven by the first approach to use a component-level perspective for combined hazard and threat analysis as a safety & security mechanism performed during concept design, popular safety analysis methods were further extended to include security analysis as well. For example, FMEA (Failure Modes and Effects Analysis) [14] was extended towards FMVEA (Failure Mode, Vulnerabilities and Effects Analysis) [16], HARA (Hazard Analysis and Risk Assessment) [20] was extended towards SAHARA (Security-Aware Hazard and Risk Analysis) [9], STPA (System Theoretic process Analysis) [4] towards STPA-Sec (Systems-Theoretic Process Analysis for Security) [21].

## 3 State-of-the-Art

In this section we provide an overview of the state-of-the-art safety and security analysis methods, presenting them in a chronological order. Our work provides a complementary analysis to the threat analysis and risk assessment methods presented in [7] by looking at the characteristics of methods that perform a combines safety & security analysis from different viewpoints as are presented in Sect. 4.

### 3.1 CHASSIS

Combined Harm Assessment of Safety and Security for Information Systems (CHASSIS) [15] was introduced in 2012 for supporting the derivation of safety and security requirements of a system. For an already developed system, design artefacts can be reused for CHASSIS as well, otherwise, if the system is in the design phase, safety and security analysis can be performed in parallel. During CHASSIS, functional requirements artefacts, such as use cases and sequence diagrams, are analyzed with HAZOP (Hazard and Operability studies) keywords. The analysis results in a set of safety and security requirements documented in UML (Unified Modelling Language). During elicitation of functional requirements, D-UCs (Diagramatic Use Cases) and T-UCs (Textual Use Cases) are used for describing users, functions and services, whereas UML sequence diagrams are used to model sequences of interactions between objects. During the elicitation of safety and security requirements, potential misuses of the system are identified in brainstorming sessions that involve domain experts as well as safety and security experts. Misuses are identified through a combination of use cases and HAZOP guide words, documented in D-MUC (Diagrammatic Misuse Cases), T-MUC (Textual Misuse Cases) and MUSD (Misuse sequence Diagrams). Failures are documented in FSD (Failure-Sequence Diagrams).

### 3.2 FMEVA

Failure Mode, Vulnerability and Effect Analysis Method [16] was introduced in 2014 as a security-oriented extension of the FMEA (Failure Mode and Effect Analysis) with the scope of unifying safety and security cause-effect analysis. Extending the analysis of a given *failure mode*, which describes how a system quality attribute fails, FMVEA proposes a *threat mode*, which describes how a security attribute of a component fails. Vulnerabilities are considered the causes for failures of security attributes. Vulnerabilities analyzed together with potential attacks form the likelihood of a threat mode. Following the system decomposition into functions, the FMVEA analysis provides a combined list of failures and threat modes together with their causes and a risk estimate. In a cyber-physical system such as a vehicle, typical functions are either processing or communication functions. As such, over time, frequently encountered failure modes for input and output have been identified [13]. Performed during system design, FMVEA provides a systematic analysis of failures and threats effects, allowing development to proceed effectively.

Because FMVEA focuses on the analysis on component functions and their interactions, it does not address the system-level functions explicitly. As a consequence, the analysis can result in a list of vulnerabilities that can be potentially exploited for each component, but may miss critical threads to the overall system.

### 3.3 SAHARA

Security-Aware Hazard Analysis and Risk Assessment technique has been introduced in 2015 [9] as a joint approach for performing safety and security analysis through a combination of HARA [20] for the automotive domain and STRIDE (Spoofing, Tampering, Repudiation, Information Disclosure, Denial of Service, elevation of Priviledges) [10]. Focusing on the automotive domain, SAHARA quantifies the probability of an identified security threat and aligns its Security Level (SecL) to the ASIL (Automotive Safety Integrity Level) defined in the HARA. As in CHASSIS, functional use-cases are analyzed, but in SAHARA, the STRIDE model is used for the identification of possible security threats. In a following paper [8] Macher et al. recommend using DREAD (Damage potential, Reproducibility, Exploitability, affected users, Discoverability) model instead of STRIDE for distributed systems.

### 3.4 STPA-Sec

System Theoretic Process Analysis - Security Method (STPA-Sec) [21] is an extension of the base STPA approach based on the STAMP (System Theoretic Accident Model and Processes) causality model. Starting from the idea that an attacker could infiltrate a system by exploring an open and undiscovered vulnerability, once the intention is there, Young and Leveson introduce the idea that the ultimate goal of security should be the assurance of maintainability of critical functions despite such intrusions [3]. Therefore, the STPA-Sec, unlike the other methodologies, does not focus on finding vulnerabilities in the individual components, but views safety as a system property and not a component property. Focusing on enforcing safety and security concerns at the system level, hierarchical control structures are created where higher-level components constrain lower-level components. As a result, the safety and security incidents are treated as a hierarchical control problem rather than a simple chain of events.

STPA-Sec works on a specification of the system's accidents and losses, along with hazards that can potentially lead to accidents. The method performs a systematic analysis according to existing functional control diagrams. Then, a set of guide words are used to reason about causes of hazards.

## 4 Applying Safety and Security Analysis Methods on Collaborating Systems

In this section, we summarize the results of comparisons resulting from application of the methods presented in the previous section on a use case from the automotive domain. A more detailed report of the method's application can be found at [6].

### 4.1 Use Case Definition

For exemplifying the above mentioned safety and security co-analysis, we refer to a use case from the automotive domain which is general enough to ease the mapping to similar domains, such as robotics, while allowing us to be specific in the explanations. Through I2V (Infrastructure-to-Vehicle) communication, a platoon leader receives information from a RSU (Road Side Unit) about an imposed speed limit. Following a PLF (Predecessor Leader Following) [22] communication typology, the platoon leader further on sends the information to the platoon members via V2V (Vehicle-to-Vehicle) communication. In a platoon, all vehicles apply the same speed and acceleration at the same time by maintaining the closest reasonable distance. This maneuver reduces air friction and implicitly reduces the fuel consumption. A detailed description of the use case can be found at [6].

Overall, during the study and applicability of the methods, multiple differentiating factors have emerged. We group these differentiating factors in the following way:

1. ***Type of Approach:*** FMVEA follows a bottom-up approach starting with system design. For performing FMVEA, a system must be divided into components such as: sensors for distance measurement, network devices, ECU controllers, etc. FMVEA maintains the focus on single-component vulnerabilities while documenting their impact on the overall system. Both SAHARA and CHASSIS, even though they do not focus on single components, are also

bottom-up approaches that base the safety and security co-analysis on functional decomposition of systems. SAHARA performs a functional decomposition of systems, whereas CHASSIS performs a behavioral decomposition. STPA-Sec, on the contrary, takes a top-down approach, starting by identifying accidents and system losses. Afterwards, the analysis is conducted with the focus on control loops between components, in order to identify if an error of control at the higher levels can potentially result in a vulnerability that can cause an accident or a loss.

2. ***Needed Artefacts:*** FMVEA needs a physical decomposition diagram of a system. CHASSIS requires various forms of use cases and sequence diagrams at the start of the analysis. SAHARA requires functional decomposition diagrams of the system. STPA-Sec needs some non-standard artefacts such as: definition of accidents or losses for the system, and eventually hazard analysis.
3. ***Creativity of Analysis Process:*** Given the creative process of analysis, one important factor for method applicability is the way in which the analysis guidelines stimulate the thought process of a safety and security analyst. For FMVEA, the guideline directs the analyst to look at the system as a composition of components and then focus on single components for threat identification. CHASSIS provides a clear direction to focus on interaction among subsystems through the system level artefacts. STPA-Sec also clearly states that the analysis should be focused on control loops. The only method where we did not see a clear direction on analysis is SAHARA. SAHARA provides a sophisticated mechanism for risk rating but how should one approach the system to identify those risks is not clear.
4. ***Application Phase:*** The phase of development, after security analysis is completed, is a critical factor for successful risk mitigation, as the latter can become expensive and more challenging when applied in the later phases. FMVEA, CHASSIS and SAHARA can commence in the system design phase. FMVEA, due to its minimal requirements, can start earlier compared to the other two. For new system development, STPA-Sec is recommended to start at the conceptual phase as one of the earliest activities.
5. ***Generated Artefact:*** FMVEA and SAHARA generate the classic quantitative tabular risk analysis summary with numbers to indicate the priority. STPA-Sec generates a functional control diagram of the system highlighting the control loops and a list of control loop issues that can result in a hazard. Among all methods CHASSIS generates the most sophisticated and graphical artefacts i.e. T-MUCs, D-MUCs and MUSDs. There is a contradicting characteristic to artefacts generated needs to be highlighted here i.e. repeatability. We think if FMVEA and SAHARA are done by multiple teams or people, the results will not vary a lot since the analysis is strictly formulated at every step. In case of CHASSIS it is partially formulated i.e. use of existing artefacts and HAZOP but still analysis is open to interpretation. STPA-Sec results of the same system by different people will probably vary most widely since it does not restrict the analysis to strict formulation.
6. ***Risk Evaluation:*** All methods presented in this paper can identify security risks associated with the system. In particular, FMVEA and SAHARA pro-

vide evaluation of risks compared to each other. SAHARA integrates ASIL levels to indicate relative weighted risks. Evaluation of risks helps in prioritizing the mitigation strategies. CHASSIS does not highlight the prioritization of risks. STPA-Sec, due to the fact that it links a hazard directly to the accidents or losses it may cause, an indirect understanding of the risk priority.

7. ***Usability:*** In terms of usability and understandability, FMVEA has an advantage, being known in many industries including embedded applications in automotive domain. Both FMVEA and SAHARA methods are well-structured to be followed by any team and recommended guide words from STRIDE help stimulate the risk analysis. CHASSIS and STPA-Sec require domain knowledge in the application area and security analysis to generate an exhaustive list of risks. The artefacts generated in STPA-Sec can be complicated to understand for someone lacking understanding of control loops. Also, because this method looks beyond the system, it requires domain expertise. In CHASSIS, the artefacts like misuse cases and misuse sequence diagrams are expressed in UML, therefore an understanding of UML notations is necessary.
8. ***Efforts:*** While no quantitative effort estimation has been performed during this study, in a report from a 2018 [2], Torkildson et al. compare efforts between FMVEA, CHASSIS and STPA-Sec for safety-security analysis of an autonomous boat quantitatively. They found out that STPA-Sec took almost double the efforts of FMVEA. CHASSIS took roughly 50% more effort than FMVEA. This is in line with our qualitative comparison with respect to the factors above i.e. STPA-Sec and CHASSIS both require a longer preparation phase and evaluation of more artefacts.

The Table 1 below provides a summary of the comparison results.

**Table 1.** Summary of comparison

| Method | FMVEA | CHASSIS | STPA-Sec | SAHARA |
|---|---|---|---|---|
| Approach | Bottom-Up | Bottom-Up | Top-Down | Bottom-Up |
| Needed Artefacts | Low Complexity: Physical Decomposition | High Complexity: US, SDs | Medium Complexity: Functional Control Diagram | Low Complexity: Functional Decomposition |
| Provides Risk Rating | Yes | No | No | Yes |
| Requires Domain Expertise | No | Yes | Yes | No |
| Enforces Creative Analysis | Yes | Yes | Yes | No |
| Repeatability of Results | High | Low | Low | High |
| Required Efforts | Low | Medium | Hign | Medium |
| Requires Special Tool Skill | No | Yes | No | No |

## 5 Conclusions and Future Work

The current paper reports on comparison results between safety and security analysis methods applied on a use case example from the automotive domain.

Newer security analysis methods base their approach on safety analysis methods, like FMEA and FTA, which were conceptualized 50–60 years earlier, when systems were predominantly mechanical and component-oriented. Pushed by technological developments, the digital systems we see today, including vehicles, are software intensive, tightly coupled, highly automated and connected. The platooning system we analyse in our paper is an example of such a system. The vehicles who themselves are a system of connected ECUs communicate with each other and RSUs to maintain constant longitudinal gap to improve fuel economy and traffic efficiency. The use case from the automotive domain is similar to various scenarios, including communicating and collaborating robots.

The methods reviewed in this paper represent a first attempt towards protecting a system from malicious attacks. However, according to Leveson [3], an attacker is very likely to penetrate a target system. In this case, mechanisms and methods should be devised for protecting the safe operation of a system in situations in which the attacker already made their way into the system. In the context of complex ecosystems, where actors and systems directly interact from design time to runtime, leaving developers the possibility to patch running systems with software updates, an emergent vulnerability of receiving malicious software through runtime updates becomes evident [11]. On top of this, vulnerabilities caused human stakeholders directly affect information security and therefore, human aspects of trust are becoming increasingly important future topics to be addressed by security experts.

For safeguarding a system operation at runtime, analysis methods traditionally implemented during the design need to be extended with runtime considerations as well. Mainly because the operational context of systems interacting in highly dynamic ecosystems cannot be foreseen during design time, current methods need to be extended with runtime considerations of environment characteristics that are impossible to predict.

**Acknowledgment.** This work was supported by the project BIECO (www.bieco.org) that received funding from the European Union's Horizon 2020 research and innovation programme under grant agreement No. 952702.

## References

1. AUTOSAR: AUTOSAR. https://www.autosar.org/. Accessed 09 June 2021
2. Erik Nilsen, T., Li, J., Johnsen, S.O., Glomsrud, J.A.: Empirical studies of methods for safety and security co-analysis of autonomous boat. Safety and Reliability-Safe Societies in a Changing World (2018)
3. Gößling-Reisemann, S.: Resilience-preparing energy systems for the unexpected. An edited collection of authored pieces comparing, contrasting, and integrating risk and resilience with an emphasis on ways to measure resilience, p. 73 (2016)
4. Ishimatsu, T., et al.: Hazard analysis of complex spacecraft using systems-theoretic process analysis. J. Spacecr. Rocket. **51**(2), 509–522 (2014)
5. ISO: ISO/SAE FDIS 21434 Road vehicles - Cybersecurity engineering. https://www.iso.org/standard/70918.html. Accessed 08 June 2021

6. Kar, S.R.: Report of Method's application (2021). https://drive.google.com/drive/folders/12c3TrLUy-xLpGQmw5zfVvUECnPtxC9h2. Accessed 08 June 2021
7. Macher, G., Armengaud, E., Brenner, E., Kreiner, C.: A review of threat analysis and risk assessment methods in the automotive context. In: Skavhaug, A., Guiochet, J., Bitsch, F. (eds.) SAFECOMP 2016. LNCS, vol. 9922, pp. 130–141. Springer, Cham (2016). https://doi.org/10.1007/978-3-319-45477-1_11
8. Macher, G., Armengaud, E., Brenner, E., Kreiner, C.: Threat and risk assessment methodologies in the automotive domain. Procedia Comput. Sci. **83**, 1288–1294 (2016)
9. Macher, G., Sporer, H., Berlach, R., Armengaud, E., Kreiner, C.: Sahara: a security-aware hazard and risk analysis method. In: 2015 Design, Automation Test in Europe Conference Exhibition (DATE), pp. 621–624 (2015). https://doi.org/10.7873/DATE.2015.0622
10. Microsoft: STRIDE (2009). https://docs.microsoft.com/en-us/previous-versions/commerce-server/ee823878(v=cs.20)?redirectedfrom=MSDN/. Accessed 09 June 2021
11. O'Neill, P.H.: How a $1 million plot to hack tesla failed (2020). https://www.technologyreview.com/2020/08/28/1007752/how-a-1-million-plot-to-hack-tesla-failed/. Accessed 08 June 2021
12. Palin, R., Ward, D., Habli, I., Rivett, R.: ISO 26262 safety cases: compliance and assurance. In: 6th IET International Conference on System Safety 2011, pp. 1–6. IET (2011)
13. Pentti, H., Atte, H.: Failure mode and effects analysis of software-based automation systems. VTT Ind. Syst. STUK-YTO-TR **190**, 190 (2002)
14. Ramaiah, B.S.M.P.S., Gokhale, A.A.: FMEA and fault tree based software safety analysis of a railroad crossing critical system. Glob. J. Comput. Sci. Technol. (2011)
15. Raspotnig, C., Katta, V., Karpati, P., Opdahl, A.L.: Enhancing chassis: a method for combining safety and security. In: 2013 International Conference on Availability, Reliability and Security, pp. 766–773. IEEE (2013)
16. Schmittner, C., Gruber, T., Puschner, P., Schoitsch, E.: Security application of failure mode and effect analysis (FMEA). In: Bondavalli, A., Di Giandomenico, F. (eds.) SAFECOMP 2014. LNCS, vol. 8666, pp. 310–325. Springer, Cham (2014). https://doi.org/10.1007/978-3-319-10506-2_21
17. SESAMO: D4.2 Integrated Design and Evaluation Methodology (2014). http://sesamo-project.eu/content/d42-integrated-design-and-evaluation-methodology. Accessed 08 June 2021
18. SESAMO: Security and Safety Modelling (2015). http://www.sesamo-project.eu/. Accessed 08 June 2021
19. da Silva Amorim, S., Neto, F.S.S., McGregor, J.D., de Almeida, E.S., von Flach, G., Chavez, C.: How has the health of software ecosystems been evaluated?: a systematic review. In: Proceedings of the 31st Brazilian Symposium on Software Engineering, pp. 14–23. ACM (2017)
20. Stolte, T., Bagschik, G., Reschka, A., Maurer, M.: Hazard analysis and risk assessment for an automated unmanned protective vehicle. In: 2017 IEEE Intelligent Vehicles Symposium (IV), pp. 1848–1855. IEEE (2017)
21. Young, W., Porada, R.: System-theoretic process analysis for security (STPA-SEC): Cyber security and STPA. In: 2017 STAMP Conference (2017)
22. Zheng, Y., Li, S.E., Wang, J., Li, K., et al.: Influence of information flow topology on closed-loop stability of vehicle platoon with rigid formation. In: 17th International IEEE Conference on Intelligent Transportation Systems (ITSC), pp. 2094–2100. IEEE (2014)

# General Track

# A Simple User Guide for Erasmus+ Teaching and Training. The Case of the University of Catania

Paola Clara Leotta(✉) and Vincenza Tutino

University of Catania, Piazza Università 2, 95100 Catania, Italy
{pcleotta,cinzia.tutino}@unict.it

**Abstract.** The ***international exchanges for teaching*** performed by lecturers of higher education institutions in any academic field, as well as the ***mobility for training*** carried out by administrative staff in the form of participation in events or observation periods in the workplace, is a strategic activity of the Erasmus+programme. The action is of great interest for the scientific community and has created great expectations, because staff mobility could well be a significant incentive not only for individual professional growth, but also a boost for the development of higher education institutions' international policies. Its impact, in fact, can reflect on the institution as a whole, contributing to performing better institutional strategies and smooth governance, as well as strengthened international cooperation. The results of Erasmus staff mobility in European universities have been examined in various studies such as the EU Erasmus+Annual Reports and the REALISE project. This paper intends to raise the awareness on the importance of these international exchanges involving higher education institution staff and to provide some guidelines with which to develop a successful activity within the next Erasmus+programme edition. An increased participation to this action will provide academic and administrative staff with valuable opportunities to develop new expertise, improve language proficiency and develop intercultural skills thus broadening professional horizon.

**Keywords:** Staff mobility for teaching and training · University welcoming strategy · Recognition and career incentives

## 1 Introduction, Approaches, Methodologies and Dataset of the REALISE Project

The results of Erasmus+(EU regulation n. 1288/2013) [1] staff mobility in European universities have been examined in the EU Erasmus+Annual Reports [2] and more specifically in the REALISE project. Some guidelines that can be useful for the next Erasmus+programme edition. (2021 Erasmus+programme guide) [3] are provided here.

---

Paola Clara Leotta is the author of Abstract, Keywords and Introduction, approaches, methodologies and dataset while Vincenza Tutino is the author of Results and Conclusion.


J. J. Gude Prego et al. (Eds.): CISIS 2021/ICEUTE 2021, AISC 1400, pp. 245–253, 2022.
https://doi.org/10.1007/978-3-030-87872-6_24

According to the REALISE project (https://realise-erasmusplus.fr/), in which the University of Catania has participated as the sole Italian representative, the Erasmus staff mobility action has created a general sense of frustration in the participating higher education institutions. This is due to the poor investment in the action by the EU and the national governments and the scarce recognition and incentives awarded by the higher education institutions. Although enough European funding has been allocated to staff mobility programmes, the impact both on individuals and institutions' international strategies is not so relevant as expected. Given that staff mobility is certainly an indirect means of enhancing students' mobility, it is unclear why it has not been fully implemented so far. In fact, the teaching and training mobility of staff has not kept pace with that of the students' mobility, and this is why the REALISE project has endeavoured to raise the awareness of the importance of these kinds of exchanges. Staff mobility is indeed a valuable opportunity for academic and non-academic staff to develop new expertise, improve their language proficiency and develop intercultural skills thus broadening their professional horizon. Therefore, it is important that universities undertake activities to boost this kind of mobility, to remove obstacles, to foster academic recognition of exchange periods abroad and to introduce reward mechanisms in staff careers.

The REALISE project was realised in two phases. The first phase included a survey for all partners' higher education institution staff and international relations offices and a series of interviews with university management, both investigations were very useful to collect data and helpful to draw institutional attention to the topic. The second phase was based on the creation of a *Toolbox* [4], including innovative practical actions, valuable to establish and provide to the large public guidelines to enhance procedures for the implementation of staff mobility. The aforementioned guide aims at identifying and developing innovative practices for the implementation of staff mobility; helping participants to obtain the maximum benefit from the action's potential; fostering the recognition of mobility in the career development, raising institutional awareness of the added value of staff mobility. The handbook is made up of four parts. The first is dedicated to the strategic level; the second is devoted to the actions that institutions can implement in order to improve the quality of staff mobility. Communication and promotion is the main issue of the third part; and, finally, the recognition and encouragement of mobility is the subject of the fourth part. It is clear that the topic of the last part is particularly strategic, because it could be one of the major levers for higher education institutions to increase the number of beneficiaries and improve the quality and impact of staff mobility.

The aim of the survey and the interviews, carried out in 10 universities located in 10 different European countries, was to test the attitudes towards Erasmus staff mobility: there were 6202 respondents to the questionnaire as against 48 interviews conducted with members of senior leadership. It has been recognised that international staff mobility bears several benefits to the individuals and institutions, including not only the enhancement of the university's educational mission but also the development of personal skills, such as reaching specific targets and problem-solving. The survey and the interviews have also identified obstacles to mobility, such as insufficient funding, difficult working conditions and responses around the promotion of the programme that indicate dissatisfaction around the visibility of opportunities and partner universities. However, it appears clear that the mobility experience is highly valued by staff with almost total agreement that their participation fully met their expectations and had an impact on their lives enabling them to discover a new European identity (Table 1).

**Table 1.** REALISE survey results (https://realise-erasmusplus.fr/sites/default/files/Key%20findings-report.pdf) [5]

| **Table 1.1. REALISE survey results: Mobility demographics and activities** |
|---|
| 1 in 3 respondents have participated in European mobility in the past 5 years; there has been a 10% increase in participation of administrative staff in the past 5 years; the number of academic staff in engineering going on mobility has halved while the amount of academic staff in Humanities and Social Sciences going on mobility has slightly increased; the vast majority of all staff who went on mobility have worked at a university for 6+ years; 2 in 3 members of staff are female; 9 in 10 staff are confident as regards their strongest foreign language |
| **Table 1.2. REALISE survey results: Management** |
| A range of factors have been found to encourage individuals' decisions to go on mobility: these include more recognition being given through pay rises and job promotion, linking mobility with university strategy and, generally, showing support for staff mobility. Other factors include simplifying administrative procedures, providing family support and sensible work arrangements, increasing the length of the visit, and raising awareness of the emotions associated with the mobility experience, such as curiosity and anxiety. The mobility factors discouraging mobility include problematic work arrangements and poor promotion: respondents lamented the lack of information about partner universities and about the programme; additionally, the key individual problem experienced associated with mobility is insufficient funds, whereas the key institutional obstacle to mobility concerns working conditions, particularly in the lack of the possibility of having a replacement at work. The vast majority of respondents are satisfied to have achieved all their set goals during mobility, and the remainder have achieved them partially. Overall, 99% of those people who went on mobility thought that their participation in this Erasmus staff mobility programme fully met their expectations |
| **Table 1.3. REALISE survey results: Promotion and dissemination** |
| The main source of information about the programme is through word of mouth while informal dissemination, such as the sharing of information about the visit with colleagues, is the most popular means for spreading the results of mobility |
| **Table 1.4. REALISE survey results: Recognition** |
| 84% of staff who know colleagues who have been on mobility recognise that Erasmus staff exchange has helped them professionally however, overall perceived recognition is rather low: barely more than 1 in 3 staff members who have gone on mobility in the past 5 years feel their experience has been highly valued and acknowledged by their institution. It has been recognised that the value of mobility is in enhancing the university's educational mission and improving society's sustainability, contributing to emotional well-being, providing opportunities for personal development and collaboration, enhancing skills, reaching specific targets (e.g. administrative staff), and favouring problem–solving. Obstacles to the recognition of mobility include administrative staff themselves not being aware of the benefits of mobility |

## 2 Results: The Case Study of the University of Catania

The case study of the University of Catania has pointed out some evidence that could be useful to share.

According to the University of Catania experience, it is clear that the prerequisite of a successful internationalisation strategy for staff mobility in a higher education institution is to appoint department internationalisation coordinators whose role is to operate synergistically within an efficient task force to pursue a coherent strategy and obtain homogenous results. Before being engaged in an exchange action, the internal status quo needs to be investigated and defined, internationalisation priorities included in the university strategic plan, so as to be aware of what the main activities and intervention areas are. Do department internationalisation coordinators have enough authority to get the department directors commitment and raise institutional awareness of the added value of staff mobility? Are there clear guidelines on administrative aspects aimed at removing the obstacles related to the organisation of the exchanges? The answer of the University of Catania to these questions has been the creation of a University Direction Cabin appointed by the Rector, including an Erasmus Institutional Coordinator, specifically concerned with Erasmus+matters. The Rector Delegates, who are the members of the University Direction Cabin, work together to accomplish the internationalisation objectives established in the university strategic plan. Moreover, each Department Director has appointed an Internationalisation and Erasmus Coordinator who works on political and strategic issues in order to implement the internationalisation strategy at a department level and whose duty is also to provide support to department staff involved in exchange actions. These two political authorities at the central and the decentralised level, operate synergistically within an Internationalisation Commission, a collegial body that holds periodical meetings to evaluate the ongoing actions and shares good practices to implement the international activities. Administrative Offices at central and departmental level ensure the operational support and manage staff mobility procedures. The result of this transversal action has been a better structured system for outgoing teaching and training staff that has obtained the general approval of the university community. This first strategic organizational effort is mandatory to reach a higher level of internationalisation and to get an increased acknowledgment by the scientific community.

In addition, the experience of the University of Catania shows that it is very important to have a clear and defined welcoming scheme for incoming staff, because receiving foreign partners in an efficient way and giving them a visibility among lecturers and administration staff is a specular way to implement outgoing staff mobility and help participants understand the great benefit they can get from the action's potential. Consequently, it will be very important to have good internal coordination among the various actors involved in international welcoming through the creation of a Permanent Working Group that includes relevant stakeholders, belonging to the central administration and the departments, both at academic and administrative level, in order to better define and eventually ameliorate staff mobility priorities. Moreover, the involvement of the university communication officers, able to use their sectorial expertise to promote staff mobility and support colleagues in charge of welcoming incoming staff within the departments is strategic. Therefore, what do higher education institutions interested in improving their staff mobility need to do? Firstly, a clear webpage with simple steps to apply for incoming mobility should be made available; secondly agreements with accommodation and service providers (hotels, hostels, on-line platforms, restaurants, tourist guides) should be signed and made well visible on the website; thirdly a welcome kit including information on courses, local maps and cultural and social opportunities offer should be created together with promotional gadgets. Finally, an annual occasion, called "Staff Week", should be planned to meet colleagues coming from partner universities in order to deal with specific issues on international relations, share best practices and open the doors to specialist areas (laboratories, observatories, historical buildings). In addition, the results of each annual action need to be evaluated by answering the following simple questions: Was our action successful? Did it correspond to the objectives of the strategic plan? After having evaluated the result that the activities achieved, an institution is ready to start a new and better performance. The University of Catania, in particular, has explored and chosen some channels to be used for the promotion of incoming staff mobility, published webpages with simple step-by-step procedures to apply for incoming mobility, created a system of on-line application and a database to register mobility flows. It has already signed several agreements for accommodation of incoming staff (hotels, hostels, online platforms) and made them visible on the website; it has also improved a successful model of "Staff Week" to concentrate the incoming exchange flows in a precise moment of the year, thus not overburdening lecturers and international relation officers with a demanding activity in addition to their normal academic and administrative commitments. The University of Catania has also appointed a cultural mediator to help non-EU guests with visa and logistic support, because the Erasmus KA107 extra-UE mobility is a new priority in the university strategic plan. It is really important for the University of Catania to make the staying of their international guests fruitful and enjoyable because it will increase the number of incoming exchanges with a consequent benefit for the students and a better ranking of the institution as a whole.

From the experience of the University of Catania it finally emerges that the most important instruments to increase the number of beneficiaries of the staff teaching and training actions and improve the quality and impact of staff mobility are undoubtedly the recognition of mobility experience and career progression opportunities. To achieve this important goal, a Human Resources Commission for the evaluation and recognition of staff mobility should be appointed. Universities should explore the current situation within their institutions: is there a career progression strategy for staff that takes into consideration international mobility experience? Is there a Human Resources Commission? A good way to go about this, could be the preparation of a short questionnaire for the staff of the institution on the level of awareness concerning staff mobility opportunities and its added value for professional career improvement. After the analysis of the data, a report could be prepared and a working group involving academic and administrative representatives in charge of preparing a document defining the actions to be taken, the procedures to be implemented and the incentives to be given at the end of the mobility organised. Moreover, a recognition table for the evaluation of mobility could be adopted and a procedure for the recognition of the international exchanges in the employees' professional dossier could be created. Specifically, at the University of Catania, a Focus Group on Staff Mobility has been created; a questionnaire has been distributed to the departments in order to collect information on best practices and critical aspects of staff

**Table 2.** Evaluation criteria for STA mobility (STA: international exchange for teaching) realise recognition table experimented on a pilot basis by the University of Catania (https://realise-erasmusplus.fr/content/handbook-good-practices)

| Mobility phase | What is evaluated | Evaluation criteria | Evaluation parameters | Score | Evaluation instrument | Who is evaluating |
|---|---|---|---|---|---|---|
| Before the mobility (38% of the total score) | Partnership evaluation | Study cycle at the hosting University | III (Phd) | 3 | Teaching programme | Sending University Commission |
| | | | II (Master) | 2 | | Sending University Commission |
| | | | I (Bachelor) | 1 | | |
| | | Didactic activity length | 6 days and more | 3 | Teaching programme | Sending University Commission |
| | | | 4/5 days | 2 | | |
| | | | 2 days | 1 | | |
| | Hosting professor/tutor evaluation | Use of innovative teaching | Syllabus/Programme | 1 | Teaching programme | Sending University Commission |
| | | | Methodology and materials | 2 | Teaching programme | Sending University Commission |
| During the mobility (48% of the total score) | Partnership evaluation | Number of students attending lectures | Up to 10<br>Between 10–20<br>Over 20 | 1 | Hosting institution questionnaire | Hosting institution |
| | | | | 2 | | |
| | | | | 3 | | |
| | Teaching evaluation | Linguistic skills<br>Performance | Lectures delivered in the hosting country language | 1 | Hosting institution questionnaire | Hosting institution |
| | | | Student evaluation (optional) | 1 | Hosting institution questionnaire | Hosting institution |

*(continued)*

**Table 2.** (*continued*)

| Mobility phase | What is evaluated | Evaluation criteria | Evaluation parameters | Score | Evaluation instrument | Who is evaluating |
|---|---|---|---|---|---|---|
| | | | | 2 | | |
| | | | | 3 | | |
| | | | Hosting university evaluation | 1 | Hosting institution questionnaire | Hosting institution |
| | | | | 2 | | |
| | | | | 3 | | |
| After the mobility (14% of the total score) | Partnership evaluation | Proposal about new mobility experiences | Visiting invitation or any other proposal of collaboration, agreement related to shared activities | 2 | Hosting institution questionnaire/invitation letter | Sending University Commission |
| | | | New STA mobility | 1 | Hosting institution questionnaire/invitation letter | Sending University Commission |
| | Outputs | Use of results at home institution | Availability to participate in welcome back seminars | 2 | Beneficiary final report | Sending University Commission |

mobility; a policy document has been prepared; a recognition table proposal has been created and experimented on a pilot basis; a promotional meeting with the various actors involved has been organised. Moreover, the University of Catania has adopted specific financial measures aimed at increasing the support to Erasmus+staff's outgoing mobility, in the hope that this economic effort, together with the measures previously described, will bring about an increase, and even double, the number of successful staff exchanges for teaching and training. The University is still in the process of appointing a permanent Human Resource Commission in charge of evaluating staff mobility performance and defining the incentives to be given. This last action is the most important for the internationalisation process but also the most critical to perform, that why the University of Catania is still on the way to define a detailed time scale to reach this ambitious objective. In fact, in the next three years the University Direction Cabin will improve a more detailed recognition strategy and the governing bodies will approve an incentives policy to support the successful departments (Table 2).

## 3 Conclusions

A recommendation document (https://realise-erasmusplus.fr/content/conclusions-policy-dialogue) [6] was elaborated by the REALISE partnership in order to maximise the impact of the Erasmus+staff mobility action both on **individuals** and **institutions**. The recommendation document contains some relevant suggestions at institutional, national and European level in the hope of their acceptance for the next Erasmus+edition in force from 2021 to 2027.

**Recommendations at Institutional Level**
Improve the institutional recognition and encouragement of staff mobility and have it formalized; increase funding; simplify the mobility requirements and application procedures; organise data collection; develop international mobility or adopt an institutional internationalisation strategy that promotes staff mobility; strengthen the communication and cooperation between the institutional units engaged in developing staff mobility; improve staff mobility promotion and dissemination, support the preparation of outgoing staff and ameliorate the welcoming of incoming staff.

**Recommendations at National Level**
Launch initiatives that support the inclusion of international mobility in the career advancement system adopted at national level; implement national funding and/or co-funding for international staff mobility activities; include international staff mobility as a criterion used for national evaluation campaigns of HEIs.

**Recommendations at European Level**
Encourage Member States by giving guidelines and incentives to improve staff mobility in accordance with what the Bologna process follow-up group on mobility and internationalisation initiated; reconsider the rules of the Erasmus+programme concerning Staff Mobility.

Overall, it can be deduced that the Erasmus teaching or training period is assessed positively by both the individual beneficiaries and higher education institutions and certainly has a positive impact on the competences acquired by lecturer and staff as well as on their informal professional development [7]. In general, the staff involved in the exchanges rate their experience very positively and the self-assessment of professor and administrative staff is supported by the positive feedback of university management, who similarly rate in several cases the competences of mobile staff as being superior when compared to those of non- mobile staff in their approach to problem solving. Moreover, the exchange experience through the Erasmus+programme helps lecturers and administrative staff in their integration in international networks and in increasing their accountability abroad. It could be argued that a short mobility period cannot have a high impact on the professional life of an experienced professor or a staff member, but the opposite seems to be true. The beneficiaries of the mobility action report a positive influence on their career prospectus, especially regarding international contacts and perspectives development. However, it is a shared opinion among the mobile staff that the support of the home institution is not so significant as they had expected. Indeed, most of them report that their Erasmus exchange is considered an individual activity only and this often means extra work without any compensation. Therefore, it is strategic for the development of the international university strategy to define a recognition policy and provide participants with career acknowledgments and financial incentives. It would be also advisable to follow the suggestions related to the REALISE project. The case study of the University of Catania could prove crucial for the reinforcement of the overall professional value of the Erasmus+exchanges as well as being a good starting point for the participation in the new Erasmus+programme edition 2021–2027.

**Acknowledgements.** The present article discusses selected outcomes of the REALISE projects in which the authors have participated respectively as the Internationalisation and Erasmus Coordinator of the Department of Educational Sciences of the University of Catania and as a member of the key-staff of the project. The whole project can be found via https://realise-erasmusplus.fr/. On the website of the REALISE project videos of some success stories are also available. Acknowledgements also go to the other members of the REALISE working group of the University of Catania: Adriana Di Stefano, Teodora Nicoleta Pascu and Maria Luisa Leotta.

## References

1. Regulation (EU) n° 1288/2013 establishing 'Erasmus+: The Union Programme for Education, Training, Youth and Sport, European Parliament and Council, Official Journal of the European Union, December 2013
2. Statistical Annex Erasmus + Annual report 2019: European Commission (2020)
3. The Erasmus+ Programme Guide 2021/2027: European Commission, March 2021
4. Lionel, C., et al.: Erasmus Staff Mobility Handbook of Good Practices, Université Paul-Valéry Montpellier 3, May 2019
5. Sara, C., Ball, S., Rachel-Naseem, N.: Report on a Comparative Analysis of Current Practices in Erasmus Staff Mobility at European HEIs. Middlesex University London, London (2018)
6. Conclusions of the Policy Dialogue of the REALISE Project. Université Paul-Valéry Montpellier 3, August 2019
7. Engel, C.: The impact of Erasmus mobility on the professional career: empirical results of international studies on temporary student and teaching staff mobility. Revue Belge de Geographie **4**, 1–16 (2010)

# Virtual Implementation of Practical Control Subjects as an Alternative to Face-to-Face Laboratory Lessons

Francisco Zayas-Gato(✉), Álvaro Michelena, Esteban Jove, José-Luis Casteleiro-Roca, Héctor Quintián, Elena Arce-Fariña, and José Luis Calvo-Rolle

Department of Industrial Engineering, University of A Coruña, CTC, CITIC, Ferrol, A Coruña, Spain
{f.zayas.gato,alvaro.michelena,esteban.jove,jose.luis.casteleiro, hector.quintian,elena.arce,jlcalvo}@udc.es

**Abstract.** The coronavirus disease (COVID-19) has led to a serious health crisis with unprecedented socioeconomic consequences. Since World Health Organization's pandemic declaration, most governments applied control policies such as lockdown, mobility restriction, social distancing and so on. In this context, adapting educational planning to that scenario carries a major challenge for the global education community. Furthermore, since higher education face-to-face laboratory practice has been temporarily cancelled, creating realistic virtual scenarios by means of modern software applications represent an interesting alternative. This practical approach can be achieved by using two interesting software tools such as Node-RED and Factory I/O. In this paper, the implementation of a virtual laboratory plant using tools mentioned above, with very similar features to the laboratory plants at the Faculty of Engineering of the University of A Coruña is proposed.

**Keywords:** Virtual laboratory · Face-to-Face · Control lessons · Education in control

## 1 Introduction

The recent burst of the novel coronavirus disease (COVID-19) has conditioned higher education scheduling and teaching methodologies. Following World Health Organization (WHO) recommendations, governments of most countries have decreed mobility restriction and social distancing to prevent virus transmission. Additionally, many university facilities have been closed in an attempt to control virus spreading. Consequently, face-to-face classes have been temporarily suspended and online teaching has become mandatory. This situation has impacted dramatically the world's educational community [6,8].

One of the most affected fields in engineering education is laboratory practice and its planning. This is a key point, since theory classes are complemented with


J. J. Gude Prego et al. (Eds.): CISIS 2021/ICEUTE 2021, AISC 1400, pp. 254–263, 2022.
https://doi.org/10.1007/978-3-030-87872-6_25

laboratory practice in most subjects. This methodology fosters collaborative learning and promotes project-based learning [2,3,7].

Generally, laboratories have dedicated specific equipment for practical lessons. However, in many cases, due to limitations with available material, it is necessary to organize working groups in order to carry out tasks. In this context, due to the infectious nature of this coronavirus, working groups have now been reduced or even suspended to prevent virus transmission. Hence, many professors strove to redesign their teaching methods, relying on modern technologies to offer an alternative to real laboratories [9,13,14].

Laboratory lessons encourage student interaction and physical equipment handling. Although online theory lessons are an indispensable alternative to face-to-face ones, accessing laboratories remotely is not common practice and not possible in many cases. Thus, an interesting alternative to real labs is the design and operation of virtual laboratories.

Nowadays, thanks to constant progress in computing and digitalization, many software tools can be used to implement virtual scenarios [4,5,16,17]. In this paper, we propose the integration of Node-RED and Factory I/O software tools to create a virtual scenario as close as possible to the real experiments and plants from the Optimization and Control Laboratory of the Faculty of Engineering, at University of A Coruña (UDC).

The paper is structured as follows: after the present introduction, the motivation of this work is described. Section 4 provides a detailed explanation about the proposed method. Then, the experiments and results are shown in Sect. 5. Finally, the conclusions and future works are listed.

## 2 Background

The Optimization and Control Laboratory is one of the laboratories of the Faculty of Engineering, at University of A Coruña (UDC). This laboratory has some equipment that emulate many real industrial processes such as temperature and water-level control loops. In it, students take their practical lessons of several courses at different levels, interact with the equipment, experiment with sensors and actuators, and test their final projects.

One of the laboratory experiments is the temperature control loop. This experiment is performed by means of an educational cube-shaped enclosure made up of methacrylate pieces. Inside it, two incandescent bulbs act as a heating element. Also, a DC (Direct Current) fan fixed on a mounting hole at one side of the enclosure act as a cooler element. An LM35 temperature sensor is used to measure the temperature. Data acquisition of this plant can be done with a USB-6008 multifunction DAQ (Data AcQuisition system) device from National Instruments (see Fig. 1).

Students can organize themselves in pairs or individually to work with the temperature control systems, because there are enough units for this purpose.

Another laboratory experiment is the tank-level control loop. In contrast to temperature control systems, there are fewer tank-level control plants available.

**Fig. 1.** Temperature control plant

So, working groups must be larger to attend all the students. In addition to the above, it is necessary to take into account a new handicap due to the COVID-19 pandemic, where social distancing must be respected [1,11,15].

## 3 Case Study

This section describes the liquid tank level system and the control loop to be implemented virtually.

### 3.1 Tank-Level Plant

As mentioned before, one of the Optimization and Control Laboratory experiments is the tank-level control loop, which is used to fix the tank liquid level.

The system (see Fig. 2) has two tanks at different levels. The lower tank is used as a water storage to fill and discharge the upper tank, which is used as a liquid level control target. A centrifugal pump fed from the lower tank is used to pump water to the upper tank. For discharging it, both a manual and a servo piloted proportional valve can be used. A variable frequency drive (VFD), connected to the pump's three-phase motor, controls the upper tank fill rate. Also, the water level in this tank is measured with an ultrasonic sensor. For the current experiment, an Arduino Uno is used as a DAQ for acquiring the process signals.

### 3.2 Control System

The control system is a software-implemented virtual controller, developed by using a PC and a DAQ (see Figs. 3 and 4). It is based on a Proportional, Integral, Derivative (PID) controller [10,12]. The set point is the desired water level, and the process value is the current water level measured in the target tank. The

**Fig. 2.** Tank-level control plant

water level is measured with the ultrasonic sensor, and its output signal is read with the DAQ. Thanks to that, the PC receives and processes the signal.

The error signal, which is the input of the virtual controller, is obtained by subtracting the process value and the set point. Finally, the control signal is calculated by the virtual controller and sent through one of the DAQ's analog outputs to the plant's VFD, closing the tank-level control loop.

## 4 Integration Approach

Since the tank-level control plant from the Optimization and Control Laboratory is not available for operation because of the COVID-19 pandemic, the use of a virtual model of the real system is proposed [16].

This section introduces the software tools used for creating a virtual tank-level control plant as well as the virtual control system.

### 4.1 Virtual Plant with Factory I/O

Factory I/O is a 3D design software for factory simulation. It is a powerful tool that can complement the automation and control engineering learning. As a user-friendly software, Factory I/O allows students to create scenarios quickly and understand factory operation from a more realistic point of view (Fig. 5).

Many of the industry's most common elements can be found in Factory I/O. Some of them are: belt conveyors, electric switchboards, elevators, palletizers,

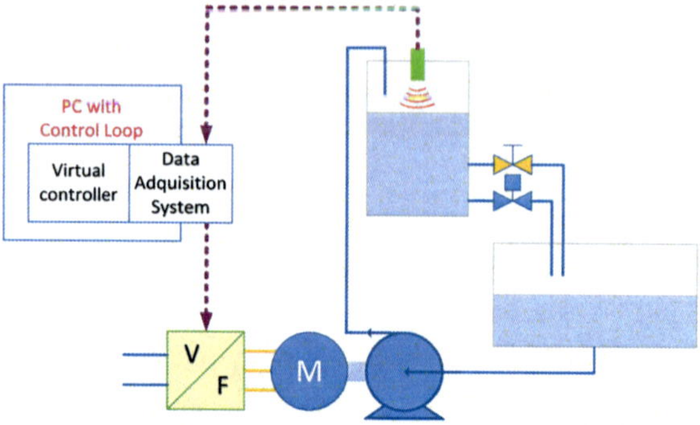


**Fig. 3.** Tank-level plant simplified scheme

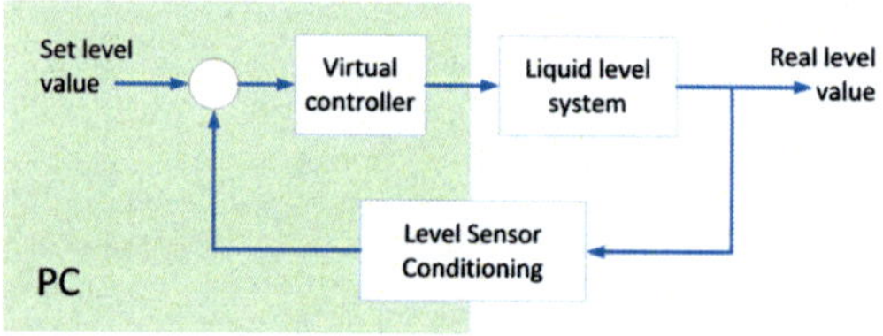


**Fig. 4.** Tank-level control system architecture

tanks, sensors, and so on. All these elements can be assembled to configure a specific industrial process and thus, build a virtual factory.

One of the most interesting scenarios is the use of Factory I/O as a validation platform for PLC-based control systems. It allows to test how an industrial process would behave before being implemented in the field. In addition, it has a wide range of industrial communication protocols, allowing connections with PLCs, microcontrollers, Single Board Computers (SBC) or even with other type of desktop applications.

The level control scene available in Factory I/O and the tank-level control plant from the Optimization and Control Laboratory, have many common features. Besides having a tank to fix the water level, the level control scene has sensors and actuators with industrial analog inputs and outputs. However, this input and output signals are not hardware available. So, an interface between the level control scene and the actual virtual controller is not possible using a hardware data acquisition card.

An alternative to a hardware data acquisition card is an industrial communication interface. Factory I/O has multiple standard communication drivers to read and write the input and output signals generated by the virtual scenes. Some of these communication drivers are Modbus TCP Client, Modbus TCP Server, OPC Client DA/UA, Siemens S7-1200/1500, and so on. Therefore, any software application with support for such communication drivers can be used as an alternative to a DAQ-based system.

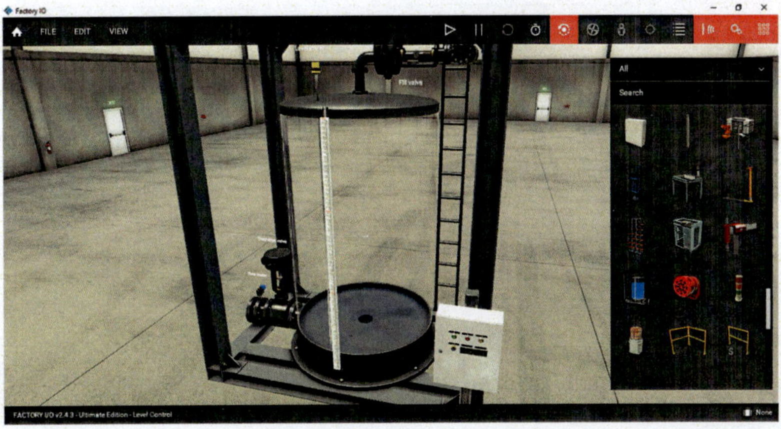

**Fig. 5.** Factory I/O level control

### 4.2 Interface and Control with Node-RED

Node-RED is a graphical programming software to connect hardware devices and also establish communications between applications. Originally developed by IBM's Emerging Technology Services team, Node-RED is now a part of the OpenJS Foundation.

Graphical programming, or, in this case, "flow-based" programming, is a way of developing applications by interconnecting blocks or "nodes" (see Fig. 6). Node-RED not only offers a palette of predefined nodes, it also has the possibility to customize nodes to perform specific functions. Each node within the application receives data through its inputs, executes actions based on that data, and sends data through its outputs. The way and order of node interconnections define the application behavior.

The Integrated Development Environment (IDE) is web server-based, so, there is no need to install any dedicated desktop application. Only a web browser is required to access the programming environment. Therefore, to develop applications for a device connected to a network, a smartphone, tablet, laptop, and so on can be used.

Node-RED makes application development easier and helps to understand the order of program execution. It can be considered as a support tool, therefore, students can focus on the subject of the course and not worry about mastering a new programming language.

The key aspect of using Node-RED for this proposal, is that it serves as the first step of the integration approach. Also, having the required communication drivers, it is possible the implementation of a virtual PID controller.

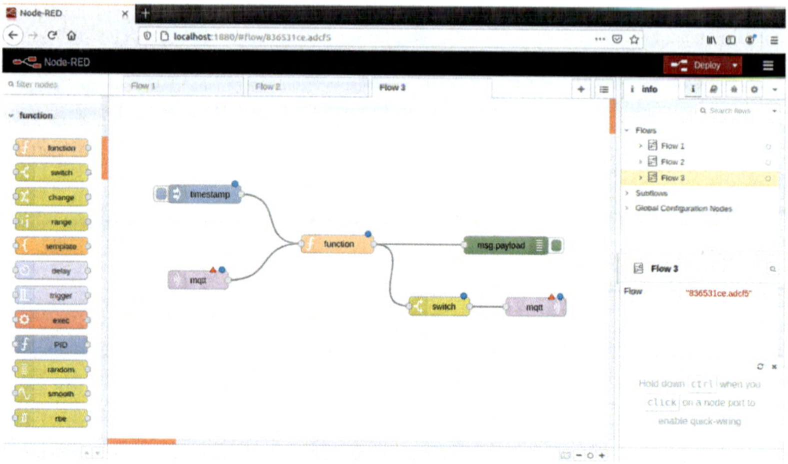


Fig. 6. Node-RED programming environment

## 5 Implemented Solution

This section describes the methods to achieve the application integration, and the implemented virtual solution as an alternative to the current real tank-level control system.

### 5.1 Communicating Applications over Modbus TCP

Modbus is one of the most popular communication protocols within the industry. After more than forty years since it was published, it is still widely used to connect millions of devices around the world. It is a de facto industry standard created by Modicon, and later adopted by many other manufacturers as it is royalty-free.

Modbus is well known for being a robust, easy-to-implement protocol. Also, it offers a client/server architecture that allows the connection of multiple devices over a fieldbus. In addition, it supports multiple versions such as Modbus RTU, Modbus ASCII, Modbus TCP, and so on. Regardless of the version used, the main goal is the data exchange between devices or applications.

Factory I/O and Node-RED have drivers to communicate over Modbus TCP. These applications can use this communication protocol to perform data exchange. Therefore, control and process value signals from the controller and the virtual plant can be sent over this link. In this case, the Modbus client side is performed by the Node-RED application and the server side by the Factory I/O scene as can be seen in Fig. 7.

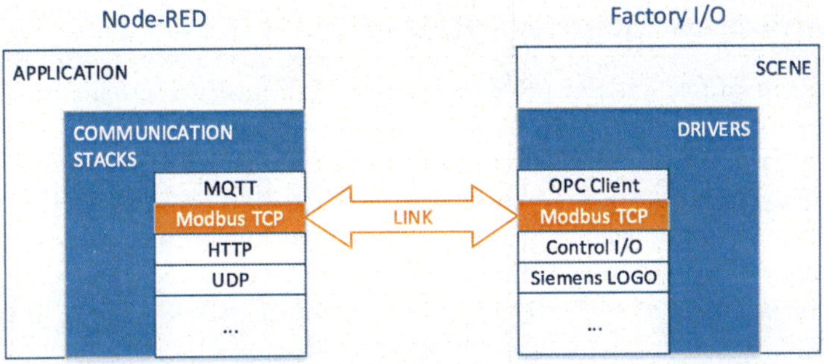


**Fig. 7.** Node-RED and Factory I/O communication link

## 5.2 PID Control with Node-RED

As stated before, the control system of the current tank-level plant is based on a Proportional, Integral, Derivative (PID) controller developed with Matlab/Simulink. Matlab/Simulink is a powerful simulation software with tools for designing and tuning PID controllers. However, it has some limitations in terms of industrial communication interfaces.

On the other hand, Node-RED has some predefined nodes that implement PID control algorithms as well as communication interface nodes. This allows the integration of communication and control into the same software suite. Similar to a Matlab/Simulink PID block, a PID control node receives the set point and the process value through its inputs, and sends the control signal through its output. Then, users can focus on learning and tuning the controller, instead of coding the PID algorithm by themselves.

The process value from the Factory I/O scene and the PID control signal from Node-RED can be exchanged using Modbus TCP interface. This makes it possible to close the control loop and run the process on a single computer (Fig. 8).

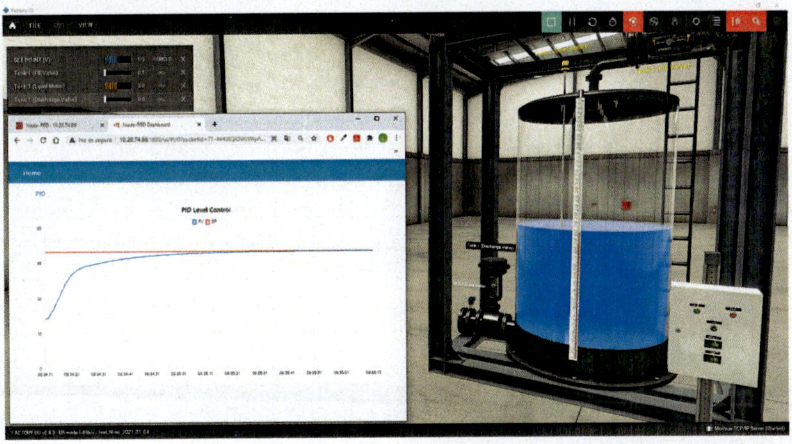

**Fig. 8.** Node-RED and Factory I/O integration

### 5.3 Visualization, Analysis and Results

The operation of the virtual plant created with Factory I/O can be displayed in real time once the control loop has been closed. Before this, it is necessary to run both applications and configure the communication interface parameters. Each application must fix its own IP address and port, as well as the required memory map for data exchange. Furthermore, it is important to verify the scale of the whole system variables to ensure right operation.

The visualization of the virtual plant response is a key aspect. In fact, one control engineering subject is the empirical tuning of PID controllers based on the time domain response parameters. Open-loop as well as closed-loop PID tuning techniques, such as sustained oscillation or relay feedback methods, are widely accepted and remain in use due to their simplicity. Tuning PID controllers with these techniques relies on accurate plotting of the plant response. Therefore, a data acquisition and visualization tool is required.

Node-RED has the node-red-dashboard subset of nodes to plot real time data. In addition to pie charts, line charts and gauges, this subset also has user interface elements such as buttons, sliders, text inputs, and so on. Combining these nodes with the control and communication features, a Complete Graphical user Interface (GUI) along with an analysis and control suite can be deployed.

## 6 Conclusions and Future Works

This work presents a solution to teach online laboratory practice, as an alternative to face-to-face classes with a real laboratory plant. Furthermore, taking advantage of Information Technology (IT) and modern software tools, students can simply design their own virtual and realistic laboratory plant, work individually, and share the results with their classmates and teacher.

Given the approach to the real case shown in the implemented solution section, it can be concluded that this proposal is presented as an interesting method to avoid a major change in scheduling as a consequence of the COVID-19 pandemic.

After testing the proposed solution, we can conclude that very good results have been obtained. Moreover, a great acceptance by students and professors has been achieved.

As future works, the virtualization of other laboratory plants and experiments like the temperature control loops, could help to mitigate the impact of the pandemic on the face-to-face modality. In addition, students also benefit from learning other software tools that could be useful in their future careers.

## References

1. Casteleiro-Roca, J.L., Barragán, A.J., Segura, F., Calvo-Rolle, J.L., Andújar, J.M.: Fuel cell output current prediction with a hybrid intelligent system. Complexity **2019** (2019)

2. Casteleiro-Roca, J.L., Javier Barragan, A., Segura, F., Luis Calvo-Rolle, J., Manuel Andujar, J.: Intelligent hybrid system for the prediction of the voltage-current characteristic curve of a hydrogen-based fuel cell. Revista Iberoamericana de Automática e Informática industrial **16**(4), 492–501 (2019)
3. Crespo-Turrado, C., et al.: Comparative study of imputation algorithms applied to the prediction of student performance. Logic J. IGPL **28**(1), 58–70 (2020)
4. Daniel, J.: Education and the Covid-19 pandemic. Prospects **49**(1), 91–96 (2020)
5. Fernández-Serantes, L.A., Estrada Vázquez, R., Casteleiro-Roca, J.L., Calvo-Rolle, J.L., Corchado, E.: Hybrid intelligent model to predict the SOC of a LFP power cell type. In: Polycarpou, M., de Carvalho, A.C.P.L.F., Pan, J.-S., Woźniak, M., Quintian, H., Corchado, E. (eds.) HAIS 2014. LNCS (LNAI), vol. 8480, pp. 561–572. Springer, Cham (2014). https://doi.org/10.1007/978-3-319-07617-1_49
6. Jena, P.K.: Impact of Covid-19 on higher education in India. Int. J. Adv. Educ. Res. (IJAER) **5** (2020)
7. Jove, E., Casteleiro-Roca, J., Quintián, H., Méndez-Pérez, J., Calvo-Rolle, J.: Anomaly detection based on intelligent techniques over a bicomponent production plant used on wind generator blades manufacturing. Revista Iberoamericana de Automática e Informática industrial **17**(1), 84–93 (2020)
8. Jove, E., et al.: Missing data imputation over academic records of electrical engineering students. Logic J. IGPL **28**(4), 487–501 (2020)
9. Jove, E., Casteleiro-Roca, J.L., Quintián, H., Simić, D., Méndez-Pérez, J.A., Luis Calvo-Rolle, J.: Anomaly detection based on one-class intelligent techniques over a control level plant. Logic J. IGPL **28**(4), 502–518 (2020)
10. Jove, E., et al.: Modelling the hypnotic patient response in general anaesthesia using intelligent models. Logic J. IGPL **27**(2), 189–201 (2019)
11. Laal, M., Laal, M.: Collaborative learning: what is it? Procedia Soc. Behav. Sci. **31**, 491–495 (2012)
12. Luis Casteleiro-Roca, J., Quintián, H., Luis Calvo-Rolle, J., Méndez-Pérez, J.A., Javier Perez-Castelo, F., Corchado, E.: Lithium iron phosphate power cell fault detection system based on hybrid intelligent system. Logic J. IGPL **28**(1), 71–82 (2020)
13. Nicola, M., et al.: The socio-economic implications of the coronavirus and Covid-19 pandemic: a review. Int. J. Surg. (2020)
14. Pérez-López, E., Atochero, A.V., Rivero, S.C.: Educación a distancia en tiempos de covid-19: Análisis desde la perspectiva de los estudiantes universitarios. Revista Iberoamericana de Educación a Distancia **24**(1), 331–350 (2021)
15. Ray, S., Srivastava, S.: Virtualization of science education: a lesson from the Covid-19 pandemic. J. Proteins Proteomics **11**, 77–80 (2020)
16. Vasiliadou, R.: Virtual laboratories during coronavirus (Covid-19) pandemic. Biochem. Mol. Biol. Educ. **48**(5), 482–483 (2020)
17. Vega Vega, R., Quintián, H., Calvo-Rolle, J.L., Herrero, Á., Corchado, E.: Gaining deep knowledge of Android malware families through dimensionality reduction techniques. Logic J. IGPL **27**(2), 160–176 (2019)

# Longitudinal Study of the Learning Styles Evolution in Engineering Degrees

Miguel A. Molina-Cabello[1,2], Karl Thurnhofer-Hemsi[1,2](✉), Enrique Domínguez[1,2], Ezequiel López-Rubio[1,2], and Esteban J. Palomo[1,2]

[1] Department of Computer Languages and Computer Science, University of Malaga, Bulevar Louis Pasteur 35, 29071 Málaga, Spain
{miguelangel,karlkhader,enriqued,ezeqlr,ejpalomo}@lcc.uma.es

[2] Instituto de Investigación Biomédica de Málaga – IBIMA, Málaga, Spain

**Abstract.** A learning style describes what are the predominant skills for learning tasks. In the context of university education, knowing the learning styles of the students constitutes a great opportunity to improve both teaching and evaluation. By using the Honey-Alonso Learning Styles Questionnaire (CHAEA), in this work, we carried out a longitudinal study of the correlation between marks and the results of the survey. The 2018/2019, 2019/2020, and 2020/2021 academic years in two different courses from Computer Sciences engineering degrees were studied. The results were analyzed in order to evaluate the impact of the adaptation to digital lessons during the last year.

**Keywords:** Learning styles · Grades · Web-based education · Higher education

## 1 Introduction

Education researchers have been searching for the factors that affect student evaluation. One of these factors is represented by the student learning styles, which refer to individual approaches or preferences used while receiving, processing, organizing, and using information [3]. They are drawn from the speculations based on behavior observation and personal statements through a variety of learning style inventories [12,13]. Thus, the Learning Style Questionnaire (LSQ) designed by Honey and Mumford [4] was derived from the Learning Style Inventory (LSI) [6] and it was later adopted by Alonso, Gallego, and Honey to the Spanish educational context [1].

Studying the learning styles involves paying attention to the characteristics, cognitive, affective, and physiological behaviors that different students can present and have a direct impact on the acquisition and learning process. This way, learning styles allow us to "conclude" how people act, and are useful for classifying and analyzing behaviors. With that in mind, different learning methodologies take into account these specific features to design the student learning process such as adaptive learning which focuses on maximizing the learning


J. J. Gude Prego et al. (Eds.): CISIS 2021/ICEUTE 2021, AISC 1400, pp. 264–273, 2022.
https://doi.org/10.1007/978-3-030-87872-6_26

performance of each student by deploying learning resources adapted to the student's characteristics [7].

The Honey-Alonso Learning Styles Questionnaire (CHAEA) is a tool designed for identifying four learning styles (activist, reflector, theorist, and pragmatist) that can be present in each student to a different degree. Activists learn by doing, while reflectors learn by observing and thinking about what is happening. Regarding theorists, they need models, concepts and facts to learn. Respecting pragmatists, they are practical people. These modalities are not exclusive so that a person can present several ones. The questionnaire consists of 80 assertions divided into four sections of 20 items corresponding to the four different learning styles. They can be utilized in any educational area at any age. Most of the research papers are focused on applying the CHAEA questionnaire to university students from many different areas, namely, engineering, sanitary, humanities, etc. [2].

The CHAEA questionnaire has been widely employed over the years. The identification of the differences between the results in the study of the learning styles preferences of teachers and postgraduate students was addressed in [11], whereas in [5] the CHAEA questionnaire, together with the ACRA scale was used as potential tools to identify the psychoeducational features of students starting a university degree. Also, a recommendation system of educational strategies based on rules according to the CHAEA questionnaire was proposed in [8]. More recently, the CHAEA questionnaire was used in [9] with engineering university students to analyze development, performance, and gender.

In this paper, a longitudinal study of the learning styles evolution by means of the CHAEA questionnaire is proposed. Concretely, the CHAEA questionnaire was used in two courses from two different computer engineering degrees from the University of Malaga during 3 academic years. Therefore, the information gathered from the CHAEA questionnaire together with the student's marks are not only used to analyze the student learning styles but also the student adaptation to the online lessons due to the pandemic. This way, this study aims to determine if the students' learning styles of a subject and their marks in the different academic years are related. If so, this would allow us in subsequent years to design the course programming adapted to the detected learning styles and thus favour the students' learning process.

The rest of this paper is organized as follows. The academic context is shown in Sect. 2. In Sect. 3 the evaluation methodologies are presented. The project results are given in Sect. 4. Finally, Sect. 5 is devoted to conclusions.

## 2 Academic Context

The University of Malaga offers a wide range of degrees from every branch of knowledge, having 59 bachelor degrees, 53 master's degrees, and 35,000 students. Therefore, diversity is not only present speaking of degrees, but also of academic years and courses. For example, whereas an engineering degree can have students with a more practical profile, students will be more thoughtful in a communications degree. However, even in the same degree, students from different academic years can undergo different profile variations.

**Table 1.** Summary of number of enrolled students (Attd.) and the participants in the survey (Resp.) among the considered subjects.

| Subject | Degree | Course | Year | Attd. | Resp. | Rate |
|---|---|---|---|---|---|---|
| TALF | Comp. Science/Comp. Engineering | 2º | 18/19 | 138 | 55 | 39.86% |
| | | | 19/20 | 135 | 58 | 42.96% |
| | | | 20/21 | 150 | 39 | 26.00% |
| TCIS | Software Engineering | 3º | 18/19 | 60 | 28 | 46.67% |
| | | | 19/20 | 70 | 42 | 60.00% |
| | | | 20/21 | 66 | 36 | 54.54% |

In this work, the information from three courses has been gathered, trying to cover a wide variety. This information is given in Table 1. Courses from the first years of a degree have been avoided since there can be an important bias in the results since many students do not be clear about which degree want to study. Likewise, a sample formed by three different courses from different degrees that belong to the Computer Sciences engineering branch has been used in order to perform a complete analysis.

On the other hand, it would be advisable to contextualize the assessment methodologies of each analyzed course briefly. Automata Theory and Formal Languages (TALF) course is one of the most theoretical courses from the computer science point of view and also has a sizeable mathematical content. The TALF assessment consists of 4 independent mid-term exams that can remove contents from the final exam. The mid-term exams have an only test about the course contents with penalization in failed questions. Students must pass all the mid-term exams to pass this course. Regarding the Computational Techniques in Software Engineering (TCIS) course, its assessment is formed by 13 laboratory practices, two mid-term exams that can remove contents, and a final exam.

In Table 2, the average and standard deviation grades of the overall group is presented for each subject and year is shown. As it can be extracted, one of the

**Table 2.** Average mark and standard deviation by differentiating between those students who have responded to the CHAEA survey, those who have not responded to the survey and all students (who has responded and who not) for each subject and course.

| Subject | Surveyed | 18/19 | 19/20 | 20/21 |
|---|---|---|---|---|
| TALF | Yes | $6.35 \pm 2.30$ | $6.47 \pm 2.13$ | $6.26 \pm 1.88$ |
| | No | $4.82 \pm 2.75$ | $5.33 \pm 2.72$ | $5.39 \pm 2.31$ |
| | Total | $5.43 \pm 2.68$ | $5.82 \pm 2.54$ | $5.62 \pm 2.23$ |
| TCIS | Yes | $8.17 \pm 0.98$ | $8.03 \pm 1.39$ | $8.27 \pm 1.40$ |
| | No | $7.93 \pm 1.24$ | $8.23 \pm 1.27$ | $8.04 \pm 1.22$ |
| | Total | $8.04 \pm 1.13$ | $8.11 \pm 1.34$ | $8.17 \pm 1.32$ |

subjects, TCIS, has outstanding qualifications since it was easy for the students. That is because its evaluation is based on practices without the need to pass a final exam. On the other hand, TALF is a complex theoretical subject that many students fail in it.

## 3 Evaluation Methodologies

There are several tools that allow to compute the learning styles of the students in an online way. In this work Moodle has been the selected software since our university uses it as learning management system [10]. This software provides many resources, the quiz module among them, which easily supports the management of a questionnaire.

At the beginning of the course, students are asked to fullfil the CHAEA questionnaire. This task is not mandatory, so that it is optional and it does not affect the final grade. Regarding the final evaluation, it is computed at the end of the course as usual.

It is known that the instructor has the possibility of selecting a simple grading or a complex one. In this work, with the objective of a clear explanation of the grading process, each student is graded with a numerical evaluation in the range between 0 and 10.

## 4 Project Results

Next, the results of the surveys and the correlation with the marks are presented for each of the analyzed subjects.

First of all, in Table 1 is presented the respondents rate with respect to the total number of students of each subject, during the last three years. Due to several students do not finish the CHAEA survey, a student is considered that has completed the survey if that survey has at least the 75% of the questions of the survey. So that it is considered as completed if it has at least 60 responded questions (CHAEA survey has 80 questions). As it can be seen, there is not a clear tendency, although the greatest differences are in the subject TALF, where there are a large number of students. The online modality of the year 20/21 has not motivated the students to complete the survey more than the past years. The equality in the number of respondents will facilitate the extraction of conclusions of the evolution among years.

The average learning style profiles for each subject are presented in Figs. 1 and 2. The colors represent the results of each year. In TALF, there is a small but not very significant change in the type of learning style. In the first two years the students seemed to be more active and theorist, while the course 20/21 pushed their reflexive side. That is, since most of the lessons were carried out in an online mode, the students tend to analyze and observe more the explanations and exercises than trying to participate actively in the lessons.

In the case of TCIS, no differences were found. This behavior might be caused due to the different evaluation system. In this subject, several small practices

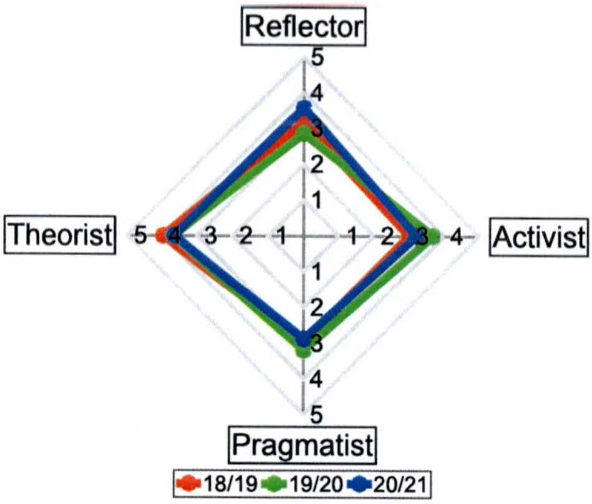


**Fig. 1.** Average learning style profile of the subject TALF.

are proposed to the students, and two mid-term exams facilitates the elimination of content. That is, the evaluation is continuous and it has not changed in the almost online year 20/21. However, the way the lessons were given in TALF have changed a bit and practice lessons have almost disappeared, provoking a small change in the learning style of the students.

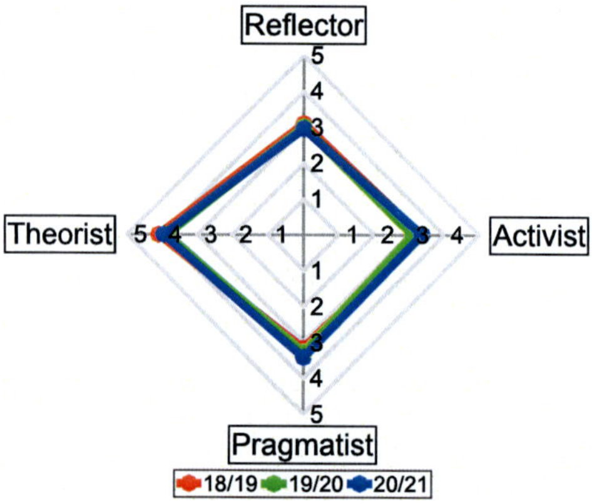


**Fig. 2.** Average learning style profile of the subject TCIS.

In order to analyze the marks, Figs. 3 and 4 exhibit the whisker plots for subjects TALF and TCIS, respectively, where the maximum whisker length has been specified as 0.75 times the interquartile range. On one hand, in Fig. 3 it can be appreciated a difference in the marks obtained by those students who completed the survey and those who did not. Except for some outlying cases, the median value is almost 2 points lower and the spread in higher. In general,

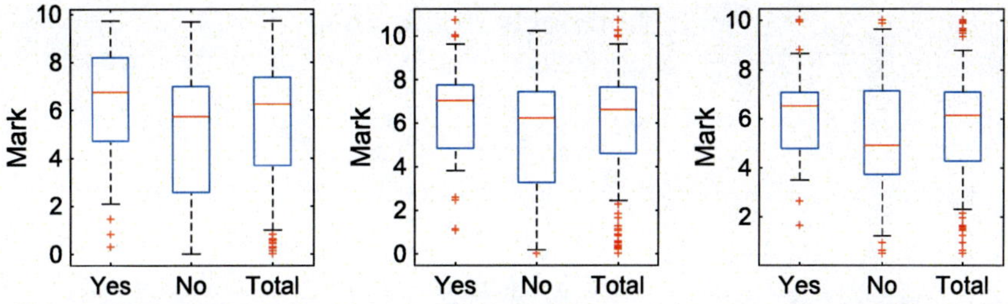


(a) 18/19 (b) 19/20 (c) 20/21

**Fig. 3.** Whisker plot of marks, by differentiating between those students who have responded to the CHAEA survey, those who have not responded to the survey and all students (who have responded and who not) for subject TALF.

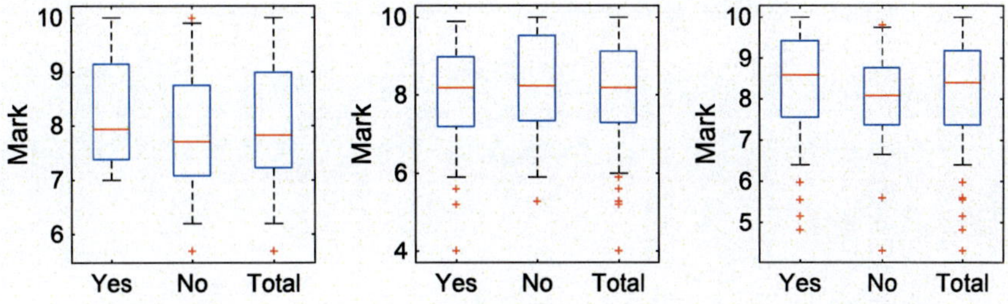


(a) 18/19 (b) 19/20 (c) 20/21

**Fig. 4.** Whisker plot of marks, by differentiating between those students who have responded to the CHAEA survey, those who have not responded to the survey and all students (who have responded and who not) for subject TCIS.

there is a great variety of marks within this set of students, which indicates that the subject is complex and needs special effort to pass it. The evolution along the years is negative, that is, the grades have got worse. The cause is unclear but it can be the decreasing level of mathematical fundamentals required for this subject.

On the other hand, TCIS presented better marks in Fig. 4, and the median and dispersion are very similar for any type of student. Nevertheless, again the students with less interest in the questionnaire are those who obtained lower marks, with the exception of the course 19/20. The overall grades are good (greater than 8) and there are few falling students.

In Table 2 the average marks can be compared. The tendency is very similar to the median represented in the whisker plots. The major discrepancies are found in TALF, where the mean values do not show a clear decrease. Even in the course 20/21, it has increased slightly with respect to two years ago. The case of the TCIS subject is analogous. The unexpected result is the 8.23 points of the course 19/20, which is higher for the students who did not complete the

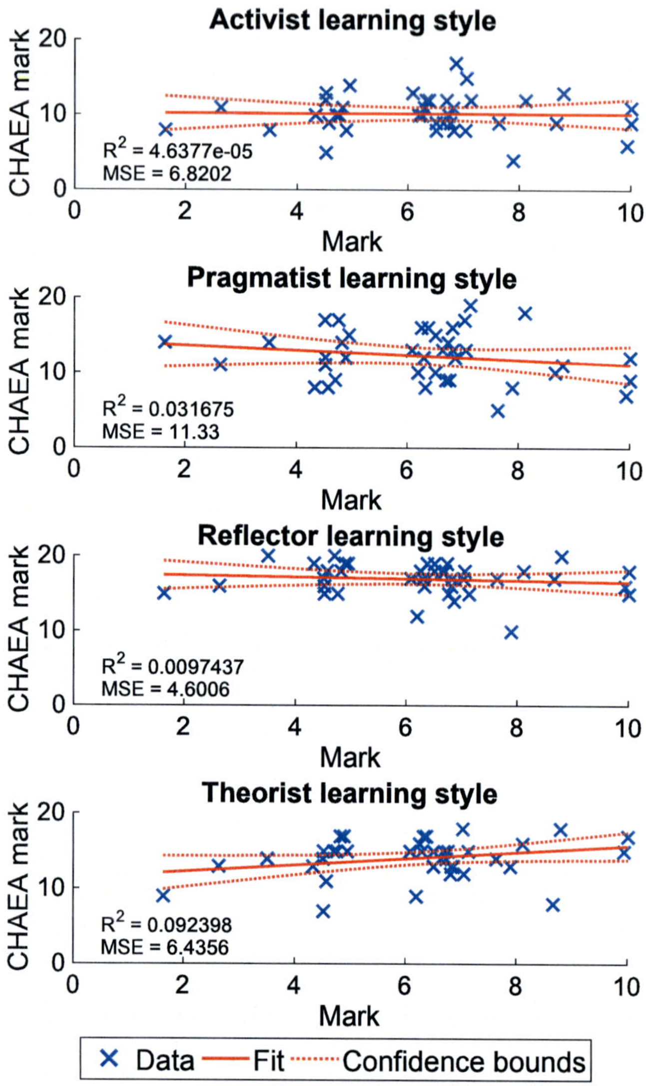


**Fig. 5.** For each learning style, the subject grades and the CHAEA marks of all participants have been analyzed with a linear regression model for subject TALF in the course 20/21.

CHAEA survey. The standard deviation of both subjects goes accordingly with what was shown in the previous figures.

By last, the correlation of the CHAEA marks and the subject grades is analyzed in Figs. 5 and 6, using the information of the last course 20/21. A linear regression was computed, using as samples each of the students of the subject. $R^2$ (where the higher, the better) and MSE (the lower, the better) measures are

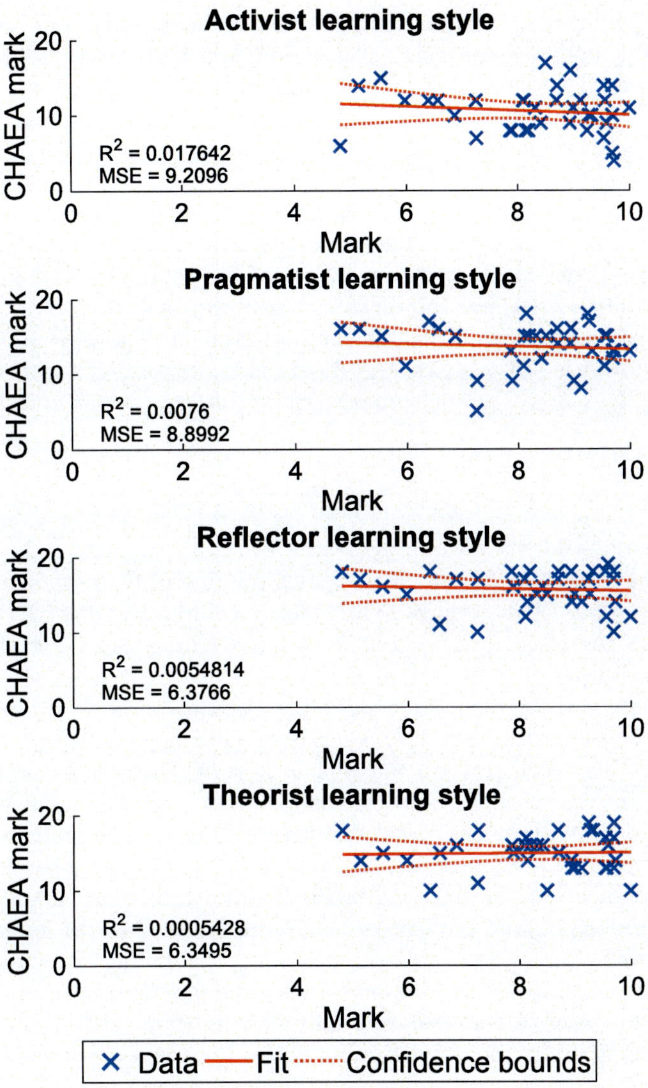


**Fig. 6.** For each learning style, the subject grades and the CHAEA marks of all participants have been analyzed with a linear regression model for subject TCIS in the course 20/21.

also provided to compare better the relation between assessment mark (which is in the range 0 and 10) and CHAEA mark (range 0 and 20) of each learning style for each student of the studied subjects. In the case of TALF, the best fits are achieved for both the reflector and theorist learning styles, where the residual values are the lowest. Actually, the best CHAEA marks are obtained in these styles, which confirm that the students are more reflexive than past years. The pragmatist style depicted the worst results, followed by the activist. Regarding the grades, it can be checked that the best students are theoretical except for one of them, who have an equilibrium in the distribution of his capacities. There are two main groups of students, one of the over 5 points (grade) and the other below. The group who passed the subject is more theoretical and active, while the other is more reflexive and a bit pragmatic. Therefore, this subject evaluation model seemed to be more adequate for those students that do not need to participate actively in the lessons. TCIS presents very similar results for the four learning styles, being the theorist and the reflector those ones with less error. No clear trend is present in any style, although the best students obtained better marks in the theorist and reflector styles.

## 5 Conclusions

The utility of the learning styles resides in the knowledge that provides to both the teachers and students on improving the learning procedure. On one side, teachers would be able to have a better understanding of their students, and to adapt their methodologies to have more efficiency and save time. On the other hand, students can identify their profile, and use strategies to face the course more appropriately.

In this work, a study over three years has been elaborated to evaluate the relationship between learning styles and grades concerning two subjects of Computer Sciences Engineering. It was found that most of the students are theorists, and in second place, reflexive. Besides, the evolution over the years and different learning contexts do not reflect a big change in their predominant learning styles. The impact of the pandemic, and therefore the change of some teaching methods, did not affect the final marks of the students, which might indicate that the methodologies and evaluation were quite adapted to the audience.

Future works intend to analyze the evolution of learning styles and marks of the students on different degrees, not only in Computer Science, where digitalization is intrinsic to the field. Actually, having more historic data would provide teachers a useful tool for designing better the whole academic course for a specific degree.

**Acknowledgments.** This work is partially supported by the University of Malaga (Spain) under grants PIE19-161, project name Cooperative Evaluation Procedures in Active Methodologies, and PIE19-219, project name Critical review of evaluation systems. The authors thankfully acknowledge the grant of the Universidad de Málaga, I Plan Propio Integral de Docencia.

## References

1. Alonso, C.M., Gallego, D.J., Honey, P.: Los estilos de aprendizaje: procedimientos de diagnóstico y mejora. Mensajero Bilbao, España (1997)

2. Escanero-Marcen, J., Soria, M., Guerra-Sánchez, M.: Estilos de aprendizaje y rendimiento académico: diferentes herramientas, diferentes resultados. Revista de la Fundación Educación Médica **21**(4), 173 (2018)
3. Göğüş, A., Ertek, G.: A scoring approach for the assessment of study skills and learning styles. Int. J. Inf. Educ. Technol. **10**(10), 715–722 (2020)
4. Honey, P., Mumford, A.: The Manual of Learning Styles. Peter Honey Maidenhead (1982)
5. Juarez Lugo, C.S., Rodriguez Hernandez, G., Luna Montijo, E.: The learning styles questionnaire CHAEA and the learning strategies scale ACRA as potential tools for academic tutoring. J. Learn. Styles **5**, 148–171 (2012)
6. Kolb, D.: Learning Style Inventory Technical Manual. McBer (1976)
7. Lai, S., Sun, B., Wu, F., Xiao, R.: Automatic personality identification using students' online learning behavior. IEEE Trans. Learn. Technol. **13**(1), 26–37 (2020)
8. Marín, P.A.R., Herrán, A.O., Darío Duque Méndez, N.: Rules based system to educative personalized strategy recommendation according to the CHAEA test. In: 2016 XI Latin American Conference on Learning Objects and Technology (LACLO), pp. 1–7 (2016)
9. Mena Lorenzo, J.L., Rodriguez-Pulido, J., Mena Lorenzo, J.A., Navarro-Guzman, I.J., Silvio Cabrera-Guzman, J.: Learning styles of engineering students: development, performance and gender. Eur. J. Educ. Psychol. **12**, 175–189 (2019)
10. Rice, W.: Moodle E-Learning Course Development. Packt Publishing Limited, Birmingham (2015)
11. Santizo Rincon, J.A., Garcia Cue, J.L., Gallego, D.J.: Two methodes for the identification of differences between studies where the CHAEA has been applied. J. Learn. Styles **1**(1), 28–42 (2008)
12. Soflano, M., Connolly, T.M., Hainey, T.: Learning styles in adaptive games-based learning. In: Herrero, Á., et al. (eds.) International Joint Conference SOCO'13-CISIS'13-ICEUTE'13. Advances in Intelligent Systems and Computing, vol. 239, pp. 709–717. Springer, Cham (2014). https://doi.org/10.1007/978-3-319-01854-6_72
13. Weinstein, C.E., Palmer, D.R., Acee, T.W.: LASSI User's Manual: For Those Administering the Learning and Study Strategies Inventory, 3rd edn. (2016)

# Peer Assessments in Engineering: A Pilot Project

Karl Thurnhofer-Hemsi[1,2](✉), Miguel A. Molina-Cabello[1,2], Esteban J. Palomo[1,2], Ezequiel López-Rubio[1,2], and Enrique Domínguez[1,2]

[1] Department of Computer Languages and Computer Science, University of Malaga, Bulevar Louis Pasteur 35, 29071 Málaga, Spain
{karlkhader,miguelangel,ejpalomo,ezeqlr,enriqued}@lcc.uma.es
[2] Instituto de Investigación Biomédica de Málaga—IBIMA, Málaga, Spain

**Abstract.** The evaluation methods employed in a course are the most important point for the students, above any other learning aspect. For teachers, this task is arduous when the number of students is high. Traditional evaluation requires the teacher to grade all the assignments and exams, while peer assessments have become a valuable tool to involve students effectively in the correction of exercises. This paper applied and analyzes the evaluation through peer review in a course of Computer Sciences Engineering. A total of six assignments and a mid-term exam were evaluated by both teachers (individually) and students (cooperatively), and the differences were discussed to extract conclusions about the viability of this evaluation model.

**Keywords:** Peer assessment · Cooperative assessment · Peer review evaluation · Web-based education · Higher education

## 1 Introduction

More and more modern methodologies are replacing traditional approaches to develop professional and personal skills in students producing highly knowledgeable individuals, stressing problem-solving in real-life contexts [10].

Assessment-based methodologies are useful for giving students valuable feedback and information on their marks. However, the number of students in higher education is higher than in primary and secondary education, which makes increases the teacher's workload.

One strategy to minimize that heavy workload is to implement peer assessment (PA), where students evaluate the work of their classmates from which an aggregate grade is obtained [1,6,17]. Such a paradigm is not only ideal for reducing the teacher's workload, but it also allows students to learn from alternative solutions to the same problems proposed by their peers [9,15]. Thus, both student interests and skills are positively affected by using this kind of methodology [4,5,7]. In fact, PA is considered by researchers as an effective pedagogical strategy involving peer evaluations with scoring methods and feedback


J. J. Gude Prego et al. (Eds.): CISIS 2021/ICEUTE 2021, AISC 1400, pp. 274–283, 2022.
https://doi.org/10.1007/978-3-030-87872-6_27

to improve learning performance [16]. Peer-assisted evaluation can be seen as a part of peer-assisted learning [2,3,18].

This strategy is also known as peer review, which can be defined as a reciprocal process whereby students evaluate and make judgments about the work of their peers and construct a written feedback commentary [9]. Therefore, students both provide feedback reviews on others' work and receive feedback reviews on their own work. This represents an important alternative to teacher feedback without increasing teacher workload. However, some drawbacks are also presented in peer reviews, such as the students' ability to produce meaningful feedback, fairness, and biases in reviewing, collusion and plagiarism, etc.

Nevertheless, most of these drawbacks can be faced through well-designed peer review tasks [13]. One of the tools to facilitate peer review is the use of a set of evaluation rules or rubrics so that students have limitations when it comes to evaluating the work of their peers, as well as they are forced to produce a numerical score for each evaluated activity [12,15]. Scoring rubrics provide a reliable and valid assessment in a structured format that makes students more confident in this task [11].

In this paper, a study of peer assessment in a course taught in four different Computer Science degrees is presented, where the marks obtained from 900 students and the teacher during 3 academic years in different assignments and mid-term exams are analyzed.

The paper is structured as follows. The academic context of this work is given in Sect. 2, whereas the evaluation methodologies are shown in Sect. 3. Section 4 provides the project results. Finally, some discussions and the conclusions are given in Sects. 5 and 6.

## 2 Academic Context

The Spanish higher education system is regulated by the Universities Organic Law (UOL). Moreover, Spain is also a member of the World Conference on Higher Education, which is sponsored by the United Nations Educational, Scientific and Cultural Organization (UNESCO) and the Bologna process.

On the one hand, the World Conference on Higher Education has the main goal to address the most important challenges in higher education with their suggested indications. Thus, different conferences have been developed in order to prepare the work of the four commissions of the world conference: relevance, quality, management and financing, and international cooperation. Also, several considerations have been performed such as the take of actions to execute projects and support international cooperation based on solidarity and the construction of an equal society thanks to research, community projects, and specialist training.

On the other hand, the Bologna process involves many European countries in order to achieve different objectives, such as the support of student mobility, the proposition of comparable teaching methodologies and evaluations, and the motivation of institutional cooperation. This process also establishes a comprehensible and comparable education system. This way, several considerations can be yielded easily such as the search for jobs. The proposed degree system is based on two levels: undergraduate (B.Sc.), and graduate (M.Sc.). In order to organize those degrees, the European Credit Transfer System (ECTS) is used, so that student interchange is achieved among other considerations.

Spanish universities can design their own degrees, but always under the umbrella of the current regulations. At this moment, three different levels compose the Spanish higher education system: undergraduate (B.Sc.), graduate (M.Sc.), and doctorate (Ph.D.). Therefore, the Spanish national common degree catalog from the previous system has been removed.

Currently, undergraduate degrees have between 180 and 240 ECTS credits (there are some exceptions with 300 or 360 ECTS credits), while most graduate programs have between 60 and 120 ECTS credits. The common undergraduate program has 240 ECTS credits and the usual graduate program has 60 ECTS credits, where the academic year has 60 ECTS credits for both undergraduate and graduate programs.

The organization of the courses has changed from the previous contents system to the competencies system with the Bologna process. Therefore, students are now given their degrees if they acquire the corresponding competencies, which have to be assessable.

The current educational innovation project has been applied to students from the Intelligent Systems course during 3 academic years. This course is taught during the fifth semester (third academic year) in 4 different degrees at the School of Computer Science at the University of Malaga (Spain), namely, the Computer Science, Software Engineering, Computer Engineering, and Health Engineering degrees. The results obtained from the peer assessment in this course are shown in the next section.

## 3 Evaluation Methodologies

There are several tools that support peer assessment in an online way. In this work Moodle has been the selected software due to our university uses it as learning management system [14]. This software facilities many resources. In particular, the workshop module, which may be considered as one of the most powerful peer assessment tool [8]. This way, the activities that students do are carried out with that workshop tool.

Intelligent Systems is the first course of the degree related to Artificial Intelligence (AI); therefore, this course deals with the fundamentals concepts of AI by introducing the diverse and wide area of AI. This is a practical course in which most of the lessons implies lab activities where students apply the theoretical concepts by managing and solving different problems. These lab sessions are complemented with some lecturing where theoretical fundamentals are presented. Lab sessions deal with simple applications of the algorithms presented in the theoretical lessons. Apart from the lab sessions (10% of the final grade), there are three assignments (each one of 30% of the final grade) about the theoretical/practical concepts: Data mining, Advanced models, and Search and Logic.

These assignments comprise the correct application and explanation of different algorithms in order to solve the addressed problems.

The procedure for a given activity or task is as follows. In the first step students do their work and submit it. Then, the teacher assigns a fixed number of submissions to each student and provide a rubric. After that, students use the rubric to evaluate their assigned submissions from other students. In order to achieve an impartial assessment, submissions are anonymous. And finally, the instructor gives a mark for each work by selecting its own evaluation and those provided by the students.

Instructor mark may be computed by using a simple grading or a more complex one. In this work, a numerical evaluation between 0 and 100 has been used as a grade. This kind of evaluation allows a clearer interpretation of the grading process.

## 4 Project Results

The innovative education project, as commented above, involves 4 courses with a total of more than 900 students during 3 academic years. All courses share a common program with the same evaluation process, which is composed by 6 mandatory assignments, a mid-term exam, and a final exam. The evaluation of the works are carried out by rubrics, which are specifically designed for each assignment. The mandatory assignments and the mid-term exam are peer reviewed by the students and professors. Due to the time limitation in the period exam, the last final exam is not peer reviewed by students, but we are studying other choices in order to include this activity in the peer review process by changing the dates and the evaluation process in the following years.

The deviation of students grade compared with teachers grade for each assignment is shown in Fig. 1. The assignments are evaluated over 100 points, and the mark differences are calculated by subtracting the teacher grade of students grade. In this sense, a negative mark difference means peer reviewers have been more strict than teacher evaluating the work. In contrast, a positive mark difference means the student has been over graded, where reviewers have shown a less rigorous evaluation than teachers.

According to Fig. 1, most of evaluation done by peer reviewers were over graded around 10% respect to the teacher grade. Interestingly, this over rating was greater in mid-term exams, as Fig. 2 shows. Furthermore, most of the outliers are located in the negative part (i.e., reviews are more rigorous) for the assignments. However, these outliers are located in the positive part for the mid-term exam. Regarding to the medians of marks, the behavior is similar: the median of the marks differences are close to 20 (over 100) for the mid-term exam, while the medians are close to zero in the rest of assignments. That is, except for the mid-term exam, the reviews were very fair on average comparing with the teacher grade.

These imbalances in the evaluation process are clearly observable in Fig. 3 where the histograms of marks differences are shown for both mid-term exam

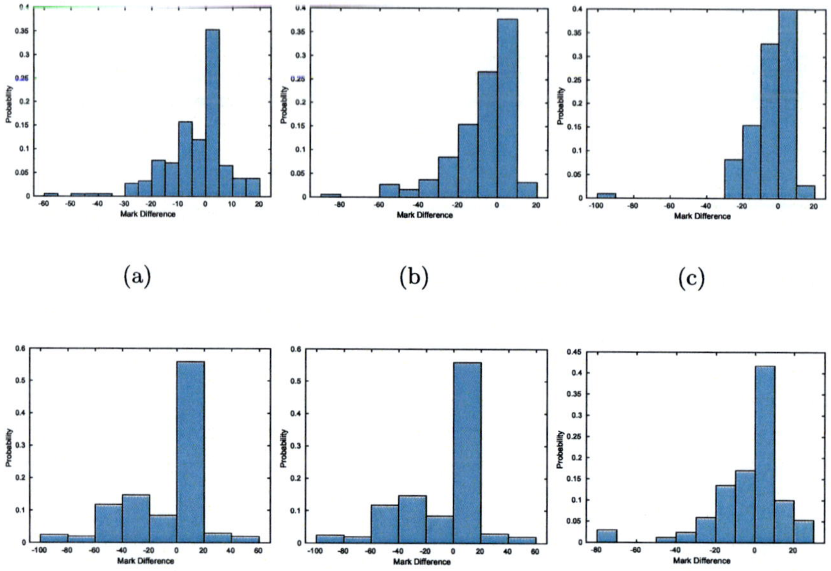

(a) (b) (c)

(d) (e) (f)

**Fig. 1.** Histogram of mark differences for several assignments.

and the rest of assignments. Possible causes of this variation are mainly related to the weight of the activity in the final grade of the course, since the contribution of the mid-term exam grade to the final grade of the course is greater than the rest of assignments due to the current evaluation procedure.

The comparison of Fig. 4 analyzes the relationship between the reviewer grades obtained by the students and the difference from the instructor's mark. The first appreciation is that the best reviewers have the lowest difference with respect to the teacher correction. Moreover, successively, these ranges are more extent for the students that obtained lower reviewer grades. It is remarkable that there are good reviewers evaluating more strictly, 40% lower than the teacher (the outlying points), which indicates they are very critical with peers. Nevertheless, the majority of the students are optimistic in their assessments.

Finally, we have compared the evolution of the reviews in peer assessment during the last three academic courses in Fig. 5. The median shows that the results are very similar for all years, despite the fact that the academic contexts are different for each course. In 2018/2019, there is exceptional stability of the mark differences around zero. However, in the following course, 2019/2020, lots of students tended to rate very strictly, having many negative differences. However, during the last year, in which the lessons were mainly given online, the evaluation of the reviewers was positive, as the median reflects. Therefore, the background of the students and of the teaching modality might affect a bit the performance of peer assessment tasks, but with no many deviations.

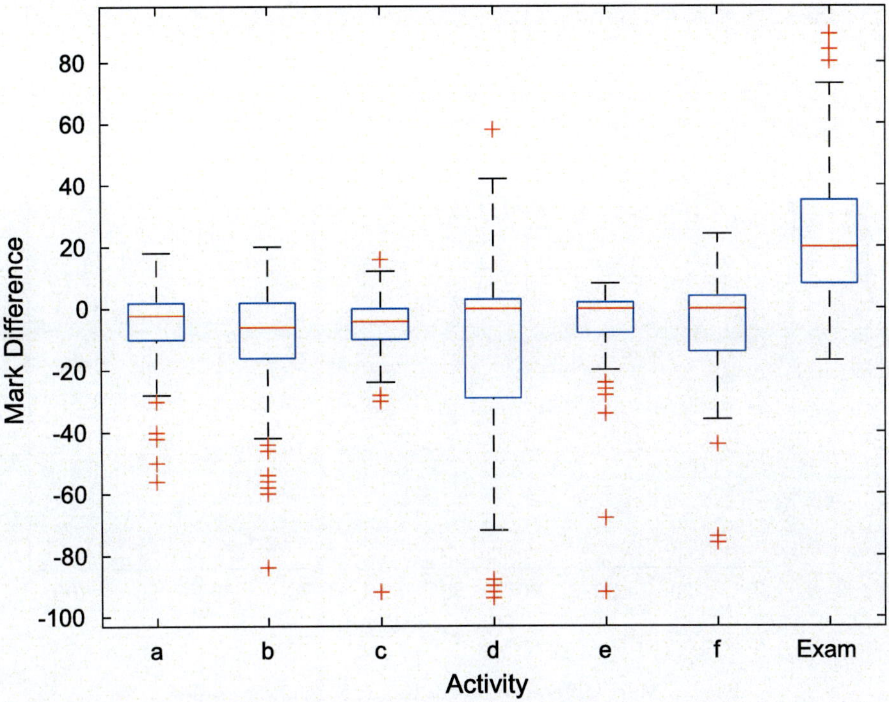


**Fig. 2.** Comparison of mark differences between the mid-term exam and the rest of assignments.

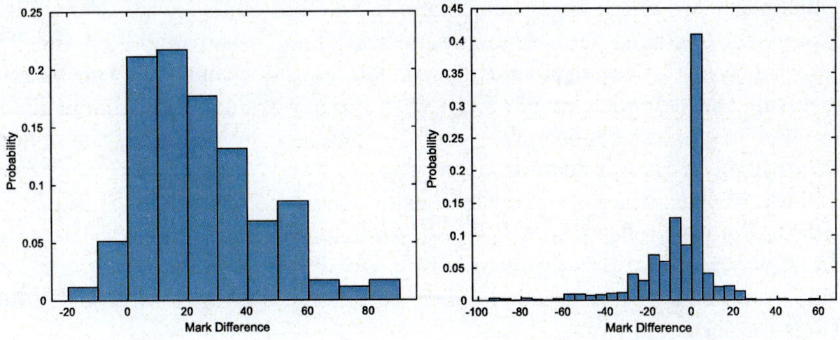


(a) Mid-term exam.

(b) Rest of activities.

**Fig. 3.** Comparing histograms between exams and the rest of assignments.

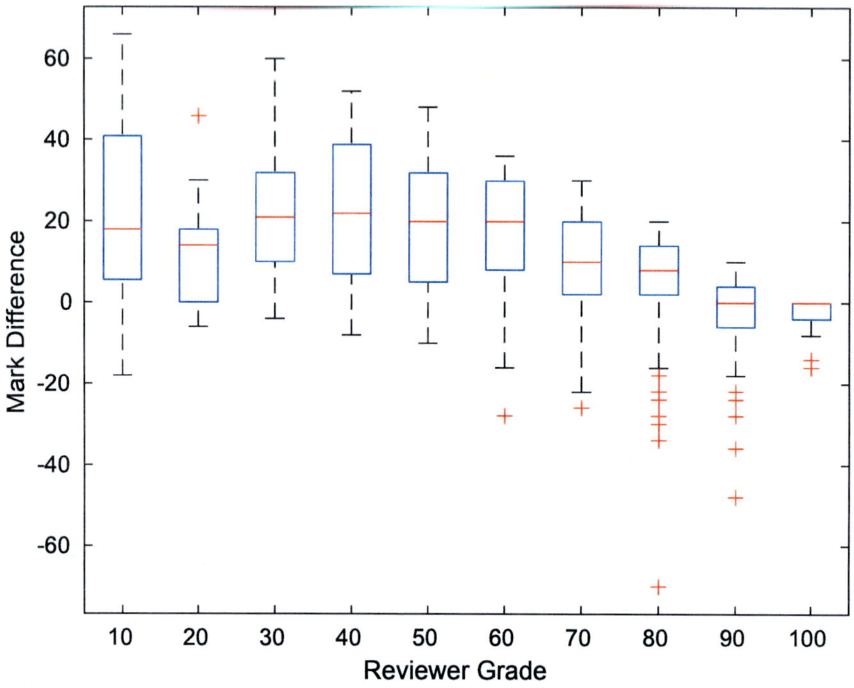


**Fig. 4.** Mark differences according the reviewer grades.

## 5 Discussion

An important point that needs to be discussed is the presence of outliers in the mark differences represented in Figs. 2, 4 and 5. As we can see in the first outcome, there are many negative outliers in the mid-course activities, while in the exam, the students tend to be optimistic. The reason might be the level of importance of the activity; that is, students consider that they can be more strict during the year with small activities since they are not determinant for the final mark. So a balance mechanism could be included by weighting the review's marks strongly in those mid-term activities.

Related to that, it is important to consider the type of students that provoke this effect. Figure 4 reflects that the best students make more restrictive assessments. A way to avoid this problem is by providing a template of a generalized solution combined with a detailed rubric to extend the range of possible valid solutions for these students.

On the other hand, when the course 2018–2019 finished, the mark difference analysis showed that there were outliers on both sides (optimistic and pessimistic), so it was difficult to determine a prevention mechanism for the next years. Actually, the following academic courses showed disparate tendencies,

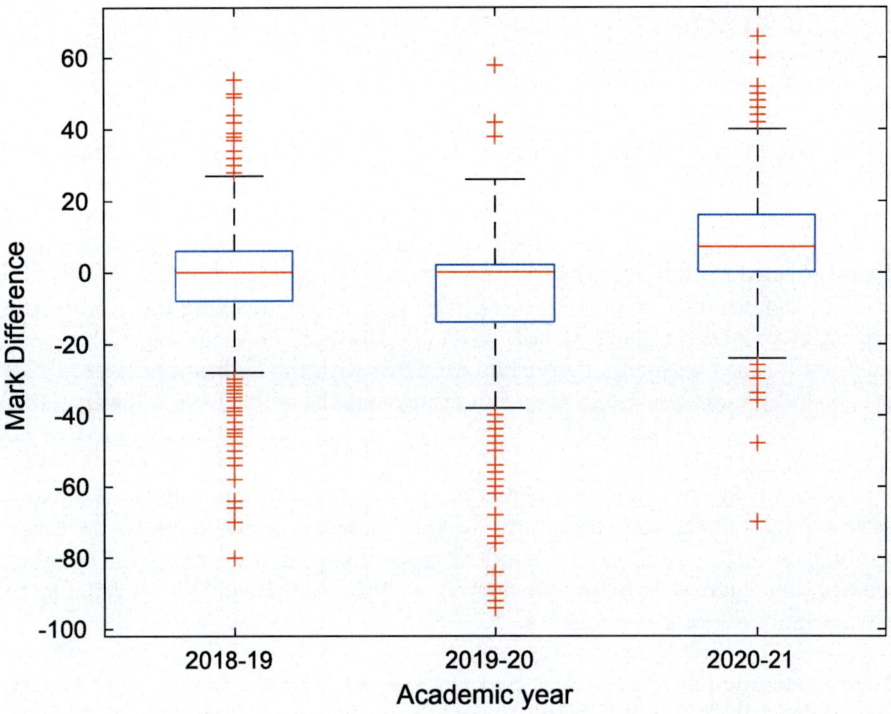


**Fig. 5.** Comparison of mark differences along academic years.

which implicate that the student context (type of students) strongly influences the peer assessment.

On the other hand, these peer assessment tasks were applied to a specific subject, which comprises a set of assignments. The type of exercise is also relevant in the results presented in this work. These assignments consisted of the application of the algorithms taught in the lessons, giving the results and an explanation of those. Thus, the evaluation by peer assessment can be completely objective since the results are fixed, and the reasoning is the same for all students. However, if the teaching objectives are different and the required activities are subjective, peer assessment could not be adequate. For example, if answers are long texts that need to be interpreted, one student may think differently than others, and the assessments would be quite different.

Therefore, this evaluation method may be only suitable for any objective evaluation. In addition, the workload is not that much when the evaluation criteria are clear. Only those sets of task assessments whose marks are scattered need to be analyzed by the teacher. If the evaluation marks are similar among the student, the teacher does not have to re-evaluate the task.

## 6 Conclusions

Cooperative evaluation allows teachers to reduce their workload and have additional points of view of the quality of the exercises done by students. The case study analyzed in this work showed a good effectiveness of the evaluation. Students typically have no more than 20% error on instructor grades, and even they tend to be very strict with their mates. Nevertheless, the importance of an exam in the overall evaluation may affect the student's grades award, since most of them provided optimist marks.

Although students reported a positive feedback about being involved in the evaluation procedure, this task still have a dependency on the teacher, that is, it must not be used without supervision. Students with the highest reviewer grades are generally more precise in the evaluation, and the worst have a lower grade, which indicates peer assessment is coherent and suitable to be included as an evaluation method.

The tendency over the last three years is stable, but the impact of remote lessons might affect the corrections of the students, which have to be taken carefully in future studies. The application of this peer review process to other courses and degrees is being considered, so more data would be available to extract more relevant conclusions.

**Acknowledgments.** This work is partially supported by the University of Málaga (Spain) under the grants PIE19-161, project name Cooperative Evaluation Procedures in Active Methodologies, and PIE19-219, project name Critical review of evaluation systems. The authors thankfully acknowledge the grant of the Universidad de Málaga, I Plan Propio Integral de Docencia.

## References

1. Brooks, C.M., Ammons, J.L.: Free riding in group projects and the effects of timing, frequency, and specificity of criteria in peer assessments. J. Educ. Bus. **78**(5), 268–272 (2003)
2. Glynn, L.G., MacFarlane, A., Kelly, M., Cantillon, P., Murphy, A.W.: Helping each other to learn - a process evaluation of peer assisted learning. BMC Med. Educ. **6**(1), 18 (2006). https://doi.org/10.1186/1472-6920-6-18
3. Hodgson, Y., Benson, R., Brack, C.: Student conceptions of peer-assisted learning. J. Furth. High. Educ. **39**(4), 579–597 (2015)
4. Hodson, R.J., Chan, W.M., Chin, K.N., Nagami, M., Suthiwan, T.: Effects of online cooperative evaluation and group rewards on high school students' argumentative writing attitudes and interaction behaviors, pp. 383–412. De Gruyter Mouton, Boston (2011)
5. Kim, C.J.: The effects of peer assessment and peer feedback in writing education for premedical students. Ewha Med. J. **40**(1), 41–49 (2017)
6. McLarnon, M.J., O'Neill, T.A., Taras, V., Law, D., Donia, M.B., Steel, P.: Global virtual team communication, coordination, and performance across three peer feedback strategies. Can. J. Behav. Sci. **51**, 207 (2019)
7. Montanero, M., Lucero, M., Fernández, M.J.: Iterative co-evaluation with a rubric of narrative texts in primary education. J. Study Educ. Dev. **37**(1), 184–220 (2014)

8. Nash, S.S.: Moodle 3.x Teaching Techniques. Packt Publishing Limited, Birmingham (2016)
9. Nicol, D., Thomson, A., Breslin, C.: Rethinking feedback practices in higher education: a peer review perspective. Assess. Eval. High. Educ. **39**(1), 102–122 (2014)
10. Palomba, Catherine A., Banta, T.W.: Assessment Essentials: Planning, Implementing, and Improving Assessment in Higher Education. Higher and Adult Education Series. Jossey-Bass (1999)
11. Panadero, E., Jonsson, A.: The use of scoring rubrics for formative assessment purposes revisited: a review (2013)
12. Panadero, E., Romero, M., Strijbos, J.W.: The impact of a rubric and friendship on peer assessment: effects on construct validity, performance, and perceptions of fairness and comfort. Stud. Educ. Eval. **39**(4), 195–203 (2013)
13. Pearce, J., Mulder, R.A., Baik, C.: Involving Students in Peer Review: Case Studies and Practical Strategies for University Teaching, pp. 1–29. Centre for the Study of Higher Education (2009)
14. Rice, W.: Moodle E-Learning Course Development. Packt Publishing Limited, Birmingham (2015)
15. Rico-Juan, J.R., Gallego, A.J., Calvo-Zaragoza, J.: Automatic detection of inconsistencies between numerical scores and textual feedback in peer-assessment processes with machine learning. Comput. Educ. **140**, 103609 (2019)
16. Rico-Juan, J.R., Gallego, A.J., Valero-Mas, J.J., Calvo-Zaragoza, J.: Statistical semi-supervised system for grading multiple peer-reviewed open-ended works. Comput. Educ. **126**, 264–282 (2018)
17. Román-Calderón, J.P., Robledo-Ardila, C., Velez-Calle, A.: Global virtual teams in education: do peer assessments motivate student effort? Stud. Educ. Eval. **70**, 101021 (2021)
18. Ross, M.T., Cameron, H.S.: Peer assisted learning: a planning and implementation framework: AMEE Guide no. 30. Med. Teach. **29**(6), 527–545 (2007)

# Game-Based Student Response System Applied to Soil Mechanics and Foundations

Maria-Victoria Requena-Garcia-Cruz(✉), Emilio Romero-Sánchez, and Antonio Morales-Esteban

Department of Building Structures and Geotechnical Engineering, University of Seville, Seville, Spain
{mrequena1,eromero13,ame}@us.es

**Abstract.** Research on student-focused approaches have been gaining importance over the years to avoid classical learning methods. In this context and linked to the development of the technology, Game-based Student Response Systems (GSRS) have been proved to enhance the students' engagement, their learning and their performance. In this study, a pilot experience has been presented using a GSRS tool called *Kahoot!* in the context of an engineering/architectural course. Several competitive games have been developed in which students answer questions interactively. The motivation and engagement of the students have been assessed by a proved scientific tool named the Keller's survey. The main objectives of this paper are: i) to assess the degree of acceptance by the students of the inclusion of game-based learning; ii) to estimate the benefits of *Kahoot!* in terms of the students' motivation and academic performance enhancement; and, iii) to verify the capacity of this method to identify learning deficiencies to improve the design of the course. The key findings have been that *Kahoot!* has increased the interest of the students in the contents of the course, enhancing their motivation and attention. The results of this research eased the assessment of the understanding of each question and the topics of the course. Therefore, GSRS enable professors to obtain feedback from students, generally and individually, and, to some extent, to adapt their teaching activities based on the students' responses.

**Keywords:** Student response system · Higher education · Performance · Motivation · STEM disciplines

## 1 Introduction

Maintaining students' attention and engagement can be difficult in Science, Technology, Engineering and Mathematics (STEM) disciplines [9]. Moreover, this worsens if the teaching method is based on teacher-centered classes, which limit the participation of the students. Consequently, they become increasingly bored, present a low academic performance and usually end up dropping out the courses [10]. Evidence of this problem can be found in the Soil Mechanics and Foundations (SMF) subject [14]. It is taught in the third year of the Degree in Fundamentals of Architecture at the University of Seville. The teaching approach of this course is based on long lectures combined with practical


J. J. Gude Prego et al. (Eds.): CISIS 2021/ICEUTE 2021, AISC 1400, pp. 284–294, 2022.
https://doi.org/10.1007/978-3-030-87872-6_28

lessons in a very limited time. Therefore, SMF students tend to engage passively with the lesson material and often struggle to understand the course's complexity and numerous concepts [13].

This reveals the need to do further research on how to enhance SMF students' academic performance as well as their motivation. In this sense, student-focused approaches have been used over the last years due to their capacity to improve students' learning compared to more traditional approaches [5]. These methodologies have recently considered technology to enhance the processes. These are called Student Response System Methodologies (SRSM). They allow creating interactive learning and facilitate the evaluation of the knowledge, the attendance and the participation of the students [15]. They mean a great opportunity to work with since students are eager to experiment with technology: they are skilled in its use and enjoy using applications and games [10].

The Game-Based Student Response Systems (GSRS) are a version of the SRSM. These methods are based on implementing educational games in the classrooms. Students compete against each other during the games, and they have found these activities to be helpful and enjoyable [12]. The GSRS are likely to: i) minimize distraction, supporting the development of the students' motivation and engagement, which in turn can influence their learning [5]; ii) provide the teacher with an easy-to-use tool to instantly evaluate the students' comprehension [4]; iii) improve the development of students' academic performance [17]; and iv), enhance the quality of the teaching and the learning beyond what is provided in conventional classrooms [10].

One of the gamification applications most preferred among students is *Kahoot!* [2]. This is a free real-time platform for game-based learning [7]. It has been used at different educational levels and its positive effects on student's learning have been proved [16]. In fact, there are several studies on analyzing the perception of students concerning the platform. According to the students' feedback, these studies concluded that *Kahoot!* can help in creating fun and active learning experiences [4]. Others have pointed out that *Kahoot!* also fosters creating engagement and motivation [10]. This game-based tool was used in several studies by higher education students. However, most of them just used the tool for certain occasions. In [1] and [4], the authors only considered one set of questions, not sufficiently proving the benefits of using the tool. In contrast, in [3], the authors presented a methodology adapted to perform several quizzes for the topics of a course. The authors suggested that the use of quizzes at the beginning of the classes was found to relatively enhance the performance of the students. Nevertheless, despite the number of studies, the authors focused only on conceptual understanding and students' perceptions. Moreover, these studies omitted scientifically assessing the students' motivation. Therefore, there is a need to further analyze whether students' learning can be improved by using this tool and method.

Consequently, this paper aims to study the impact that GSRS tools can have on the learning performance of higher education students. They have been applied to the SMF course owing to the low academic performance and motivation of its students. Different types of quizzes have been designed and applied during eight weeks using *Kahoot!*. The motivation and the engagement of the students have been assessed by a proved scientific tool. The main objectives of this paper are: i) to assess the degree of acceptance by students of the inclusion of game-based learning; ii) to estimate the benefits of *Kahoot!*

in terms of students' motivation and academic performance enhancement; and, iii) to verify the capacity of this method to identify learning deficiencies to improve the design of the course.

## 2 Method and Materials

### 2.1 Research Approach

This research aims to contribute to a better understanding of the benefits of using GSRS in learning. Also, it intends to gauge the extent to which the use of *Kahoot!* can enhance the student's academic performance, their learning experience and their motivation. Therefore, several research questions (RQ) have been proposed to address the research goals:

RQ1: Can SRSM improve the students' performance in SMF?

RQ2: How does the application of GSRS affect the students' motivation?

RQ3: What is the effectiveness of the reminders at the beginning of the classes?

RQ4: Can SRSM improve the performance of the professor and the design of the courses?

RQ1 and RQ4 have been evaluated by means of *Kahoot!* quizzes filled in by the students. RQ2 has been assessed through the Instructional Materials Motivation Survey (IMMS) developed by Keller [8]. RQ3 has been appraised by performing the quizzes at the beginning of the classes.

### 2.2 Procedure

The SMF subject is taught over 15 weeks. The teaching is concentrated during four hours per week with a 30-min break. The classes are designed following the classical teaching method. This is based on teacher-centered classes of long lectures (2 h) combined with practical lessons (2 h) (Fig. 1). Therefore, the participation of the students is limited.

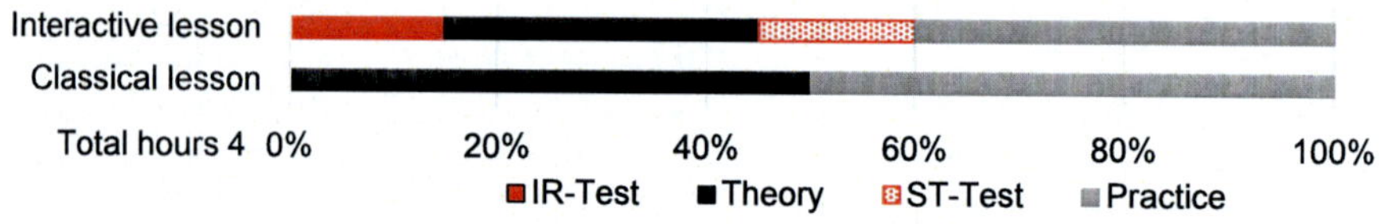


**Fig. 1.** Conceptual design of the lessons following the new interactive teaching method and the classical teaching method.

In this paper, this method has been improved by implementing two types of interactive *Kahoot!* quizzes (Fig. 1): initial reminders (IR) and standard (ST). The IR tests are performed at the beginning of the lessons to assess the level of understanding of the students of the previous week's lesson. In order to assess the effects of these IR quizzes, the students have been divided into two groups. Therefore, each group is to perform the IR every two weeks. The ST tests are carried out right after the theory lessons. In this case, it is intended to analyze the level of understanding of the students just after the teacher's presentation.

## 2.3 Curriculum Material

**Sample**
The sample was composed of 31 students, 12 men and 19 women. Most of the students attended the course for the first time, only four re-attending. In order to analyze the effectiveness of the IR, the students have been divided into two groups denoted as A and B. Each group is composed of 18 and 13 members, respectively. The division has been automatically obtained by the University of Seville platform.

**Learning Environment**
The research has been conducted on the SMF course during the academic year 2020/2021. The subject is taught in the third year and in the second semester of the Degree in Fundamentals of Architecture at the University of Seville. In order to enhance the students' attention, engagement, learning and performance, a GSRS has been implemented. This has been applied during eight weeks of the course, which corresponds to the teaching of some of its most theoretical topics. Therefore, eight different quizzes have been created according to the topics, which have been: 1) Physical properties of the soil; 2) The water in the soil. Effective pressure theory. Superficial tension in the soil; 3) Compressibility of the soil; 4) Compaction of the soil. Deformation of the unsaturated soil; 5) Shear strength of the soil; 6) Identification of the soil. In situ geotechnical tests; 7) Tensions of the soil. The elastic behavior of the soil; and, 8) Assessment of the settlement of the soil.

**Instruments**
Two different instruments have been used to carry out this experiment. The quizzes have been performed with the game-based learning tool *Kahoot!*. This has been selected due to: i) there being no need to install additional programs to use it; ii) it is a freeware; and, iii) it is the most preferred software among students. The quizzes have been designed by the authors of the research and have ten questions. Eight quizzes have been developed corresponding to each of the selected topics of the course. For each question, the students must select the correct solution from among four options. In some cases, more than one option can be correct. The questions can require the students to answer a theoretical concept, to perform short numerical calculations or to identify something from images. Each response has a limited time of 30 s. The quizzes are not anonymous in order to individually assess each performance of the students. Therefore, the students need to be identified with a nickname. The teacher displays the questions on the screen and the students proceed to answer as correctly and quickly as possible using their own device. They obtain a score based on the number of correct answers and the speed in their response. At the end of the game, the three students with the highest scores are revealed.

The students' motivation has been assessed by Keller's IMMS. This is a scientifically-validated questionnaire of 36 items which has been widely used [11]. It measures the four dimensions of the motivation: attention, relevance, confidence and satisfaction. This survey has been filled in by the students through anonymous online forms [6] at the end of the quizzes (Week 8).

## 3 Analysis of the Results

In this section, the results of this experiment are shown and analyzed. In Fig. 2, it can be observed that, generally, the scores of the IR tests have been higher than those obtained for the ST tests, except for Topics 7 and 8. A decreasing tendency of the results of the tests can be observed, which increases in the case of the IR tests. This can be due to several aspects: i) the students at the beginning of the course are more motivated; ii) the complexity and amount of concepts linearly increases during the development of the course; iii) Topics 5 to 8 are more practical than theoretical; iv) the students do not have to pass theoretical exams for Topics 7 and 8; and, v) during the first weeks of the course, the students were asked to elaborate summaries of the contents of the lessons. It can also be observed that Topic 7, regarding the elastic behavior of the soil, has had worst results. Also, significant differences can be found concerning the performance of the students. Generally, Group A has had better results than Group B.

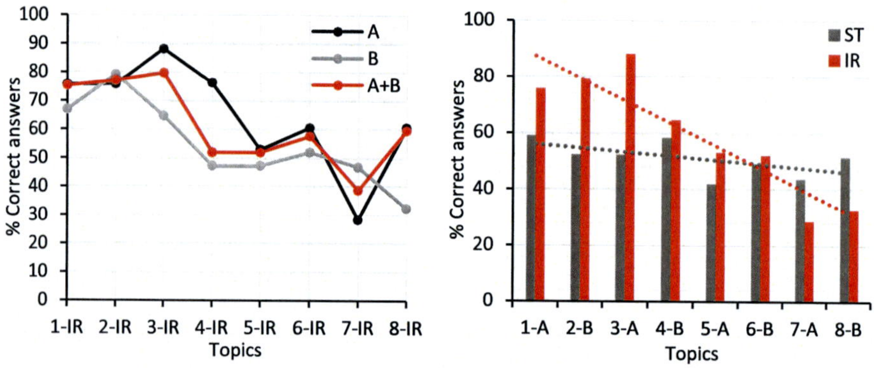


**Fig. 2.** Percentage of correct answers for the IR (left) and for each of the ST and IR (right).

Figure 3 shows the percentages of correct answers obtained for each of the questions and for each of the types of quizzes. As previously mentioned, it has been found out that Group A usually obtained better results than Group B. Also, the questions with lower scores can be intuitively assessed. Therefore, these results enable the teacher to identify the deficiencies in the understanding of the concepts. It can be observed that, generally, the score of each of the questions improves after the IR tests with the exception of Topics 7 and 8. Overall, the concepts that have not been completely understood have been: Topic 2, Q10; Topic 5, Q1, Q3, Q10; Topic 6, Q7; Topic 7, Q2, Q3, Q4, Q5, Q6; and, Topic 8, Q2, Q7 (See Table 1 from the Appendix).

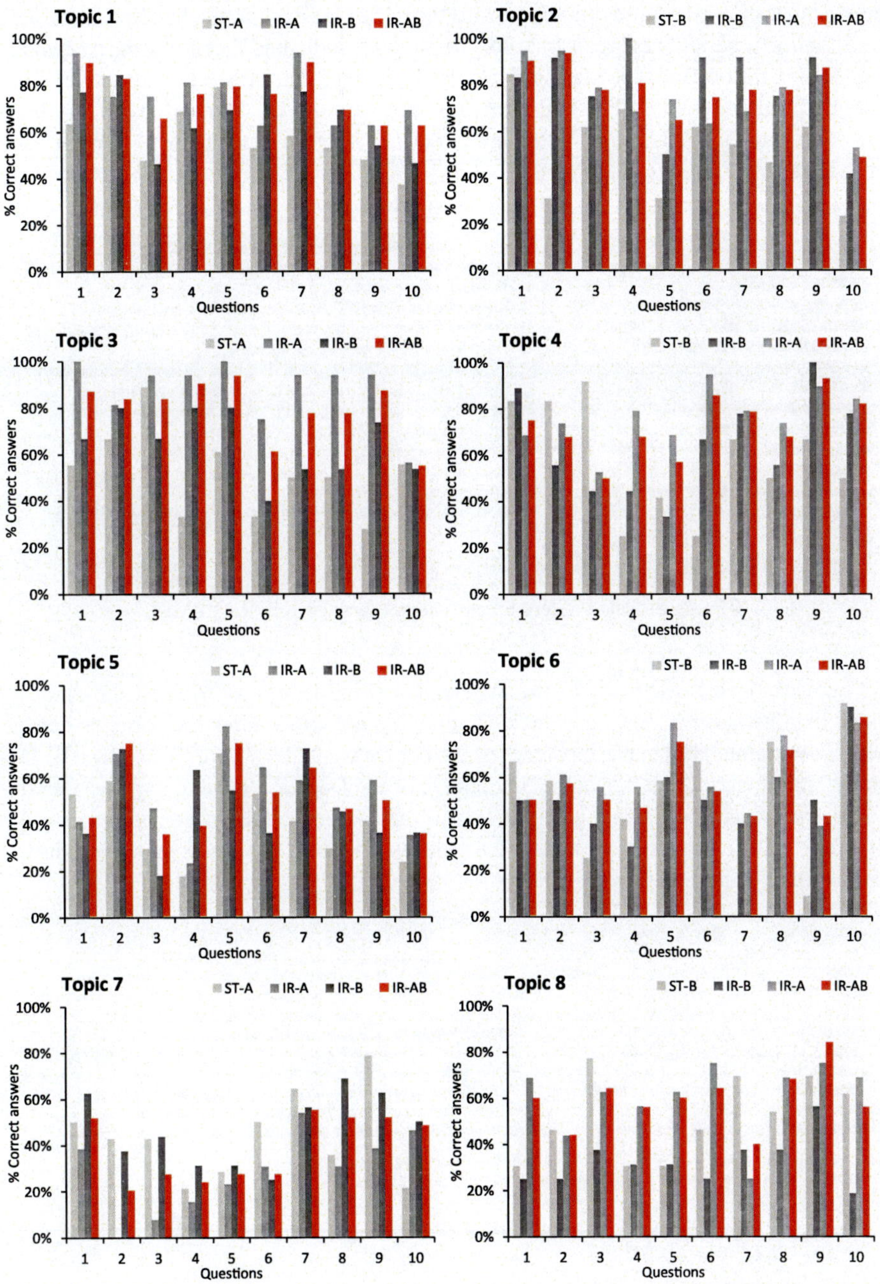


**Fig. 3.** Percentage of correct answers for each of the questions of the tests and types of quizzes.

The results obtained from the application of Keller's IMM to assess the students' attention, relevance, confidence and satisfaction are shown in Fig. 4, 5, 6 and 7. Overall, the students showed a good attitude toward the use of GSRS into the lessons. Regarding the attention assessment (Fig. 4), half of the students indicated that the *Kahoot!* quizzes got their attention from the start and they also thought that they were attractive, well-designed and enjoyable. They considered as well that the data showed in the questions was clear and helped them maintain their attention. However, they pointed out that there were too many contents in the course.

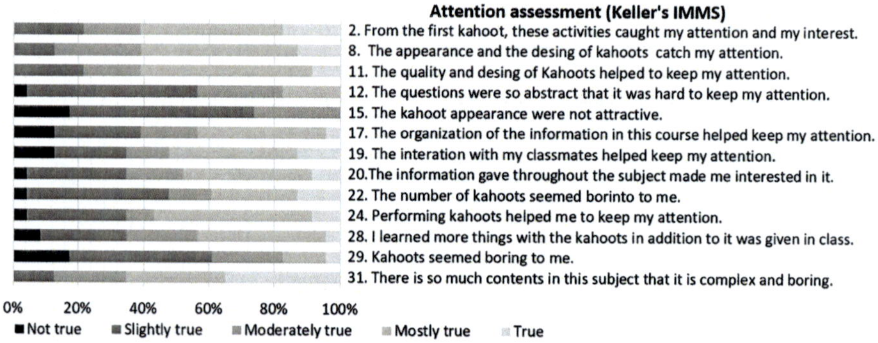


**Fig. 4.** Students' attention assessment through Keller's IMMS.

Concerning the relevance of the experiment (Fig. 5), almost 90% of the students stated that the quizzes were relevant since they did not know about the contents before. Also, 74% pointed out that the content of the course was relevant for their interests. It is important to remark that this is the first course of the Degree related to soil and foundations issues. It has also been found that 78% of them felt successful when completing the tests and 70% felt that the content of the quizzes is related to aspects explained during the theoretical and practical lessons.

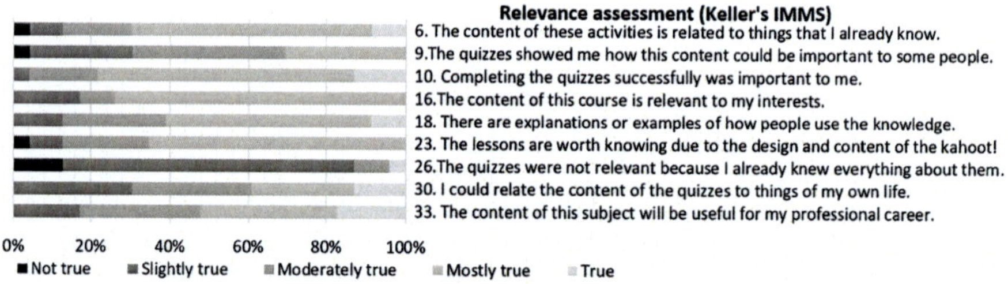


**Fig. 5.** Students' relevance assessment through the Keller's IMMS.

According to the students' feedback regarding their confidence (Fig. 6), it can be seen that half of them thought that the quizzes were going to be easy to fill in. Moreover, 65% understood what they had to do right after the introduction to the activities. However, 57% stated that the material was more difficult to understant than what they had thought at first. Also, 80% of them highlighted that as there was too much information, it was difficult to remember everything. Nevertheless, only 30% indicated that they could not completely understad the contents of the course.

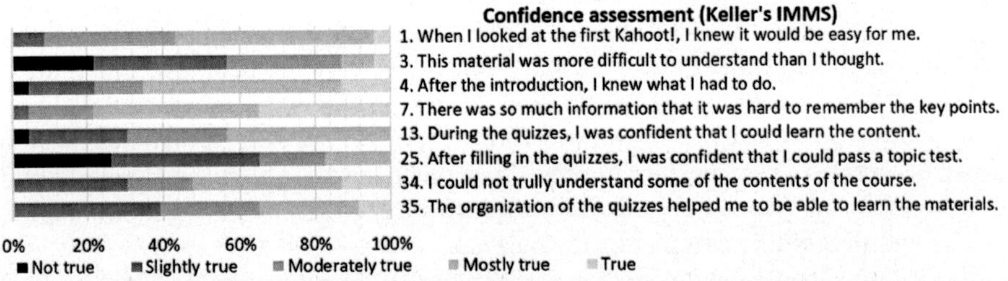


**Fig. 6.** Students' confidence assessment through Keller's IMMS.

Regarding the satisfaction assessment (Fig. 7), 60% were satisfied after finishing the quizzes. Moreover, 55% enjoyed working with *Kahoot!*, pointing out that this tool made them want to know more about the course. Moreover, most of them indicated that revealing the winners at the end of the quizzes made them feel rewarded.

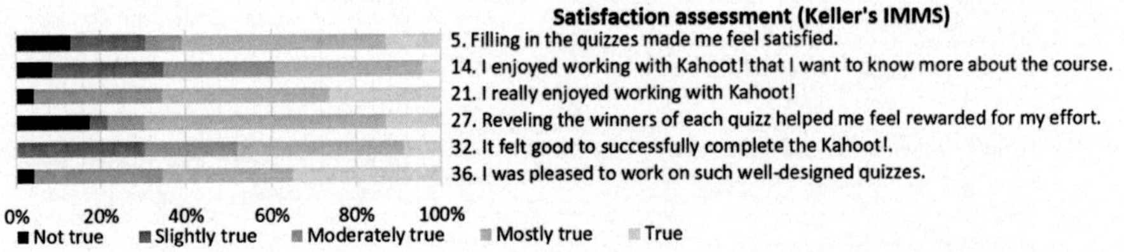


**Fig. 7.** Students' satisfaction assessment through Keller's IMMS.

# 4 Conclusion

The main finding of this pilot experiment is the potential that technology has to improve the learning experience, engagement and attention of students. In the case of SMF, the students' performance has been relatively improved by using the GSRS due to several factors. However, this has been mainly limited to the increasing complexity and number of concepts of the topics during the development of the course. It has been found that making students elaborate summaries for each of the topics has helped them to increase their scores. Generally, students had better scores in the IR quizzes than in the ST tests. It has therefore been proved that IR can somehow improve the students' understanding. Also, significant differences can be noted in the performance of the students. Generally, Group A has had better scores than Group B.

The results of this research eased the assessment of the understanding of each question and topics of the course and the identification of the deficiencies. In fact, in this case, it has been observed that the topics with the worst level of understanding have been those related to the elastic behavior of the soil and the assessment of settlements. Therefore, the GSRS enable professors to obtain feedback from students, generally and even individually. Moreover, GSRS help them, to some extent, to adapt their teaching activities based on the students' response to quizzes.

*Kahoot!* has increased the interest of the students in the contents of the course, enhancing their motivation and attention. According to the students' feedback from Keller's IMMS, it can be concluded that *Kahoot!* got their attention and that they thought these games were attractive and enjoyable. However, they considered that there are too many concepts in the course. Nevertheless, they stated that the quizzes were relevant for their interests and they felt confident when playing the games. Also, most of them felt good about successfully completing the quizzes.

Since this is the first attempt of the authors, further research is needed on this topic. This work could be improved by performing statistical analyses, comparing the results with record of previous years or control groups or increasing the size of the sample.

## 5 Appendix

**Table 1.** Questions for each of the topics.

| | | | |
|---|---|---|---|
| T1 | Q1. What are the Atterberg's limits? | T5 | Q1.What is the Mohr-Coulomb failure criterion? |
| | Q2.What is the porous index? | | Q2.Ways to apply the force in shear direct tests |
| | Q3.Which specific weights do you know? | | Q3.Parameters obtained from drained triaxial tests |
| | Q4.Determination of humidity, apparent and solid weight | | Q4.Equivalency of the simple compression test |
| | Q5.Use of the granulometric test | | Q5.Number of types of shear direct tests |
| | Q6.Difference between sediment and sieve tests | | Q6.Number of samples needed in triaxial tests |
| | Q7.What is the consistency of a soil? | | Q7.Meaning of *UU* in a test |
| | Q8.Use of the Casagrande's plasticity classification | | Q8.What is dilatancy? |
| | Q9.What are Atterberg's limits? | | Q9.What is the deviation tension in a triaxial test? |
| | Q10.Parameters defining the graduation of soils | | Q10.Identify of the tests for each of the plots |
| T2 | Q1.Select the correct effective pressure relation | T6 | Q1.How is $N_{spt}$ calculated? |
| | Q2.What type of soils are isotropous? | | Q2.What is a dynamic and continuous penetration test? |
| | Q3.What is the effective preasure? | | Q3.Differences between DPSH and DPH tests |
| | Q4.When does siphoning in soils happen? | | Q4.When is a reject attained in DPSH and SPT tests? |
| | Q5.When does suction happen in soils? | | Q5.How is the water table measured? |
| | Q6.What is the water table? | | Q6.What is obtained from a cone penetration test? |
| | Q7.What type of soil is more permeable? | | Q7.Drawbacks of the plate load test |
| | Q8.What is the specific submerged weight? | | Q8.Differences between DPH and Boreholes tests |
| | Q9.What is the piezometric height? | | Q9.Advantages of the plate load test |

*(continued)*

**Table 1.** *(continued)*

| | | | |
|---|---|---|---|
| | Q10.Select the correct value of neutral pressure | | Q10.Identify the tests for this figure |
| T3 | Q1.Which part does not belong to the oedometer (*OED*)? | T7 | Q1.What is a semi-space of Boussinesq? |
| | Q2.What is the preconsolidation pressure? | | Q2.Semi-space of Boussinesq and rigid base. Differences |
| | Q3.What is the name of this curve? | | Q3.What is a model of elastic stratum over a rigid base? |
| | Q4.A soil that has never had higher effective pressures is… | | Q4.Ocurrence of uniform / elastic distribution of tensions |
| | Q5.If a soil has had higher effective pressures, it is called… | | Q5.Use of Steinbrenner's method |
| | Q6.Identify the equation with the plot | | Q6.Parameters important in a semi-space of Boussinesq |
| | Q7.What are $C_s$ and $C_c$? | | Q7.Important parameters in the elastic models of soils |
| | Q8.Determination of the drainage path | | Q8.Distribution of tensions for a flexible footing |
| | Q9.In this relation, what is $E_m$ and $E$? | | Q9.Determination of the pressure under a point below |
| | Q10.How is the consolidation grade defined? | | Q10.Determination of $I_r$ using Fadum's abaqus |
| T4 | Q1.Most used to determine the compaction of the soil | T8 | Q1.Type of soil characterized by instant settlement |
| | Q2.Differences between normal and modified proctor test | | Q2.What happens in a primary consolidation settlement? |
| | Q3.Difference between the Lambe and free swelling tests | | Q3.In which soil will a consolidation settlement occur? |
| | Q4.For which tests should unaltered samples be used? | | Q4.What is the angular distortion? |
| | Q5.Identify the tests to which these curves belong to | | Q5.What is an instant settlement? |
| | Q6.Curves are obtained from the underload flood test | | Q6.*OED* method. Type of soil to calculate settlements |
| | Q7.What expansiveness tests are qualitative? | | Q7.Elastic method. Type of soil to calculate settlements |
| | Q8.What is a qualitative expansiveness test? | | Q8.Parameters needed in elastic models |
| | Q9.What is the meaning of an expansive soil? | | Q9.Value of the maximum distortion according to CTE |
| | Q10.What is the active layer of a clay? | | Q10.Why is the *OED* method not used for sands? |

## References

1. Amaro-Mellado, J.L., Antón, D., Pérez-Suárez, M., Martínez-Álvarez, F.: Game-based student response system applied to a multidisciplinary teaching context. In: Martínez Álvarez, F., Troncoso Lora, A., Sáez Muñoz, J.A., Quintián, H., Corchado, E. (eds.) CISIS/ICEUTE -2019. AISC, vol. 951, pp. 329–339. Springer, Cham (2020). https://doi.org/10.1007/978-3-030-20005-3_34
2. Bicen, H.: Determination of university students' most preferred mobile application for gamification. World J. Educ. Technol. Current **9**, 18 (2017)

3. Cantero-Chinchilla, F.N., Díaz-Martín, C., García-Marín, A.P., Estévez, J.: Innovative student response system methodologies for civil engineering practical lectures. Technol. Knowl. Learn. **25**(4), 835–852 (2019). https://doi.org/10.1007/s10758-019-09410-z
4. Castro, M.J., et al.: Impact of educational games on academic outcomes of students in the Degree in Nursing. PlosOne. **14**(7), e0220388 (2019)
5. Felszeghy, S., et al.: Using online game-based platforms to improve student performance and engagement in histology teaching. Med. Educ. **19**, 273 (2019)
6. Google Inc.: Google Forms( 2021). https://www.google.es/intl/es/forms/about/. Accedido: 05 mayo 2021
7. Kahoot! Play Kahoot (2021)! https://kahoot.it/. Accedido 04 mayo 2021
8. Keller, J.M.: Motivational Design for Learning and Performance: The ARCS Model Approach. Springer, US (2010)
9. Kuenzi, J.J., Matthews, C.M., Mangan, B.F.: Science, Technology, Engineering, and Mathematics (STEM) Education Issues and Legislative Options, United States (2006)
10. Licorish, S.A., Owen, H.E., Daniel, B., George, J.L.: Students' perception of Kahoot!'s influence on teaching and learning. Res. Pract. Technol. Enhanc. Learn. **13**(1), 1–23 (2018). https://doi.org/10.1186/s41039-018-0078-8
11. Loorbach, N., Peters, O., Karreman, J., Steehouder, M.: Validation of the instructional materials motivation survey (IMMS) in a self-directed instructional setting aimed at working with technology. Br. J. Educ. Technol. **46**(1), 204–218 (2015)
12. Rahmahani, D.: The effect of gamified student response system on students' perception and achievement. Int. J. Eng. Pedagog. **10**(2), 45–59 (2020)
13. Requena-Garcia-Cruz, M.V.: Propuesta de innovación docente en arquitectura mediante el aprendizaje basado en problemas. In: Medina, E.N., Porlán, R. (eds.) Ciclos de mejora en el aula. Año 2019. Experiencias de innovación docente de la Universidad de Sevilla, pp. 449–468. Editorial Universidad de Sevilla (2020). https://doi.org/10.12795/9788447221912.019
14. Requena-Garcia-Cruz, M.V., Morales-Esteban, A.: Classroom improvement cycle in architecture by means of problem-based learning. In: Herrero, Á., Cambra, C., Urda, D., Sedano, J., Quintián, H., Corchado, E. (eds.) ICEUTE 2020. AISC, vol. 1266, pp. 377–387. Springer, Cham (2021). https://doi.org/10.1007/978-3-030-57799-5_39
15. Turan, Z., Meral, E.: Game-based versus to non-game-based: the impact of student response systems on students'achievements. Engage. Test Anxiet. Inf. Educ. **17**(1), 105–116 (2018)
16. Wang, A.I., Tahir, R.: The effect of using Kahoot! for learning – A literature review. Comput. Educ. **149**, 103818 (2020)
17. Wichadee, S., Pattanapichet, F.: Enhancement of performance and motivation through application of digital games in an English language class. Teach. English with Technol. **18**(1), 77–92 (2018)

# Longitudinal Study of Grades for the Industrial Electronics and Automation Engineering Degree Programme

José Antonio López-Vázquez[1], Elena Arce[1](✉), María Isabel Fernández-Ibáñez[1], José Luis Casteleiro-Roca[1], Francisco Zayas[1], and Andrés Suárez-García[2]

[1] Department of Industrial Engineering, University of A Coruña, CTC, CITIC, Ferrol, A Coruña, Spain
{jose.lopez,elena.arce,isabel.fibanez,jose.luis.casteleiro, f.zayas.gato}@udc.es

[2] University Defense Center at Spanish Naval Academy, Marín, Pontevedra, Spain
andres.suarez@cud.uvigo.es

**Abstract.** The low academic performance rates in the majority of the STEAM (Science, Technology, Engineering, Arts and Math) degrees, is an issue that influences not only the degree quality rates, but also the teaching staff assessments. In this study, 254 graduates of the Industrial Electronics and Automation Engineering degree were monitored over eight academic years. For this purpose, individual academic records were compiled and the individual grades were analyzed. The results showed that there were inter-course differences. Significant differences were also observed in the grades for different subjects. Statistical analysis results also showed that in subjects with low performance rates less than 20% of grades exceeded pass mark. Subjects where final exam does not have a 100% weight in the qualification, performance rates are higher. In addition, significant differences were found between the grades of compulsory and the optional subjects.

**Keywords:** Academic performance · Electronic and automation engineering · Qualifications · Longitudinal study

## 1 Introduction

Royal Decree 1393/2007 of 29 October 2007, which establishes the organization of the official university education, requires the existence of an internal quality assurance system (IQAS) in the faculties or colleges for the verification and accreditation of official university degrees. This Royal Decree (RD) establishes a series of performance indicators for subjects and degrees. However, neither RD 1393/2007 nor RD 861/2010, which amends the previous one, establishes any reference value for these indicators. In the absence of reference values, the IQAS usually take the historical mean value as the standard.

In addition, the European Standards and Guidelines for Quality Assurance of Higher Education establish the degree's verification report establishes standard performance


J. J. Gude Prego et al. (Eds.): CISIS 2021/ICEUTE 2021, AISC 1400, pp. 295–304, 2022.
https://doi.org/10.1007/978-3-030-87872-6_29

values as a quality criterion. When subject's results do not meet the standard value set out, the teaching staff responsible for the subject must draw up a self-report explaining the results.

Academic results are evaluated on the basis of three indicators: success rate, performance rate and evaluation rate [1, 2]. The performance rate is defined as the ratio of the number of credits passed by students enrolled to the total number of credits enrolled. The success rate represents the proportion number of credits passed by students enrolled divided by the total number of credits for which students took an exam. The assessment rate is a percentage representing the number of credits for which students took an exam divided by the total number of credits enrolled. Therefore, the indicators of the IQAS reports depend on the qualifications' distribution. In addition, academic results are also used in the DOCENTIA programme for the assessment of teaching performance.

Thus, student performance rates are key factors in quality ratios (both at the overall university, degree and subject level. Performance ratio is lower at engineering degrees, than in other areas of knowledge. Thus, in the 2018–2019 academic year, the average performance rate in Spanish university degree studies was 78%: the minimum value was in Engineering and Architecture area (67.5%) and the maximum in Health Sciences area (84.7%) [3]. The same performance is observed in the University of A Coruña (UDC). In the 2018–2019 academic year, the average performance rate in UDC degree was 75.8%: the minimum value was in Engineering and Architecture (68.9%) and the maximum in Health Sciences (91.1%) [4].

Figure 1 shows a comparison of the evolution of the performance rate indicator in the Spanish University System (SUS) with the Industrial Electronics and Automation Engineering (IEAE) degree. The results for IEAE at the UDC are lower than those of the SUS.

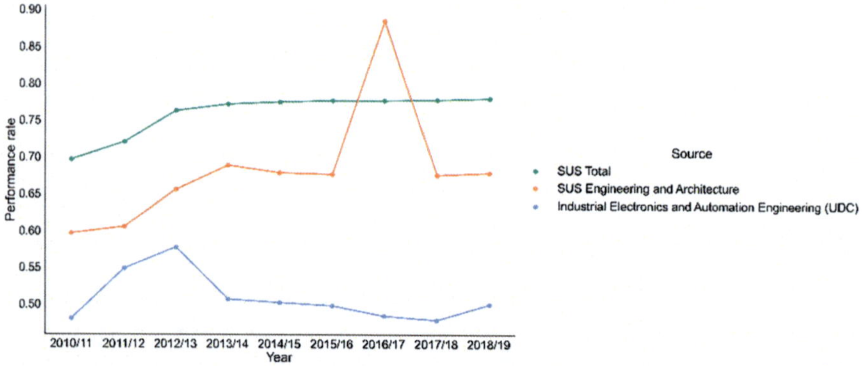


**Fig. 1.** Evolution of performance rates.

Therefore, it is necessary to analyze academic performance on the basis of the grades distribution. This will allow specific improvement actions to be proposed. The scientific literature has focused on the analysis of student performance with two objectives: to generate indicators to measure the quality of lecturing and to make predictions of future performance [5–7]. This study presents the results of a longitudinal analysis with a sample of 254 students in the IEAE degree over 8 academic years. The results show that in the second year of the degree the grades are higher than in the first and third year. Differences between grades depending on the nature of the subject (compulsory or optional) were found.

## 2 Methodology

The first step to analyze academic performance is to define it. The study of Dorta, et al. [8] was taken as a reference. In this work, student performance is analyzed on the basis of the subjects passed grades. That is, the qualifications are greater than or equal to 5. Grades are a quantitative variable, which takes values in the interval from 5 to 10.

This study analyzed the records of 254 students who completed their studies in IEAE at the UDC over 8 academic years. The dataset consisted of the individual grades. The individual academic records were accessed. The academic records only contain passed grades, so our database lacks information on grades lower than 5 (fail).

81% of the students were male, while 19% were female. The low presence of women is characteristic of STEM disciplines. The 94% of students accessed from Baccalaureate, only 6% accessed from Vocational Training.

The average access grade was 6.29 and the average grade of the degree was 6.28. Students took an average of 6.78 years to complete their studies. 1.57% of students completed their studies within the stipulated years, while 98.43% of students required more years. Specifically, 7.48% of students require another academic year to achieve the degree and 90.94% require two or more. Tables 1 and 2 show the subjects analyzed.

**Table 1.** Statistical measures of students' grades in compulsory subjects.

| Subject | Year | Mean | Median | Mode | SD | Max. range | Variation coefficient |
|---|---|---|---|---|---|---|---|
| Physics | 1st | 6.22 | 5.50 | 5.50 | 1.19 | 9.60 | 19.13% |
| Mathematics I | 1st | 5.83 | 5.50 | 5.50 | 0.91 | 9.20 | 15.61% |
| Circuit theory | 1st | 5.93 | 5.50 | 5.00 | 1.08 | 10.00 | 18.21% |
| Computer science | 1st | 6.61 | 6.10 | 5.50 | 1.24 | 10.00 | 18.76% |
| Engineering drawing | 1st | 6.16 | 6.00 | 5.50 | 0.98 | 9.20 | 15.91% |
| Electronics technology | 1st | 6.18 | 5.50 | 5.50 | 1.22 | 10.00 | 19.74% |
| Statistics | 1st | 5.87 | 5.50 | 5.00 | 1.00 | 10.00 | 17.04% |
| Chemistry | 1st | 6.85 | 7.00 | 5.50 | 1.37 | 10.00 | 20.00% |

*(continued)*

**Table 1.** (*continued*)

| Subject | Year | Mean | Median | Mode | SD | Max. range | Variation coefficient |
|---|---|---|---|---|---|---|---|
| Mechanical systems | 1st | 5.71 | 5.50 | 5.00 | 0.95 | 10.00 | 16.64% |
| Automatic regulation | 2nd | 5.87 | 5.50 | 5.00 | 1.04 | 9.50 | 17.72% |
| Analogue electronics | 2nd | 6.35 | 6.00 | 5.50 | 1.32 | 10.00 | 20.79% |
| Digital electronics | 2nd | 6.11 | 5.65 | 5.00 | 1.22 | 10.00 | 19.97% |
| Electronic systems design | 2nd | 6.80 | 6.60 | 5.50 | 1.51 | 10.00 | 22.21% |
| Electrical systems | 2nd | 6.08 | 5.80 | 5.50 | 0.99 | 10.00 | 16.28% |
| Industrial programming | 2nd | 5.99 | 5.50 | 5.00 | 1.10 | 9.50 | 18.36% |
| Oil-pneumatic systems | 2nd | 7.09 | 7.00 | 7.00 | 1.21 | 9.50 | 17.07% |
| Mathematics II | 2nd | 6.54 | 6.20 | 5.50 | 1.38 | 10.00 | 21.10% |
| Industrial informatics | 3rd | 5.86 | 5.50 | 5.00 | 1.05 | 10.00 | 17.92% |
| Industrial automation | 3rd | 5.94 | 6.00 | 6.00 | 0.66 | 9.30 | 11.11% |
| Electronic instrumentation | 3rd | 5.96 | 5.50 | 5.00 | 1.14 | 10.00 | 19.13% |
| Technical office | 3rd | 6.30 | 6.15 | 7.00 | 0.89 | 9.00 | 14.13% |
| Business administration & industrial organization | 3rd | 6.51 | 6.20 | 5.00 | 1.19 | 9.60 | 18.28% |
| Power electronics | 3rd | 6.37 | 6.00 | 5.00 | 1.45 | 10.00 | 22.76% |
| Energy technology | 3rd | 6.20 | 5.85 | 5.00 | 1.15 | 9.50 | 18.55% |
| Final degree project | 3rd | 7.69 | 7.50 | 7.00 | 1.13 | 10.00 | 14.69% |

## 3 Results

A total of 41 subjects were analyzed. Of these, 25 were compulsory (Table 1) and 16 were optional (Table 2). The Coefficient of Variation (CV) results (Table 1) showed that the Industrial Automation grades (CV = 11.11%) were homogeneous. In other words, the assessment system used in this subject did not discriminate the level of achievement reached by the students, as there was no variability in the grades. On the other hand, Power

**Table 2.** Statistical measures of students' grades in optional subjects.

| Subject | N | Mean | Median | Mode | SD | Max. range | Variation coefficient |
|---|---|---|---|---|---|---|---|
| Industrial drawing | 59 | 5.88 | 5.80 | 5.00 | 0.79 | 8.50 | 13.45% |
| Polymer physics | 148 | 5.99 | 5.50 | 5.50 | 1.09 | 10.00 | 18.13% |
| Computer process control | 26 | 6.58 | 6.00 | 6.00 | 1.32 | 10.00 | 20.14% |
| Thermal machines | 19 | 6.65 | 7.00 | 5.00 | 1.34 | 9.00 | 20.13% |
| Heat transmission | 58 | 6.89 | 7.00 | 5.50 | 1.47 | 10.00 | 21.31% |
| Procurement management | 53 | 6.41 | 6.70 | 5.00 | 1.08 | 8.50 | 16.84% |
| Chemistry extension | 118 | 7.00 | 7.50 | 7.50 | 1.37 | 10.00 | 19.52% |
| Composite materials technology | 17 | 6.04 | 5.50 | 5.00 | 1.13 | 9.00 | 18.79% |
| Electrical energy billing | 51 | 6.66 | 6.00 | 5.00 | 1.41 | 9.00 | 21.18% |
| Environmental engineering and management | 155 | 6.72 | 6.80 | 5.50 | 1.09 | 9.00 | 16.25% |
| Electronic control of electrical machines | 32 | 6.37 | 6.00 | 6.00 | 0.87 | 8.50 | 13.67% |
| Home automation | 133 | 6.64 | 6.60 | 7.50 | 1.11 | 9.30 | 16.78% |
| English | 190 | 6.60 | 6.50 | 5.50 | 1.28 | 10.00 | 19.43% |
| Air conditioning systems | 22 | 6.10 | 5.75 | 5.00 | 1.27 | 9.50 | 20.75% |
| Advanced control techniques | 69 | 7.21 | 7.00 | 7.00 | 0.67 | 8.50 | 9.32% |

Electronics was the subject with the greatest dispersion of marks. Same analysis was carried out on the optional subjects (Table 2). Advanced Control Techniques, Industrial Drawing and Electronic Control of Electrical Machines grades were homogeneous. The optional subjects' marks showed greater dispersion than those for compulsory ones. The violin plots showed this homogeneity (Fig. 5). The kernel density shows how grades accumulate around a particular range. Because not all students attended all the optional subjects, Table 2 includes a column (N) with the sample size. The grade of 5 was the most frequently. 5 points was the mode value for 11 of the 26 subjects analyzed. More than 40% of the subjects had no record of 10-point grades.

The influence of subject type on grades was analyzed. The results showed a significant effect, $F(1, 506) = 32.47, p < .001$, for the factor subject type (compulsory vs. optional), which explains 6.1% of the variance, $eta^2 = .061$. Specifically, in the optional subjects (M = 6.57), students obtained higher evaluation than in the compulsory ones (M = 6.28).

A repeated measures mean comparison analysis was also performed. Results showed significant differences in the average grade for the year factor (intra factor), $F(2, 506) = 21.56, p < .001, \eta^2 = .146$. A posteriori contrast, with Bonferroni's level of protection (.05/3 = .017), showed that the average evaluation was significantly higher in the second year (M = 6.35) than the first (M = 6.15) and third year (M = 6.16). The grades of the second-year subjects were compared with each other. A repeated measures mean comparison analysis was performed and the results showed significant differences in the average grade for the subject factor (intra factor), $F(7, 1771) = 34.62, p < .001, \eta^2 = .120$. A posteriori contrasts, with Bonferroni's level of protection (.05/8 = .006), showed that the average evaluation for the subjects Electronic Systems Design (M = 6.80), Oil-Pneumatic Systems (M = 7.09) and Mathematics II (M = 6.54) were significantly higher than the other subjects of the second year grades (Fig. 3). In other words, the higher grades in three compulsory subjects cause the second year to be different from the first and third. The grades for the whole degree are grouped in a very narrow range of values. Thus, a small variation in a few subjects in a course can cause the significant difference observed. The syllabus for these three subjects were reviewed. The assessment systems in "Oil-Pneumatic Systems" and "Electronic Systems Design" included problem-based learning methodologies.

The grades distributions for both compulsory and optional subjects were analyzed (Fig. 2, 3, 4 and 5). There was no overall homogeneity for the grading patterns. However, when density functions were analyzed, subjects with worst performance rates could be identified. The data profile showed clustering in the bottom, for subjects with low grades. For instance, this pattern was identified in Mechanical Systems (Fig. 2). Interquartile range was between 5 and 6 points (75%). Grades above 7.5 points were considered outliers, in this profile subjects. Chemistry and Computer Science were first year subjects with higher performance rates (Fig. 2). These subjects' grades distribution was much more evenly distributed than ones for subjects with lower results. This indicates the existence of a bias in the grades. The distribution in some subjects such as Industrial Automation (Fig. 4) was striking. The grades were grouped by ranks. This probably reflects a binary system of assessment and grading. The Final Degree Project, which had the highest grade average, showed a unique pattern. Grades were grouped around 7, being residual for the extreme grades. This subject showed outliers below the lower limit. Figure 4 showed how average values were between 5.5 and 8 points for subjects with highest marks (i.e. Oil-Pneumatic Systems and Electronic Systems Design).

Figure 5 showed that higher density in the "A" category (distinction), for the optional subjects. This fact was atypical for compulsory subjects. Final Degree Project marks

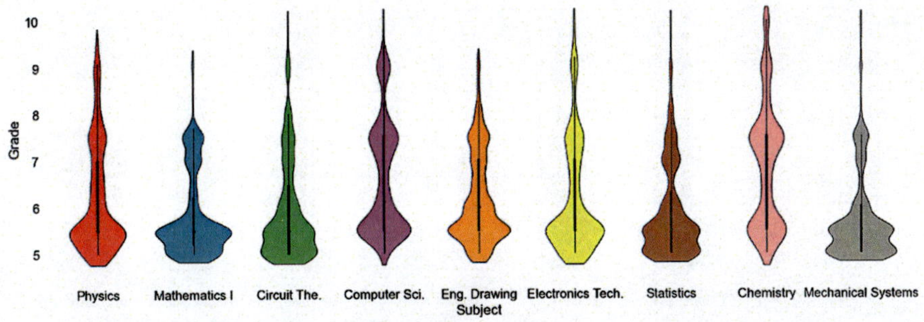


**Fig. 2.** Violin plots for grade distribution in first-year compulsory subjects.

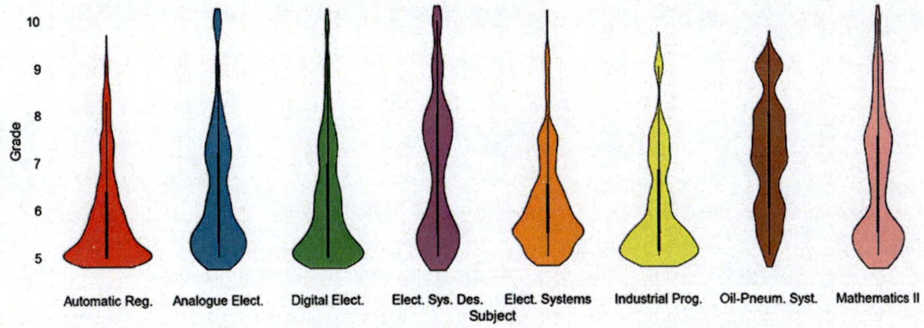


**Fig. 3.** Violin plots for grade distribution in second-year compulsory subjects.

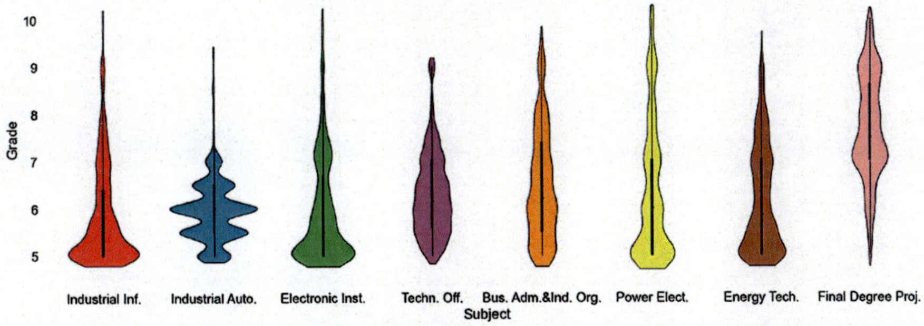


**Fig. 4.** Violin plots for grade distribution in third-year compulsory subjects.

distribution was grouped around the "B" and "A" levels. Once again, the grade distributions showed the influence of the assessment systems. Thus, the grades were grouped by sections (e.g. English, Thermal Machines, Chemistry Extension).

### 3.1 Analysis of the Subjects with the Worst Rates of Performance

Subjects with the lowest average grades were analyzed (Fig. 6). Mathematics I was a first-year subject taught on an annual basis. Various teaching methodologies were applied

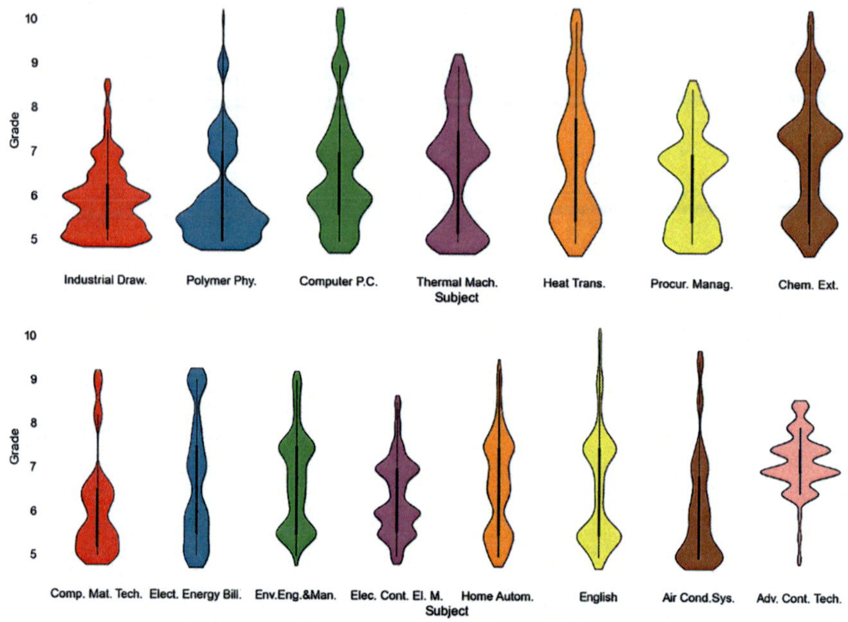


**Fig. 5.** Violin plots for grade distribution in optional subjects.

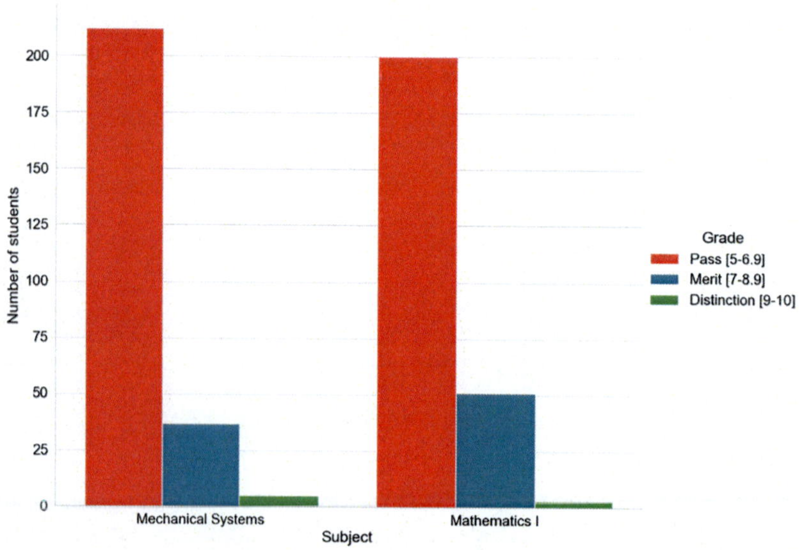


**Fig. 6.** Summary of grades in Mechanical Systems and Mathematics I.

on Mathematics I: Introductory activities, ICT practices, workshops, concept maps, problem solving and master class. Two assessment strategies were used in Mathematics I: Objective test (90%) and Workshop (10%). Mechanical Systems is a first-year subject taught in the second term. Master class was the only teaching methodology included in the syllabus. The evaluation strategy was a final exam (100% of the qualification). The syllabus did not specify details of the assessment test. In both Mathematics I and Mechanical Systems, the percentage of passing grades [5–6.9] is above 80% (Fig. 6).

On the contrary, the final exam has a weight of less than 70% of the qualification in subjects with highest performance rate (e.g. Computer Science or Chemistry). This enhances continuous assessment. The student works progressively on the competences and the teaching staff can better monitor the teaching-learning process.

The quality reports made by the teaching staff (academic years 2016/2017, 2017/2018 and 2018/2019) were also analyzed. The reports did not contain an analysis of results. The teaching staff responsible for first-year subjects requested that the grades comparison should be made on a year-by-year basis. Actually, the grades were compared at the global level of the degree. The reports indicated as possible causes of poor performance: the greater difficulty of some subjects, low class attendance, the high number of non-attendance students at final exams (especially in the second round), the absence of prior knowledge. The reports also contain proposals for improvement: reducing syllabus content, establishing "key" subjects, reducing competences, introducing partial tests or planning laboratory practices that motivate students.

## 4 Conclusions

254 records of students who complete their studies in IEAE degree over 8 academic years were analyzed. The average access grade was low in comparison with other degrees. The access grade is a predictor of academic performance. This partly explains the low performance rates in the first year subjects.

The nature of the subject (optional vs. compulsory) explains more than 6% of the variance of the grades. Grades in optional subjects are significantly higher than in compulsory ones. No homogeneity was found in grade distribution patterns, but subjects with worst performance rates were identified. 75% of marks were grouped around 5-points (pass) in subjects with the worst performance rates. Performance is higher in the second-year. This is due to the influence of three subjects.

The subjects with the worst performance rates were analyzed. The assessment strategies were focused exclusively on a final examination in these subjects. The IQAS reports were reviewed. The low marks were endorsed to the difficulty of the content, the low attendance rate and the high number of non-attendance students at final exams. The proposals for improvement do not include changes in assessment strategies or teaching methodologies. The proposals for improvement included in the IQAS committee reports focused on the revision of syllabus and creation of practical bulletins. None of the reports include proposals for actions to detect real training deficiencies, so that teachers can reinforce those concepts that students claim to be unfamiliar with and thus improve performance rates. Therefore, it is proposed to the IQAS committee of the PUC to conduct personal interviews with the faculty involved. This measure could help to discover and overcome the problem of low grades. The interviews can also be extended to the student body.

Therefore, the teaching staff should be encouraged to think deeply on the assessment system and the distribution of grades. This task should be focused both on exploring new teaching and assessment methodologies. Thus, continuous assessment should be strengthened. In addition, efforts should be made to detect students' educational shortcomings and to design programs to improve academic performance.

## References

1. IESE: Business school. University of Navarra. Accreditations. ANECA [Online] (2021)
2. ACSUG: Agency for Quality Assurance in the Galician University System. Guide to the evaluation of university degrees external to the Spanish university system [Online] (2018)
3. Spanish Ministry of Universities: Facts and Figures on the Spanish University System. Publication. General Technical Secretariat of the Spanish Ministry of University, Madrid (2021)
4. University of A Coruña: Transparency portal. Academic information [Online]. Available: https://www.udc.es/es/transparencia/resultados/informacion-academica/.
5. Crespo-Turrado, C., et al.: Comparative study of imputation algorithms applied to the prediction of student performance. Logic J. IGPL **28**(1), 58–70 (2020)
6. Miguéis, V.L., Freitas, A., Garcia, P.J., Silva, A.: Early segmentation of students according to their academic performance: a predictive modelling approach. Decis. Support Syst. **115**, 36–51 (2018)
7. Larson, L.M., Pesch, K.M., Surapaneni, S., Bonitz, V.S., Wu, T., Werbel, J.D.: Predicting graduation: the role of mathematics/science self-efficacy. J. Career Assess. **23**(3), 399–409 (2015)
8. Guerra, R., Marrero, I., Abdul-Jalbar, B., González, R., Darias, N.: Un modelo predictivo del rendimiento académico a partir de las calificaciones de bachillerato y PAU. In: Navarro, A.V. (ed.) De los procesos de cambio al cambio con sentido, pp. 119–136. Vicerrectorado de Docencia. Universidad de La Laguna. Servicio de Publicaciones de la Universidad de La Laguna (2019). https://doi.org/10.25145/b.innovaull.2019.009

# Analysis of Students' Online Interactions in the Covid Era from the Perspective of Anomaly Detection

José Otero(✉), Luciano Sánchez, Luís A. Junco, and Inés Couso

Oviedo University, Oviedo, Spain
{jotero,luciano,lajunco,couso}@uniovi.es

**Abstract.** During the pandemic, most of the teaching has been done online. The lack of face-to-face interaction has many undesirable effects, including students being less focused, not receiving feedback on how they are approaching the current topic/task, and an increased risk of cheating. It is expected that those students with similarly graded assignments/exams would have similar interactions during online teaching sessions. The opposite is an anomaly, for better or worse. It is possible to find out if assignments/exams are legitimate by using anti-plagiarism tools or by carefully examining submissions, but it is time-consuming and only protects against one type of fraud. In this paper, we propose to apply anomaly detection techniques to the students' interactions to reduce the number of assignments/exams that need to be checked against fraud.

**Keywords:** Fraud detection · Anomaly detection · Class participation · Online teaching

## 1 Introduction

Students' interaction helps the teacher to know whether their form of communication is being successful, whether the students are staying involved and to predict whether each individual student will perform well.

In face-to-face teaching, many students adopt a passive attitude [14], and this behaviour is exacerbated in online teaching. The Covid-19 pandemic has worsened the situation, as in many cases teaching has had to adapt to an online format with little preparation time [3,10]. The need for tools that reinforce interaction between teachers and students is evident.

Numerous studies have been published about participatory learning, such as [13] and [15], where participation is studied from the perspective of the learner.

This work has been funded by the Ministry of Science and Innovation of the Government of Spain and by the FEDER funds of the European Community through projects with codes PID2020-112726RB-I00 and TIN2017-84804-R.


J. J. Gude Prego et al. (Eds.): CISIS 2021/ICEUTE 2021, AISC 1400, pp. 305–314, 2022.
https://doi.org/10.1007/978-3-030-87872-6_30

Both studies conclude that the students would prefer to be encouraged to participate (when they are expected to do so) with clear messages. In [13] it is also indicated that the learner would like not to be asked embarrassing or awkward questions. Studies [7] and [18] conclude that collecting feedback and giving credit for student participation positively influences it. Reference [21] is a study focused on Information Technologies students that reaffirms the above: the students participate more if their effort is recognized.

Learning Analytics is the use of data about learners in order to improve learning and teaching [5,9]. The category of systems devoted to provide services related with online teaching is Learning Management Systems (LMS). These systems provide tools to create activities to be completed online and the students' interactions are routinely logged. An active area of research is predicting the students' performance using educational data, including selecting the attributes and metrics that are better suited for the task [4,16]. Focused on Computer Science teaching, in [17] a particular kind of online activity (source code submissions) is analysed to predict the pass rate of online courses.

From the cited works, seems pretty clear that students' online activity logging yields valuable insights about their future performance and encourages them to keep focus during teaching.

In contrast to other studies such as [4] or [16], in which the aim is to predict the students' grade using different indicators, in this study the aim is to detect anomalous students. We will understand that a student is "anomalous" if there are significant differences between the ordinary evaluation of their activities and the grade that would have been assigned by algorithms such as [11] or [17]. In other words, we seek to identify students who, having attended all classes with good results, have received a low grade (or the opposite) in order to provide them with a personalized follow-up.

In the following section, we first explain the collected data and the computed metrics from raw data. Then we introduce feature selection techniques focusing in the chosen one. This is followed by the proposed strategy for outlier analysis. Below, we detail the experimental setup, number of students, number of responses, method parameters, etc. We continue with the results of the study. The paper ends with the obtained conclusions and future work.

## 2 Data Acquisition and Processing

### 2.1 Activity Logging and Metrics Extraction

At this moment, most of the face to face activities have been replaced with online versions and thus we need an LMS or a custom tool to record the students' interactions. The requisites that we need to be fulfilled are:

1. Simple enrolling/authentication.
2. On the fly correct answer setting.
3. Some kind of graphical aggregated display of the answers.
4. Full activity logging.

At this moment, the most popular LMSs are Google Classroom [23] and Moodle [22]. In Table 1 these systems' features are summarized. There is a lack of key features in these systems, thus we have developed an online replacement. The questions are posed orally using Microsoft Teams and the answers are collected using a custom online application built in Python using `http.server` module. We want the lectures to be as close as possible to the usual ones and thus the questions are unplanned and related with the topic being explained at any given moment. The log of the app looks like this:

```
2020-10-06 12:10:37,222 - ip=X.X.X.X module=0 user=professor answer=C1
2020-10-06 12:10:40,551 - ip=X.X.X.X module=0 user=XXX answer=R1
2020-10-06 12:10:45,717 - ip=X.X.X.X module=0 user=XXX answer=R1
```

Each question log begins with a line with the professor as user along with the correct answer. Then the students' answers are logged. For each user, the date/time, IP address, module and user name along with his/her answer are also logged. With this information we compute all the metrics commented in Subsect. 2.2.

**Table 1.** Popular LMS vs. our tool requisites.

| Tool | Requisite 1 | Requisite 2 | Requisite 3 | Requisite 4 |
|---|---|---|---|---|
| Google classroom | | | ✓ | |
| Moodle | | | ✓ | ✓ |
| Our tool | ✓ | ✓ | ✓ | ✓ |

## 2.2 Metrics and Data Preprocessing

The catalogue of metrics computed from the activity logging for each student is:

- Total number of answers by the students.
- Average delay in answering the first response since the beginning of the session.
- Fisher's five number summary of the answer delay from the first answer among the students: for each question, the first answer is taken as a reference and subsequent student's answers delays are measured from that reference. This set of numbers summarizes the shape of distribution of the students' answering delays.
- Number of joined sessions.
- Percentage of sessions where the student answers all the questions.
- Percentage of sessions where the student answers 50% of the questions or more.
- Number of times that the student's answer is the same as most students.
- Number of times that the student answer is the same as the minority of the students.

We have devised these twelve metrics from the raw data logging grounded on our own experience in teaching and our previous work [17]. Other works in this area such as [20] make use of similar metrics related with time stamps, number of answers and so on.

This dataset has been preprocessed to remove redundant attributes by means of a feature selection [1]. In this preliminary work, we have chosen the importance feature ranking embedded in Extra-trees algorithm [8] and simply select the best $n$ variables to work with. This allows us to fine tune the optimal number of variables without running again the feature selection procedure.

### 2.3 Anomaly Detection

Our problem fits in the Anomaly Detection field, finding entities such that their behaviour is different from the expected one with the available data [2]. Some works use the term Outlier Detection instead [24].

At this moment, due to the massively online teaching and testing, we are interested in finding students with *tentative* marks that are unexpected given the logged online interactions during the whole term. In this way, the amount of submitted tasks/exams subjected to an exhaustive review against cheating and plagiarism will be greatly reduced. Anomaly detection methods are classified in several categories according to different criteria [24]. In our case, there is no spatial or temporal local scope, we want to label or score the candidates for an exhaustive review, the data are not labelled and we have no information about the data distribution. Thus, we need a global, labelling (or a scoring one if a suitable threshold for normalities is known), unsupervised (or semi-supervised) non-parametric method.

We have chosen *i*Forest (Isolation Forest) [12], a method based on isolating the data points using a recursive partitioning (a tree) and taking as normality measure a function of the averaged path length (over a forest) from the root node to the terminating node. The shorter this averaged path for a datum is, the more likely is for it to be an anomaly. The anomaly score is computed as $s(x,n) = 2^{-\frac{E(h(x))}{c(n)}}$, where $h(x)$ is the average length of the path for $x$, $n$ is the number of external nodes and $c(n)$ is the averaged path length of failed Binary Search Tree searches.

### 2.4 Outlier Analysis Discussion

In our case, outliers are students with unexpected scores, given the data logged regarding the online sessions. Thus, we need to compare each outlier student metric with the ones obtained by the inliers in order to ascertain if that metric value is a fraud evidence or not. For the whole set of inlier students, we compute bootstrap [6] confidence intervals for the mean $[\hat{\theta}^*_j, \hat{\theta}^*_k]$, where $\hat{\theta}^*_j$ and $\hat{\theta}^*_k$ are the jth and kth quantile of the bootstrap mean estimate, $j = \alpha/2 \times B$ and $k = (1-\alpha/2) \times B$. The confidence level is $\alpha$ and the number of samples is $B$. In this work we compute confidence intervals for the mean with levels 70, 80, 90,

95, 99 and 99.5. For each student/metric, we check if it falls within any of the confidence intervals computed with inliers data. If it does so, that metric is not unusual, otherwise it is. If the metric is unusual, we analyse if it is above the higher limit or below the lower limit of the wider confidence interval. Depending on the metric, better scores than expected might be justified or not. For instance, if we are analysing the first answer delay, a value above the higher limit of the widest confidence interval for this metric means that the outlier student usually joins late the online session. The opposite happens when the first answer delay is below the lower limit of the widest confidence interval for this metric, in this case the outlier student usually joins the online session earlier than their peers, this is a sign of engagement, interest in the subject, etc. that justifies a score higher than expected. The opposite happens with the number of attended sessions, in this case higher values are better.

## 3 Experimental Set-Up

In this work, we collected data from 69 students from the same module during the first term of 2020–2021 course. We chose the "Systems Administration" module for the preliminary study shown in this work. We focused in online laboratory sessions, where the students were asked to perform several tasks using an operating system installed in a virtual machine. The number of online sessions was 18 and a total of 68 questions were asked during these sessions. The number of questions for each online session is not evenly distributed, ranges from 1 to 13. Answering the questions was not mandatory.

At the beginning of the session, they were asked if they had joined the online activity and after that the professor asked the students several questions. We haven't logged the questions, just the answers.

In this work we set the parameters of the different algorithms to the typically recommended values found in the corresponding Python modules documentation.

As we stated earlier, we used random forest feature ranking to select the best features. The only parameter set for this method is the number of estimators, 200.

As mentioned in Sect. 2.3, the outlier detection method that best suits our needs is *i*Forest. The configuration of this method is as follows:

- Number of estimators: 200
- Number of samples to train each estimator: minimum of 256 and number of samples.
- Contamination: auto, computed as in [12].

The number of repetitions for bootstrap confidence interval estimation is 1000.

## 4 Results

Random forest feature importance ranking brings out the results in Fig. 1. From the collected set of twelve features, we have selected the first eight, in decreasing importances these are: average first answer delay (6), percentage of times that the student's answer is the less common (11), number of attended sessions(7), number of answered questions (0), 25% quantile delay (2), percentage of times that the student's answer is the most common (10), minimum delay (1), 75% quantile delay (4). In Fig. 2 a PCA (Principal Component Analysis) plot of the selected features with outliers (yellow) and inliers (dark) highlighted is shown, with an anonymized label added. As can be seen, outliers are isolated around the tightly clustered inliers as expected. From 69 students, we have identified 16 outliers; these are the candidates to an exhaustive check against plagiarism and cheating, a 76.81% reduction. The staff review has found eight anomalous results, all of them within the 23.2% of detected outliers.

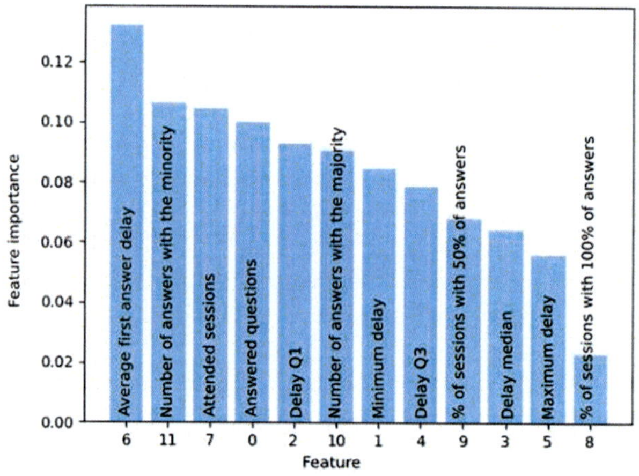


**Fig. 1.** Feature importances.

In Table 2 the obtained metrics for each outlier along with the confidence interval for each metric is shown. Those cases where the metrics do not fit within the confidence interval are marked with an asterisk and the widest confidence interval is shown, otherwise the tightest confidence interval that contains the metric is shown along with the confidence level instead of an asterisk. At the end of each row, the scores for four assignments that were used to evaluate the students are found.

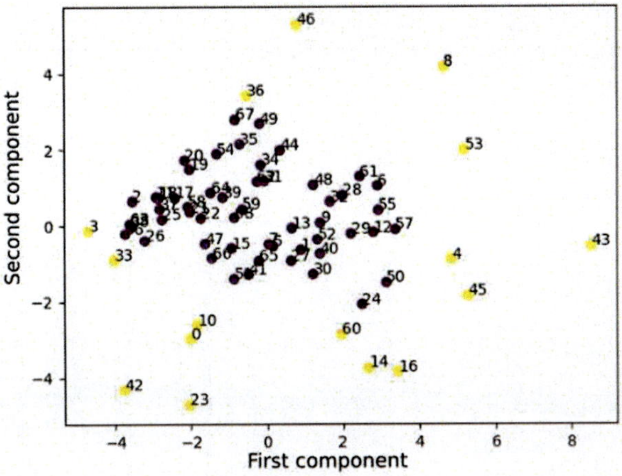


**Fig. 2.** PCA representation of outliers (yellow) and inliers (dark) labelled with an anonymous identifier. (Color figure online)

In the following, we will carefully analyse two of the outliers' data in order to explain how the proposed data pipeline processing allows us to give an exhaustive follow-up to those students with abnormal metrics.

- Student 46: the scores of this student abruptly increase from the first half of the term to the second half. The average first answer delay for this student (metric 6) is greater than the right side value of the widest confidence interval for this metric. This means that this student usually joins the online activity late. Conversely, his/hers number of answers (metric 0) is above the right side of the confidence interval for this metric, this means that the degree of engagement is high. The same happens with the number of attended sessions (metric 7), they attended most of the sessions. If we check the metrics related with the distribution of answers delays, we can see that they usually are low, this means that they is usually focused during the online sessions. In this case, the diagnosis arising from these data is that this student worked hard after the low midterm scores, it is identified as an outlier because other students with a similar pattern have failed.
- Student 8: this student also has a sudden increase of scores in the second half of the term. In this case, seems that the first answer delay is small (metric 6), he/she usually joins the online session promptly. Other delay metrics (2, 1, 4) are also quite low. The number of answers (metric 0) is low, as the number of attended sessions (7), this is unexpected because the assignments and exams are related with the online sessions: this student submitted assignments and exams need to be checked carefully.

**Table 2.** Outliers selected features vs inliers confidence intervals. The last four columns are assignments/exams grades.

| Outlier | 6 | 11 | 7 | 0 | 2 | 10 | 1 | 4 | | | | |
|---|---|---|---|---|---|---|---|---|---|---|---|---|
| 0 | 0.995,391.792 (390.211,737.31) | *,0.333 (0.375,0.709) | 0.8,0.5 (0.49,0.667) | *,21.0 (21.588,38.176) | *,11.84 (14.016,18.665) | *,0.429 (0.43,0.782) | 0.9,5.482 (5.378,8.816) | *,113.041 (36.485,73.399) | 9.7 | 9.1 | 9.0 | 9.0 |
| 3 | 0.99,701.169 (403.422,723.774) | *,0.952 (0.375,0.709) | *,0.889 (0.386,0.758) | *,45.0 (21.588,38.176) | *,10.99 (14.016,18.665) | *,0.857 (0.43,0.782) | 0.95,5.092 (5.071,9.228) | 0.7,44.576 (42.972,57.709) | 9.3 | 10.0 | 8.0 | 8.5 |
| 4 | *,331.227 (390.211,737.31) | *,0.143 (0.375,0.709) | *,0.111 (0.386,0.758) | *,6.0 (21.588,38.176) | 0.9,17.363 (14.857,17.664) | *,0.119 (0.43,0.782) | *,16.968 (4.214,10.229) | 0.99,38.952 (37.433,72.019) | 5.0 | 6.2 | 6.0 | 6.5 |
| 8 | *,182.832 (390.211,737.31) | *,0.143 (0.375,0.709) | *,0.333 (0.386,0.758) | 0.995,22.0 (21.588,38.176) | *,10.178 (14.016,18.665) | 0.99,0.452 (0.452,0.769) | *,1.288 (4.214,10.229) | 0.7,47.77 (42.972,57.709) | 0.0 | 6.0 | 8.5 | 9.0 |
| 10 | *,915.617 (390.211,737.31) | 0.99,0.667 (0.387,0.689) | 0.99,0.722 (0.405,0.745) | 0.7,30.0 (27.059,33.353) | 0.95,17.765 (14.613,17.943) | 0.7,0.643 (0.555,0.683) | *,13.337 (4.214,10.229) | *,159.323 (36.485,73.399) | 9.25 | 8.41 | 8.5 | 7.0 |
| 14 | *,339.291 (390.211,737.31) | *,0.095 (0.375,0.709) | *,0.111 (0.386,0.758) | *,6.0 (21.588,38.176) | 0.9,17.544 (14.857,17.664) | *,0.143 (0.43,0.782) | *,17.089 (4.214,10.229) | *,76.915 (36.485,73.399) | 8.6 | 7.1 | 6.0 | 6.5 |
| 16 | *,948.977 (390.211,737.31) | 0.7,0.524 (0.479,0.602) | 0.95,0.444 (0.441,0.706) | *,16.0 (21.588,38.176) | *,22.421 (14.016,18.665) | *,0.31 (0.43,0.782) | *,3.953 (4.214,10.229) | *,329.086 (36.485,73.399) | 6.9 | 5.1 | 5.0 | 6.0 |
| 23 | *,1931.784 (390.211,737.31) | 0.7,0.524 (0.479,0.602) | 0.9,0.667 (0.464,0.686) | *,20.0 (21.588,38.176) | 0.99,18.333 (14.173,18.446) | *,0.405 (0.43,0.782) | *,2.989 (4.214,10.229) | *,178.695 (36.485,73.399) | 9.2 | 9.5 | 9.0 | 9.0 |
| 33 | *,785.07 (390.211,737.31) | *,0.762 (0.375,0.709) | 0.99,0.722 (0.405,0.745) | *,39.0 (21.588,38.176) | *,12.444 (14.016,18.665) | 0.99,0.762 (0.452,0.769) | 0.99,4.608 (4.483,9.854) | 0.7,43.865 (42.972,57.709) | 10.0 | 8.4 | 10.0 | 8.0 |
| 36 | *,919.631 (390.211,737.31) | 0.99,0.667 (0.387,0.689) | *,0.833 (0.386,0.758) | *,42.0 (21.588,38.176) | *,8.96 (14.016,18.665) | *,0.857 (0.43,0.782) | *,1.145 (4.214,10.229) | 0.7,46.656 (42.972,57.709) | 3.87 | 7.16 | 9.0 | 8.5 |
| 42 | 0.9,451.436 (450.851,661.649) | 0.9,0.619 (0.44,0.639) | *,0.778 (0.386,0.758) | 0.9,35.0 (25.353,35.059) | *,23.861 (14.016,18.665) | 0.7,0.619 (0.555,0.683) | 0.95,5.127 (5.071,9.228) | *,80.714 (36.485,73.399) | 9.6 | 10.0 | 9.0 | 9.0 |
| 43 | *,87.623 (390.211,737.31) | *,0.048 (0.375,0.709) | *,0.056 (0.386,0.758) | *,3.0 (21.588,38.176) | *,31.083 (14.016,18.665) | *,0.048 (0.43,0.782) | *,24.459 (4.214,10.229) | 0.7,44.332 (42.972,57.709) | 1.5 | 5.3 | 6.0 | 6.5 |
| 45 | *,319.4 (390.211,737.31) | *,0.095 (0.375,0.709) | *,0.111 (0.386,0.758) | *,6.0 (21.588,38.176) | *,20.642 (14.016,18.665) | *,0.119 (0.43,0.782) | *,18.935 (4.214,10.229) | 0.7,57.154 (42.972,57.709) | 5.8 | 5.1 | 6.0 | 6.5 |
| 46 | *,899.283 (390.211,737.31) | *,0.952 (0.375,0.709) | *,0.889 (0.386,0.758) | *,47.0 (21.588,38.176) | *,9.961 (14.016,18.665) | *,1.0 (0.43,0.782) | *,0.718 (4.214,10.229) | 0.7,44.415 (42.972,57.709) | 3.6 | 3.7 | 7.0 | 8.0 |
| 53 | 0.99,690.029 (403.422,723.774) | 0.995,0.381 (0.375,0.709) | 0.8,0.5 (0.49,0.667) | *,19.0 (21.588,38.176) | 0.995,18.454 (14.016,18.665) | *,0.405 (0.43,0.782) | *,12.324 (4.214,10.229) | *,81.006 (36.485,73.399) | 2.2 | 3.6 | 8.0 | 8.5 |
| 60 | 0.7,532.356 (484.674,618.101) | *,0.286 (0.375,0.709) | *,0.278 (0.386,0.758) | *,18.0 (21.588,38.176) | *,31.913 (14.016,18.665) | *,0.381 (0.43,0.782) | *,15.542 (4.214,10.229) | 0.7,55.664 (42.972,57.709) | 7.12 | 7.33 | 8.0 | 8.5 |

## 5 Conclusions and Future Work

We have developed a tool to comprehensively record the students' interactions during online teaching sessions. Some of the metrics obtained from the registered data hold valuable information about the students' engagement and attention that are related with their performance, but those metrics have to be carefully selected. Unexpected grades (anomalies) can be seen as cheating or plagiarism indicators helping to reduce the workload of thoroughly checking each submission against fraud. Comparing individual metrics values of the outliers with the inliers ones, we can ascertain if an unexpectedly high score is justified or not. In this preliminary study our algorithm concludes that 76.8% of the students do not have anomalous behaviour. An independent, manually performed review has found eight anomalous results, all of them within the 23.2% of detected outliers. In any case, the sample size is not large enough to claim that these results are an adequate estimate of the precision and recall values.

In future work, we will study the temporal evolution of the metrics in the analysis. In addition, we believe that the application of Explainable Artificial Intelligence techniques [19] to this problem will make the results of the analysis more useful to the teacher: we need not only to identify candidates for a thorough review, but also to understand why they are candidates for it. Finally, after this academic year ends, we will have registered the interactions of the students in other modules and groups. We plan to verify if the findings of this preliminary study hold for these other sets of data.

## References


1. Bolón-Canedo, V., Sánchez-Maroño, N., Alonso-Betanzos, A.: A review of feature selection methods on synthetic data. Knowl. Inf. Syst. **34**, 483–519 (2013). https://doi.org/10.1007/s10115-012-0487-8
2. Chandola, V., Banerjee, A., Kumar, V.: Anomaly detection: a survey. ACM Comput. Surv. **41** (2009). https://doi.org/10.1145/1541880.1541882
3. Crick, T., Knight, C., Watermeyer, R., Goodall, J.: An overview of the impact of COVID-19 and "emergency remote teaching" on international CS education practitioners. Presented at the SIGCSE 2021 - Proceedings of the 52nd ACM Technical Symposium on Computer Science Education (2021). https://doi.org/10.1145/3408877.3439680
4. Dhankhar, A., Solanki, K., Dalal, S., Omdev: Predicting students performance using educational data mining and learning analytics: a systematic literature review. In: Raj, J.S., Iliyasu, A.M., Bestak, R., Baig, Z.A. (eds.) Innovative Data Communication Technologies and Application. LNDECT, vol. 59, pp. 127–140. Springer, Singapore (2021). https://doi.org/10.1007/978-981-15-9651-3_11
5. Du, X., Yang, J., Shelton, B.E., Hung, J.-L., Zhang, M.: A systematic meta-review and analysis of learning analytics research. Behav. Inf. Technol. **40**, 49–62 (2021). https://doi.org/10.1080/0144929X.2019.1669712
6. Efron, B., Tibshirani, R.J.: An Introduction to the Bootstrap. CRC Press, Boca Raton (1994)

7. Foster, L.N., et al.: Increasing low-responding students' participation in class discussion. J. Behav. Educ. **18**(2), 173–188 (2009). https://doi.org/10.1007/s10864-009-9083-8
8. Geurts, P., Ernst, D., Wehenkel, L.: Extremely randomized trees. Mach. Learn. **63**, 3–42 (2006). https://doi.org/10.1007/s10994-006-6226-1
9. Hantoobi, S., Wahdan, A., Al-Emran, M., Shaalan, K.: A review of learning analytics studies. Stud. Syst. Decis. Control **335**, 119–134 (2021). https://doi.org/10.1007/978-3-030-64987-6_8
10. Iglesias-Pradas, S., Hernández-García, Á., Chaparro-Peláez, J., Prieto, J.L.: Emergency remote teaching and students' academic performance in higher education during the COVID-19 pandemic: a case study. Comput. Hum. Behav. **119** (2021). https://doi.org/10.1016/j.chb.2021.106713
11. Keržič, D., Aristovnik, A., Tomaževič, N., Umek, L.: Assessing the impact of students' activities in e-courses on learning outcomes: a data mining approach. Interact. Technol. Smart Educ. **16**, 117–129 (2019). https://doi.org/10.1108/ITSE-09-2018-0069
12. Liu, F.T., Ting, K.M., Zhou, Z.-H.: Isolation-based anomaly detection. ACM Trans. Knowl. Discov. Data **6**, 1–39 (2012). https://doi.org/10.1145/2133360.2133363
13. Loftin, C., Davis, L.A., Hartin, V.: Classroom participation: a student perspective. Teach. Learn. Nurs. **5**(3), 119–124 (2010). https://doi.org/10.1016/j.teln.2010.02.004
14. Abdullah, M.Y., Bakar, N.R.A., Mahbob, M.H.: The dynamics of student participation in classroom: observation on level and forms of participation. Procedia Soc. Behav. Sci. **59**, 61–70 (2012). https://doi.org/10.1016/j.sbspro.2012.09.246
15. Mustapha, S.M., Rahman, N.S.N.A., Yunus, M.M.: Perceptions towards classroom participation: a case study of Malaysian undergraduate students. Procedia Soc. Behav. Sci. **7**, 113–121 (2010). https://doi.org/10.1016/j.sbspro.2010.10.017
16. Namoun, A., Alshanqiti, A.: Predicting student performance using data mining and learning analytics techniques: a systematic literature review. Appl. Sci. (Switz.) **11**, 1–28 (2021). https://doi.org/10.3390/app11010237
17. Otero, J., Junco, L., Suárez, R., Palacios, A., Couso, I., Sánchez, L.: Finding informative code metrics under uncertainty for predicting the pass rate of online courses. Inf. Sci. **373**, 42–56 (2016). https://doi.org/10.1016/j.ins.2016.08.090
18. Paff, L.A.: Does grading encourage participation? Evidence & implications. Coll. Teach. **63**(4), 135–145 (2015). https://doi.org/10.1080/87567555.2015.1028021
19. Roscher, R., Bohn, B., Duarte, M.F., Garcke, J.: Explainable machine learning for scientific insights and discoveries. IEEE Access **8**, 42200–42216 (2020). https://doi.org/10.1109/ACCESS.2020.2976199
20. Riestra-González, M., del P. Paule-Ruíz, M., Ortín, F.: Massive LMS log data analysis for the early prediction of course-agnostic student performance. Comput. Educ. **163**, 104108 (2021). https://doi.org/10.1016/j.compedu.2020.104108
21. Sixsmith, A., Dyson, L.E., Nataatmadja, I.: Improving class participation in it tutorials and small lectures. Presented at the ACIS 2006 Proceedings - 17th Australasian Conference on Information Systems (2006)
22. Wild, I.: Moodle 3.x Developer's Guide. Packt Publishing Ltd. (2017)
23. Zhang, M.: Teaching with Google Classroom. Packt Publishing Ltd. (2016)
24. Zimek, A., Filzmoser, P.: There and back again: outlier detection between statistical reasoning and data mining algorithms. WIREs Data Mining Knowl. Discov. **8** (2018). https://doi.org/10.1002/widm.1280

# Low Cost Three-Phase Motor Speed Control System Design for Educational Laboratory Practices

Álvaro Michelena, Francico Zayas-Gato, Esteban Jove(✉), José-Luis Casteleiro-Roca, Héctor Quintián, and José Luis Calvo-Rolle

Department of Industrial Engineering, University of A Coruña, CTC, CITIC, Ferrol, A Coruña, Spain
{alvaro.michelena,f.zayas.gato,esteban.jove, jose.luis.casteleiro,hector.quintian,jlcalvo}@udc.es

**Abstract.** This document presents the design, construction and commissioning of a low cost test bench for monitoring and controlling the speed of a three-phase motor. This test bench will complete the offer of industrial models available in the Optimization and Control Laboratory of the Faculty of Engineering at University of A Coruña (UDC), in which students will be able to carry out process control procedures applying different theoretical concepts acquired throughout the degree course.

**Keywords:** Test bench · Three-phase motor · Variable frequency drive

## 1 Introduction

Formerly, technical engineering university degrees had a lot of theoretical teaching. This fact caused that many students, once they had finished their studies, and despite having a great theoretical knowledge, found difficulties to put it into practice in industrial situations. However, recent changes in university curricula have changed this situation [5].

Currently, after the Bologna declaration was established, the practical lessons represent a mandatory part of the university subjects. The implementation of this practical sessions in the university requires that the universities have specific laboratories with technical material and equipment, where the students can carry out their tasks [11].

In the field of engineering, these laboratories incorporate some educational plants that try to simulate industrial processes. However, many of the technical equipment of these laboratories are very expensive, so universities can only have a few units, or, in some cases, may not have that specific equipment [1,7].

However, to solve this problem many universities create their own industrial systems implementing low-cost devices to reduce the cost of its manufacture. In this way, and due to its simple design and its reduce price, these models can be replicated easily to facilitate that all the students could carry out their tasks.


J. J. Gude Prego et al. (Eds.): CISIS 2021/ICEUTE 2021, AISC 1400, pp. 315–324, 2022.
https://doi.org/10.1007/978-3-030-87872-6_31

In this way, this article presents the design, assembly and commissioning of a low-cost test bench for monitoring and control of a three-phase motor [4,12,13].

The paper is structured as follows: after the present introduction, the background and the motivation of this work are described. Section 3 provides a detailed explanation about the proposed method. Then, the experiments and results are shown in Sect. 4. Finally, the conclusions and future works are listed.

## 2 The Control Engineering Laboratory

The Optimization and Control Laboratory is one of the laboratories in the Faculty of Engineering, at University of A Coruña. This laboratory has some different educational plants that emulate some industrial processes where the students can implement control algorithms. Examples of these processes are for instance temperature, level or pressure control loops.

### 2.1 Laboratory Background

As mentioned above, one of the systems at this laboratory is the temperature control loop which is used by students for accomplishing temperature regulation and control exercises. This system is made with methacrylate box, which emulates a stove, where two incandescent bulbs are used as a heating elements. In addition, this system includes in one of sides a DC fan for making disturbances [9,16].

A level control system is used for regulating the liquid level in a tank. For this, the plant is based on two tanks at different levels. The lower tank is the liquid storage for feeding the centrifugal pump, which propels the water to the upper tank. Whereas, the upper tank is where the water level is controlled. For measuring the water level at the upper tank, an ultrasonic sensor S18UUA from Banner is used [3,15].

Furthermore, the tanks are connected through a pipe with a manual or servo piloted valves. The pumped liquid flow depends on the centrifugal pump speed that is controlled with a Variable Frequency Drive (VFD) from Schneider.

### 2.2 Motivation and Objectives

The temperature control system is relatively a slow process where the variable to be controlled, the temperature, varies slowly in both the heating and cooling processes. Furthermore, in the case of the level control plant, the system also is a relatively slow process, although faster than the temperature loop. For this reason, and with the aim of training with different kinds of industrial processes, the design and assembly of a new plant with a faster performance process is proposed.

As explained above, it was decided to design a speed control system for a three-phase engine. The speed variable to be controlled varies with a fast dynamic. The following features should be considered the system design:

- The system should be as simple as possible, minimizing the mechanical parts of the system to avoid possible mechanical problems.
- The design must be flexible and modular too, in which modifications can be made without the need for major changes.
- All necessary security measures should be included to make the handling of the system as safe as possible.
- Finally, the design should be economical using low-cost hardware whenever possible.

In general, the main goal of the project is to design a simple, economical, safe, and modular system for the speed control of a three-phase motor.

## 3 Proposal

This section describes the proposal for the speed control system of a three-phase motor. Although some alternatives have been taken into account, analyzing their advantages and disadvantages, this research will mainly discuss the selected design.

In general, the system includes the next main elements: three-phase motor and loads, variable frequency drive, speed sensor and DAQ. In addition, other components are added to the design to ensure proper and safe operation. The general scheme of the system is shown in the Fig. 1.

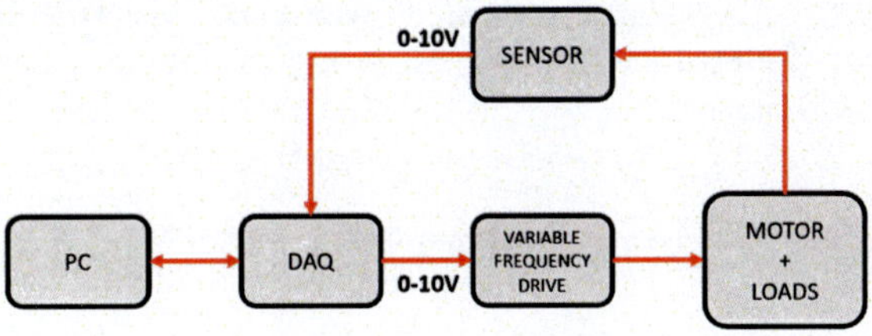


**Fig. 1.** System overview

The speed sensor is the device for obtaining a voltage proportional to the rotational speed of the system (motor and loads). This voltage is the analog input signal to the DAQ, which converts it into a digital value with which the control algorithm, which is execute on a computer, works. This algorithm generates output values that are sent to the DAQ, where they are converted to analogue values with which the VFD works. Depending on the output value of the control algorithm, the VFD will receive a DC proportional signal (0–10 V) and will generate the right three-phase motor supply, for controlling the motor rotational speed.

### 3.1 Motor-Loads-Sensor Arrangement

One of the main aspects of system design is to select a layout for the motor, loads and sensor. Different alternatives were analysed for the arrangement of these elements, but the layout that simplified the mechanical couplings of the system was selected. Therefore, the solution adopted was to place the load system coupled to the front axle of the motor, while the speed sensor would be located at the rear.

As a result of this arrangement, the chosen motor should be one in which the shaft must be accessed from the rear. One of the main advantages of this arrangement is that the axis of rotation of the system is limited to the motor shaft itself, so no extension or shaft alignment processes are required.

### 3.2 Motor Selection and Load System Design

Once the three main components of the system has been selected, a load system is designed. One of the factors to be considered is that the loads should be replaceable, with the aim of small disturbances can be applied to the system.

Firstly, different available alternatives on the market were studied, such as the use of electromagnetic brakes, or a generator on which electrical loads (such as power resistors) could be connected. However, all of this proposals were discarded because this solutions cannot generate an inertial system which is necessary to obtain overshoots responses. Therefore, the proposed solution was to design the load system.

Before of designing the load system, it was necessary to select the motor. Then, to know the dimensions of the front axle for making a design that could be coupled to it is required.

The following motor characteristics were considered:

- As a result of the selected arrangement, the motor must have access to its shaft from the rear.
- An economical and low-maintenance three-phase motor should be sought. For this reason the selected motor should have a squirrel cage rotor.

Considering the limitations, a three-phase CEMER IE1-MSE802-4 motor from the manufacturer Cosgra was selected. This motor is characterised by being a three-phase asynchronous motor with a power of 1 HP and a nominal speed of 1500 r.p.m., with a squirrel cage rotor and with a lower price than others with the same characteristics.

This model, like most of the electric motors, features a shaft extension at the rear to attach the motor's cooling system, usually consisting of plastic blades coupled to the rear axle. This system incorporates a metal protection casing. Once the electric motor used has been selected, the load system can be designed. The proposed design consists of 2 parts: the load discs and the coupling of the load discs to the motor shaft.

The loads correspond to metal discs of different thicknesses and weights, while the coupling is designed to be attached to the motor shaft and serve as a support for the load discs. The main advantages of this design are that one or more discs can be included to the system to simulate different loads, and that the loads can be replaced easily by unscrewing the wingnuts. In addition, this design offers the possibility of adding unbalanced loads that generate disturbances in the system. Finally it is important that the discs are made of heavy materials such as stainless steel or titanium, in order to be able to brake the system.

### 3.3 Selection and Arrangement of the Speed Sensor

Currently, there are many different speed sensors like optical encoders, magnetic encoders, tachogenerators, etc. However, in order to limit the system mechanical parts and simplify the design, the considered alternatives were magnetic encoders because, this kind of sensor, do not need contact with the shaft to measure its rotational speed.

For the system, an incremental magnetic encoder of 20 pulses per revolution was selected since it was low cost and had an adequate resolution for the designed system. Incremental encoders generates a pulse train and, knowing the time between samples, the quantity of detected pulses and the sensor resolution, the rotational speed can be determined.

The chosen model uses two Hall sensors for obtaining the quadrature output. To activate the Hall sensors and generate the output pulses a 10-pole magnetized disc is used. For measuring the rotational speed, the magnetic disc must be coupled to the motor shaft. In this way, when the magnet rotates, the Hall sensor are activated and deactivated producing the quadrature output signal.

For using this sensor it was necessary the design of a bracket for its right positioning. The design of this housing is proposed to be manufactured in plastic by means of a 3D printer to reduce its cost. In addition, for a easy sensor electrical connection, a small bracket is designed to hold a 4-pin male connector, which will be fixed to the casing of the ventilation system. The casing and the connector bracket designs are shown in Fig. 2.

(a) Design (b) Real model

**Fig. 2.** Casing and connector bracket

For the right measurement of the speed sensor, it is important to place the magnet disc correctly centred on the motor's rotation axis. For this reason, a small bracket is also designed to ensure that the magnet is correctly centered on the motor shaft. This bracket is shown in Fig. 3.

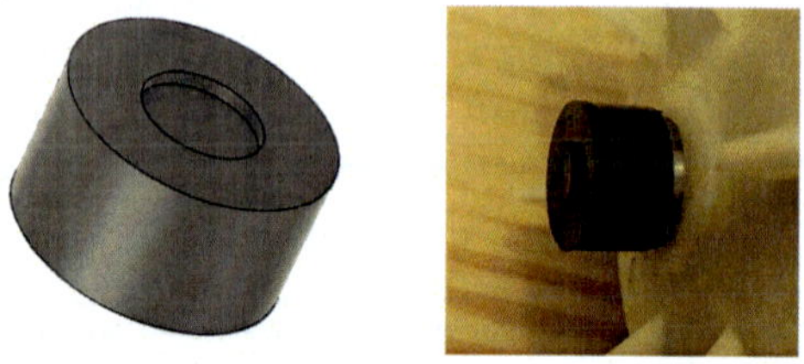

(a) Design (b) Real Model

**Fig. 3.** Magnet disc bracket

One of the features of the system its flexibility. For this reason, the sensor output signal was adapted to DC 0–10 V for its reading from a DAQ. For the sensor output conditioning and to obtain proportional voltage to the rotational speed, the adaptation signal system is needed.

The Arduino Uno microcontroller will be programmed to count both the rising and falling edges of the two output channels of the sensor for a specific sample time. In this way, knowing the number of detected edges and the sensor resolution, the speed value is calculated.

### 3.4 Variable Frequency Drive (VFD) Selection

Another of the main devices in the system is the Variable Frequency Drive, that generates the three-phase motor supply, with a variable frequency and voltage depending on the value of the analogue input signal (0–10 V).

Nowadays, the VFD are standardised and there are not significative differences between the models. Therefore, for the system designed, any VFD, that is adapted to the electrical motor characteristics, could be used. For the system, the Schneider ATV12HU075M2 model has been used, since the UDC had some units available, and, in addition, this model is used on the level control systems with a right, safe and simple functionality.

### 3.5 Data Acquisition Card (DAQ)

Data Acquisition Card is the used device for converting the system analogue signals to digital values and digital signals to analogue (0–10 V). Initially, some models of data cards were analyzed, but due to the high price of the commercial ones, other lower-cost alternatives were considered, based on microcontrollers and low-cost hardware. For these reasons it was decided to use an Arduino Uno together with a conditioning circuit for both the input and output signals. The Arduino is programmed to read an analogue voltage and convert it to digital value when it is necessary. For the connection between the computer and the DAQ, the Arduino serial port was used.

### 3.6 Safety Devices

For the system design, it is necessary to analyze the risks or hazards that may occur when it is put into operation. In this way, two kinds of problems were analyzed:

- Electrical hazards at connections of the different devices.
- Mechanical risks caused by rotating the load system at high speed.

In order to minimize these risks, both electrical connection and mechanical safety devices were implemented. Due to the simplicity of the proposed design, the mechanical risks are focused on the possibility of the user coming into contact with the moving parts when they are in motion, and on the parts projections (loading and coupling). For these reasons, and taking into account the possible trajectories of the elements that could be projected and the possibility of replacing the load discs, it was proposed the design of a protection box which is presented in Fig. 4.

(a) Desing (b) Real Model

**Fig. 4.** Complete system

The protection box is designed to be made of polycarbonate, a transparent plastic material with a high impact resistance. Also the design includes a small door in the front side to access to the load system and replace the discs. This door includes a limit switch to stop the system if the door is open.

On the other hand, to minimize electrical hazards, a circuit breaker was added to protect the system from overcurrents and short circuits. In addition, it was designed a housing to protect the user from the electrical connections and for simplifying the connections. All housing are designed to make the power connections via cable glands, while the control signal connections are made with banana connectors (the sensor connection is made via a 4-pin connector).

The following housings are designed for the system:

- Circuit breaker contact protection housing.
- VFD contact protection housing.
- Sensor output signal conditioning circuit protection housing.
- DAQ protection housing.

The final design and the real system could be seen in Fig. 5.

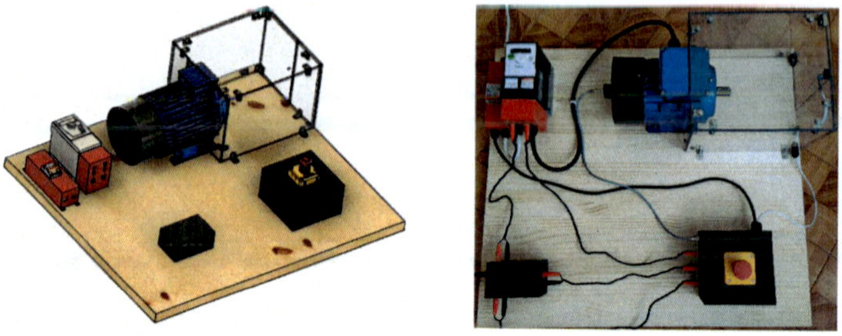

(a) Desing (b) Real Model

**Fig. 5.** Complete system

## 4 Results

Once the system was designed, it was built, assembled, and commissioned. Initially, a testing was made for the right operation checkup. These tests were able to detect small faults that were easily solved with simple settings. Special care was taken during of the speed sensor placement (incremental magnetic encoder), because a wrong positioning could produce measurement errors.

Also, it was necessary to configure the different parameters of the VFD. Once all the system's devices had been checked, some operational tests were made. For this purpose, a speed control algorithm based on a PI controller was programed in Matlab.

The system was tested for different speeds and the obtained results are presented in Fig. 6.

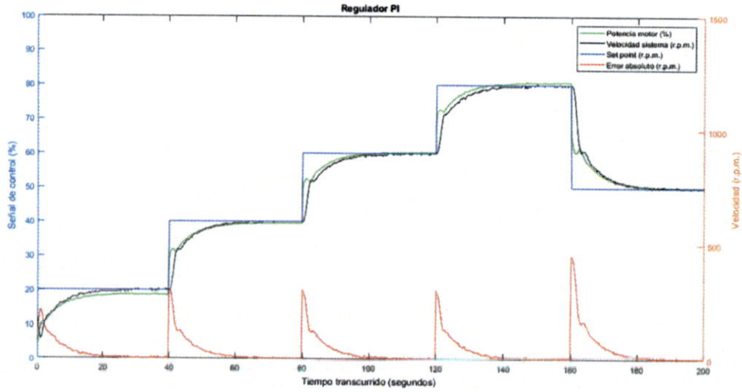


**Fig. 6.** System control with PI-controller

The system absolute error is shown in red, the setpoint value in blue, the control signal sent to the VFD in green and the system speed (feedback signal) in black. As can be seen, the system behavior is similar to a first order system.

In conclusion, it should be noted that the system designed works correctly, and the achieved response behaves like a first-order system response.

## 5 Conclusions and Future Works

This paper describes the design of a speed control system of a three-phase motor for different coupled loads. This laboratory plant gives a new fast dynamics system that completes the industrial models of the Optimization and Control Laboratory at the Faculty of Engineering at the University of A Coruña. In this way, the laboratory incorporates slow dynamics systems (temperature control), medium speed systems (level control) and systems with a fast dynamic (speed control). Therefore, students will be able to have an approach of the typical processes that can be found actually in the industry, and put into practice concepts and knowledge acquired in the degree.

On the other hand, and due to the current pandemic, the remote laboratory possibility is being considered [2,6,8,10,14]. With it, students can connect, by internet, to the different systems without the laboratory presence, thus facilitating practices in the different control systems, improving the quality of online teaching.

**Acknowledgements.** CITIC, as a Research Center of the University System of Galicia, is funded by Consellería de Educación, Universidade e Formación Profesional of the Xunta de Galicia through the European Regional Development Fund (ERDF) and the Secretaría Xeral de Universidades (Ref. ED431G 2019/01).

## References

1. Aláiz-Moretón, H., Castejón-Limas, M., Casteleiro-Roca, J.L., Jove, E., Fernández Robles, L., Calvo-Rolle, J.L.: A fault detection system for a geothermal heat exchanger sensor based on intelligent techniques. Sensors **19**(12), 2740 (2019)
2. Alaiz-Moretón, H., et al.: Bioinspired hybrid model to predict the hydrogen inlet fuel cell flow change of an energy storage system. Processes **7**(11), 825 (2019)
3. Casteleiro-Roca, J.L., Barragán, A.J., Segura, F., Calvo-Rolle, J.L., Andújar, J.M.: Fuel cell output current prediction with a hybrid intelligent system. Complexity **2019**, 1–10 (2019)
4. Casteleiro-Roca, J.L., Javier Barragan, A., Segura, F., Luis Calvo-Rolle, J., Manuel Andujar, J.: Intelligent hybrid system for the prediction of the voltage-current characteristic curve of a hydrogen-based fuel cell. Revista Iberoamericana de Automática e Informática industrial **16**(4), 492–501 (2019)
5. Crespo-Turrado, C., et al.: Comparative study of imputation algorithms applied to the prediction of student performance. Logic J. IGPL **28**(1), 58–70 (2020)
6. Domínguez, M., Reguera, P., Fuertes, J.J.: Laboratorio remoto para la enseñanza de la automática en la universidad de león (españa). Revista Iberoamericana de automática e informática industrial **2**(2), 36–45 (2010)

7. Fernández-Serantes, L.A., Estrada Vázquez, R., Casteleiro-Roca, J.L., Calvo-Rolle, J.L., Corchado, E.: Hybrid intelligent model to predict the SOC of a LFP power cell type. In: Polycarpou, M., de Carvalho, A.C.P.L.F., Pan, J.-S., Woźniak, M., Quintian, H., Corchado, E. (eds.) HAIS 2014. LNCS (LNAI), vol. 8480, pp. 561–572. Springer, Cham (2014). https://doi.org/10.1007/978-3-319-07617-1_49
8. Gonzalez-Cava, J.M., Reboso, J.A., Casteleiro-Roca, J.L., Calvo-Rolle, J.L., Méndez Pérez, J.A.: A novel fuzzy algorithm to introduce new variables in the drug supply decision-making process in medicine. Complexity **2018**, 1–15 (2018)
9. Jove, E., Alaiz-Moretón, H., García-Rodríguez, I., Benavides-Cuellar, C., Casteleiro-Roca, J.L., Calvo-Rolle, J.L.: PID-ITS: an intelligent tutoring system for PID tuning learning process. In: Pérez García, H., Alfonso-Cendón, J., Sánchez González, L., Quintián, H., Corchado, E. (eds.) SOCO/CISIS/ICEUTE -2017. AISC, vol. 649, pp. 726–735. Springer, Cham (2018). https://doi.org/10.1007/978-3-319-67180-2_71
10. Jove, E., et al.: Missing data imputation over academic records of electrical engineering students. Logic J. IGPL **28**(4), 487–501 (2019). https://doi.org/10.1093/jigpal/jzz056
11. Jove, E., Casteleiro-Roca, J.L., Quintián, H., Méndez-Pérez, J.A., Calvo-Rolle, J.L.: Virtual sensor for fault detection, isolation and data recovery for bicomponent mixing machine monitoring. Informatica **30**(4), 671–687 (2019)
12. Jove, E., Casteleiro-Roca, J.L., Quintián, H., Méndez-Pérez, J.A., Calvo-Rolle, J.L.: Anomaly detection based on intelligent techniques over a bicomponent production plant used on wind generator blades manufacturing. Revista Iberoamericana de Automática e Informática Industrial **17**(1), 84–93 (2020). https://doi.org/10.4995/riai.2019.11055
13. Jove, E., Casteleiro-Roca, J.-L., Quintián, H., Méndez-Pérez, J.A., Calvo-Rolle, J.L.: Outlier generation and anomaly detection based on intelligent one-class techniques over a bicomponent mixing system. In: Martínez Álvarez, F., Troncoso Lora, A., Sáez Muñoz, J.A., Quintián, H., Corchado, E. (eds.) SOCO 2019. AISC, vol. 950, pp. 399–410. Springer, Cham (2020). https://doi.org/10.1007/978-3-030-20055-8_38
14. Jove, E., Casteleiro-Roca, J.L., Quintián, H., Méndez-Pérez, J.A., Calvo-Rolle, J.L.: A new method for anomaly detection based on non-convex boundaries with random two-dimensional projections. Inf. Fusion **65**, 50–57 (2021)
15. Jove, E., Casteleiro-Roca, J.L., Quintián, H., Simić, D., Méndez-Pérez, J.A., Calvo-Rolle, J.L.: Anomaly detection based on one-class intelligent techniques over a control level plant. Logic J. IGPL **28**(4), 502–518 (2020)
16. Vega Vega, R., Quintián, H., Calvo-Rolle, J.L., Herrero, Á., Corchado, E.: Gaining deep knowledge of android malware families through dimensionality reduction techniques. Logic J. IGPL **27**(2), 160–176 (2019)

# Evaluation of E-learning Criteria to Improve the Academic and Functional Results of PhD Studies

Bruno Baruque-Zanón[1](✉) and Ana María Lara-Palma[2]

[1] Computer Science Department, University of Burgos, Avd. Cantabria s/n., 09006 Burgos, Spain
bbaruque@ubu.es

[2] Management Engineering Department, University of Burgos, Avd. Cantabria s/n., 09006 Burgos, Spain
amlara@ubu.es

**Abstract.** The European Commission in its Digital Education Action Plan (2021–2027) sets the objective of readjusting education and training to the digital age. It highlights two main guidelines to reach this goal. Firstly, to promote the development of a high-performing digital education ecosystem and, secondly, to enhance digital skills and competences for the digital transformation. Inspired by this transformative spirit, the University of Burgos (UBU) proposes a roadmap to set up a digital educational model around e-learning as a key factor. In the academic year 2015–2016 the institution implemented for the first time an online Engineering Degree. Since then, sixteen official degrees have been developed to shape the map of the online academic offer in the institution (five Online Degree studies, four Online Master studies, seven Streaming Master studies). The objective of this paper is to evaluate the e-learning criteria to improve the academic and functional results of PhD studies. The methodological approach consists in a quantitative evaluation (eighteen items in rubric format) that gather interest data in e-learning approach with the Intranet Tool. The concluded results allow stablishing foresight and actions for being a reference in e-learning education in PhD studies at UBU.

**Keywords:** PhD studies · E-learning · International studies · On-line learning tools

## 1 Introduction

After 20 years since the implementation of the European Higher Education Area (EHEA), universities have incorporated a set of improvements in order to achieve the objectives and guidelines of Bologna, Prague and Berlin. The European Association for Quality Assurance in Higher Education (ENQA) has highlighted on numerous occasions the high degree of commitment of European cooperation to improve teachers training and to engage the students to the stage of education.


J. J. Gude Prego et al. (Eds.): CISIS 2021/ICEUTE 2021, AISC 1400, pp. 325–335, 2022.
https://doi.org/10.1007/978-3-030-87872-6_32

However, when everything pointed to a stabilization of the actions for improvement, the year 2020 has proven to be a challenge for the functioning of every academic institution with a global pandemic caused by the Covid-19 virus. Along with this unforeseen scenario, new threats arise putting in danger all the goals reached to date. It is needed to act immediately to continue maintaining the quality ratios and standards of the basic principles of the educational system.

Learning is the engine of any educational system. However, learning is not limited solely to the action of harboring knowledge. Learning should feature knowledge transfer, competitiveness improvement, satisfaction and progress. Based on Bologna Process Implementation Report [1], learning is the result of a process where knowledge is a statement of what the individual knows, understands and is able to do on completion of work. Moreover, a report from the Conference of Spanish University Rectors, CRUE [2], considers learning as a compendium of instrumental, cognitive, social and interpersonal skills, work planning and results orientation, systemic skills and self-determination, values and motivational aspects.

One of the most relevant contributions that the educational system has made to maintain learning levels during the pandemic has been the incorporation of e-learning into the work roadmap for the academia community. This technique, widely used in some institutions, has been introduced in the teaching of regular classes (going from face-to-face to virtual or streaming modality), third-cycle studies (facilitating collaborative repositories) and remote meetings (favoring internationalization).

Consequently, the pandemic has not implied a setback in the way of transferring knowledge, quite the contrary. New technologies and investment in technical support have given a boost to how the academic community interacts.

In the following subsection, we tackle the concrete aspects of the e-learning context in PhD studies, in a particular Spanish example, the University of Burgos.

## 2 E-learning Context in Third Cycle (PhD) Studies

The new reports from the European Commission show a glossary of new future areas of improvement, such as, degree structures, quality assurance and recognition, social dimension, internationalization and future developments [3]. The action plan called e-Europe focuses, among others, on e-learning research initiatives on Third Cycle (PhD) Programs.

The fundamentals of this plan comprise a large set of initiatives [4] to ensure that all people related to education have access to equipment, IT, multimedia and Internet access. It also proposes to train teachers in order to integrate these instruments into their educational method, develop high-quality European content in the multimedia field, and accelerate the link with online education and training systems.

Making a temporary review of the literature on the dimension that encompasses e-learning (online learning through collaborative technologies), Table 1 shows a taxonomy and the collective benefits derived from its usability.

Authors' framework cover a range of overlapping perspectives noting that e-learning taxonomy brings together a series of standpoints: learning units (contents and knowledge), social work (collaborative interaction) and technology (learning platforms).

Regarding the benefits obtained from applying the e-learning methodology, authors identify a significant percentage of improvements in learning and interaction motivated by the use of technology. All of them complement each other and enable an in-depth critical analysis of the boundaries of e-learning as a resource to enhance learning.

**Table 1.** Review of the literature on the dimension that encompasses e-learning (taxonomy and benefits)

| Literature review | E-learning taxonomy | Derived e-learning benefits |
|---|---|---|
| Kumar, P., Saxena, C., Baber, H. (2021) [5] | Learning platforms: widely spread, flexible, easily accessible, perpetually open, eliminates barriers of distance, exponential enrollment of students | – Qualitative learning<br>– Satisfaction of learners |
| Rababa, N. (2021) [6] | Teacher-centered and memory-based education to student-centered education (collaboratively work, build their knowledge, foster critical thinking) | – Significant improvement in the higher thinking skills<br>– Remove many social and cultural barriers |
| Al-Fraihat, D., Joy, M., Masádeh, R., Sinclair, J. (2020) [7] | Direct result of the integration of technology and education | – Perceived usefulness<br>– Perceived satisfaction and use |
| Papi, C. (2019) [8] | Needings for move forward obstacles and levers in digital technology support | – Techno-pedagogical practices and activities<br>– Leadership |
| Aparicio, M., Bacao, F., Oliveira, T. (2016) [9] | Typology of e-learning system´s services (including users) | – Integration of learning strategies, technologies and stakeholders |
| Wang, T. H. (2014) [10] | Assessment-centered e-Learning system | – Personalized e-Learning material<br>– Improvement of misconceptions (especially for low-level prior knowledge)<br>– Enhance learning |
| Bernard, R. M., Abrami, P. C., Borokhovski, E., Wade, C. A., Tamim, R. M., Surkes, M. A., Bethel, E. C. (2009) [11] | Technology and achievements | – Student interaction-Linear relationship between IT and outcomes |

(*continued*)

**Table 1.** (*continued*)

| Literature review | E-learning taxonomy | Derived e-learning benefits |
|---|---|---|
| García, F. J. (2007) [12] | Lifelong learning with storage facilities for use and reuse contents and learning units | – Permanent learning process<br>– Qualifies persons to acquire and update their knowledge, skills and abilities in various stages and in various learning contexts |

Although much of e-learning benefits have been clearly identified in bachelor and master studies, it is needed to analyze them also in regard to the third cycle studies. Quoting a reflection in PhD studies [13], "learning proceeds amidst what candidates do, what peers do, what advisor do, what academics responsible for professional learning and leading and researching in the academy do".

## 2.1 Doctoral Program Review. A Particular Case Study (UBU)

In the academic year 2018–2019, 86,619 students enrolled on third grade studies (RD 99/2011) at the Spanish universities. Specifically, 14,005 of them were listed in the disciplines of engineering and architecture. Figure 1 shows the distribution of the number of students enrolled in RD 99/2011 PhD by disciplines and gender [14].

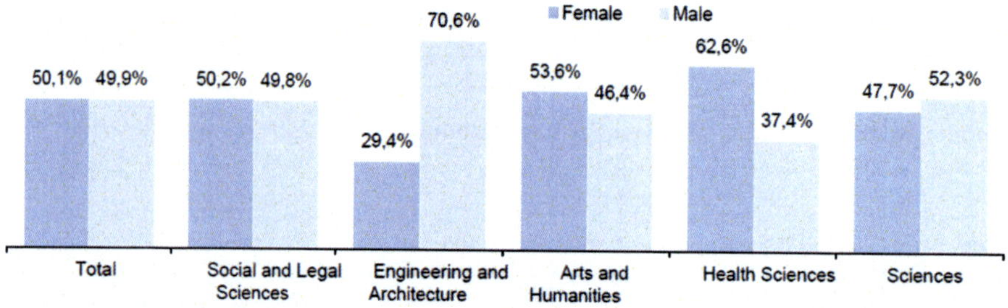


**Fig. 1.** Distribution of the Number of Students Enrolled in RD 99/2011 PhD in Spain (2018–2019 academic year)

Out of the total enrolled, 26.2% of the students were foreigners and 73.8% were Spanish. Fifty Schools of Doctorate offer their academic Programs for national and international participants. In Castilla y León, the number of foreign PhD students was 1,689, which represents 33.9% of the total number of students enrolled in the autonomous communities of the country. Since the early twentieth century, the University of Burgos has been evolving promoting attention in high standards and rankings of learning, specifically in technology-enhanced learning. In this regard, improvements in equipment, networks, streaming/virtual communication and corporate platforms show the most outstanding advances. This strategy has positioned the institution at the rankings indicated in Table 2. According research status (international doctorate), the institution occupies the 12$^{th}$ position out of 49 public universities evaluated [15]. More specifically, the Engineering Program Doctorate data reflects a 6.7% of international thesis [16].

**Table 2.** Research ranking by contribution at UBU

| Source | Research ranking | Contribution |
|---|---|---|
| U-Ranking 2019 | 10th position<br>9th position | Number of high level impact publication (Q1)<br>Research and development |
| Fundación CyD | 9th position | Number of high level impact publications (Q1) |
| U-Multirank 2019 | 2 out of 77 | Interdisciplinary publications |
| | 14 out of 49 | Students mobility |
| | 12 out of 49 | International PhD |

As far as PhD studies are concerned, e-learning is not only a strategic factor for continuous improvement and academic excellence, but the paradigm of collaborative knowledge sharing at a critical moment such as the one brought by the pandemic. The objective of this paper is to carry out an evaluation of e-learning criteria to improve the academic and functional results of PhD studies. The methodological approach consists in a quantitative evaluation (eighteen items in rubric format) that gather interest data in e-learning approach with the Intranet Tool.

## 3 Summative Dimensions of E-learning. A Criteria Rubric for PhD Studies

Based on the Bologna Process Implementation Report, The European Students' Union (ESU) [1] defines student-centered learning as "both a mindset and a culture [...] characterized by innovative methods of teaching which aim to promote learning in communication with teachers and other learners and which take students seriously as active participants in their own learning, fostering transferable skills such as problem-solving, critical thinking and reflective thinking".

E-learning must gather relevance in this worldwide pandemic crisis and serve for these above mentioned purposes. PhD support has to provide content resources, learning paths, guiding principles, assessment activities, synchronous and asynchronous seminars [17], IT platforms, social collaborative presence and metacognitive engagement.

Our contribution highlights a rubric valuing eighteen items codified in e-learning approach with the Intranet Tool (ease of use, tech support/help availability, hypermediality, accessibility, mobile design, privacy, social presence, teaching presence and cognitive presence). In this sense, the results of this study will show how close the university is to the Bologna commitments.

The rubric aims to be a scoring guide and it is based on [18] (Table 3 brings together the description of the parameters that make up the rubric) and own elaboration adapted to PhD Studies.

Drawing on the rubric, next section shows findings of the methodological approach (empirical study).

## 4 Analysis of Survey Results

The rubric shown in Sect. 3 has been published as a web form in order to let the PhD students of the University of Burgos anonymously access it and send their opinion about each item.[1]

Table 4 shows the most interesting details of the student sample. It includes the opinion of a 33% of students of the "Doctoral Program in Law Economics and Social Sciences" and a 66% of the "Doctoral Program in Industrial Technologies and Civil Engineering", being these two the more related to the authors and therefore with a biggest opportunities to reach students. It is interesting also to note that there were more international than national students that answered the questionnaire. The main reason could be that these kind of students are those who need to make a more extensive use of online tools for communication with their supervisors and are, therefore, more involved in the improvement of those tools. Finally, the number of years spent at PhD studies is evenly distributed among recently enrolled students and those who spent more time in their PhD studies.

Regarding the results, each item of the rubric was given a range of 1 to 3, *"1"* meaning a clear dissatisfaction with the valued aspect of the system, and *"3"* meaning the most favorable opinion on the aspect. The result obtained for each item is shown in Fig. 2, Fig. 3 and Fig. 4.

The most favorable valued items are those related to the accessibility to the intranet, along with the data privacy and security of this system. More precisely, the item "Operating System" has a mean value of $2.6 \pm 0.47$, being the highest valued; while the items "Access" and "Data Privacy" have both a mean value of $2.3 \pm 0.47$. On the other side of the spectrum, the lowest rated items are the "Ease of Use", with a mean value of $1.5 \pm 0.5$, then "Customization" with a mean value of $1.6 \pm 0.48$ and finally "Hypermediality", "User Focused" and "Enhancement of Cognitive Tasks" all three with a mean value of $1.6 \pm 0.5$.

This means that, while the system deals satisfactorily with technical aspects available in modern software (especially those systems used by public services such as public education or health systems), it lacks on other aspects, such as ease of use or customization of the user experience, in which the user has many examples in private systems; where the user experience and UI design are preferred over other considerations.

Finally, the last two questions, which provided a more detailed scale (1 to 4) for two important dimensions of the doctoral studies, gave the results in Fig. 5. The mean of the "Collaborative Research" has a value of $1.5 \pm 0.5$ and the "Dissemination of Knowledge", a mean value of $1.66 \pm 0.74$, ranking quite low in relation with other aspects. This also points in a similar direction: while the on-line platform is perceived as offering the basic functionality and complying with privacy requirements, it lacks in aspects of user communication and cooperation, providing a quite limited freedom for the students to cooperate and disseminate results by means of this platform.

[1] A copy of the form can be found at: https://forms.gle/fFevDqsk8kNfQonA7.

**Table 3.** Rubric taxonomy summary [18] and own elaboration

| Ratio | E-learning taxonomy | Derived PhD e-learning benefits |
|---|---|---|
| Ease of use | The ease of use criterion focuses on design characteristics that contribute to user-friendliness and intuitive use | – Tools are perceived more intuitive, easy of use and offers guidance through user engagement<br>– *Students* optimize time and questions |
| Tech support | Tech support focuses on optimal support localized, up-to-date, responsive to users´ needs and timely | – Enhance campus-based technical support with robust support from the platform<br>– *Students* perceive quality, advanced technology and autonomous learning |
| Hyper-mediality | Hypermediality is achieved by providing multiple forms of media as well as the ability to organize lessons in a non-sequential way | – Encourage instructors and students to engage with and communicate through different forms of media in a flexible, nonlinear fashion<br>– *Students* learn in a visual context with a more assistive technology |
| Accessibility standards | An e-learning tool should adhere to mandated requirements for accessibility, legislative accessibility standards and generally accepted guidelines | – Provide information to meet accessibility standards<br>– *Students* find a placeholder for all services |
| Accessibility participation | E-learning tools that address the needs of diverse users and include broader understandings of literacies and student capabilities | – *Students* develop their research work on global spectrum platforms (educational inclusion and equal opportunities) |
| Technical | Selecting an e-learning tool that can be installed and run on up-to-date versions | – *Students* do not find incompatibilities in the operating system |
| Mobile design | For e-learning tools accessed using a mobile device. Any e-learning tool that accesses the internet should offer an offline mode | – *Students* are able to expand access independently for those who have limited or intermittent connectivity<br>– *Students* save time, are open to work in a big window time and timetable is flexible |

*(continued)*

**Table 3.** (*continued*)

| Ratio | E-learning taxonomy | Derived PhD e-learning benefits |
|---|---|---|
| Privacy | Privacy shows the IP policies of the e-learning tool and the user's control over how content is shared | – Users maintain their IP rights and are able to exercise full control over how their content is made public<br>– *Students* feel confident with intellectual properties, online communities and ethical policies |
| Social presence | Social presence provides synchronous and asynchronous channels for knowledge exchange establishing a sense of community between users | – Instructors are encouraged to design learning activities to build a sense of community<br>– *Students* feel willing to participate more actively |
| Teaching presence | Tool provides easy-to-use features that enhance an instructor's ability to effectively engage in facilitation activities. Alignment between intended learning outcomes and learning activities and assessments | – Effective teaching provides timely input, information, feedback, thinking skills, modelling inquiry and demonstrating cognitive engagement<br>– Students find a clear and simple assessment mode/method that examines their progress |
| Cognitive presence | Select technologies that redefine tasks adding functional value for finding answers to a complex problem. Directly links formative feedback and metacognitive practice | – Enhances learning. Instructors support higher-order learning tasks. Formative feedback to modify thinking or behaviour for improved learning |

**Table 4.** Details of the sample of students on the presented study

| PhD program | | Nationality | | Years of studies | |
|---|---|---|---|---|---|
| Economics and law | Industrial engineering | German | Spanish | Less than 3 | 3 or more |
| 33% | 66% | 66% | 33% | 50% | 50% |

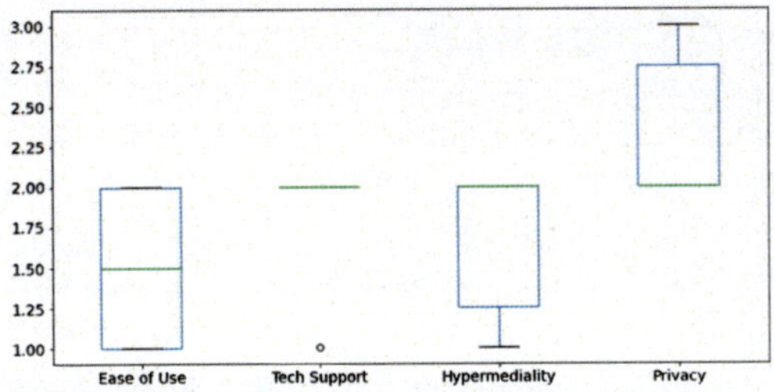


**Fig. 2.** Boxplot of answer distributions for the items 1 to 4 of the rubric

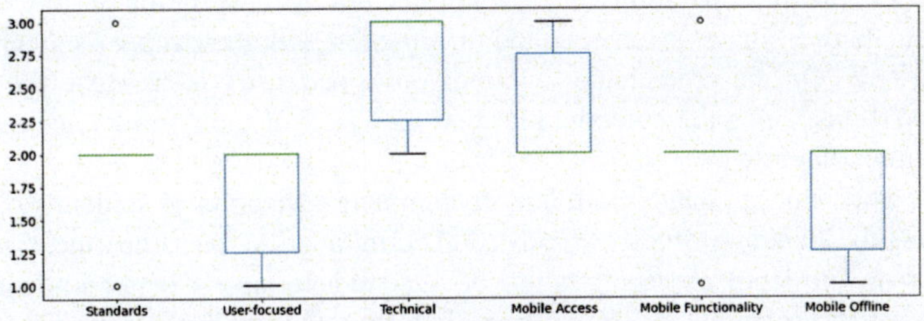


**Fig. 3.** Boxplot of answer distributions for the items 5 to 10 of the rubric

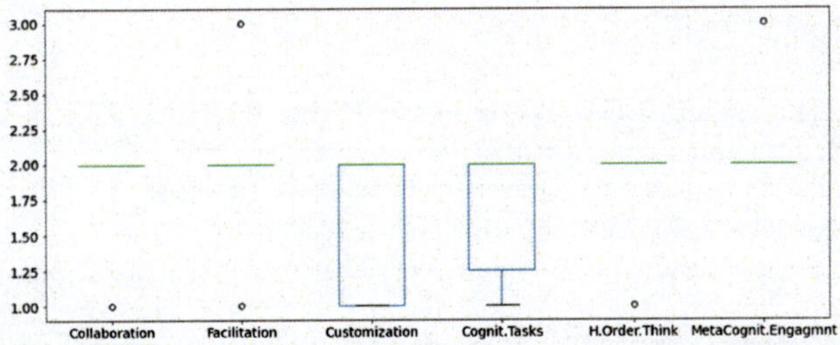


**Fig. 4.** Boxplot of answer distributions for the items 11 to 16 of the rubric

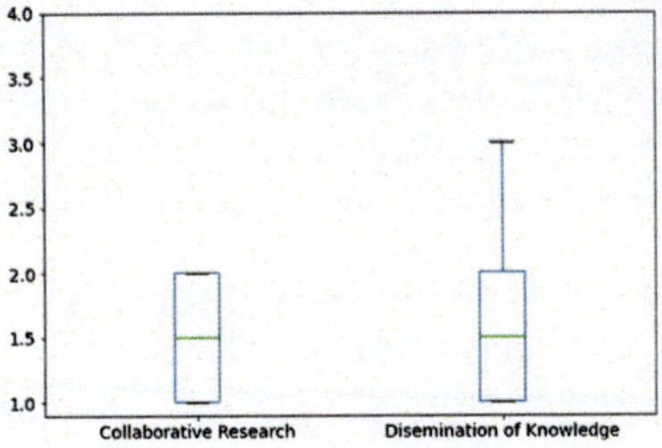


**Fig. 5.** Boxplot of answer distributions for corresponding to items 17 and 18 of the rubric

## 5 Conclusions and Future Work

The contribution presents the general framework of the third cycle studies in the Spanish education system, focusing on how the e-learning modality could be leveraged for this kind of studies. This might be especially critical in situations like the current COVID-19 pandemic situation; but could be also interesting to be used in highly on-line connected societies, such as our own.

As an analysis of a concrete instance, the results of the survey performed on a sample of PhD students on the University of Burgos is included. In this aspect, the main conclusions to highlight are twofold: on one hand, there is a good perception of the students about the interoperability and data protection included in the on-line platform that enables them to share documents and information with their supervisors and other university staff. On the other hand, it seems quite necessary to improve the current available systems, since the students feel that they lack of tools to interact with each other in a more fluent way.

Future work would include both a more complete collection of student's opinions, by widening the student profiles with other PhD programs. At the same time, this would enable a more in-depth characterization of our students and a better knowledge of the perceived shortcomings of the on-line platform, along with their requirements of collaboration opportunities with their peers on the beginning of their research career.

## References

1. European Commission/EACEA: The European Higher Education Area in 2020. Bologna Process Implementation Report (2020)
2. Hernández, J., Pérez, J.A.: La Universidad Española en cifras 2013/2014, 2014/2015, 2015/2016, 2016/2017, 2017/2018. CRUE Universidades Españolas
3. https://eacea.ec.europa.eu/national-policies/eurydice/content/third-cycle-phd-programmes-79_en
4. Aznar, I.: La metodología Blended e-learning aplicada a la formación universitaria como respuesta a los planteamientos de reforma establecidos en la declaración de Bolonia: studio de un caso. REIFOP Revista Electrónica Interuniversitaria de Formación del Profesorado 8(4) (2005)
5. Kumar, P., Saxena, C., Baber, H.: Learner-content interaction in e-learning – the moderating role of perceived harm of COVID-19 in assessing the satisfaction of learners. Smart Learn. Environ. **8**(1), 1–15 (2021). https://doi.org/10.1186/s40561-021-00149-8
6. Rababa, N.: The effect of e-learning in developing high thinking skills. Int. J. Data Netw. Sci. **5**(1), 43–46 (2021)
7. Al-Fraihat, D., Joy, M., Masádeh, R., Sinclair, J.: Evaluating e-learning systems success: an empirical study. Comput. Hum. Behav. **102**, 67–86 (2020)
8. Papi, C.: Towards a generalization of digital technology in education? Digit. Soc. Socio-polit. Issues **1**, 179–190 (2019)
9. Aparicio, M., Bacao, F., Oliveira, T.: An e-learning theoretical framework. Educ. Technol. Soc. **19**(1), 292–307 (2016)
10. Wang, T.H.: Developing an assessment-centered e-learning system for improving student learning effectiveness. Comput. Educ. **73**, 189–203 (2014)
11. Bernard, R.M., et al.: A meta-analysis of three types of interaction treatments in distance education. Rev. Educ. Res. **79**(3), 1243–1289 (2009)

12. Peñalvo, F.J.: eLearning: covering lifelong in the European Space for higher education. Int. J. Contin. Eng. Educ. Life-Long Learn. **17**(6), 403–405 (2007)
13. Manidis, M., Addo, R.: Learning a practice through practice: presenting knowledge in doctoral spoken presentations. Stud. Contin. Educ. **39**(3), 235–250 (2017)
14. Subdirección General de Actividad Universitaria Investigadora de la Secretaría General de Universidades. Ministerio de Universidades. Datos y Cifras del Sistema Universitario Español. Publicación (2019–2020)
15. UBU Homepage: https://www.ubu.es/portal-de-transparencia/resultados/posicionamiento-en-rankings-de-universidades
16. Baruque-Zanón, B., Lara-Palma, A.M.: International PhD Quality Assessment: Growing Awareness of Transnational Projection. In: Herrero, Á., Cambra, C., Urda, D., Sedano, J., Quintián, H., Corchado, E. (eds.) ICEUTE 2020. AISC, vol. 1266, pp. 132–141. Springer, Cham (2021). https://doi.org/10.1007/978-3-030-57799-5_14
17. eCampusAlberta: eLearning Rubric. a resource for quality online curriculum development (2017)
18. Anstey, L.M., Watson, G.P.L.: A rubric for evaluating e-learning tools in higher evaluation. EDUCAUSE Rev. (2018)

# Learning by Doing in Higher Education: An Experience to Increase Self-learning and Motivation in First Courses

Oscar Fontenla-Romero, Francisco Bellas(✉), Noelia Sánchez-Maroño, and J. A. Becerra

Escuela Universitaria Politecnica, University of A Coruña, Ferrol, Spain
{oscar.fontenla,francisco.bellas,noelia.sanchez, jose.antonio.becerra.permuy}@udc.es

**Abstract.** This article presents an experience of application of the Learning by Doing methodology through a Project Based Learning (PjBL) activity in the "Computer Science" compulsory subject, which belongs to the first year of the Electrical Engineering degree at the University of A Coruña. The goal was to increase the students' self-learning, leading to a lower degree of plagiarism and a higher motivation. The main contribution of this experience is on the introduction of a real problem to be fully solved in a first year compulsory subject. The results obtained during two academic years, applying this learning by doing methodology, showed an increase in student motivation to reach a value of 94% and also a 60% reduction of similarity between submitted programs. In addition to describing the experience and analyzing the results, the didactic impacts are discussed in a more global way.

**Keywords:** Project-based learning · Learning by doing · Educational robotics

## 1 Introduction

Project based learning (PjBL) is a teaching methodology belonging to what is called "active learning", which is based on the fact that the concepts are acquired or reinforced by the students as they are required to solve a problem in the real world [2]. This concept is opposed to the "passive learning" approach of the traditional teaching, supported by master classes where students are receptors of information that has to be later memorized [6].

PjBL has been applied worldwide in primary, secondary and higher education, during the last 20 years with general successful results [5,7]. In [3], authors perform a meta-analysis comparing the effects of PjBL versus traditional teaching on student academic achievement, showing that PjBL had a more positive impact on students' academic achievement than traditional instruction did. Moreover, it was shown the benefits of fostering students' innovation by supporting their autonomy during learning tasks. However, PjBL is far from being


J. J. Gude Prego et al. (Eds.): CISIS 2021/ICEUTE 2021, AISC 1400, pp. 336–345, 2022.
https://doi.org/10.1007/978-3-030-87872-6_33

an official methodology in education, but an optional approach that teachers implement on their own.

The implantation of PjBL in higher education is "natural", because its main features like problem-solving, team work or autonomous learning are supposed to be critical skills to prepare students for the labour market [3]. Although the number of research papers analyzing the formal consequences of applying PjBL in higher education is lower than in pre-university education [4], most of them are focused on experiences belonging to optional subjects in the last years of technical degrees, where specific problems can be proposed to students, but with low generalization towards other disciplines [1,9].

The current work shows that PjBL can be also highly beneficial when applied in compulsory subjects belonging to the first years of university degrees. This type of subject is less "flexible" than the optional ones of the last years, mainly because they cover basic concepts and topics and thus students could not understand how these concepts could be applied in a practical problem. Moreover, the temporal organization of such subjects is typically more adjusted (similar to pre-university education), so the PjBL activities have to be more controlled and, consequently, the autonomy of students decreases. The main contribution of this experience is not on the introduction of PjBL at this level, but on the application of a learning by doing methodology in a basic subject, in this case through a robotic problem in the real world that students had to solve. Therefore, students had to focus their work on providing a reliable solution, which was the main evaluated aspect, leading to an indirect learning of the basic programming skills, as they were required to solve the task.

The results obtained during two academic years are described and analyzed, showing a very important increase in their self-learning and motivation. Although the experience presented here has been carried out in a computer science subject of a technical degree, the conclusions obtained from its application can be generalized to other compulsory subjects simply by adapting the practical activity to the specific field.

## 2 Description of the PjBL Activity

In the new activity, small robots are incorporated in the challenge to solve, and a learning by doing methodology is proposed. The general idea is that students have to program a device that interacts with the real world, i.e. outside the digital environment of the computer, and closer to their specialization as engineers of the industrial sector. In addition, the solution is mainly evaluated from a practical perspective, considering the proper response of the robot in the task. The abilities required to obtain such a reliable solution are obtained by students as required to achieve the goal, without specific training on them from the teachers.

With this approach, the aim was to achieve the following didactic objectives related to the skills of the subject:

- Delve into key aspects of operation of the computer programs, such as performance, robustness and reliability. In order to solve the challenge in a contin-

uous fashion, the students had to focus their attention in these aspects, less treated in the subject until the moment, but very relevant for their training.
- Increase the proactivity of students when solving problems that may arise during the execution of the proposed project, as they can observe the robot operating in the real world and the errors that it commits.
- Improve the motivation of students by using a device of great current popularity and interest for their future work.

These objectives were aligned with the issues drawbacks in the previous years, and their global goal was to increase the self-learning capabilities of the students.

In order to carry out the activity, 12 mBot robots[1] were acquired. This robot is one of the most economic of the market and it is based on an arduino board and free software.

The methodology applied to introduce the activity in the subject was the same in both degrees (Electrical and Industrial Electronic and Automatic Engineering), but in a different semester:

1. Carrying out a face-to-face practical session, of two hours duration, with the objective of introducing to students the fundamentals of the mBot hardware and programming.
2. Accomplishment of the programming project with the robot. The students were organized in pairs and they should carry out the challenge in a maximum period of 5 weeks. For it, they should request the robots to the teachers of the subject and to use them in the college, not being possible to take them outside.
3. Performing a face-to-face defense. At the end of the 5-week period, students had to submit the C code of their solution to the project, and a defense of this solution in person with the teachers was carried out. They had to explain the details of the developed program and, more important, show the practical response of the robot in different trials.
4. Carrying out a competition between the different groups, in which the one with the best performance obtained an extra score in the evaluation of the practice. This activity was open for public audience at the college, so that any other student or teacher could attend it.

With this methodology, the main goal was to increase the motivation of the students towards the subject by introducing the challenge with the robot. In addition, the competition was supposed to favor this motivation because they had to show publicly their work and compare it with others. As commented above, we intended to achieve a lower number of plagiarism in the task too. The use of physical devices that may differ slightly from one another was key to achieving this objective, because the same program could not work. Moreover, the value of the competition result in the students' score was introduced to face this goal of promoting original solutions. Finally, but as expected, this methodology allowed to develop also relevant skills in the engineering field: the ability to work in a team and the ability to proactively solve problems.

[1] https://www.makeblock.com/mbot.

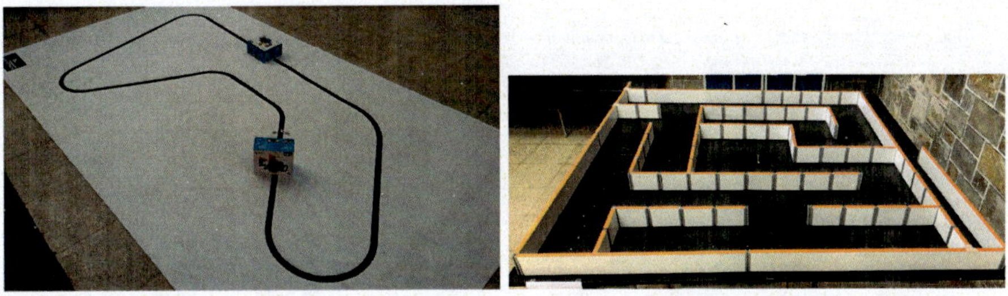

**Fig. 1.** Experimental setups to carry out the projects

As an example of the specific type of challenges that the students must solve, Fig. 1 shows the working environments proposed in courses 2017/18 and 2018/19. The left image shows a circuit created with a dark line on a white background. The objective of the program to be developed was to achieve that the robot, in a totally autonomous way, follows the dark line, as fast as possible, performing a complete lap. With the aim of increasing the difficulty of the challenge, we randomly placed 2 obstacles on the route that the robot had to circumvent. In the competition among students, it was valued the time required to perform a complete lap and the robustness of the solution, that is, the number of times the goal was achieved from a set of 3 consecutive attempts. Specifically, the programming and defense had a 70% value of the grade, while the competition and robustness was the remaining 30%.

The right image in Fig. 1 shows another of the real environments used in the challenge. In this case, the environment is a maze, and the objective of the developed program was to guide the robot towards the center of the maze, with independence of the origin point. Again, in the competition, the time employed to achieve the goal and the number of times that it was achieved in 3 consecutive attempts were considered in the evaluation, using the same values than in the previous year.

As it can be observed in the evaluation criteria, the practical issues of the solution were very relevant. Although teachers did not provide specific training regarding how to make the solution more robust and reliable, it was a key aspect in the grade, so students had to obtain these skills in an autonomous way. The consequence of this learning by doing methodology was that they realize how relevant is an efficient and simple programming in order to achieve such robustness in the validation stage.

Once the activity was completed, students conducted an anonymous survey to get feedback on the activity. The development of the activity took place during the 2017/18 and 2018/19 academic years, so it was possible to analyze its evolution according to the feedback provided. The results of these surveys are shown and discussed in the following section.

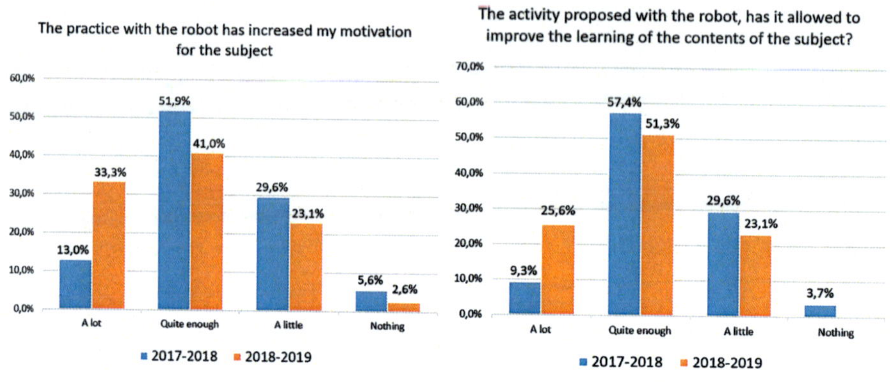


**Fig. 2.** Questions 1 and 2 of the student survey

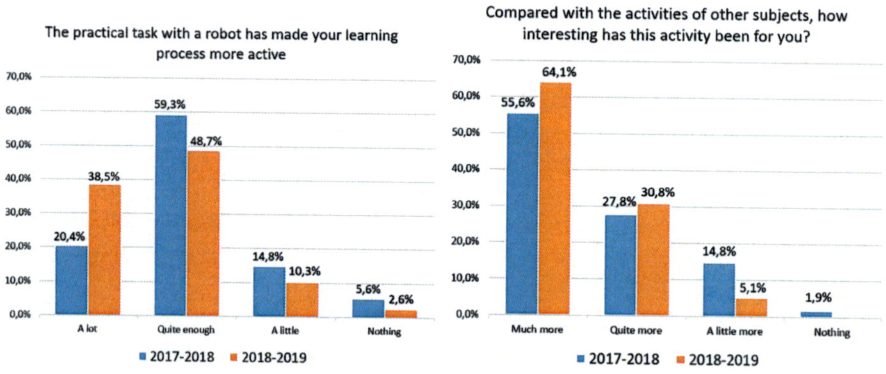


**Fig. 3.** Questions 3 and 4 of the student survey

## 3 Results

In order to objectively evaluate the results of this activity, some surveys were made, which were answered by the students at the end of the activity during the 2017/18 and 2018/19 school years. Figures 3, 4, 5 and 6 include the answers in each of the questions, as well as a comparison between them in both academic years.

First of all, it is important to clarify that, during the 2018/19 academic year, certain improvements were made to the activity over the original approach. Specifically, the students' progress was monitored at an intermediate stage, with the aim of ensuring constant development over time and avoiding the accumulation of tasks on dates close to the delivery of the program. In addition, motivated by the student's feedback about the activity, technical improvements were made to the robots in order to reduce the time needed for the students to adjust the solution, like the inclusion of a second ultrasonic sensor.

As it can be seen in the four figures, the perception of the activity in the 2018/19 school year (orange color) was better than in the previous year (blue color), so we can conclude that the modifications made resulted in an improvement in the motivation of the students. This type of continuous improvement is necessary in an PjBL methodology, especially in the first years of its implementation, since it is extremely complicated to predict all the problems that students may encounter in an open work approach like this.

Analyzing in detail the results in Fig. 2 left, it can be seen that in both academic years, a remarkable degree (quite enough or a lot) of improvement was achieved in student motivation, reaching 64.9% of students in the 2017/18 academic year and 74.3% in 2018/19. Moving to Fig. 2 right, it can be observed how the majority of students perceived that the activity was relevant to improve the learning of the subject's contents (quite enough or a lot). This is an interesting result, because the main subject contents were the fundamentals and formal aspects of programming, while the objective of the activity was to make the robot perform as well as possible. Hence, students' impression was that they improved their programming skills although no explicit mention was done by teachers in these aspects.

The question shown in Fig. 3 left aimed to reflect how using a real robot made students' learning more pro-active, that is, they had to take decision on their own to obtain the solution. Again, we can observe that 79.7% in 2017/18 and 87.5% in 2018/19 detected an improvement in this aspect. Finally, if we take a look at the results shown in Fig. 3 right, we see how the activity was highly interesting for them as compared to other subjects (mainly compulsory too). Consequently, the motivation and interest for the subject was increased, as expected, while the learning of the fundamental concepts was maintained and even, improved in some aspects.

On the other hand, the goal of decreasing similar or plagiarized work was also achieved. To evaluate this objective, a software tool was used that allows, in an automatic and objective way, to detect similarities in the source code of programs. The software tool used in this study was MOSS [8], developed by Stanford University. As commented in Sect. 2, before the implementation of this activity with the robot, students already developed a PjBL activity, but focused on creating a program to solve a more traditional computer science project. The top part of Fig. 4 contains the results of the detection of plagiarism while using this previous approach, in the course 2016/17 of the Electrical Engineering Degree, for 15 submitted programs (note that the same program may appear several times with different degrees of plagiarism, and other could not be shown).

Accordingly, the bottom part of Fig. 4 contains the percentage of lines of code considered identical among the 12 programs delivered in the 2018/19 academic year in the same degree, using now the robot activity. As it can be seen, only 3 programs were detected with some similar lines and a maximum of 20% similarity. Analyzing Fig. 4, it is clear that the improvement in this aspect was very remarkable, as it significantly reduces the percentage of maximum similarity (from 84% to 20%). With a level of 84% similarity between two documents

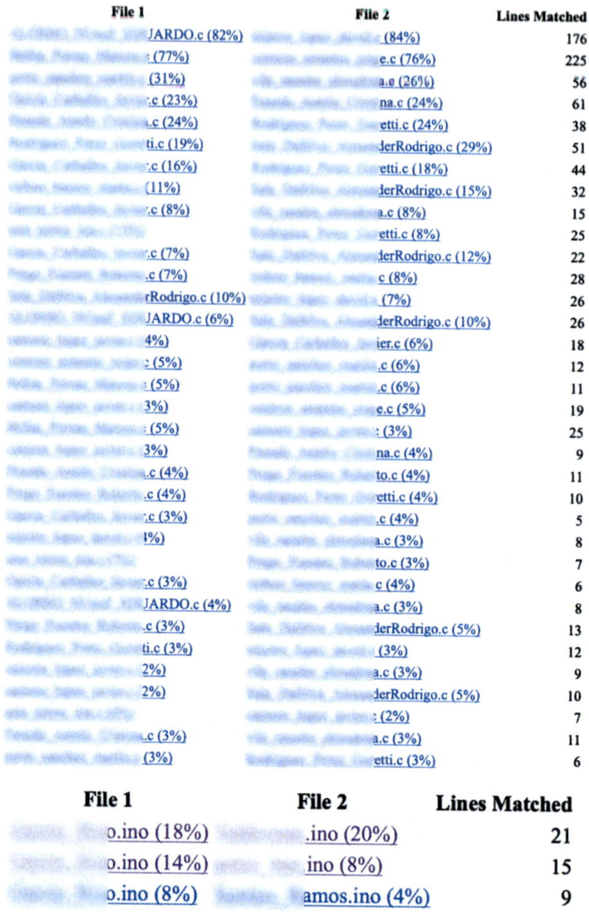

| File 1 | File 2 | Lines Matched |
|---|---|---|
| JARDO.c (82%) | (84%) | 176 |
| (77%) | e.c (76%) | 225 |
| (31%) | a.c (26%) | 56 |
| .c (23%) | na.c (24%) | 61 |
| .c (24%) | etti.c (24%) | 38 |
| ti.c (19%) | derRodrigo.c (29%) | 51 |
| .c (16%) | etti.c (18%) | 44 |
| (11%) | derRodrigo.c (15%) | 32 |
| .c (8%) | a.c (8%) | 15 |
| | etti.c (8%) | 25 |
| .c (7%) | derRodrigo.c (12%) | 22 |
| .c (7%) | c (8%) | 28 |
| rRodrigo.c (10%) | (7%) | 26 |
| JARDO.c (6%) | derRodrigo.c (10%) | 26 |
| 4%) | ier.c (6%) | 18 |
| (5%) | .c (6%) | 12 |
| (5%) | .c (6%) | 11 |
| 3%) | e.c (5%) | 19 |
| (5%) | (3%) | 25 |
| 3%) | na.c (4%) | 9 |
| .c (4%) | to.c (4%) | 11 |
| .c (4%) | etti.c (4%) | 10 |
| .c (3%) | .c (4%) | 5 |
| 4%) | a.c (3%) | 8 |
| | to.c (3%) | 7 |
| .c (3%) | c (4%) | 6 |
| JARDO.c (4%) | a.c (3%) | 8 |
| .c (3%) | derRodrigo.c (5%) | 13 |
| ti.c (3%) | (3%) | 12 |
| 2%) | a.c (3%) | 9 |
| 2%) | derRodrigo.c (5%) | 10 |
| | (2%) | 7 |
| .c (3%) | a.c (3%) | 11 |
| (3%) | etti.c (3%) | 6 |

| File 1 | File 2 | Lines Matched |
|---|---|---|
| o.ino (18%) | .ino (20%) | 21 |
| o.ino (14%) | ino (8%) | 15 |
| o.ino (8%) | amos.ino (4%) | 9 |

**Fig. 4.** Percentage of lines of code detected as identical by the MOSS plagiarism detection tool for courses 2016/17 (top) 2018/19 (bottom) in the Electrical Engineering Degree (image blurred due to privacy)

(source code) is obtained a fairly clear evidence of a case of plagiarism while the level of 20% is quite reasonable taking into account that the challenge posed is common to all groups of students and, therefore, there will be parts of the code, which correspond to more fundamental parts of programming, whose similarity is inevitable.

Figure 5 shows a similar comparison but, in this case, for the Industrial Electronic and Automatic Engineering degree, before and after applying the new PjBL activity (top and bottom part). Although the difference is not so high, it is again clear that the number of programs with any similarity has been drastically reduced. Finally, Fig. 6 shows the average score (grade) of the PjBL activity and the percentage of submitted programs from year 2010/11 until 2018/19,

| File 1 | File 2 | Lines Matched |
|---|---|---|
| ego.c (38%) | (48%) | 84 |
| ro.c (30%) | istina.c.c (24%) | 84 |
| o_Javier.c (15%) | el.c (17%) | 52 |
| o_Javier.c (12%) | c (8%) | 29 |
| (11%) | sabel.c (12%) | 53 |
| erto.c (14%) | c (14%) | 47 |
| .c (11%) | .c (10%) | 23 |
| (14%) | .c (10%) | 44 |
| .c (10%) | c (12%) | 48 |
| : (6%) | (4%) | 20 |
| c (8%) | (11%) | 30 |
| erto.c (11%) | ro.c (5%) | 92 |
| .c (12%) | (4%) | 28 |
| c (10%) | ı.c (9%) | 17 |
| c (9%) | os.c (10%) | 22 |
| c (9%) | os.c (10%) | 31 |
| c (7%) | os.c (10%) | 28 |
| (9%) | ateo.c (5%) | 27 |
| .c (7%) | (9%) | 22 |
| c (7%) | istina.c.c (7%) | 22 |
| varo.c (7%) | (8%) | 18 |
| .c (9%) | o_Javier.c (6%) | 24 |
| erto.c (8%) | c (4%) | 17 |
| c (7%) | lro.c (9%) | 26 |
| : (4%) | el.c (7%) | 17 |
| c (6%) | ateo.c (4%) | 11 |
| varo.c (6%) | .c (5%) | 24 |
| : (4%) | (5%) | 23 |
| ego.c (6%) | (3%) | 33 |
| varo.c (6%) | rgio.c (13%) | 15 |
| c (6%) | (5%) | 31 |
| ro.c (7%) | sabel.c (6%) | 21 |
| c (6%) | el.c (6%) | 16 |
| (7%) | ı.c (6%) | 15 |
| o_Javier.c (5%) | (2%) | 18 |
| (6%) | perto.c (6%) | 66 |
| (4%) | ateo.c (3%) | 2 |

| File 1 | File 2 | Lines Matched |
|---|---|---|
| ain.ino (5%) | (14%) | 41 |
| (26%) | .ino (14%) | 23 |
| (12%) | ino (5%) | 10 |
| no (5%) | .ino (6%) | 11 |

**Fig. 5.** Percentage of lines of code detected as identical by the MOSS plagiarism detection tool for the activity of courses 2015/16 and 2017/18 in the Industrial Electronic and Automatic Engineering Degree (image blurred due to privacy)

that is, before and after applying the learning by doing methodology. It can be observed that the average score (6.59 vs 6.41) was more or less the same, meaning that the difficulty of the activity was similar. But the number of submitted programs increased from 58.5% up to 66%, which objectively reinforces the idea that students found this new approach more motivating in the last years.

| | Academic degree | Academic year | Average score of the project | Percentage of students who submmited the project |
|---|---|---|---|---|
| Academic years using the previous project (without mBot) | EE | **2010/11** | 6,75 | 62,2% |
| | IEAE | **2010/11** | 6,28 | 75,0% |
| | EE | **2011/12** | 5,82 | 50,0% |
| | IEAE | **2011/12** | 6,99 | 71,0% |
| | EE | **2012/13** | 6,09 | 56,1% |
| | IEAE | **2012/13** | 7,93 | 61,0% |
| | EE | **2013/14** | 6,52 | 53,4% |
| | IEAE | **2013/14** | 5,91 | 56,3% |
| | EE | **2014/15** | 6,67 | 54,5% |
| | IEAE | **2014/15** | 8,28 | 63,6% |
| | EE | **2015/16** | 7,38 | 51,4% |
| | IEAE | **2015/16** | 7,12 | 62,2% |
| | EE | **2016/17** | 5,18 | 44,1% |
| | | **Mean:** | 6,69 | 58,5% |
| | | | | |
| Academic years using the new project approach with mBot | IEAE | **2016/17** | 6,49 | 66,7% |
| | EE | **2017/18** | 6,33 | 74,3% |
| | IEAE | **2017/18** | 5,35 | 58,3% |
| | EE | **2018/19** | 6,78 | 58,3% |
| | IEAE | **2018/19** | 7,60 | 72,3% |
| | | **Mean:** | 6,51 | 66,0% |

**Fig. 6.** Percentage of students participating in PjBL activity and score achieved. IE stands for Industrial Engineering and IEAE for Industrial Electronic and Automatic Engineering

## 4 Conclusions

The main contribution of this work is on the introduction of a learning by doing methodology in a first year compulsory subject of higher education. It was carried out by means of a PjBL activity where the challenge to be faced by students included a real device, in this case, a robot, and where the main evaluated aspect was its practical response. As a consequence, basic programming skills were acquired as they were required to solve the task properly. Based on the answers to the surveys performed by students, it can be concluded that the methodology has proved positive in improving students' motivation. Moreover, the results have been progressively improved from one academic year to another, based on the feedback provided by the students in the first surveys:

- In academic year 2017/18, 94.4% of the students showed an increase in their motivation at some level.
- In the 2018/19 academic year, 97.4% of students reported increased motivation at some grade level.

On the other hand, a very important reduction, contrasted by the use of the MOSS tool, has been achieved in the level of plagiarism of the activities delivered by the students. This result reinforces the main conclusion of this experience, that the methodology has been very successful in terms of increasing the students' self-learning capacities. These conclusions can be extrapolated to other similar subjects, especially compulsory and first year courses. The aim would be involving students in the realization of projects with real application in their degree, not only in the specific subject. It will imply to carefully select the application field and the challenge, so to solve it, students have to acquire fundamental skills of the subject, although it is not explicit in the instructions or training. In addition, the evaluation of the activity must be based on the result, but taking care that such result is a consequence of a proper skill application. Finally, an open competition or exhibition is proposed to show the solution, the reduction of plagiarism is clearly reduced.

Regarding future work, as soon as the pandemic situation allows to carry out group activities, our intention is to continue with this activity and improving it: new robotic challenges, more configuration options, and others.

**Acknowledgment.** The authors wish to acknowledge the support received from the CITIC research center, funded by Xunta de Galicia and European Regional Development Fund by grant ED431G 2019/01. We also would like to acknowledge to the Escuela Universitaria Politécnica of the University of A Coruña for the support for the acquisition of the robots and the experimental setups of the challenges.

## References

1. Balve, P., Albert, M.: Project-based learning in production engineering at the Heilbronn learning factory. Proc. CIRP **32**, 104–108 (2015). https://doi.org/10.1016/j.procir.2015.02.215
2. Blumenfeld, P.C., Soloway, E., Marx, R.W., Krajcik, J.S., Guzdial, M., Palincsar, A.: Motivating project-based learning: sustaining the doing, supporting the learning. Educ. Psychol. **26**(3–4), 369–398 (1991). https://doi.org/10.1080/00461520.1991.9653139
3. Chen, C.H., Yang, Y.C.: Revisiting the effects of project-based learning on students' academic achievement: a meta-analysis investigating moderators. Educ. Res. Rev. **26**, 71–81 (2019). https://doi.org/10.1016/j.edurev.2018.11.001
4. Guo, P., Saab, N., Post, L.S., Admiraal, W.: A review of project-based learning in higher education: student outcomes and measures. Int. J. Educ. Res. **102**, 101586 (2020). https://doi.org/10.1016/j.ijer.2020.101586
5. Holm, M.: Project-based instruction: a review of the literature on effectiveness in prekindergarten through 12th grade classrooms. Rivier Acad. J. **7**(2), 1–13 (2018)
6. Jensen, A.A., Stentoft, D., Ravn, O.: Interdisciplinarity and Problem-Based Learning in Higher Education Research and Perspectives from Aalborg University: Research and Perspectives from Aalborg University. Springer, Heidelberg (2019). https://doi.org/10.1007/978-3-030-18842-9
7. Lasauskiene, J., Rauduvaite, A.: Project-based learning at university: teaching experiences of lecturers. Proc. Soc. Behav. Sci. **197**, 788–792 (2015). https://doi.org/10.1016/j.sbspro.2015.07.182
8. Stanford University: MOSS: a system for detecting software similarity. http://theory.stanford.edu/~aiken/moss/
9. Yadav, A., Shaver, G.M., Meckl, P.: Lessons learned: implementing the case teaching method in a mechanical engineering course. J. Eng. Educ. **99**(1), 55–69 (2010). https://doi.org/10.1002/j.2168-9830.2010.tb01042.x

# Experience of a Lifetime: Erasmus+ Traineeship Mobility on a Wind Energy Project

Benoit Barbe[1](✉), Segundo Esteban[2], Matilde Santos[3], and Christophe Pasquier[4]

[1] Polytech Clermont, 63100 Aubière, France
benoit.barbe@etu.uca.fr
[2] Faculty of Physics, University Complutense of Madrid, 28040 Madrid, Spain
sesteban@ucm.es
[3] Institute of Knowledge Technology, University Complutense of Madrid, 28040 Madrid, Spain
msantos@ucm.es
[4] Electrical Engineering Department, Polytech Clermont, 63100 Aubière, France
christophe.pasquier@uca.fr

**Abstract.** International mobility enriches both, educational institutions and students that enjoy it. In this paper, the process of getting an Erasmus+ traineeship and the benefits derived from it are presented. This work shares the experience of a group of French students who have enjoyed an internship at a Spanish university in 2021. The importance of preparing the stay and the procedure to be carried out from both educational institutions, the one of origin and the receiving, is shown. For the successful of this mobility, it is also key to define a project that motivates the students and suits their background and skills. In this case, a project focused on renewable energies has been proposed. It consists on the design of a floating wind turbine prototype. Clean energy has proven to be very attractive to students and it is a way to discover new vocations for young researchers.

**Keywords:** Internationalization · Transnational · University · Education · Erasmus · Renewable energies · Wind energy project

## 1 Introduction

The internationalization of a university has been always seen as something beneficial for the institution [1]. But there are actions that promote this internationalization that have a direct impact on the students, and therefore benefit and enrich both, the education institution and the applicants [2]. This is the case of the mobility actions that are carried out among students from European universities under different agreements. This transnational aspect, so important in university education, has been stopped by the COVID'19 pandemic in many places [3, 4]. However, in some countries this possibility has remained open.

A growing literature addresses the experiences of international students in higher education [5]. This work shares the experience of a group of French students who have enjoyed a Erasmus+ traineeship at a Spanish university in 2021. Specifically, a group of


J. J. Gude Prego et al. (Eds.): CISIS 2021/ICEUTE 2021, AISC 1400, pp. 346–355, 2022.
https://doi.org/10.1007/978-3-030-87872-6_34

students from the Polytech Clermont have spent a four-month stay (from May to August) at the Complutense University of Madrid, in order to collaborate on a research project.

The framework of this mobility is the Erasmus+ traineeship program of the European Union. ERASMUS is the acronym of the name of the programme: European Community Action Scheme for the Mobility of University Students. The purpose of the ERASMUS Programme is to further student mobility within Europe [6]. The Erasmus+ traineeship program is offered for students currently enrolled in higher education institutions in Programme countries at Bachelor and Master level as well as for doctoral candidates. These opportunities are also open to recent graduates. This specific program aims at not only to improve communication, language and inter-cultural skills, but also soft skills highly valued by future employers [7].

Throughout this paper, the importance of preparing the stay and the procedure to be carried out from both educational institutions, the one of origin and the host, is shown. For the successful of the exchange, it is also key to define the project in which the students are going to work on, and this project should suit their background and skills. In this case, the project aimed to be a motivational one. It is focused on renewable energies, something the attracts the students and connects with their ideals of a cleaner world. It is aligned with one of the main topics that the 2021–2027 Erasmus programme is fostering, the green and digital transitions [7]. Specifically, the design of a floating wind turbine prototype is proposed.

The structure of the paper is as follows. Section 2 describes the procedure to apply for the Erasmus grant from the student point of view, and the advantages and difficulties they may face. Section 3 presents the policy of the destination university. In Sect. 4 the project and the tasks assigned to the student are explained. The paper ends with the conclusions and future works.

## 2 The University of Origin

International mobility is a commitment by the Network Charter's Polytech. It can take the form of a study visit or a work placement. The destinations concern all the countries of the world. The Polytech schools have a dense network of international relations which enables them to offer various possibilities.

The Polytech network actively participates in European programs dedicated to education and especially the Erasmus+ program. "Meeting the challenges of tomorrow", motto of the Polytech network, means for some students to get out of their knowledge field, out of their home country, their habits, their comfort, to eventually find a new vision of how technology and innovation can improve the way we live.

The policy of the Polytech school in Clermont (Fig. 1) drives students to do an internship abroad during the degree, for a minimum of three months. The conditions are the same as for an internship in France. The Polytech advises the students to enjoy this mobility before the last year of the degree, so they may gain international experience while studying. This way they can also check one's ability to adapt before joining a larger mobility program or getting a job abroad [8].

Engineering students are encouraged to personally approach potential host companies or universities. Just in case of difficulty, they will turn to the relevant departments at Polytech Clermont (International Relations Department) and/or to the University (UCA).

**Fig. 1.** Polytech Clermont (France).

### 2.1 A Personal Experience: A Long Process to Get There

As the first co-author of this paper, I will share my experience.

This internship was, first of all, an opportunity to confront myself with the difficulties of working on an international project in a foreign country. Of course, it was also a way to improve my English and the official language of the destination, in this case, Spanish. But what most matters, working with a foreign team collaborating in a real research project was really exciting. This definitely enriches the curriculum and, more important, our preparation for the professional life.

The student, the own higher education institution and the receiving organisation must sign a Learning Agreement for Traineeships to ensure a transparent and efficient preparation of the exchange abroad. This document sets out rights and responsibilities of the various parties, a detailed programme of the traineeship, information on insurance, and how your traineeship will be recognised upon successful completion.

The procedure to get all the paperwork done depends on the different ways each University and country has to solve the bureaucracy.

In our case, we, the students of Engineering, have to make the first move. That is, we must contact the host company or university and to define the topic of the internship. Then, it must be validated by the corresponding Department. Internships abroad have to be justified and the work done presented on our return to France, to be evaluated.

This financial support is available to all students. However, it is essential to pay attention to the selection criteria. The grant application files are received by the International Relations Office in Nice. In this case, I was able to open an Erasmus+ application which would allow me to obtain the funding if I proved a certain level of English knowledge and submitted the completed documentation on time (Fig. 2).

In a second step, I went to the Auvergne-Rhône-Alpes region in which I study, where I was able to obtain financial aid on the condition that I submit a complete file on my return from my internship on the different activities I have carried out. And finally, I went to the city of Chamalières, where I live, to find out if I was eligible for an international mobility grant. The essential criterion was that I had to justify my schooling for the year 2020/2021 and to hand in a completed file on my return to France, as for the financial aid from the Auvergne-Rhône-Alpes region.

The last phase of this long process is to contact a suitable destination university. You have to choose among all the universities that have an Erasmus agreement with the Polytech School. University Complutense of Madrid (Spain) was a possible option. Even more, you must contact someone at the Complutense University that is willing to receive students and who can offer a project for the internship.

POLYTECH CLERMONT-FERRAND

Academic year: 2020-2021
(Année universitaire)

Internship agreement between

1 - THE EDUCATIONAL OR TRAINING INSTITUTION (L'ETABLISSEMENT D'ENSEIGNEMENT OU DE FORMATION)

Name (Nom) POLYTECH CLERMONT-FERRAND.
Address (Adresse) Campus des Cézeaux - 2 avenue Blaise Pascal
TSA 60206 - CS 60026 63178 AUBIERE Cedex FRANCE
Tél: 33 4 73 40 77 01
Represented by (agreement-signing party) (Représenté par (signataire de la convention)) . Monsieur Christian LARROCHE
Capacity of the representative (Qualité du représentant) Director
email stages.polytech@uca.fr

2 - HOST ORGANIZATION (L'ORGANISME D'ACCUEIL)

Name (Nom) UNIVERSIDAD COMPLUTENSE DE MADRID
Address (Adresse) Avda. de Séneca
2 Ciudad Universitaria
28040 MADRID
Espagne
Represented by (agreement-signing party) (Représenté par (nom du signataire de la convention)):
Fatima MARTIN HERNANDEZ
email : vdalls@ucm.es
Location of internship (if different from that of the organization) (Lieu du stage (si différent de l'adresse de l'organisme)):

3 - THE INTERN (LE STAGIAIRE)

Last name (Nom) BARBE First name (Prénom) BENOIT

TITLE OF INTERNSHIP OR TRAINING COURSE TAKEN AT THE INSTITUTION OF HIGHER EDUCATION (INTITULE DE LA FORMATION OU DU CURSUS SUIVI DANS L'ETABLISSEMENT D'ENSEIGNEMENT SUPERIEUR ) Génie Electrique Quatrième année

SUBJECT OF INTERNSHIP (SUJET DE STAGE) : Theoretical and practical study on a prototype : analysis and vibration control of floating wind turbines (FloatWind)
Dates: From (Du) 19/04/2021 To (Au) 09/08/2021
Representing a total duration of ....4 months
(Représentant une durée totale de) (Nombre de Semaines / de Mois / de Jours (rayer la mention inutile) )
and corresponding to (Et correspondant à) ............ actual hours of attendance at the host organization (Heures de présence effective dans l'organisme d'accueil)
Distribution, in case of discontinuous attendance : ........... Number of hours per week or hours per day (cross out the inappropriate item)
(Répartition si présence discontinue) (nombre d'heures par semaine ou nombre d'heures par jour (rayer la mention inutile))
Comments (Commentaire):

SUPERVISION OF INTERN BY THE EDUCATIONAL INSTITUTION (ENCADREMENT DU STAGIAIRE PAR L'ETABLISSEMENT D'ENSEIGNEMENT)
First and Last name of academic advisor (Nom et prénom de l'enseignant référent) LENGAGNE Sébastien
email: sebastien.lengagne@uca.fr

SUPERVISION OF INTERN BY THE HOST ORGANIZATION (ENCADREMENT DU STAGIAIRE PAR L'ORGANISME D'ACCUEIL)
Full name of training supervisor (Nom et prénom du tuteur de stage):
Matilde SANTOS
email msantos@ucm.es

La Région Auvergne-Rhône-Alpes
Clermont Auvergne Institut National Polytechnique
Erasmus+
F AUBIERE04

**Certificat de présence**
***Attendance form***

Document à faire établir dès le début de la formation ou du stage, sans rature ni surcharge

Partie à remplir par l'étudiant

Nom Barbé
Prénom Benoît

Partie à faire compléter et signer par le responsable (ou son représentant) de l'établissement ou de l'entreprise d'accueil à l'étranger

Je soussigné(e) (Nom et prénom du signataire) Matilde Santos Peñas
*I, the undersigned*
Fonction : Prof. Dr. and Supervisor
*Role (Job title)*
Etablissement / Entreprise : University Complutense of Madrid
*Institution / Company*
Pays : Spain
*Country*
Certifie que (nom de l'étudiant) : Benoît Barbé est présent dans mon établissement.
*Certify that (name of student) is present in my establishment*

**Durée prévisionnelle du stage ou de la formation**
*Estimated time of this training period*
du (JJ/MM/AA) : 19 / 04 / 2021 au 09 / 08 / 2021
*From (starting date) - Day / Month / Year to (ending date) - Day / Month / Year*

Et - dans le cadre d'un stage - perçoit une indemnisation/rémunération d'un montant mensuel net de ....0€.... (Préciser la devise : € - $ - £ - autre).
*And - within the framework of an internship – perceives a compensation/payment for a monthly amount of: ............ (Specify the legal currency: € - $ - £ - another one)*

Fait à Madrid le 19/04/2021
*Place Date*

Cachet de l'établissement
*Stamp of the establishment*

FACULTAD DE INFORMÁTICA

S'il n'y a pas de tampon, établir ce certificat sur papier à en tête de l'établissement ou joindre à ce certificat signé une carte de visite du signataire contresignée.
*If there is no stamp to establish this certificate on paper at the head of the establishment or to join to this signed certificate a business card of the signatory countersigned*
Signature manuscrite :

Page 2 sur 5

**Fig. 2.** Internship agreement (left) and Erasmus agreements (right).

In my case, I was offered a project that I found very interesting. It was focused on renewable energies. It was the design of a prototype of a wind turbine and its control. I decided to take it as it sounded very attractive. I though my background would allow me to work on it, particularly if I address it form the electronic point of view. I hoped to fulfill the expectations, though the topic was quite difficult and complex. But this opportunity was an ideal starting point to promote what I believe in: to contribute to develop new and clean ways to generate electricity.

## 2.2 Pros and Cons

Once I have decided to go abroad with this Erasmus+ traineeship, I reflected on the pros and cons.

- As a student of a technical degree, working in Madrid will allow me to experience the innovation cycle from the inside, to join technical meetups, and stay up to date with the new developments in a current and innovative field.

- In addition, to learn new ways of working, different techniques, and to develop some skills and learn new concepts in a foreign country can pave the way for fundamental changes in one's home country, in the professional developments. It is a fact that for most companies, a work experience abroad may be a capital asset of qualification to get good qualified jobs.
- Besides, the Complutense University was a very prestigious one, and very well known. That added an extra value to my choice.
- Last, but not least, if one wants to further develop foreign language skills, immersion in another country is a way to get that. Not only the official language of the country, Spanish in this case, but also English, as in the working environment we usually use the latter.

Among the disadvantages or risks of studying abroad, we can mention the following. First, the expenses. Living in another country may not be cheap, and the funding received are small. The accommodation but also the often very expensive insurance in the host country, the price of air tickets, and in general, the standard of living of the foreign country, may make your financial situation difficult. In any case, it is usually possible to apply for several types of grants, depending on the destination, the program you join, the type of mobility, and the duration of the stay.

But in summary, I think that enjoying a mobility program between different universities is an opportunity to be seized. Moreover, the recently published figures on international students around the world show that this practice is growing because of the benefits you get.

## 3 From the Host University

Spain is full of assets that make it one of the most attractive countries in terms of education, living environment or culture. Choosing Spain is a guarantee of personal fulfillment in many respects, in one of the most welcoming countries in Europe.

In fact, this country has a world-renowned level of higher education, especially university. Some of the oldest universities in the world are located there. Not in vain it is one of the most requested destinations by international students [9].

In addition, the Spanish educational system is based on a network of public and private institutions that have made it rank 12th in terms of internal spending on research and development. The government helps and has policies to facilitate access to research. Studying in Spain is less expensive compared to other destinations and there are many scholarships available.

Spanish is the second mother tongue in the world by number of speakers, only behind Mandarin Chinese. It is the third language in the global score of speakers after English and Chinese. Today 7.6% of the world's population is Spanish-speaking. Studying in Spain is, therefore, a guarantee to improve your level of Spanish and access many job opportunities with this language.

### 3.1 The Destination University

At an educational level, one of the goal of this Erasmus mobility is to initiate links between the two universities involved, so as to be ready to establish, afterwards, real long term collaboration between both institutions. In this case, they were the Polytech INP Clermont (Fig. 1) and Complutense University of Madrid (Fig. 3), that was the destination University chosen. The Complutense University offers top quality teaching and facilities.

The applicant had the chance to debate with the lectures at France the convenience of joining this university for a 4-month internship. Then he had to search for contacts within UCM. After a couple of emails, the candidate understood that the UCM, particularly the professor he contacted, was in line with his expectations. Even more, she was open to welcome foreign students despite the health context we are currently experiencing due to the COVID pandemic.

Particularly, the agreement was set with the Faculty of Physics, although in this case the supervisor of the exchange was at the Computer Sciences Faculty (Fig. 4). That is why the student was assigned an office at this Faculty, on the second floor, shared with other undergraduate and postgraduate students. That was also part of the benefits of this traineeship.

**Fig. 3.** University Complutense of Madrid, campus. Spain.

**Fig. 4.** Physics (left) and Computer Sciences (right) faculties.

The host institution was also interested because this way it will count on students with a suitable background to contribute to the research that was going on at different

Departments and Faculties. Besides, the integration of foreign students contributes to the dynamism and internationalization of the Complutense University.

They checked my academic and professional background, as I had to send my CV. As the project I was going to be involved was about renewable energies, they considered my knowledge and experience sufficient. Indeed, I had already approached this field of energy, so to join it and try to contribute was something feasible. I would be able to understand the specific and technical terms and the main goals.

## 4 Joining a Research Project on Renewable Energies

As a student at Polytech Clermont Ferrand, I wanted to continue going deep into the field of energy, with which I have been somehow related through my studies and experience. To that end, I actively pursued an energy-related end-of-college internship.

During my last professional experiences, I had the perspective of developing skills in this area. In fact, I have worked in the rehabilitation of the Guy Thomas hospital in Riom (France). It had suffered a flood in one area, and I helped design a backup method of electricity distribution in the building, according to French standards. This work experience allowed me to achieve some experience in the energy grid. Indeed, I was able to understand how the facilities are connected to the electrical grid and to get hold of the technical vocabulary of this field. Therefore, the next step was to study other forms of energy, specifically renewable energies, which seemed to be very efficient and at the cutting edge.

This experience abroad could also be a great opportunity to collaborate in an international project in a new field for me, wind energy and, more specifically, floating wind turbines.

### 4.1 The Project of the Traineeship

In the area of renewable energies, wind energy is generating great expectations in terms of energy production, with high efficiency, but also causing a relatively important environmental impact. One solution proposed to alleviate the visual and noise impact of the on-land turbines is the installation of offshore wind turbines (OWT), and in particular, in the deep sea (floating offshore wind turbines, FOWT).

The floating turbines, which are quite recent and are now being developed all over the world, can be installed far from the coast, at great depths. But they need stable and flexible foundations. They present many interesting technological challenges [10].

Due to their situation on the sea surface, with strong winds, waves and currents, they are subject to intense loads and undesirable vibrations. Hence, the ability to reduce the impact of loads is extremely important as it allows for increased reliability and possibly the use of lighter and cheaper structures. But, on the other hand, the potential for wind energy is increased because the typical wind speed is above that of its land-based counterpart and the turbulence rate is lower.

In this research project, that was already going on, the control of floating wind turbines was addressed. All types of control that play a role in the performance of these renewable energy systems are studied, but the control of the blades to maintain the desired output power while reducing the vibrations is highlighted. The stability of the FOWT is a specific and important issue that the on-land ones do not present, and that even the bottom-fixed off-shore wind turbines (attached to the sea ground) do not have to such an extent.

The research includes the modelling of these complex, non-linear and unstable systems. These models are necessary in order to design the control strategies. And not only we are interested in the wind turbine model but in the environment, mainly waves and wind. It also deals with the control of the wind turbine, to optimize the energy that can be produced by applying intelligent techniques to the multivariable control of these devices [11].

Some of this goals have been already addressed, thought they are still open to research and to new finding. But to progress in the research, it was necessary to have simulation and real model where to test the control strategies proposed. Numerical tools have been applied to solve the complex dynamics of the floating devices but it was also quite necessary to have experimental installations where we can test the analytical and simulation developments [12, 13]. So, we decided to design a prototype of a floating wind turbine, to test different control algorithms and to measure the vibrations.

## 4.2 The Floating Wind Turbine Prototype

A small-scale turbine would allow us to simulate the behaviors of full-scale systems (prototype) in such a way that the real values can be obtained from the values measured on a model by applying identified scaling factors. Thus, the project requires the design and implementation of a small scale replica of a floating turbine.

In order to make this prototype, it was first necessary to find out about the various platforms that currently exist for the floating turbines, and decide which one we were going to implement. After checking the four most common stabilization strategies, we decided to use barge-type floating wind turbines. The overturning torque is compensated by the high buoyancy of the system. The structure is fixed to the sea ground with mooring lines (Fig. 5, the one on the left).

**Fig. 5.** Four types of floating wind turbine. Barge-type on the left.

Then we have to select the electronics component that were necessary to assemble the FOWT. First, the rotor with the blades. We have to research on internet and ask different providers because we need it to be angle to change the angle, in order to test the pitch control. After this searching, we bought the Dancing Wings Hobby kit, with two blades and a mechanical system that allows the system to set the angle. The rotor should be attached to a tube, to emulate the tower of the wind turbine.

An Arduino board was also bought to process the signals that are send to the servomotor that moves the blades. The Elegoo Arduino kit provides different sensors that were quite useful to measure different signals.

The Matlab software package is used to development a graphical interface to make easier for the user to control the inclination of the blades. The goal is to test different pitch control algorithms, so it was important to connect the PC with the control algorithms and the motor of the blades. The MPU-9250 connects the microcontroller via an I2C interface and has free libraries for the Arduino platform, allowing us to simply integrate it into any project.

A platform was built. It is just an horizontal surface where to fix the tower and the rotor. A system of four spring will be used to simulate the oscillations of the waves. In order to analyze the influence of the waves on the structure, an ultrasonic sensor was added. It allows us to measure the space between the platform and the seabed so, to measure the high of the waves.

This platform is equipped with an IMU sensor. This will allow us to measure the movement of the platform in real time from. Using Matlab software, we will analyze the platform's movements with the creation of three graphs: accelerometer, gyroscope, magnetometer. These will allow us to study in detail the influence of each parameter (pitch control, wind direction, wave movement).

## 5 Conclusions

This paper is focused on the experience of an Erasmus+ traineeship. It aims to show the benefits of enjoying this international experience, that in this particular type of programme, aims at getting the student collaborate in a research project.

Advantages of this mobility are shown, not only from the student point of view but also from the receiving university. The difficulties of the long administrative process that has to be carried out to get all the paperwork solved are also shared. As a conclusion, the involvement of both universities, host and origin, and the corresponding supervisors, is key to the success of these exchanges.

In addition, the importance is selecting a motivational project has been highlighted. In this case, besides being related to an attractive topic, such as wind energy, it is allowing the beneficiary to put into practice the knowledge he has acquired along his degree on different technological subjects (mechanical, electronics, computation, etc.). Working on a real problem opens a door to what research is and how to address it. Another conclusion that can be drawn is that it is important to find topics that motivate the students to get them involved, and to consider their background so they can address it successfully.

The result is quite positive. Although in this case the mobility is still going on, so far it has provided many benefits to both parts.

As a suggestion for future experiences of this type, we would advise the applicants to pay attention to the selection criteria, and to the people in charge to deal with the students' questions without delay.

**Acknowledgement.** This work was partially supported by the Spanish Ministry of Science, Innovation and Universities under MCI/AEI/FEDER Project number RTI2018-094902-B-C21.

## References

1. Ciesielkiewicz, M., Bonilla, C., Santos Peñas, M.: The Acquisition of Competences in Transnational Education Through the ePortfolio. In: Herrero, Á., Cambra, C., Urda, D., Sedano, J., Quintián, H., Corchado, E. (eds.) ICEUTE 2020. AISC, vol. 1266, pp. 75–83. Springer, Cham (2021). https://doi.org/10.1007/978-3-030-57799-5_8
2. Zagirnyak, M., Gladyr, A., Nozhenko, V.: Results and problems of international academic exchanges. In: 2020 IEEE Problems of Automated Electrodrive. Theory and Practice (PAEP), pp. 1–4. IEEE (2020)
3. Gabriels, W., Benke-Åberg, R.: Student exchanges in times of crisis. Erasmus Student Network Aisbl, Europe. https://tinyurl.com/5n773k2 (2020)
4. Koris, R., Mato-Díaz, F., Hernández-Nanclares, Núria.: From real to virtual mobility: Erasmus students' transition to online learning amid the COVID-19 crisis. Eur. Educ. Res. J. **20**(4), 463–478 (2021). https://doi.org/10.1177/14749041211021247
5. Huisman, J., Vlegels, J., Daenekindt, S., Seeber, M., Laufer, M.: How satisfied are international students? The role of town, gown and motivations. Compare: J. Comp. Int. Educ., 1–19 (2021)
6. Teichler, U.: Student mobility in the framework of ERASMUS: findings of an evaluation study. Eur. J. Educ. **31**(2), 153–179 (1996)
7. Erasmus+ Programme: https://ec.europa.eu/programmes/erasmus-plus/about_en
8. Soares, M.E., Mosquera, P.: Linking development of skills and perceptions of employability: the case of Erasmus students. Econ. Res.-Ekonomska Istraživanja **33**(1), 2769–2786 (2020)
9. Breznik, K., Skrbinjek, V.: Erasmus student mobility flows. Eur. J. Educ. **55**(1), 105–117 (2020)
10. Tomás-Rodríguez, M., Santos, M.: Modelling and control of floating offshore wind turbines. Rev. Iberoamericana Autom. Inform. Ind. **16**(4), 381–390 (2019)
11. Sierra-García, J.E., Santos, M.: Performance analysis of a wind turbine pitch neurocontroller with unsupervised learning. Complexity **2020**, 1–15 (2020). https://doi.org/10.1155/2020/4681767
12. Sánchez, C., Muñoz de la Peña, D., Gómez-Estern, F.: Automated generation of control design benchmark problems for computer-assessed education with Doctus. Rev. Iberoamericana Autom. Inform. Ind. **17**(1), 1–9 (2020). https://doi.org/10.4995/riai.2019.11243
13. Lerma, E., Costa-Castelló, R., Griñó, R., Sanchis, C.: Tools for teaching digital control in engineering degrees. Rev. Iberoamericana Autom. Inform. Ind. **18**(2), 189–199 (2021)

# Social Perceptions of Science and Technology Among Students of Early Childhood and Primary Education

Roberto García-Morís[1(✉)] and Javier Alfonso-Cendón[2]

[1] Department Specific Didactics, University of A Coruña, Campus de Elviña, 15008 A Coruña, Spain
roberto.Garcia.moris@udc.es
[2] Department of Mechanical, Computing and Aerospace Engineering, University of Leon, Campus de Vegazana s/n, 24071 Leon, Spain
javier.alfonso@unileon.es

**Abstract.** The objective of this article is to analyse social perceptions of Science and Technology among trainee pre-school and primary school teachers. The study will pay particular attention to how perceptions vary according to the training course studied and to gender. To reach these objectives we used a questionnaire based on Spain's Survey of Social Perceptions of Science and Technology Survey (SSPST), a study that has taken place every two years since 2002. We gathered responses from (N = 383) young people in professional training preparatory to careers teaching in either pre-school or primary school settings. The interest in focussing on such a narrow section of the population stems from the fact that, in order to address the gender gap still found in science and technology professions, it is necessary start with schools. To this end, we must first consider teacher training, hence the need to investigate trainee teacher's perceptions of science and technology. Preliminary results show that while perceptions do not seem to vary as a function of training course (pre-school or primary), there do appear to be differences depending on participants' gender.

**Keywords:** Gender gap · Science · Technology · Teacher training · Pre-school education · Primary education

## 1 Introduction and Rationale

Many studies both in Spain and internationally show a distinct gender gap in the field of Science and Technology. In general, findings show women representing around 30% of the research community, although in Spain this figure has hovered around 38% for the last decade [1]. This gender gap continues to exist, although, in Spain, it has been slowly narrowing such that in the most recent report on this topic, *Científicas en cifras 2021* [8], it was found that representation of women in research had risen to 40%, had reached equality (52%) in Public Administration, was at 43% in Higher Education, while in the business sector it was at 31% (ibid, p.90).


J. J. Gude Prego et al. (Eds.): CISIS 2021/ICEUTE 2021, AISC 1400, pp. 356–361, 2022.
https://doi.org/10.1007/978-3-030-87872-6_35

The Survey of Social Perceptions of Science and Technology [3] is a study that has been carried out regularly in Spain since 2002 by the department of Scientific Culture and Innovation at the Spanish Foundation for Science and Technology (*Fundación Española para la Ciencia y la Tecnología*, FECYT). This study, which has taken place every two years since 2002, makes an in-depth analysis of the relationship between science, technology, and society by examining perceptions among the general population. All the data and the questionnaires completed for every edition of the survey are publicly available. In this way, we have been able to use a modified version of this instrument to investigate a narrower section of the population: trainee teachers preparing for careers in either pre-school or primary school teaching, that is, future teachers of science and technology in these two first phases of education.

Childhood is an important stage with respect to developing interest in scientific careers, hence the importance of fostering new teachers' skills in science and technology [5]. This has particular significance in the present context where, year on year, young people show ever less enthusiasm for science and technology subjects [6, 7], despite the innate tendency of youngsters towards discovery and scientific experimentation.

The report entitled "Girls' and women's education in Science, technology, engineering and mathematics (STEM)", [4] highlights the findings of numerous studies identifying the disadvantages faced by girls in STEM due to factors linked to socialisation and educational practices. The educational system, schools, educational materials, among other things, all effect the level of interest shown by girls for these subjects. Another influence appears to be having female teachers in these subjects, which can lead to girls having greater confidence in their possibilities of success.

As a result of these considerations, our work focuses on a group of student teachers (N = 383) training in pre-school or primary education with the aim of uncovering their perceptions concerning science and technology, subjects that they will need to teach as non-specialist educators. Teacher training courses for these phases of education are dominated by women students and this is especially the case for pre-school where over 90% of student teachers are female. In this first report, we will focus particularly on analyses according to the type of training course (pre-school or primary education) and according to gender. To gather our data, we used a questionnaire based on specific questions extracted from the 2018 SSPST survey.

## 2 Objectives

This work seeks to address one general objective (GO) and two specific objectives (SO).

- GO: to analyse social perceptions of Science and Technology among trainee teachers in pre-school and primary education.
- SO-1: to investigate differences in perceptions depending on participants' course of studies (either pre-school or primary).
- SO-2: to investigate differences in perceptions depending on participants' gender.

Our initial hypothesis is that, while differences may exist between trainees on the two different courses, these will not be significant. In contrast, we expect that gender-disaggregated data will reveal differences, in line with international studies and reports that have analysed the gender gap, particularly in the field of science and technology.

## 3 Methodology

Data for this investigation was gathered using a semistructed questionnaire comprising 7 questions taken from the 2018 Survey of Social Perceptions of Science and Technology (SSPST). More specifically, we selected questions Q1, Q2, Q8A, Q8B, Q22, Q23 and Q25 from the SSPST and, for our purposes renumbered them 1 to 7. We also collected participants' sociological data, such as age, gender, and previous and current studies (Fig. 1).

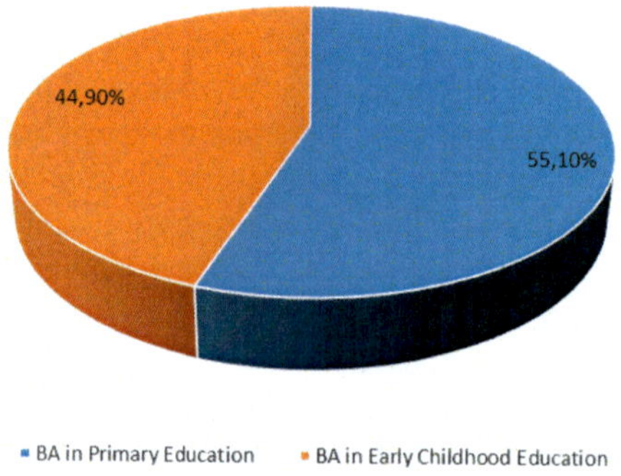


**Fig. 1.** Distribution of participants by degree

We obtained a total of 383 completed questionnaires, 16% from men and 84% from women. Participants all attended one of two universities: the University of A Coruña in the northwest of Spain or the University of Cordoba in the south. The median age of participants was 21 years. In this work we will be presenting the results obtained from questions 3, 4, 5, and 6 (using our numbering system). These questions concerned the value given to different professions (Q3); opinions on what is scientific or not (Q4); and participants' assessment of their scientific understanding and their level of scientific education (Q5 and Q6).

## 4 Results

Question 3 asked participants to rank the following professions: doctor, scientist, engineer, judge, journalist, businessperson, teacher, member of the clergy, and politician. The most highly ranked profession, on a scale of 0–5, was that of doctor which received a median score of 4.92. This was a higher score than teaching, although, as might be expected here, this profession also received a high score: 4.89, which put it in second place. The professions of scientist and engineer were also highly regarded, being placed in third and fourth position with median scores of 4.85 and 4.52, respectively. This agrees with the SSPST 2018, which showed the same scoring pattern for professions in their survey population. We did not find any significant differences when the data were analysed in terms of gender, indeed, we found that median scores were almost identical among men and women in our study. This is also in line with SSPST findings, although this survey did find differences with respect to gender when data was also grouped according to age. However, these differences were only apparent in age cohorts different to that studied here which concerned a university population [9].

Question 4 asked participants to rate certain professions according to how scientific they felt them to be. Rating was done on a scale from 1 (not at all scientific) to 5 (very scientific). The profession that was considered most scientific was that of physicist (4.81), followed by doctor (4.73) and psychologist (4.21). Other professions such as physiotherapist, sociologist, economist, and homeopathist were rated, in list order, between 4 and 3. There were no large differences between responses **according to degree**, but gender did appear to have an influence. Men rated the professions of physicist and doctor as more scientific (4.92 and 4.9 respectively) than did women (4.79 and 4.67). However, although some scores are very close, women rated several professions as more scientific than men: psychologist (4.23 women, 4.11 men); sociologist (3.86 women, 3.823 men); economist (3.37 women, 3.23 men); and even homeopathist (3.36 women, 3.25 men). In contrast, the profession of physiotherapist was given a median score of 4.02 by men and 3.87 by women. These patterns in the analysis according to degree revealed that pre-school trainees scored professions similarly to women respondents while primary trainees response pattern more closely resembled that seen in men's responses. This is probably simply a reflection of the fact that pre-school trainees are more likely to be female.

Question 5 asked about participants' perceived level of scientific understanding and answers show variation according to gender. In response to the statement: "Science subjects have always given me trouble", 26.09% of women participants moderately agreed compared to only 16.39% of men. This trend is also apparent in the data obtained for the statement: "Science is so specialised I struggle to understand it". Here, 33.23% of women were in moderate agreement compared to 24.59% of men. Responses to the statement: "Knowing about science and technology is important to my daily life", were similar among men and women in all five possible answers to this question except in the case of "disagree" where 8.2% of men gave this response compared to only 1.55% of women.

With respect to question 6, concerning the level of science and technology education participants' felt they had received. Here, there were few differences between participants in terms of their degree with 63.51% of primary school trainees and 61.06% of pre-school trainees reporting that they had received a standard level of education in this area. Of primary trainees, 20.85% felt they had a low level, and this percentage was higher, 24.42%, among pre-school trainees. Thus, in general terms, pre-school trainees rated themselves as having a lower level of science and technology education than primary school trainees. Looking at the data with reference to gender, women felt that they had received low levels of education in this field to a greater degree than did men: 22.98%

compared to 19.68%. This goes some way to explaining the differences seen in terms of course studied since pre-school trainees were predominantly female. Thus, we can say that there is a gender-dependent difference in how participants viewed their science education.

## 5 Conclusions

The questions analysed in this article concerned participants' assessment of certain scientific professions and participants' own scientific understanding and education. The results from this initial study show certain variation in the social perception of science and technology among a corpus of trainee teachers and differences are most marked with respect to participants' gender [9, 10].

Although there are some differences between perceptions depending on the course studied by our sample of trainees, these appear to be due to the fact that our pre-school corpus was predominantly female compared to the more gender-equal corpus in primary education. Thus, the main differences lie in the gender-disaggregated data and while these were only small with respect to participants' assessments of scientific professions, major differences were seen in terms of participants' perceptions of their own scientific education. In particular, women were found to be more likely to rate their level of scientific education as "low" than our male participants [1, 9, 10], in contrast to the results of other studies [11].

Perceptions of science and technology among teachers of infant and primary education play an essential role in awakening children's interest in science and providing them with positive scientific role models from an early age [12]. This study's use of the general population Survey of Social Perceptions of Science and Technology (SSPST) to analyse and compare attitudes specifically among trainee pre-school and primary teachers thus represents a key starting point for future research into the improvement, advancement and innovation of teacher training in this area [7].

**Acknowledgements.** This work would have not been possible without the full cooperation of the students and faculty from the University of A Coruña and University of Córdoba who kindly participated in this experience.

## References

1. Puertas, F.: El papel de las mujeres en la Ciencia y la tecnología. Santillana (2019)
2. Ministry of Science and Innovation: Científicas en Cifras 2021. Observatorio Mujeres, Ciencia e innovación (2021)
3. Ministry of Science and Innovation: Percepción Social de la Ciencia y la Tecnología en España. Informes y Publicaciones. https://icono.fecyt.es/informes-y-publicaciones/percepcion-social-de-la-ciencia-y-la-tecnologia-en-espana, 10 June 2021
4. FECYT: Fundación Española para la Ciencia y la Tecnología. https://www.fecyt.es/. 10 June 2021
5. DigitalES: El desafío de las vocaciones STEM. Por qué los jóvenes españoles descartan los estudios de ciencia y tecnología. DigitalES (2019)

6. Solbes, J., Montserrat, R., Furió, C.: El desinterés del alumnado hacia el aprendizaje de la ciencia: implicaciones en su enseñanza. Didáctica de las ciencias experimentales y sociales **21**, 91 (2007)
7. FECYT: ¿Cómo podemos estimular una mente científica? Estudio sobre vocaciones científicas. Obra Social "la Caixa", FECYT y everis (2015)
8. UNESCO: *Descifrar el código: la educación de las niñas y las mujeres en ciencias, tecnología, ingeniería y matemáticas (STEM)*. UNESCO (2019)
9. Ortega, E.: De la brecha de género al análisis interseccional de disciplinas STEM. Percepción social de la Ciencia y la Tecnología 2018. FECYT, pp. 189–201 (2019)
10. Vázquez-Cupeiro, S.: Ciencia, estereotipos y género: una revisión de los marcos explicativos. Convergencia **22**(68), 177–202 (2015)
11. Gómez-Escobar. A., Fernández-Cézar, R.: ¿Afecta una intervención didáctica a la percepción del género en la ciencia de futuros docentes? In: Innovación Docente e Investigación en Educación. Avanzando el proceso de enseñanza-aprendizaje. Dykinson, pp. 123–133 (2020)
12. Solbes, J.: Les empremtes de la ciència. Alzira, Ed Germania (2002)

# SPECIAL SESSION: Sustainable Personal Goals: Engaging Students in their Learning Process

# Wind Turbines Control Optimization: A Problem-Driven Proposal to Learn Genetic Algorithms

Jesús Enrique Sierra-García[1](✉) and Matilde Santos[2]

[1] Electromechanical Engineering Department, University of Burgos, Burgos, Spain
jesierra@ubu.es
[2] Institute of Knowledge Technology, Complutense University of Madrid, Madrid, Spain
msantos@ucm.es

**Abstract.** Engineering students increase its motivation when they face exciting challenges. In this sense, renewable energies may be a motivational and attractive topic as it is seen as a contribution to a cleaner world. Specifically, wind energy plays nowadays a key role in the sustainability of the worldwide electric grid. But wind turbines (WT) are complex devices that need control systems to maintain the output power around the rated value. Control engineering students address this problem with a theoretical background on regulation, but they usually do not know how to improve the standard solutions. Artificial intelligence techniques, and specifically genetic algorithms (GA), have not received the due attention in electronics and industrial engineering yet, despite its utility as optimization tool. This evolutive optimization tool can be exploited to tune the parameters of the controllers and thus to optimize the operation of WTs. In this work, a problem-driven didactic proposal that can help student to learn optimized control techniques on WTs is presented. The students must work with a mathematical model of a WT. Once they have understood its performance, they design a PID controller to stabilize the output power. This regulator is manually tuned so they are aware of the complexity of this parameters adjustment for complex systems. Genetic algorithms are then presented as a powerful optimization tool to help them in this task. The proposal has been designed with the aim of achieving a set of didactic competences and learning objectives. To achieve it, a set of concepts are defined and scheduled in several sessions. In addition, a practice script has been develop to guide the students through the learning process.

**Keywords:** Control · Tuning · Wind turbines · Optimization · Genetic algorithms · Education

## 1 Introduction

Engineering students are motivated by problems that connect with their lofty ideals [1, 2]. It is well known that young people get excited with examples of real world applications that they found useful [3, 4]. This is even more true if the application contributes to


J. J. Gude Prego et al. (Eds.): CISIS 2021/ICEUTE 2021, AISC 1400, pp. 365–373, 2022.
https://doi.org/10.1007/978-3-030-87872-6_36

get a cleaner environment and to decelerate the climate change. That is why renewable energies can be considered an area of experimentation that encourages students to apply their knowledge in a practical and useful way [5].

Particularly, wind energy is a key pillar to achieve the Paris Climate Goals and the Sustainable Development Goals (O. D. S.) of the 2030 Agenda [6]. Indeed, it is a vital resource to guaranty the sustainability of the electrical grid. The seventh goal of the 2030 Agenda claims that it is necessary to ensure access to affordable, reliable, sustainable and modern energy for all. Besides, wind energy is a mature technology; hence, real examples of the performance of wind turbines (WT) are available to be presented and explained to students. Still, there are many engineering challenges related to wind turbines that must be addressed and are worthy to bring up [7–9].

To further develop the control theory discipline it is key to incorporate new researchers who are willing to innovate and apply recent methodologies and techniques, particularly in this field of wind energy. Intelligent techniques have been already succesfully applied to WT control [10–12]. One of these techniques is genetic algorithms (GA), a Soft Computing technique based on the evolution of species [13]. The term soft computing was proposed by Lotfi A. Zadeh, the father of fuzzy logic. Soft computing is a collection of methodologies that aim to exploit tolerance, imprecision and uncertainty of information to achieve tractability, robustness, and low cost solutions for complex problems [14]. Among all the concepts encompassed by Soft Computing, it is worth mentioning: neural networks, fuzzy logic, and genetic algorithms. The latter, GA, is often studied in computer science degree. However, despite being a useful tool to optimize all type of engineering processes, it is not well known neither explained in electrical, electronics, telecommunications, or industrial engineering degrees. As an optimization tool, GA can be used to fine tune controller parameters and thus to optimize the performance of WTs [15]. A good way to incorporate these computational intelligence tools in the engineering studies is to propose a real control problem and use these techniques to solve it. Today, the WT control problem is a motivating challenge to learn new control concepts and promote STEM vocations between the students.

In this work, a problem-driven didactic proposal that can help student to learn optimized control techniques on WTs is presented. The students must work with a mathematical model of a WT. Once they have understood its performance, they design a PID controller to stabilize the output power. This regulator is manually tuned so they are aware of the complexity of this parameters adjustment for complex systems. Genetic algorithms are then presented as a powerful optimization tool to help them in this task.

In order to guide the students through the learning process, a script with several questions and practical exercises has been prepared. These questions help them identify and focus on the key concepts. The proposal has been designed with the aim of achieving a set of didactic competences and learning objectives. To achieve it, a set of concepts are defined and scheduled in several sessions.

The rest of the paper is structured as follows. The didactical approach is described in Sect. 2. Section 3 details the script that students have to carry out to learn the main control concepts, with practical exercises. The paper ends with the conclusions and future work lines.

## 2 Didactic Proposal

Students must study the mathematical equations that describe the behavior of a wind turbine. The model represents the dynamics of a small on-land wind turbine, so they can understand how it works based on their internal variables [16]. The next step will be to design a controller to get the wind turbine producing the maximum energy. These goals require a good knowledge of STEM subjects such as mathematics, physics, mechanics, engineering, computer sciences, etc.

The project is divided into three main tasks: to understand how wind turbines work by simulating the wind turbine mathematical model; to design and manually tune a PID control for that WT; and finally, to apply genetic algorithms to optimize the performance of the PID controller. This didactic practical proposal is developed based on a set of contents, competences, objectives, and an evaluation procedure, described below.

### 2.1 Competences

The didactic proposal has been designed to achieve a set of specific competences and a set of cross competences.

- Specific competences:
  - Ability to model and represent physical systems by their differential equations and to simulate the evolution of the state variables and outputs with some software package or language programming such as Matlab/Simulink [17] or Phyton.
  - Ability to use different input signals such as steps, ramps and sinusoidal signals in an effective way in order to obtain some specifications of the time and frequency response: rise time, settling time, cut-off frequency … of the wind turbines.
  - Ability to determine the control goals for a wind turbine and set these objectives in the controllers.
  - Ability to design simple control systems for wind turbines based on the PID control law, and to tune the parameters of the controllers to obtain the desired response.
  - Ability to configure and execute genetic algorithms to optimize the parameters of the control systems for the wind turbine using Matlab/Simulink.

- Cross competences [18]:
  - Logical thinking and ability to think in a systematic way, based on facts and evidences. The analysis of how the controllers modify the response of the real systems help develop this logical thinking.
  - Teamwork or the ability of a team to work collaboratively to achieve a common objective in the most efficient way. In this case, the common goal is the optimization of the control of the wind turbine.
  - Troubleshooting. Search of causes of problems, critic analysis and resolution of engineering problems. The practice with simulation computational tools develops these competences.

– Search and management of information. To acquire abilities related with the use of computer programs to search, classify, sort and manage scientific information, especially in the fields of wind energy and control engineering.

### 2.2 Objectives

- To handle the mathematical equations that govern the operation of wind turbines and identify the relevant variables.
- To identify the problems that the controllers solve.
- To compare open loop and close loop systems. To identify the differences between manipulated variables and controlled variables.
- To simulate dynamic systems given by their differential equations.
- To classify and recognize SISO, MIMO, first order, second order and higher order systems. To identify the differences between these systems.
- To obtain the time and frequency response of a system.
- To design and tune a PID controller.
- To identify the phases and elements of the genetic algorithms.
- To configure and run genetic algorithms to optimize the performance of the controllers.

### 2.3 Contents and Planning

The proposal is divided into three main modules. First, an on-land wind turbine is mathematically modelled, and it is simulated with a computational tool. After this, a PID regulator is designed to control the wind output power. Finally, this controller is optimized by genetic algorithms. Each one of these modules is divided into several sessions so the students can understand and handle the main concepts related to the wind turbine control in an organized and systematic way. Each main concept is assigned some exercises to be solved, in order to test if students have understood it and they are able to put the acquired knowledge into practice. These exercises help focus the learning process and the information searching.

- Playing with wind turbines (Exercises 1–3)

  – Session 1: Operation of a wind turbine. Space of states. SISO and MIMO systems. Sensors and actuators. Mathematical model of a wind turbine. Main variables. Equations.
  – Session 2: Simulation of the wind turbine model. Different conditions. Different wind speed and profiles.

- Controlling the wind turbines (Exercises 4–5)

  – Session 3: Control systems: manipulated and controlled variables. Close loop control. Time response and specifications.
  – Session 4: The PID controller. Tuning the three gains of a PID. Effects on the response.

   - Session 5: Application of a PID regulator to a wind turbine.

- Optimizing the controllers (Exercises 6–13)

   - Session 6: Optimization. Cost function. Restrictions. Variables. Solvers. Genetic algorithms.
   - Session 7: Optimization of PID controllers by genetic algorithms. Application on wind turbines.
   - Session 8: Multi-objective optimization. Pareto front. Application on wind turbines.

### 2.4 Evaluation

In order to guide and orientate the students, a script with the practical exercises has been prepared to help the student better understand the problem (Sect. 3). This script presents several exercises on the main control concepts. Besides, the results of the exercises can be used to evaluate if the student has reached the required level of knowledge and also as a feedback about what aspects need to be reinforced.

At the end of the training the students must summarize all the information in a document with the format of a scientific paper. This helps them to familiarize with this type of documents and also to motivate them as it makes them feel closer to the research community. In addition, to write their finding is a good way to summarize the knowledge they have acquired and to show the results. They must provide the files with the Simulink simulation models and the figures.

The evaluation is divided into two parts: the scientific paper and the practical exercises. The preparation of the scientific paper is valued the 30%. This includes the writing, the structure, the correct use of grammar, and the correct use of the references. The other 70% is obtained by the correct implementation of the practical exercises. In this part all the exercises are equally weighted.

## 3 Script of the Practical Exercises

The use of wind energy plays a key role in the sustainability of the electrical grid worldwide. Its deployment grows day by day. However, in order to maximize the efficiency of these energy sources it is necessary to control the power energy they produce, harvesting the wind. The following terms and the corresponding mathematical model will introduce you in the world of wind turbines from the simulation and control perspective.

Current in the armature:

$$\dot{I}_a = \frac{1}{L_a}\left(K_g \cdot K_\phi \cdot w - (R_a + R_L)I_a\right) \tag{1}$$

Where $L_a$ is the armature inductance (H), $K_g$ is a dimensionless constant of the generator, $K_\phi$ is the magnetic flow coupling constant (V·s/rad), $R_a$ is the armature resistance, $R_L$ is the resistance of the load (Ω). As it is well known, in an electric generator, the current depends on the variation of the magnetic flow and this variation depends on the angular velocity. This equation is key to transform the rotor mechanical energy into electrical energy.

Tip-speed ratio:

$$\lambda = (w \cdot R)/v_{ef} \tag{2}$$

Where $w$ is the angular rotor speed (rad/s), $R$ is the radius or blade length (m), $v_{ef}$ is the effective wind velocity in the blades (m/s) and $\lambda$ is the tip-speed ratio, which is dimensionless. The tip-speed ratio is an auxiliary signal that helps us interpret how different the rotor speed is from the wind speed.

Power coefficient:

$$C_p(\lambda_i, \theta) = c_1\left[\frac{c_2}{\lambda_i} - c_3\theta - c_4\theta^{c_5} - c_6\right]e^{-\frac{c_7}{\lambda_i}} \tag{3}$$

$$\lambda_i = \left[\left(\frac{1}{\lambda + c_8}\right) - \left(\frac{c_9}{\theta^3 + 1}\right)\right]^{-1} \tag{4}$$

The values of the coefficients $c_1$ to $c_9$ that define the power coefficient $C_p$ depend on the characteristics of the wind turbine. These equations show that the angle of the blades, $\theta$, can be used to modulate the amount of wind energy that is effectively transformed to mechanical energy. When the angle is in feather position, that is, 90°, this percentage is minimum and the blade acts as a brake.

Angular rotor speed:

$$\dot{w} = \frac{1}{2 \cdot J \cdot w}\left(C_p(\lambda_i, \theta) \cdot \rho\pi R^2 \cdot v_{ef}^3\right) - \frac{1}{J}\left(K_g \cdot K_\phi \cdot I_a + K_f w\right) \tag{5}$$

Where $J$ is the rotational inertia (kg.m$^2$), $R$ is the radius or blade length (m), $\rho$ is the air density (kg/m$^3$), $K_f$ is the friction coefficient (N.m/rad/s), and $v_{ef}$ is the effective wind velocity in the blades (m/s). This equation shows how the wind velocity modifies the angular rotor speed, and thus the wind energy that is obtained.

Pitch angle:

$$\ddot{\theta} = \frac{1}{T_\theta}\left[K_\theta\left(\theta_{ref} - \theta\right) - \dot{\theta}\right] \tag{6}$$

Where $\theta_{ref}$ is the pitch reference, $K_\theta$ and $T_\theta$ are dimensionless parameters of the pitch actuator. The input of the wind turbine is the pitch reference; therefore, this is the output of the controller, or manipulated variable. This equation shows how the pitch reference updates the pitch angle. The power constraints of the pitch actuator make that the new pitch reference is not instantaneously transformed into a new pitch angle in the blade.

Output power:

$$P_{out} = R_L \cdot I_a^2 \tag{7}$$

The controller must try to stabilize the output power of the wind turbine around its nominal value, thus this can be considered the controlled variable. Real values of the parameter for a 7 kW wind turbine can be found in [19].

**Exercise 1.** *Try to understand what these equations represent. Identify the manipulated variables, the controlled variables and the disturbances of the system. Identify the sensors and the actuators of the system. Identify the state variables. Is the system SISO or MIMO? Why? Search for the parameters of a real turbine.*

**Exercise 2.** *Implement the model of the wind turbine in Matlab/Simulink. Try different initial conditions for the state variable with wind speed at 0 m/s. What happens?*

**Exercise 3.** *Set the wind speed to 12.5 m/s. Set the pitch reference to 0, set the reference to pi/8, set the reference to pi/4. Is the system linear? Why? Is it possible to reach the same conclusion looking at the mathematical expressions? How?*

**Exercise 4.** *Try a control architecture in Simulink with a P regulator. Set the wind speed at 12.25 m/s and the power reference to 7kW. Study the time response of the system. Is the value of the output power 7kW? What is the stationary error? How can you reduce it? What is the overshoot? How can you reduce it?*

**Exercise 5.** *Now, improve the controller tuning the integral and the derivate gains. Apply the PID to the wind turbine and show the results. Test the controller with different wind values. Does the performance of the controller depend on the wind range? Why? Comment the performance of the wind turbine with this controller. Save the constants that provide the best results.*

As it has been experienced, the process of tuning a PID for a complex and non-linear system such as WT is not a trivial task. Genetic algorithms can help us in this process. They are very useful to optimize all type of processes. To do it, it is necessary to define a cost function, that is, the function that we want to minimize/maximize. The individual that will evolve to the best solution is formed with the variables that will be modified every iteration, to eventually optimize the cost function. The genetic algorithm will search the optimum individual by genetic operators [13, 20].

**Exercise 6.** *Can the wind turbines control be considered an optimization problem? How? What would the cost function be? What would the constraints be? If this problem were addressed by genetic algorithms, how would the individual be?*

Matlab provides an optimization tool to carry out this process. It is included in the Optimization Toolbox. It can be executed by commands or by an user interface. This user interface is shown by the command optimtool. Different solvers are available, among others, genetic algorithms.

**Exercise 7.** *Execute the command optimtool. Investigate in the documentation how the function cost, constraints and variables must be specified.*

**Exercise 8.** *Find the minimum of the function* $f(x_1, x_2) = 100\left(x_2 - x_1^2\right)^2 + (1 - x_1)^2$ *in the set* $x_1^2 + x_2^2 \leq 4$ *by using the command optimtool. This function is the Rosenbrock's function and has a minimum at (1,1). Represents this function in a plot. Try different solvers to optimize the function.*

**Exercise 9.** *Implement a function in Matlab to model the cost function defined in the exercise 6. This function must call the Simulink block defined in the Exercise 2.*

**Exercise 10.** *Optimize the wind turbine controller by the optimtool command and the cost function that has been implemented in the exercise 9. Try different wind profiles.*

**Exercise 11.** *Compare the performance of the optimized controller and the controller of the Exercise 5. What controller works better? Why?*

In order to prevent the degradation of the systems and their actuators, it is a common practice in control engineering to consider the control effort as a metric to be optimized. There are different ways to measure the control effort but it is generally estimated by,

$$C_{eff} = \int u(t)^2 dt \tag{8}$$

Where $u(t)$ represents the output of the controller.

**Exercise 12.** *How can we calculate the control effort of the pitch control of a wind turbine? Modify the simulation block to generate an output that indicates the control effort.*

**Exercise 13.** *Can the wind turbine control problem be considered as a multi-objective problem? How? Modify the cost function of the Exercise 6 to simultaneously consider the power error and the control effort.*

## 4 Conclusions and Future Works

As the 2030 Agenda claims, it is necessary to enhance international cooperation to facilitate access to clean energy research and technology, including renewable energy, energy efficiency and advanced and cleaner fossil-fuel technology, and promote investment in energy infrastructure and clean energy technology [6]. Therefore, it is essential to incorporate new researchers who innovate and apply recent methodologies and techniques in this field.

A way to encourage a greater number of engineering students to work on this field is to propose them topics that connect with their interests and values. The introduction of the WT control problem is a motivating, challenging and relevant problem. It is undoubtedly a good starting point for boosting these research vocations.

In this work, a motivational proposal based on clean energy for education in control in engineering degrees is presented. First the wind turbines are introduced as the system to be studied. A mathematical model of a real small wind turbine is presented. The mechanical and electrical differential equations are given and explained. These equations are simulated using a computational tool. This way the students can experiment and understand the operation of wind turbines. Then they can address the second task, to design a controller to stabilize the output power at its rated value. Finally, students are introduced into the optimization field by understanding how genetic algorithms work and applying them to optimize the designed PID controller.

The didactic proposal is developed in order to achieve a set of competences and objectives. The main control concepts are identified and scheduled. In addition, a script with a set of practical exercises has been developed to guide the student through the learning process.

As future works, we plan to evaluate this proposal with students and to see if they are really motivated to learn with this real application. So far we have already confirmed that clean energy is a motivational topic that students chose for their final degree project. Other future works include to design and apply yaw control and to extend the model to offshore wind turbines.

**Acknowledgement.** This work was partially supported by the Spanish Ministry of Science, Innovation and Universities under MCI/AEI/FEDER Project number RTI2018-094902-B-C21.

## References

1. Martin, A.J.: Examining a multidimensional model of student motivation and engagement using a construct validation approach. Br. J. Educ. Psychol. **77**(2), 413–440 (2007)
2. Starr, C.R., Hunter, L., Dunkin, R., Honig, S., Palomino, R., Leaper, C.: Engaging in science practices in classrooms predicts increases in undergraduates' STEM motivation, identity, and achievement: a short-term longitudinal study. J. Res. Sci. Teach. **57**(7), 1093–1118 (2020)
3. García-Álvarez, F.M., Santos, M.: Technology-enhanced learning by experiments with real systems. In: 14th International Technology, Education and Development Conference, pp. 272–280 (2020)
4. Ouariachi, T., Wim, E.J.: Escape rooms as tools for climate change education: an exploration of initiatives. Environ. Educ. Res. **26**(8), 1193–1206 (2020)
5. Ulazia, A., Ibarra-Berastegi, G.: Problem-based learning in university studies on renewable energies: case of a laboratory windpump. Sustainability **12**(6), 2495 (2020)
6. United Nations. https://sdgs.un.org/2030agenda (2021). Accessed 07 Jan 2021
7. Tomás-Rodríguez, M., Santos, M.: Modelling and control of floating offshore wind turbines. Revista Iberoamericana de Automática e Informática Industrial **16**(4), 381–390 (2019)
8. Galán-Lavado, A., Santos, M.: Analysis of the effects of the location of passive control devices on the platform of a floating wind turbine. Energies **14**(10), 2850 (2021)
9. Sierra-García, J.E., Santos, M.: Lookup table and neural network hybrid strategy for wind turbine pitch control. Sustainability **13**(6), 3235 (2021)
10. Sierra-García, J.E., Santos, M.: Improving wind turbine pitch control by effective wind neuro-estimators. IEEE Access **9**, 10413–10425 (2021)
11. Sierra-García, J.E., Santos, M.: Exploring reward strategies for wind turbine pitch control by reinforcement learning. Appl. Sci. **10**(21), 7462 (2020)
12. Sierra-García, J.E., Santos, M.: Performance analysis of a wind turbine pitch neurocontroller with unsupervised learning. Complexity (2020)
13. Zotes, F.A., Penas, M.S.: Multi-criteria genetic optimisation of the manoeuvres of a two-stage launcher. Inf. Sci. **180**(6), 896–910 (2010)
14. Zadeh, L.A.: Soft computing and fuzzy logic. IEEE Softw. **11**(6), 48–56 (1994)
15. Lara, M., Garrido, J., Ruz, M.L., Vázquez, F.: Adaptive pitch controller of a large-scale wind turbine using multi-objective optimization. Appl. Sci. **11**(6), 2844 (2021)
16. Mikati, M., Santos, M., Armenta, C.: Electric grid dependence on the configuration of a small-scale wind and solar power hybrid system. Renew. Energy **57**, 587–593 (2013)
17. Matlab/Simulink: Mathworks. https://uk.mathworks.com/products/ (2021). Accessed 04 Jan 2021
18. Ciesielkiewicz, M., Bonilla, C.F., Santos Peñas, M.: The Acquisition of Competences in Transnational Education Through the ePortfolio. In: Herrero, Á., Cambra, C., Urda, D., Sedano, J., Quintián, H., Corchado, E. (eds.) ICEUTE 2020. AISC, vol. 1266, pp. 75–83. Springer, Cham (2021). https://doi.org/10.1007/978-3-030-57799-5_8
19. Mikati, M., Santos, M., Armenta, C.: Modelado y simulación de un sistema conjunto de energía solar y eólica para analizar su 458 dependencia de la red eléctrica. Revista Iberoamericana de Automática e Informática Industrial **9**(3), 267–281 (2012)
20. Santos, M., Cantos, A.J.: Classification of plasma signals by genetic algorithms. Fusion Sci. Technol. **58**(2), 706–713 (2010)

# Development of Competencies Applying the Project Method. Application in Environmental Engineering

Hernando Avella Forero[1], Néstor Rafael Perico-Granados[2], Pedro Mauricio Acosta Castellanos[1,3](✉), Araceli Queiruga-Dios[3], and Heidy Madeleine Arévalo Algarra[4]

[1] Universidad Santo Tomás, Tunja, Colombia
{hernando.avella,pedro.acosta}@usantoto.edu.co
[2] Corporación Universitaria Minuto de Dios, Bogotá, Colombia
nestor.perico@uniminuto.edu.co
[3] Universidad de Salamanca, Salamanca, Spain
queirugadios@usal.es
[4] Tunja, Colombia

**Abstract.** For the development of society, the training of upright professionals is required, especially in environmental matters. Therefore, it is necessary to apply training processes that involve the construction of competencies, to face the solution of the problems that currently arise. For this, academic spaces are required with methodologies that apply and execute human and disciplinary competencies; one of these methodologies is the POPBL (Project-Oriented Problem-Based Learning). This research will establish environmental education as a strategy to solve the generation of solid waste, applying the POPBL method and the development of human and professional skills and competencies. The projects were developed and applied by students from the Faculty of Environmental Engineering of the Santo Tomas University, in the first, sixth, and seventh semesters in the 2019-II period, guided by the teacher and where the main topic was environmental education and solid waste management. The researcher applied qualitative and quantitative data collection instruments to determine the state of the competencies and strengthened in the academic period through the POPBL. It was evident that the students at the beginning of this research were not clear about how these competencies were developed and their answers were according to their expectations of application and generation. As the application of research projects and teacher interventions explained the development of competencies, the students were able to relate their training process with the development of competencies and solve current environmental problems.

**Keywords:** Environmental education strategies · Project-Oriented Problem-Based Learning · Waste disposal · Solid waste · Waste source separation


J. J. Gude Prego et al. (Eds.): CISIS 2021/ICEUTE 2021, AISC 1400, pp. 374–383, 2022.
https://doi.org/10.1007/978-3-030-87872-6_37

## 1 Introduction

Based on the need that currently arises in the social context on environmental education in solid waste management, the high consumption of products and the lack of knowledge of the proper management of these wastes, lead the world population to face an environmental problem, concerning the irrational generation of organic and inorganic waste, a product of the exaggerated consumption of human beings. This has caused contamination of water sources, effects on the ozone layer, deforestation of forested areas, and extinction of fauna and flora [1]. At a global level 2010 million tons of solid waste are generated annually, this due to the rapid urbanization that occurs mainly in the cities of the world, being the continents of Asia and Europe where this type of waste is most generated [2]. Between the years 1950 and 2017, 9.2 million tons of plastic was produced, that is, more than one ton for each person who currently lives on earth, with single-use products and packaging being the most consumed, and at least 10% of this plastic was recycled [3]. In the last decades, humanity accelerated the production of waste, 2,100 million tons are produced per year [4]. In Colombia, the generation of waste is approximately 11.6 million tons annually, where part of these solids can be reused [5]. About 12 million tons of garbage are generated per year and of this only 17% is recycled [6]. Due to this problem, where the statistics are conclusive and in which decisive actions are required, there is a lack of competent professionals to solve this need, and higher education today faces an interesting stage, with some uncertainties and complexities. One of the causes is globalization and the challenges that universities face, given that globalization implies taking advantage of important opportunities, but in the same way, challenges concerning the traditional values of universities such as autonomy, academic freedom, research, student work, and evaluation that are threatened by this [7]. This has generated an impact on the training of engineers, especially because it is trained for competitiveness and not in a solidarity education, although there are approaches for training in values [8, 9]. As the training process occurs traditionally, competencies that the student can develop through strategies such as research projects are limited [10]. In Colombia, the problem of higher education is in slow changes in the face of social, economic, cultural, scientific, and political phenomena. Also, the crisis in aspects such as research and the decline in quality, which is reflected in the lack of skills such as leadership [8]. The lack of competencies in professionals in Colombia concerning their social and political commitment and low comprehensive training is another factor that affects the education of engineering students [11].

From these problems presented (generation of solid waste without its adequate disposal and separation, the lack of environmental education in society, and professional education with few skills to act in the face of real problems) lies the need to intervene society, though, of an environmental education that impacts on the management of solid waste generated by humans, involving methodologies such as POPBL to apply environmental projects aimed at solving this problem, training not only professionals but also people committed to social transformation [11]. The POPBL is a methodology in which the student appropriates his learning from the formulation of research questions, motivated by his natural curiosity. The Teacher acts as a guide and supervisor within the research process. The results of the process are finally exposed to a specific audience. This methodology is characterized by having a greater understanding and deeper learning [12]. In Sect. 2 we detail the methodology used within the selected groups of the environmental engineering program of the Universidad Santo Tomas in Colombia, in Sect. 3, we show the different results obtained disaggregated by competencies and subjects. Finally, we propose a discussion in Sect. 4, where we present the point of view on the observed, especially emphasizing the remarkable of this research.

## 2 Methodology

Carrying out this project is based on the Problem-Based Learning - Project-Oriented method, this being a goal-oriented procedure. For this, there are two research approaches: quantitative and qualitative, where the quantitative measures and the qualitative comprise [13]. Within the research, the two approaches were handled, which is defined as a mixed-method, which represents systematic, empirical, and critical processes, involving data collection and analysis, with an integrated discussion to make inferences from the investigated information and have a better understanding of study work [14].

This research was carried out in the Environmental Engineering program of the Santo Tomas University in Colombia, South America. Environmental engineering in Colombia is an undergraduate degree, which usually lasts 10 semesters or 5 years[15]. The research was carried out with 59 students, who were taking the courses of introduction to environmental engineering in the first semester, solid waste management in the sixth semester, and environmental management and processes in the seventh semester, all carried out in the year 2019-II. The students from the areas of introduction to engineering, solid waste management, and environmental management and processes, were the informants and participated with the answer to the questionnaires. The techniques and instruments used were the documentary review, observation, interviews, and surveys.

The methodology consisted of the documentary review on the subject to be worked on and based on this and the problems constructed, the instruments were established (survey, interview, and observation grid), validated by a committee of experts, composed of a civil engineer, a master's degree in education and a doctor of educational sciences. A pilot survey was applied to determine if the instrument was understood by the students and corrections were made to it, on the assessment of the studied competencies. From this test, the survey and initial interview were applied to the students of the first, sixth, and seventh semesters, with a total of sixty-five students. Working groups were created, of two or three students, on average, with problems with their surroundings, for projects in environmental education. The research was done over 4 years, but the instruments were applied in the 2019-II semester. The teacher was the guide of the projects in the planning, execution and results obtained phases. Finally, at the end of the semester, interviews and final surveys were applied to the students, evaluating through the observation grid the oral presentation and their written work, the projects of each group of students .

From the tabulation of the results, their analysis was made and the conclusions and recommendations of the research project were obtained. For the development of the research, three phases are required (see Fig. 1). In the first, information was consulted about the topic to be worked on, the methodology to be used was established, determining the instruments that allow the research to collect information. In the second phase, the instruments were applied to the participating population, at the beginning and end of the academic semester. On the part of the students, work teams were formed for the application of the research projects, where they were guided by the teacher in doubts about their application. Finally, the final phase was the analysis of the results obtained from phase two, with the confrontation of these with the main authors on the subject, in the same way, the conclusions obtained from the research will be specified.

The human and professional competencies identified and worked on during the research were autonomy, leadership, teamwork, reflection, critical thinking, oral and written communication, teamwork, reflection, critical thinking, and expertise. An initial survey was applied to find a diagnosis on the state of the competencies in the students and specific corrections to the applied formats. The type of research sampling is non-probabilistic for convenience because it allowed the selected sample to be given intentionally by the researcher, given the closeness of the researcher to the sample to control the training variable. The applied techniques and instruments began with a documentary review and continued with observation to see the environment in which the researcher obtained the information. The interview and survey allowed finding the answers before and after the process. They provided the information for the triangulation to obtain conclusions. The ABP-OP was developed in research groups, where students selected solid waste management in different social spaces in their environment.

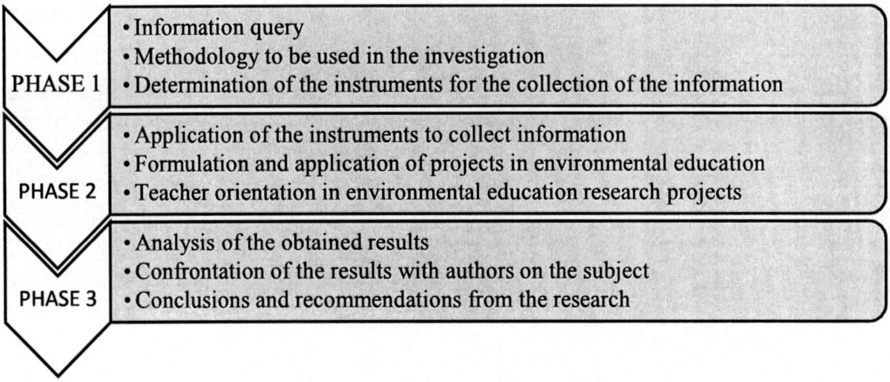


**Fig. 1.** Methodology applied to research.

## 3 Results

The results are given from the applied instruments, specifically; surveys, interviews, and observation grid. In the same way, the human competencies that were developed with the students were identified, based on the area that solid resources involve within their topics to work on. In the initial and final survey applied to the students, the changes in each of the competencies were recorded in Table 1 and the strengthening of the same was observed due to the application of the ABP-OP and the teacher's explanations about each competition in various academic spaces.

**Table 1.** Percentage of competencies development.

| Competencies | Percentage of increase | |
|---|---|---|
| | First semester | Sixth and Seventh Semesters |
| Autonomy | 19% | 41% |
| Critical thinking | 26% | 3% |
| Expertise | 6% | 7% |
| Teamwork | 67% | 44% |
| Oral communication | 39% | 3% |
| Written communication | 45% | 15% |
| Leadership | 25% | 7% |
| Reflection | 5% | 1% |

Among the results of the interviews, there is a summary matrix of student opinions regarding each of the evaluated competencies, which is shown in Table 2.

The observation grid allowed evaluating the progress of the competencies by the teacher, each competence was evaluated from items related to each competence. This was applied in the evaluation of the written projects and presentations of each subject on their research projects in environmental education. Table 3 shows an example of the observation grid applied for competencies such as autonomy and teamwork within the exhibition and written work.

**Table 2.** Final interview matrix.

| Introduction to Engineering | Solid waste management | Environmental Management of Processes |
|---|---|---|
| •Autonomy:Development process in personal matters<br>•Critical thinking: In aspects about class methodologies in the training process.<br>•Expertise: Reinforcement of theory with practice.<br>•Teamwork: Identified as necessary in professional life.<br>•Oral and written communication: Strengthened with activities such as essays and reports, as well as presentations and transmission of ideas.<br>•Project method: They use it to materialize ideas.<br>•Leadership: Lack of initiative in their work groups.<br>•Reflection: Contextualization of their training with reality | •Autonomy: Decision making in research projects.<br>•Critical thinking: Faced with the knowledge transmitted to them, in addition to the use of other sources (books).<br>•Expertise: Sometimes the relationship between theory and practice is not given.<br>•Teamwork: Strengthened through activities in the formative process.<br>•Written and oral communication: Use of technical language in writing and exposition. Shyness one of the factors to decrease.<br>•Project method: For skills development.<br>•Leadership: Listening to the opinions of colleagues for projects.<br>•Reflection: Proposal of real problems and solution, from your knowledge | •Autonomy: It grows from the freedom to work on problems in its community.<br>•Critical thinking: Facing your professional training.<br>•Expertise: The practice is more relevant during the training process.<br>•Teamwork: Fundamental in professional work.<br>•Oral and written communication: The oral expression of their ideas, what is written is not given in an adequate way.<br>•Project method: Applied in classroom projects and other areas.<br>•Leadership: Strengthened in the training process. It occurs depending on the work group.<br>•Reflection: Faced with environmental problems, with possible solutions. |

## 4 Discussion

The students were able to identify the competence of autonomy as optimal in their training process as environmental engineers, according to the applied surveys, it allows them to debate with their classmates, acquire knowledge by themselves, they manage to apply it in their training process and personal growth mainly in decision making. Competition increased by 19% in first semester students and in the other semesters analyzed the increase was greater, specifically 41%. This competence is formed in the displacement of the state of heteronyms in the early ages of the human being to decision-making, where the actions of the subject are following the law that is established in their society [16, 17], reflecting itself in the decision-making and judgment-making processes, through free decisions or elections that establishes rules for itself [18].

**Table 3.** Observation grid results.

| Competencies | Evaluation or observation results analyzed areas (observation grid) | | |
|---|---|---|---|
| | Introduction to engineering | Environmental management and processes | Solid waste |
| Showing/exposition work | | | |
| Autonomy | | | |
| The student answers the questions asked without the help of his classmates | Mostly YES | YES | YES |
| Dominate the subject that is exposed | YES | YES | YES |
| Reflects an investigation on the subject | YES | YES | YES |
| Written work | | | |
| Teamwork | | | |
| Organization and presentation of work are observed appropriately | YES | YES | YES |
| YES, one of the groups needs to improve writing and presentation | YES | YES | YES |

The teamwork competence, when measured through the survey, allows us to see that the first semester students considered this competence very important for their training process with a percentage of 67%, in the case of the other semesters with a percentage of 44%. Through teamwork, the students unanimously expressed that it is important to apply it in their training processes, because it allows them to share ideas or opinions, learning from their peers, being fundamental in their professional learning, evidenced by this in the interviews. This competence allows the student to develop skills and abilities, including setting group goals, integration capacity, conflict resolution, responsibility, planning, task coordination, and time management, to achieve an objective interactively with their group. work and the environment around you [19–22]. For its part, reflection is reinforced in academic spaces through real problems in the environment such as the need for environmental education to reduce solid waste. A shortcoming that occurs in students is that they do not question the procedures in laboratories or the field, and in several cases, the information given by the teacher, indicates that the student does not inquire or investigate these processes before performing them in practice. Reflective thought is born from situations that represent a dilemma and which supposes alternatives [23]. This is linked to economic, technological, political, scientific, and social development, where educational centers require the guidance and training of professionals who can provide solutions to social problems and their construction [24]. The decision-making processes in human beings sometimes lack logic and are taken spontaneously, implying intuitive knowledge that is developed through lived experience [25]. Critical thinking was strengthened with the application of this in their research projects and in their day-to-day training process in acquiring knowledge, it allowed students to be critical of their environment. A negative aspect that is observed in the opinion of the students is that they feel fearful of giving their opinions in the face of negative technical aspects that they observe and do not provide a solution for those shortcomings that they observe, expressing this in the interviews. Therefore, within the applied survey, a significant 26.5% increase in proficiency was found, mainly in first semester students, in sixth and seventh-semester students the increase was 3%. This competence helps them in the proper treatment of information, which is increasingly growing and important in society, analyzing statements and making value judgments based on criteria that are observed in the context [26]. Therefore, the student thinks critically when he bases his ideas and responds to situations that are presented to him, building his reasoning beyond abstract skills [27], where the teacher as an educational actor influences the development of critical thinking of students [28].

Concerning oral and written communication, it was observed that written communication is more difficult to develop and that they have little ability to express ideas or make themselves understood through this means. In oral expression, the nervousness and lack of confidence of the students are evident when expressing their ideas and in some cases making their ideas understood, but this is more developed than the written one, evidencing the above in interviews and surveys. Oral and written communication is key in the education of students and future professionals, who must be competent in the use of their mother tongue, through the exercise and development of forms of oral expression (conversations, debates, academic presentations, or professional) and written expressions such as writing texts with coherence, adequacy, and correctness of texts [29]. Communicating appropriately unites the student's abilities and skills, to build their knowledge, through the understanding, analysis, synthesis, and assessment of their ideals using expressions through oral and written messages [30, 31]. The use of theory and practice for the construction of knowledge is known as expertise. Within the themes given to students within their training process, it is observed that the student shows that the combination of theory and practice is given optimally, as evidenced in surveys and interviews. One of the shortcomings that are observed is that in some of the classes, especially in the upper semesters, the theory does not agree with the practice that the students carry out, affirmed by them in interviews. The competition was improved between 6 and 7% in the semesters analyzed. This allows the construction of knowledge thanks to the balanced combination between practice and theory, which reflects the expertise, skill, and ability in people, in what they learn and where this has greater opportunities and results in its application to communities [32]. In conclusion, repeating an activity is not enough to be an expert, it requires theoretical knowledge, where the student sets challenges with more and more obstacles, where the student looks for a way to overcome them [33, 34]. In the first semester, autonomy was improved by 19%, and in the other semesters analyzed by 41%. This indicates that the first semester students at the end of their project were more aware of how autonomy grows since at the beginning of the research they gave opinions of how it should be, without an appropriate experience of it. Teamwork, on the other hand, was one of the competencies that they affirmed was necessary within their training processes since it allows them as competencies to function as future professionals appropriately with their colleagues or colleagues. The percentage of improvement in this competence was 67% in first semester students.

## References

1. Hernández, S., Corredor, L.: Reflexiones sobre la importancia económica y ambiental del manejo de residuos en el siglo XXI, Rev. Tecnol. **15**(1), 57–76 (2016). https://dialnet.unirioja.es/servlet/articulo?codigo=6041529
2. Mundial, B.: Los desechos: un análisis actualizado del futuro de la gestión de los desechos sólidos, Banco Mundial (2018). https://www.bancomundial.org/es/news/immersive-story/2018/09/20/what-a-waste-an-updated-look-into-the-future-of-solid-waste-management
3. Heinrich Böll Stiftung y Brealk free from Plastic. Atlas del plástico. Datos y cifras sobre el mundo de los polímeros sintéticos, 2ª. Creative Commons, El Salvador (2019)
4. Revista Semana. Crisis mundial por la basura: solo el 16% de los desechos son reciclados. Revista Semana - Sostenible. 09 Jul 2019. https://sostenibilidad.semana.com/medio-ambiente/articulo/crisis-mundial-por-la-basura-solo-el-16-de-los-desechos-son-reciclados/44932.
5. Monterrosa, H.: Colombia podría aprovechar 40% de las toneladas de residuos que genera anualmente. La Republica (2019). https://www.larepublica.co/responsabilidad-social/colombia-podria-aprovechar-cerca-de-40-de-los-116-millones-de-toneladas-de-residuos-que-genera-al-ano-2813141. Accessed 01 Aug 2020.
6. Revista Semana: El 78% de los hogares colombianos no recicla. revista semana - sostenible, ene. 03 (2020)
7. Lopez, F.: Tendencias de la educación superior en el mundo y en américa latina y el caribe. Avaliação Rev. da Avaliação da Educ. Super. **13**(2), 267–291 (2006). https://doi.org/10.1590/S1414-40772008000200003.
8. Capote, G.E., Rizo, N.R., Bravo, G.: La formación de ingenieros en la actualidad. una explicación necesaria. Rev. Univ. y Soc. **8**, 21–28 (2016)
9. Perico-Granados, N., Dávila, M., Murcia, K., Arevalo, H.: Construcción de competencias humanas en Ingeniería. Hojas y Habl., **19**, 99–121 (2020). https://doi.org/10.29151/hojasyhablas.n19a6
10. De Zubiría, J.: La vigencia de las ideas de Alberto Merani en psicología y educación (2018). https://fliphtml5.com/pjcep/qymc/basic. Accessed 08 Dec 2020
11. Perico-Granados, N.R., Yaneth Galarza, E., Diaz Ochoa, M.L., Arévalo Algarra, H.M., Perico-Martinez, N.R.: Guía práctica de investigación en ingeniería: apoyo a la formación de docentes y estudiantes. Corporación Universitaria Minuto de Dios, Bogotá UNIMINUTO (2020)
12. Bell, S.: Project-Based Learning for the 21st Century: Skills for the Future. Clear. House A J. Educ. Strateg. Issues Ideas **83**(2), 39–43 (2010). https://doi.org/10.1080/00098650903505415
13. Niño, V.: Metodología de la investigación. Diseño y ejecución. Ediciones de La U, Bogotá D.C (2011)
14. Hernández Sampieri, R., Fernandez, C., Baptista, M.P.: Metodología de la investigación., Sexta edic. McGraw-Hill Education, México (2014)
15. Castellanos, P.M.A., Queiruga-Dios, A., Encinas, A.H., Acosta, L.C.: Environmental education in environmental engineering: analysis of the situation in Colombia and Latin America, Sustainability (2020). https://doi.org/10.3390/su12187239

16. Piaget, J.: El criterio moral en el niño, traducción al español de Nuria Vidal. Editorial Fontanella S.A, España (1974)
17. Galindo, J.: Sobre la noción de autonomía en Jean Piaget., Educ. y Cienc. **15**, 23–33 (2012). https://doi.org/10.1097/00042192-199704040-00073
18. Sieckmann, J.-R.: El concepto de autonomía, Doxa. Cuad. Filos. del Derecho. **31**(31), 465–483 (2008). https://doi.org/10.14198/doxa2008.31.28
19. Steven, M., Campion, M.: The knowledge, skill, and ability requirements for teamwork: implications for human resource management. J. Manage. **2**(20), 503–530 (1994)
20. Aguado, D., Arranz, V., Valera, A.: Desarrollo de la competencia transversal trabajo en equipo mediante contenidos elearning: una ayuda oara la insercion laboral. Relada **2**(4), 104–111 (2010)
21. Barraycoa, J., Lasaga, O.: La competencia de trabajo en equipo: más allá del corta y pega. Vivat Acad., **0**(111), 66–70 (2015). https://doi.org/10.15178/va.2010.111.65-69
22. Perico-Granados, N.R., Umba, M., Tovar-Torres, C., Reyes, C.: Proyectos educativos para estudiantes de educación básica en Colombia: estrategia de aprendizaje en matemáticas. Rev. Venez. Gerenc. **25**(92), 1741–1757 (2020)
23. Dewey, J.: Cómo pensamos. Nueva exposición de la relación entre pensamiento reflexivo y proceso educativo. Paidos, Barcelona (1989)
24. Sánchez, E., López, M.: El contexto actual de la formación de profesionales reflexivos. Rev. Temas **7**, 45–55 (2013). http://revistas.ustabuca.edu.co/index.php/TEMAS/article/view/573
25. Schön, D.: El profesional reflexivo, cómo piensan los profesionales cuando actúan., 1 Edición. Paidós, Barcelona España (1998)
26. Patiño, H.: El pensamiento crítico como tarea central de la educación humanista, DIDAC. Univ. Iberoam., pp. 10–17 (2014)
27. Suárez, J., Martín, J., Villaveces, L., Pabón, D.: Pensamiento crítico y filosofía. Un diálogo con nuevas tonadas. Universidad del Norte, Barranquilla (2018)
28. Tamayo, O., Zona, R., Loaiza, Y.: El pensamiento crítico en la educación. Algunas categorías centrales en su estudio. Rev. Latinoam. Estud. Educ. **11**(2), 111–133 (2015)
29. Cantú, L.: Roque, M.C.: Comunicación para ingenieros. Grupo Editorial Patria, México (2014)
30. Cid, A., Merchand, T.: La expresión oral y escrita, habilidad indispensable en la formación de ingenieros. ANFEI (2014)
31. Carreño, P.: La comunicación oral y escrita en la formación de ingenieros. Pedagog. y curriculo **13**(26), 146–152 (2012). https://doi.org/10.21500/01247492.1289
32. Dewey, J.: Naturaleza humana y conducta. Fondo de cultura económica, Mexico (2014)
33. Acosta, D., Vasco, C.: Habilidades, competencias y experticias: más allá del saber qué y el saber cómo. Corporación Universitaria Unitec, Univesidad de Manizales y Fundación Centro Internacional de Educación y Desar-rollo Humano (Cinde), Bogotá, D.C (2013)
34. Garcia, L., Mercado, C., Trespalacios, A.: La experticia en la planeación a través de la estrategia de núcleos temáticos integradores. Universidad del Norte (2020)

# A Systematic Review of the Inclusion of Education for Sustainable Development in Environmental Engineering

Pedro Mauricio Acosta Castellanos[1(✉)], Lina T. Gonzalez[1], and Araceli Queiruga-Dios[2]

[1] Universidad Santo Tomás, Bogotá, Colombia
{pedro.acosta,lina.gonzaleza}@usantoto.edu.co
[2] Universidad de Salamanca, Salamanca, Spain
queirugadios@usal.es

**Abstract.** Sustainable development (SD) is a great global commitment in economic, social, and environmental terms of the century XXI. This model integrates conventional developmental models that seek economic growth, but at the same time encourages human development, all within the framework of respect and sustainable use of natural resources. This integrative work is a challenge that must involve all levels of society. Higher education is not out of this line, since universities must strive for knowledge, research and innovation to achieve or at least promote sustainable development. For achieving this, there is a model called Education for Sustainable Development (ESD). Environmental engineering is a branch of engineering that seeks through the technique to solve environmental problems that involve engineering development. ESD has become an indirect measurement instrument for the inclusion of SD in curricula, which demonstrate how up-to-date the study plan is. With this research, based on a systematic literature review (SLR), it was sought to answer some research questions about the degree of inclusion of ESD in the environmental engineering curriculum and to recognize the success stories that are documented in the Science Direct, Scopus, and ERIC databases. It was possible to identify, among other things, the most promoted competencies within ESD and the techniques and tools used by universities to promote SD in this engineering curriculum.

**Keywords:** Education for sustainable development · Educational strategies · Environmental engineering · Environmental education · University education · Systematic review

## 1 Introduction

Today's world has an environment of globalization, made up mainly of capital and technological innovation [1]. In this context, environmental problems are increasingly recurrent and have been generated as a consequence of changes in life cycles, loss of biodiversity, overexploitation, and imbalance in the irrational use of natural resources.


J. J. Gude Prego et al. (Eds.): CISIS 2021/ICEUTE 2021, AISC 1400, pp. 384–393, 2022.
https://doi.org/10.1007/978-3-030-87872-6_38

In the same way, the predominant global development model is based mainly on excessive economic production standards, achieving unlimited development [2]. Thus, the protection and conservation of natural resources have taken a "back seat".

The current environmental crisis is caused by anthropogenic factors [3], for this reason science and technology are possible sustainable tools to solve this crisis [4]. The lack of economic resources in some societies pressures the depletion of natural resources, in the same way, does the standard of living of some developed societies generate high rates of consumption and environmental depredation [3]. It has been recognized that these crises are not only problems of innovative development, but are also the result of a lack of environmental knowledge and behavior [5]. The only viable solution to face and overcome difficulties is governed by a challenge of optimization and adjustment of a better quality of education in environmental terms [6].

Facing environmental, economic, technological, and social challenges is not easy since it is necessary to promote the acquisition of values, behaviors, and collective environmental awareness, where natural resources are assured and protected in the best way [7]. Environmental education (EE) deals with solving anthropogenic environmental problems, depending on the emergence of activities and behaviors of a responsible nature with the environment [3].

From another point of view, sustainable development implies a high level of personal and humanistic training. In this way, Education for Sustainable Development proposes an educational transformation and a moral change, achieving social and political action in favor of sustainability. The institutional commitment establishes a strategic vision of sustainability that must be embraced in higher education institutions. In other words, formal environmental teaching reinforces the inclusion of sustainability, not only at the student level but also at the teacher level, where research encourages the development of active knowledge in favor of SD [8].

ESD refers to a restructuring approach to teaching in higher education institutions, providing answers to current problems in society [9]. This approach is based on the complementation of the engineering curricula, to provide ponderable knowledge, values, and behavior models necessary today to form an environmentally sustainable society. Therefore, a proper understanding of ESD is essential, focused on the incorporation of a teaching process based on active learning and critical or systemic thinking, through dynamic methods that motivate students to acquire knowledge, skills, strategies, and methodologies to forge a sustainable future [10].

The main task of universities is to train students in order to promote and improve quality in technology and innovation. So that these innovations are friendly with the environment. Environmental engineers deal with structures, equipment, and systems, which are designed to protect, improve the quality of the environment, health, and public welfare [11]. In the specific case of Environmental Engineering, the incorporation of ESD into the curricular plans allows a methodological reorganization, as a fundamental support tool for both teachers and the student body, spreading a disciplinary verticality committed to SD [12]. Cultural, social, economic, and biological diversity motivate a change in the conception of social and professional responsibility, where the learning and capacity of the competencies or skills that are required increases, that is, not only an individual change but also social [8].

Bearing in mind the importance of ESD within universities, and in environmental engineering, the following research questions were posed to establish a real state at the research level about the inclusion of ESD within this engineering:

*Q1: What are the competencies that are proposed or used to achieve SD in environmental engineering programs?*
*Q2: What strategies do universities use to promote sustainable development and include the Sustainable Development Goals (SDG) in their work?*
*Q3: What are the methodologies that are commonly used to include ESD in the teaching of Environmental Engineering in higher education?*

## 2 Methodology

To solve the research questions, a systematic literature review was used, aimed at the inclusion of Education for Sustainable Development in Environmental Engineering.

To solve the research questions a systematic literature review (SLR) was used. The questions aimed at the inclusion of Education for Sustainable Development in Environmental Engineering. Likewise, it was sought to determine the value of ESD and the influence it has to establish an ideal development for the optimal behavior of environmental educational learning that the academic community should incorporate in universities. It was sought to determine the influence that ESD has on the learning of environmental behavior in universities.[13]. The protocol was developed under the standards of the PRISMA statement to respond in detail to the research questions [14]. The search of the publications was done from 1987, when the Sustainable Development was welcomed from the Brundtland Report by the ONU [15], until 2021.

The databases used were: ERIC, because it contains abundant data on teaching, training, and education [16]. Science Direct because it is focused on science and engineering [17], and Scopus, for being recognized as the greatest quality scientific database [18].

The keywords that were included in the search were "*Education for Sustainable Development*" or "*Environmental Engineering*", with their acronyms and synonyms. It was defined only to include studies or research from journals and book chapters; in English or Spanish languages. The summary of identification and selection is illustrated in Fig. 1.

## 3 Results

In the sustainable development objectives, universities are the fundamental axes that promote a change in social behavior through a technological, environmental, and economic commitment, which they consolidate from the application and understanding of ESD. For universities, to involve cross-sales human competencies is not unknown, and it is increasingly common to involve the social responsibility issues in the study plans of different disciplines. Engineering is not out of these processes, since its main goal is comprehensive student development [19]. In this sense, for ESD to be viable in university education, especially in Environmental Engineering, there are different approaches to

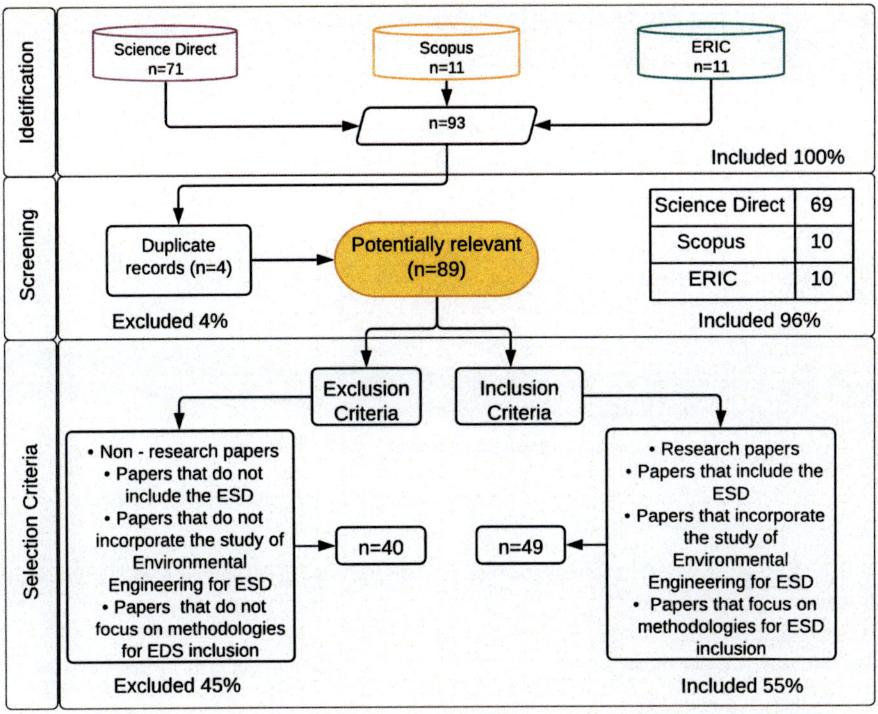


**Fig. 1.** Review scheme adapted from the methodological flow suggested by PRISMA.

evaluate teaching strategies, methods, or approaches that are based not only on cognitive learning but also on social-emotional and behavioral learning [20].

According to the results obtained, ESD has had a greater presence since 1996. There is a marked influence on the number of publications thanks to the decade for ESD proposed by the UN, this period is between 2005 and 2015 [21]. It is in this period that most publications are related to ESD and environmental engineering.

On the other hand, concerning the tools used for the promotion and inclusion of ESD in environmental engineering, diversity can be observed, where no one specifically predominates. Tables 1 and 2 show the publications of each of the databases, which include the implementation of assessment tools and learning platforms or pedagogies included in the curricula as a solution to better implement ESD in Environmental Engineering. We found some interesting tools such as the Staunch system, which assesses sustainability in universities. This tool can be used in the curricula or as environmental auditing instruments in universities [22]. Another innovative tool is Knowledge, Experience, and Self-directed learning (KES). In an ideal KES cycle, students gain knowledge and experience in their first three years of study. These early years are characterized by formal teaching and learning activities with a self-directed approach. In the last years, the students are directed to lead their learning [23].

**Table 1.** Tools that assist the inclusion of ESD in the environmental engineering curriculum.

| Tool | Percentage of papers that includes it |
|---|---|
| Implementation of a cyber-learning platform with modules and web courses | 12% |
| Collaborative research projects | 16% |
| Incorporation (KES) cycle and professional fields in the study plans | 10% |
| Implementation of an Environmental Management System (EMS) | 4% |
| Using the C-map method | 4% |
| Using the STAUNCH system | 8% |
| Audit Instrument for Sustainable Higher Education (AISHE) | 8% |

**Table 2.** ESD pedagogical methodologies and models, included in the environmental engineering curriculum.

| Methodology or pedagogical model | Percentage of articles that includes it |
|---|---|
| Sustainable virtual education | 22% |
| Innovation and implementation of green or ecological engineering and eco-design | 33% |
| Technology or communication plans | 27% |
| Using basic science and engineering science to solve social problems | 22% |
| Implementation of other ways of thinking | 14% |
| The Young Teachers Program (YMP) | 10% |
| Novel teaching methods | 12% |
| Implementation of the Capstone project | 2% |
| Partial reprogramming of the curricular contents | 12% |

### 3.1 Applied Competencies in Environmental Engineering

The economic and social council of the UN accepted ESD as a strategy, where it should be considered by the member states, this invitation not only raises the standards of curriculum development but also strengthens the capacity of individuals, groups, communities, organizations, and countries to pass judgments and elections in favor of sustainable development. It should be noted that self-criticism and ethics are two instruments that help to achieve sustainable balance from the society-nature relationship [24].

Table 3 shows the different competencies included in environmental engineering, to improve the understanding and promotion of ESD in the curricula. At this point it should be clarified that competencies are considered as the capacity for collective analysis in different domains (environment, society, technology, and economy), that is why decision-making focuses on knowledge that improves in one way or other systems thinking [20].

**Table 3.** Results of the types of competencies used to achieve ESD.

| Competence | Percentage of items |
|---|---|
| Anticipated self-knowledge | 12% |
| Design capabilities of ethical-environmental systems | 12% |
| Ability to solve environmental, social, and economic problems | 9% |
| Regulatory competences | 9% |
| Understanding of a professional and ethical responsibility | 10% |
| Professional and ethical planning skills | 11% |
| Work in the interdisciplinarity | 19% |
| Critical thinking | 13% |
| Teamwork | 5% |

It is important to consider all these competencies as preliminary axes for the integration of knowledge and the management of complexity to formulate judgments that allow discerning a reflection or view of reality on the social and ethical responsibilities of the engineer [25].

### 3.2 University Strategies to Promote SD

The SD has been adapted by some universities and adjusted to their realities without losing its meaning or essence [26]. For this, different strategies have been sought and specifically in environmental engineering, 7 were found that are used to promote SD. This means that, although ESD is not formally or explicitly mentioned within the investigations, the impulse of SD itself makes it a form of ESD. In this sense, the most predominant strategy within universities is to promote attitudes and behaviors based on environmental education with 21%, followed by the development of plans and projects with 18%, the change of curricular structures with 14%, the development of environmental and social reasoning programs 13%, the promotion of systemic thinking 12%, social research 11% and different undefined environmental strategies 11%.

### 3.3 Specific Activities Around ESD

Among the activities that universities use the most, it was possible to show that practices are used to a greater extent to encourage the so-called "learning by doing" which helps in an integral way to articulate the curriculum and with this approach in a more understandable way ESD concepts and guidelines [27]. The most used activities in ESD within environmental engineering are; technological projects in environmental sustainability with 21%, followed by 13% by student lectures or presentations, specific courses in ESD or SD with 13%, 8% are research workshops, in the same way 8% corresponds to participation in virtual or face-to-face forums, 7% to seminars, 6% corresponds to academic outings, in the same way 6% is for technical internships, likewise international internships correspond to 6%, role plays 6%, 2% correspond to video analysis or films, like 2% for the realization of concept maps, distance courses correspond to 1% and surveys and interviews with 1%.

### 3.4 University Methodologies that Promote Sustainable Development

Through the articles that were analyzed by the quality evaluation, the different methodologies that can be applied within the Environmental Engineering program were made known, in order to achieve the application and teaching of ESD. 21% of the articles are related to the structuring of an institutional plan where SD is promoted and strengthened, 16% correspond to green engineering programs, 15% to research programs on sustainable issues, 13% correspond to sustainable technology projects, 10% refer to performance analysis models of sustainable systems and tools, 9% correspond to models for economic decision-making and evaluation of sustainable projects, 8% attend the case study modules and finally the sustainability perspectives within educational approaches 8%.

## 4 Discussion

The environmental and social challenges of century XXI, show the need for a change in lifestyle in society. This is why education plays an important role in achieving the formation of balanced societies based on the pillars of the economy, technology, and the environment [28]. However, the first efforts to incorporate education on environmental issues in formal and informal educational settings under these precepts were not the best, due to the lack of will and institutional support [29]. The result of the SLR indicates that university curricula for environmental engineering require greater incorporation of ESD. Implementation must go hand in hand with participatory teaching and learning methods [28]. ESD promotes the adoption of competencies such as critical thinking, the elaboration of hypotheses for the future, and the collective adoption of decisions, these in one way or another reinforce the systemic change from formal education to quality education [30].

The United Nations, especially in the period between 2005 and 2014, has promoted the formal inclusion of ESD in higher education so that in this way universities are promoters of SD. Curricula must adopt a reflective nature around sustainability as a field or a concept, through an interdisciplinary, diverse, and multimodal learning pedagogy [26]. Universities must also understand the importance of implementing evaluations or management plans that help determine the degree of incidence or understanding of the competencies, methods, and strategies that contribute to the implementation of sustainable projects [31].

Students or future engineers must have the necessary confidence in the consideration and evaluation of the sustainability of each of the socially implemented projects, to solve the problems of the environmental crisis [10]. It must be accepted that it is not an easy task to sensitize environmental engineering students about sustainability in the short, medium, and long term. Indeed, this is why final semester students must demonstrate their sustainability analysis capabilities before graduating [13].

Higher education standards around the world take into account the SD, accrediting entities in engineering such as ABET mention the importance of environmental sustainability in engineering [32].

We find that Europe, North America, and Asia are the regions that most understand the dimension of the importance of ESD within institutional curricular plans. The strategies used by universities that promote the inclusion of ESD in Environmental engineering to achieve sustainability, lie in 21%, confirming that the attitude and behavior of environmental education are important. At the same time, the social, economic, cultural, and political dimensions have less weight, which affects a short or biased understanding of SD. In other words, the pillars (society, economy, and environment) of ESD remain on the sidelines in the understanding of most students of environmental engineering, so the integration of competencies, strategies, and methodologies may be the answer. to the correct and formal application of ESD [33].

Finally, the pedagogical approach continues to be a problem, since it depends on the degree of inclusion of SD in the curriculum. One possible effective response is green engineering research or projects, which should build on society and foster relevant activities with transformative goals [34].

## References

1. Sauvé, L.: Environmental education between modernity and postmodernity: searching for an integrating educational framework. Can. J. Environ. Educ. **4**, 9–35 (1999)
2. Gijón Cuello, A.: Problemas ambientales y Educación Ambiental en la escuela. Cent. Nac. Educ. Ambient. 1–24 (2003)
3. Mora Penagos, W.M.: Educación Ambiental y educación para el desarrollo sostenible ante la crisis planetaria: demandas a los procesos formativos del profesorado. Cent. Investig. y Desarro. Científico – Univ. Dist. Fr. José Caldas. 26, 7–35 (2009)
4. Herrera Travieso, D.M.: Science, technology and environmental health. Inst. Nac. Hig. Epidemiol. y Microbiol. 1–19 (2007)
5. Acosta Castellanos, P.M., Queiruga-Dios, A., Hernández Encinas, A., Acosta, L.C.: Environmental education in environmental engineering analysis of the situation in Colombia and Latin America. Sustainability **12**(18), 7239 (2020). https://doi.org/10.3390/su12187239
6. Martínez Castillo, R.: La importancia de la educación ambiental ante la problemática actual. Rev. Electrónica Educ. **14**(1), 97–111 (2010)
7. Sánchez, J., Domínguez, R., Sunkel, O.: Recursos naturales, medio ambiente y sostenibilidad: 70 años de pensamiento de la CEPAL, 1st ed., vol. 1. Comisión Económica para América Latina y el Caribe, Panamá (2019)
8. Ramírez Sánchez, R.: La educación superior para el Desarrollo Sostenible. Univ. Veracruzana (Gestion la Int. la Vincul. y la Cult.), pp. 4–22 (2017)
9. Martínez Iñiguez, J.E., Tobón, S., Romero Sandoval, A.: Problemáticas relacionadas con la acreditación de la calidad de la educación superior en América Latina. ALEPH 17(0), 1–18 (2017)

10. Litzner Ordóñez, L.I., Ries, W.: Education for Sustainable Development in the context of higher education in Bolivia. Perceptions of university professors. Ediciones Univ. Salamanca, pp. 149–173 (2019). https://doi.org/10.14201/teri.19037
11. Moghaddam, M.A., Taher-shamsi, Maknoun, R.: El papel de la educación en ingeniería ambiental en desarrollo sostenible en Irán. Emerald **5**(2), 123–130 (2007). https://doi.org/10.1108/14676370710726599
12. William, R., Avendaño, C.: La Educación Ambiental (EA) como herramienta de la responsabilidad social (RS). Red Rev. Científicas América Lat. el Caribe, España y Port., pp. 94–115 (2012)
13. Gonzalo Muñoz, V., Sobrino Callejo, M.R., Benítez Sastre, L., Marín Coronado, A.: Revisión sistemática sobre competencias en desarrollo sostenible en educación superior. Rev. Iberoam. Educ. **73**, 85–108 (2017)
14. Moher, D., Liberati, A., Tetzlaff, J., Altman, D.G.: Preferred reporting items for systematic reviews and meta-analyses: the PRISMA statement. PLoS Med. **6**(7), e1000097 (2009). https://doi.org/10.1371/journal.pmed.1000097
15. Larrouyet, C.: Desarrollo sustentable : origen, evolución y su implementación para el cuidado del planeta. Univ. Nac. Quilmes, Secr. Posgrado (Repositorio Inst. Digit. Acceso Abierto) (2015)
16. Chacón Alvarado, L.: ERIC Educational Resources Information Center. Esc. Bibl. Doc. e Inf. **8**(1) 26–34 (2006)
17. Codina, L.: Science Direct: base de datos y plataforma digital de Elsevier. Plataformas académicas, pp. 1–5 (2018)
18. Codina, L.: Scopus: el mayor navegador científico de la web. El Prof. la Inf. **14**(1), 1–6 (2005)
19. Rosa Ruiz, D., Giménez Armentia, P., De la Calle Maldonado, C.: Education for Sustainable Development: the role of the university in the 2030 agenda (transformation and design of new learning environments). Rev. Prism. Soc. **25**, 180–202 (2019)
20. Sahakian, M., Seyfang, G.: Una revisión de la enseñanza del consumo sostenible: de la construcción de competencias al aprendizaje transformador. Rev. Prod. mas Limpia, 1–23 (2018). https://doi.org/10.1016/j.jclepro.2018.06.238
21. Casalvázquez Hernández, G.: Educación para el Desarrollo Sostenible>>: una nueva mirada para abordar los problemas desde la educación permanente. Educ. y Innovación, 1–10 (2015)
22. Lozano, F.J., Lozano, R.: Developing the curriculum for a new bachelor's degree in engineering for sustainable development. J. Clean. Prod. **64**, 136–146 (2014). https://doi.org/10.1016/j.jclepro.2013.08.022
23. Padfield, R., Papargyropoulou, E., Baharun, S., Noor, M.J.M.M., Kinoshita, C.: 'Ningen Ryoku': the Japanese way in inculcating human skill into engineering education. Procedia – Soc. Behav. Sci. **56**, 369–376 (2012). https://doi.org/10.1016/j.sbspro.2012.09.665
24. Jickling, B.: ¿Un futuro para la sostenibilidad? Div. Artes y Ciencias, pp. 2–7 (2000). https://doi.org/10.1023/A:1005211410123
25. Perpignan, C., Yacine, B., Robin, V., Eynard, B.: Perspectiva de la educación en ingeniería para el desarrollo sostenible: una evaluación de la madurez de las habilidades técnicas avanzadas y transversales en el diseño ecológico. Elsevier, pp. 1–6 (2020). https://doi.org/10.1016/j.procir.2020.02.05
26. Melanie DuPuis, E., Ball, T.: Cómo no qué: enseñar la sostenibilidad como proceso. *ciencia, práctica y política* **9**(1) 64–75 (2013). https://doi.org/10.1080/15487733.2013.11908108
27. Lehmann, M., Christensen, P., Du, X., Thrane, M.: Problem-oriented and project-based learning (POPBL) as an innovative learning strategy for sustainable development in engineering education. Eur. J. Eng. Educ. **33**(3), 283–295 (2008). https://doi.org/10.1080/03043790802088566
28. Vásquez Vargas, M.J.: Educación para el desarrollo sostenible (EDS). Friedrich Eberrt Stift. Am. Cent. **7**, 1–12 (2014)

29. Sauve, L., Brunelle, R., BerrymaN, T.: Educar para el deba a el debate Políticas nacionales y educación ambiental. Reflexiones para una agenda Ambient. **7**(20–21), 74–87 (2006)
30. Radits, F.: Competencies for ESD (Education for Sustainable Development) Teachers, 12th ed., vol. 7, no. 3, MDPI AG, Bruselas (2008)
31. McCormick, K., et al.: Informe de iniciativas educativas Educación para el desarrollo sostenible y la Programa de Maestros Jóvenes. J. Clean. Prod. **13**, 1108–1111 (2005)
32. Sánchez-Carracedo, F., Lopez, D., Mart, C., Vidal, E., Climent, J.: La matriz de sostenibilidad: una herramienta para integrar y evaluar la sostenibilidad en las tesis de licenciatura y maestría de los títulos de ingenierí. Sustentabilidad (MDPI) 1–23 (2020). https://doi.org/10.3390/su12145755
33. Thürer, M., Tomasevic, I., Stevenson, M., Qu, T., Huisingh, D.: A Systematic Review of the Literature on Integrating Sustainability into Engineering Curricula. Elsevier, pp. 608–617 (2018)
34. Bascop, M., Perasso, P., Reiss, K.: Systematic review of education for sustainable development at an early stage: cornerstones and pedagogical approaches for teacher professional development. Sustentabilidad (MDPI) **11**, 2–16 (2019). https://doi.org/10.3390/su11030719

# Bioengineering as a Strategy for the Consolidation of Innovation and Technological Development Processes in the Central Zone of Colombia

Oscar Leonardo Pulido Sainea, Yuddy Alejandra Castro Ortegón(✉), and Pedro Mauricio Acosta Castellanos

Saint Thomas University, Tunja, Colombia
{oscar.pulidos,yuddy.castro,pedro.acosta}@usantoto.edu.co

**Abstract.** Bioengineering is a science that brings together the principles of biology and the tools that engineering can offer, so that different products can be created that are useful, economically viable and tangible. Thanks to the above, it can be recognized that this combination of topics favors the inclusion of environmental, agricultural and livestock aspects, since that is where the analysis of this research will be focused, which favors the recognition within this study of bioengineering as an independent variable. In Colombia, bioengineering does not have great significance or relevance because there are very few mechanisms that can help it to be applied in all activities in the national territory and therefore no significant emphasis has been placed on the aforementioned issues. For this reason, the research is directed to the Central Region of Colombia, which is known for being one of the sectors with the most abundance in biodiversity and has one of the largest territorial areas, basing its economy on agricultural issues. In accordance with these references and taking into account that bioengineering is a discipline that is oriented towards development and innovation, it is intended to have as a mediating effect its inclusion for the dependent variables already mentioned. The dependent variables oriented to development and innovation are fundamental to evaluate the current state of the Central Region of Colombia, made up of the departments of Bogotá D.C., Cundinamarca, Boyacá, Meta and Tolima. According to the above, current research is placed towards an analysis of the application of bioengineering in general terms, highlighting the importance of making a proposal that frames its scope and its inclusion in livestock, agricultural and environmental issues, favoring the advance of this region with adaptable and friendly technologies with the environment and the local population. Finally, it is essential to frame that the research was developed in a bibliographic way and with a theoretical analysis, according to the above it was possible to demonstrate the value and interest that bioengineering has today in daily activities, in development and innovation for the region central Colombia.

**Keywords:** Bioengineering · Innovation · Development · Technological · Science · Agriculture · Environment


J. J. Gude Prego et al. (Eds.): CISIS 2021/ICEUTE 2021, AISC 1400, pp. 394–401, 2022.
https://doi.org/10.1007/978-3-030-87872-6_39

## 1 Introduction

Bioengineering aims to be able to replicate or imitate biological systems so that in this way, beneficial products can be created, where the processes that you want to intervene or that you want to replace can be controlled and modified, in such a way that the processes can be increased. Mechanical, interacting with biology, natural resources and technology. On the other hand, and covering issues of environmental relevance, bioengineering has fundamental characteristics in terms of the genetic modification of plants and microorganisms that interact in the environment and it is there where Bio process engineering and Biotechnology are correctly linked [1].

The overwhelming development of bioengineering in countries that are considered as developed has been extensive and ambitious, as it warns of a marked trend in monopolization processes that refer to this new knowledge. In addition to this, the countries considered as third world countries have an impassive perspective in the face of this advance because a network is generated, which, within a few years, will deprive them of the enjoyment of products and new technologies because that is where an alternative is generated for the groundbreaking events of science. Worldwide, bioengineering encompasses different academic, investigative, theoretical and practical processes, for this reason the disciplines that structure and make it up are defined as: General Basic Sciences, Interdisciplinary Basic Sciences, Complementary Discipline, Fields of applicability [1].

The focus of the study is located in the Central Region of Colombia because this is a territory of great strategic importance for this country, not only because of its population and economic importance, but also because of other environmental dimensions, since it is there where it concentrates a large part of the different moorland ecosystems of the national territory. On the other hand, it is considered to be the first territory on a regional scale that can have a characteristic institutional framework for its management, because approximately similar figures can be created such as those managed in the Caribbean and the Pacific [2].

The departments that make up the Central Region of Colombia are Boyacá, Cundinamarca, Meta, Tolima and the capital of the country Bogotá D.C. These members can come to understand the importance of the large scale that the region has in the development of their territories, in addition to being able to establish the associative capacity that they have between them to be able to achieve positive efforts and relationships in such a way that they acquire mutual gains [2].

It is important to highlight that the Independent variable that we want to associate for this research is Bioengineering, as it has become an extremely important subject of study, in addition to serving as a guide for the establishment of new theoretical and practical projects. On the other hand, and by making the appropriate inclusion of the variables, the dependent appears, who for this case are innovation and development, which are main characteristics that favor the central territory of Colombia, since there are clear needs regarding the aspects livestock, agriculture and environment [3].

This is why the focus of the research is based on being able to identify the importance of bioengineering as a strategic discipline for the different consolidations of development and innovation techniques compared to the processes that the central zone of Colombia has.

## 2 Objectives

### 2.1 General Objective

Identify the importance of the application of bioengineering as an strategic method for the consolidation of innovation processes and technological development in the central region of Colombia.

### 2.2 Specific Objectives

- Evaluate the scope of bioengineering with respect to the fulfillment of the Sustainable Development Goals (SDG) in Colombia.
- Analyze the current state of bioengineering at the national, regional and local levels regarding environmental, livestock and agricultural issues.
- Propose potentialization ideas that contribute to the improvement of technologies through innovation and development in the face of environmental, agricultural and livestock issues in the region zone of Colombia.

## 3 Methodology

The methodology of the study is based on the search for the proposed objectives, in such a way it is established that the study will have a descriptive and explanatory approach, where an analysis that has a scope beyond the respective description is involved. of phenomena or concepts or that can also be established in relationships where the occurrences of the phenomena can be explained. On the other hand, it is important to mention the conditions in which it can be manifested, thanks to the relationship of two variables, for this case we want to show the links of the independent variable called bioengineering and the dependent variables that are considered as development and innovation, with the inclusion of a study area that is the Central Region of Colombia, [4].

It is available to work from a design that does not involve experimental methodologies, because this study will not have the conditions to intentionally vary the independent variable (Bioengineering) so that some type of effect can be generated on the de pendent variables (Development and Innovation). Thanks to the aforementioned, it can be established that it is an investigation that is oriented to the observation of the situations that already exist, in order to be able to establish the respective correlation that exists between the variables mentioned above [4].

The research approach will highlight the qualities that can be adapted from a purely theoretical research regarding bioengineering and its related, since it will allow the integration of qualitative data that evidences each of the study processes that is carried out for the consolidation of this project, in such a way that different themes can be integrated by means of bibliographic instruments that are related to the processes that involve innovation and development, aimed at agricultural, livestock and environmental conditions of the central region. It is necessary to emphasize that these qualitative data will contribute to the collection of information regarding the terms that establish the meanings that are shared in the research, thanks to this a fundamental dynamic is generated where the connotations for each area are shared in such a way that the different actors influence the environmental, social and economic dynamics of the central zone of Colombia. On the other hand, it can be established that the approach to shared meanings is framed in the state of the art, which is considered a process of social formation since much of the information collected is part of the interactions of people in said country area of the art, which is considered a process of social formation since much of the information collected is part of the interactions of people in said country area [4].

Finally, it is important to mention that this research was developed by collecting information regarding Bioengineering in the world, in Colombia and especially in the department of Boyacá. This analysis was carried out by searching for documents, articles, reports, databases, academic papers, journals and, among others, that were related to descriptive processes, training processes and general interest regarding the issues of the Bioengineering, as it is the basis of this study and is part of the new methods on the positive changes that can be adapted for the present and the future of the planet that this branch of education and exploration can offer us to implement their academic capacities, theoretical, practical and innovative in agriculture, the environment and livestock issues, because the aforementioned aspects are similar to the purposes of Bioengineering. On the other hand, emphasis was placed on data collection for the Central Region, thanks to the particularities mentioned above and thus carry out the characterization of the community. For this reason, a data analysis will be generated as it seeks to make a relationship between bioengineering, the environment, agricultural and livestock issues for said department. In such a way as to respond to the problems raised in this study. Thanks to the particularities that were mentioned above and, in this way, carry out the characterization of the community. For this reason, a data analysis will be generated asit seeks to make a relationship between bioengineering, the environment, agricultural and livestock issues for said department. In such a way as to respond to the problems raised in this study. Thanks to the particularities that were mentioned above and, in this way, carry out the characterization of the community. For this reason, a data analysis will be generated as it seeks to make a relationship between bioengineering, the environment, agricultural and livestock issues for said department. In such a way as to respond to the problems raised in this study (Fig. 1).

## 4 Analysis and Results

### 4.1 Analysis

A starting point of the research was framed from the study of biotechnology as a basis for the analysis of the implementation of bioengineering in the Central Region of Colombia, taking into account that the first includes methodological, procedural, applicative and purposeful components, that form an essential complement for the activities that are to be applied with respect to bioengineering, being directly related to each other.

It is important to mention that within the investigations that were carried out to cover the document, a great and valuable similarity was found between the independent variable that is bioengineering and biotechnology, since they are disciplines that include science, technology and engineering processes, since They are mechanisms adaptable to the events of living beings and especially for human beings, in addition to

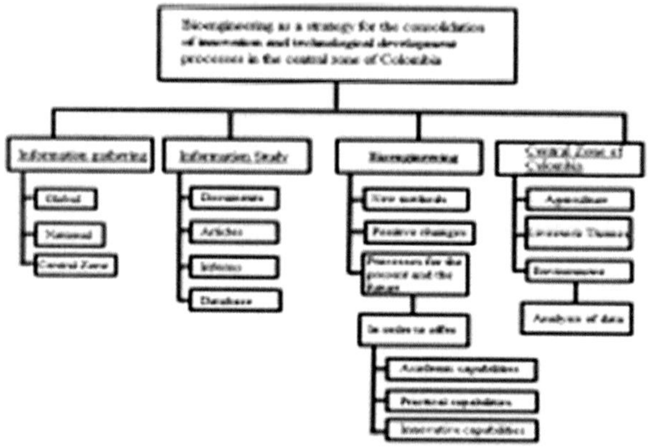

**Fig. 1.** Methodology of bioengineering as a strategy for the consolidation of innovation and technological development processes in the central zone of Colombia. Source: Authors

benefiting dependent variables such as development and innovation. Thanks to its wide range of ideas, considerable solutions can arise from problems that are evident in the central zone of Colombia, in accordance with the SDGs, the PND and the PDD, in reference to the positive conditions of agriculture, its livestock capacities and environment. Environmental.

Within the analyzes that were carried out regarding the investigations, it is important to mention that Bioengineering and Biotechnology are disciplines that, if applied correctly for each of the processes, will be very beneficial, thanks to the fact that they cover science topics with a mix technology and engineering. The phases in which it can be used is very important, since the research that was adapted to this document gave a.

Due to the results mentioned in the research, it can be highlighted that the independent variable (Bioengineering) is an important base for the central region of Colombia, since it intervenes in a positive, purposeful and effective way in all the innovation and development processes that In the case of the research, are the dependent variables, in this way the interpretation and correlation that these two variables have is correctly established, since all the activities that are carried out in the central area of Colombia and the scope that are carried out are evidenced. may have on the applicability of these two variables.

Timely result in relation to the problems that were evidenced and the possible solutions that these can give to agriculture, the environment and livestock systems. We can highlight that the PRAE play a very important role in the inclusion and consolidation of Bioengineering and Biotechnology in the academic processes that are ad dressed for the improvement and progress of the central area of Colombia, thanks to the fact that in the development objectives The themes of improving education are immersed in the development plan of this country, because bioengineering is a discipline that encompasses science and engineering issues and will contribute to the promotion of education in this sector.

### 4.2 Results

Currently, Colombia takes an average of eight (8) years for all the sustainable development objectives to be met. Among the advances that stand out in the SDGs in relation to Bioengineering and Biotechnology, the following characteristics and projects can be mentioned:

**SDG Four (4) Quality Education**

The goals that Colombia is drawing up to comply with this SDG have fundamental characteristics for the disciplines of bioengineering and biotechnology, of which the following can be highlighted:

4.7 Education for global citizenship

4B Expand Higher Education Scholarships for Developing Countries

These two aspects mentioned above stand out because they help new disciplines appear and gain strength so that more young student applicants can take advantage of these types of opportunities. Being Bioengineering a mix between science and engineering, it becomes a very attractive educational program to be able to study it.

**SDG Eleven (11) Sustainable Cities and Communities**

The goals that are set for objective number eleven (11) and that are important on the issues of bioengineering and biotechnology are:

11.5 Reduce the adverse effects of natural disasters

11.6 Minimize the environmental impacts of cities

11.7 Build green, safe and inclusive public spaces

Due to the aforementioned, it is emphasized that these aspects are a direct part of Bio engineering and Biotechnology because they can be used to implement correctly, in education, planning, development and innovation processes.

According to the National Planning Department regarding the characteristics of transport, progress has been made in the different construction of the road infrastructure that is intended for public transport systems and integrated mass transport systems in twelve (12) cities of the country. This means firstly that the accessibility so that people can be transported to educational centers do so in an adequate and positive way, that is why the inclusion of Bioengineering benefits from having this type of aspects since the inclusion of ease of acquiring knowledge regarding this discipline [5].

On the other hand, it is highlighted that within the results evidenced in the National Development Plan established for the years 2018–2022 in Colombia, it is highlighted that all the characteristics that are going to be mentioned below are relevant with respect to the themes of bioengineering and biotechnology.

In the first instance, it is evident that Colombia occupies the position number forty (40) among all the countries that emit the most greenhouse gases. On the other hand, and with regard to the dependent variables in relation to livestock, agricultural and environ mental issues, it is shown that the country has a low performance regarding the use of land and water, since it consumes 2.8 times more materials premiums than the OECD average [6].

Along with the aforementioned characteristics, the importance of bioengineering applied to all environmental processes is also highlighted, which is why it is important to highlight that in Colombia in the last six years an area of forests equivalent to 926 thousand fields have been lost. of soccer, this as a result of bad processes in environ mental zones, in addition to the fact that a large amount of loss is found in the central region of Colombia, but it is there where bioengineering appears as a mediating entity, to propose ideas for improvement of the affected sectors and help reduce the impacts that are generated by bad habits. In another vein we can also mention that there are more environmental shortcomings that can be improved and mitigated through the applicability of the branches of bioengineering [7].

"Therefore, universities by training in the undergraduate careers should promote environmental care and responsible development with nature, this can be done by one of the two known paths, be it Environmental Education or the Sustainable Development Goals" [7].

"Globalization constituted as a transforming element of current dynamics of society, mediated by the new communication and information technologies, "has the capacity to function daily as a unit in time and space in a planetary sphere" [8].

**Ideas Proposal**

It is important to be able to implement ideas and economic strategies from bioengineering for the productive sectors, so that they can be more sustainable and innovative, in this case we can show the involvement of the two variables mentioned above, in such a way that the environmental impacts can be, providing a circular economic approach. On the other hand, it is important to emphasize that through the mechanisms that bio engineering can provide, it will be possible to stop deforestation and other environmental impacts, through controls and territorial delimitations to generate new economic opportunities in livestock, agricultural and environmental systems, forming communities with sustainable assets.

Finally, it is of utmost importance to be able to strengthen higher education institutions to include bioengineering as an academic program and in this way, it will also contribute to strengthening other types of environmental organizations, including research processes, public and private management, at the same time. that is including the processes of education and dialogue regarding environmental issues and of course bioengineering.

## 5 Conclusions

- The inclusion of Bioengineering at the national level and especially in all the activities that take place in the central region is fundamental, thanks to the fact that this is a discipline that contributes to improvement and development processes, in the face of positive factors, since Being a specialty that has a com bination of scientific, technological and engineering variables, it can be correctly coupled to agricultural, livestock and environmental factors.
- Bioengineering is a very complete discipline since it has great relevance and similarity with biotechnology, these two form an essential complement and benefit both academic, economic and environmental activities.
- Bioengineering has great academic opportunities, it is necessary to promote this type of educational programs, since people with efficient professional pro files can be

trained, regarding the topics that it can cover, as already mentioned above, the union of branches science and engineering can encompass many current global issues and, in this way, contribute to the proportion of mediating and mitigating ideas.

## References


1. Fuentes, O.: Genesis y evolucion de la bioingenieria, p. 8
2. Pinto, A.L.G., Bejarano, J.H.S.: El sistema de ciudades en la Región Administrativa y de Planeación Especial, Región Central (RAPE-RC), Colombia. Rev. Ciudad. Estados Polít. **3**(3), Art. no. 3 (2016)
3. Martínez, C.C., et al.: El sistema de ciudades de la región central: configuración y desafíos, p. 50 (2016)
4. Creswell, J.W.: Research Design: Qualitative, Quantitative, and Mixed Methods Approach, 4th edn. (2013)
5. Departamento Nacional de Planeación: Ciudades y comunidades sostenibles - La Agenda 2030 en Colombia - Objetivos de Desarrollo Sostenible. Ciudades y comunidades sostenibles - La Agenda 2030 en Colombia - Objetivos de Desarrollo Sostenible (2019). https://www.ods.gov.co/es/goals/ciudades-y-comunidades-sostenibles
6. Departamento Nacional de Planeación: Plan Nacional de Desarrollo 2018–2022 "Pacto por Colombia, pacto por la equidad". https://www.dnp.gov.co/DNPN/Paginas/Plan-Nacional-de-Desarrollo.aspx
7. Castellanos, P.M.A., Encinas, A.H., Dios, A.Q., Ortegon, A.C.: Analysis of environmental sustainability educational approaches in engineering education. In: 2020 15th Iberian Conference on Information Systems and Technologies (CISTI), Sevilla, Spain, pp. 1–5, June 2020. https://doi.org/10.23919/CISTI49556.2020.9140919
8. Ortegón, A.C., Ortegón, D.C., Castellanos, P.M.A.: Medio ambiente y sostenibilidad: con la mira en los ods; capítulo 1: "El Jardín de Michuá" Ejemplo de Nueva Ruralidad para la Recuperación, Preservación y Consolidación de los Valores Culturales y Ambientales en la Zona De Influencia de la Vereda el Resguardo del Municipio de Chiquinquirá – Boyacá – Colombia (2020)

# Challenges and Competences in Master Degrees from a Comprehensive Security Perspective

Javier Espinosa García[1(✉)], Luis Hernández Encinas[1,2], and Alberto Peinado Domínguez[3]

[1] Institute of Physical and Information Technologies (ITEFI), Spanish National Research Council (CSIC), Madrid, Spain
{javier.espinosa,luis}@iec.csic.es
[2] Higher Technical School of Engineering (ICAI), Comillas Pontifical University, Madrid, Spain
[3] E.T.S. Ingeniería de Telecomunicación, Universidad de Málaga, Málaga, Spain
apeinado@ic.uma.es

**Abstract.** Nowadays, the security threats faced by society in general, and organizations in particular, are reaching unprecedented highs in terms of complexity and sophistication. Organizations must face up to new challenges in order to protect their assets (e.g., personnel, goods, information, or reputation, among others), so they need new solutions for the immediate future to achieve their business objectives efficiently and effectively. Right now, corporate security and cybersecurity within organizations are dependent on themselves and complementary at the same time, so the coordination of their activities is crucial. This joint action for a comprehensive security requires a training and educational process that provides necessary knowledge and skills. This article highlights the limitations of the current master's degrees in cybersecurity from the perspective of comprehensive security and proposes a new master's degree focused on it as a solution, which takes advantage of the skills of current degrees, and is aligned with the needs of the European Higher Education Area. This master's degree promotes sustainability through efficiency and rationalization on the general asset protection process, improving the security culture inside the organization, reducing the number of security incidents, and increasing the reputation and image of the organization.

**Keywords:** Comprehensive security · Master's degree · Security education

## 1 Introduction

We are living in a digital and diverse society [1], so almost all human activities are threatened by a growing number of potential security risks [2], both physical and cyber. This shows that security controls to protect assets by removing or avoiding dangers or damages are crucial and significant.

Nowadays, nobody doubts that companies and institutions are increasingly becoming complex [3], both from an organic and functional point of view. Organizations move in a global and highly dynamic context where technologies play a decisive and crucial role. In fact, information is one of the key assets to protect within an organization and it


J. J. Gude Prego et al. (Eds.): CISIS 2021/ICEUTE 2021, AISC 1400, pp. 402–414, 2022.
https://doi.org/10.1007/978-3-030-87872-6_40

is commonly accepted that lies in electronic media that need to be physically protected, while the cybersecurity of the physical medium that contains the information must be guaranteed. These measures are interdependent and complementary [4]. In other words, it would not be possible to guarantee the cybersecurity of an asset without a suitable physical security, and vice versa. Terms such as integral, convergence, or fusion spread in the context of security, proving that physical and cyber security complements perfectly throughout their entire value chains, providing more benefits than drawbacks.

Comprehensive security (hereafter, CS) is a multidisciplinary concept aligned with current needs and trends [5, 6]. CS covers the most relevant fields of security, both strategic, tactical, and operational ones, aimed at preventing and avoiding any damage, danger or risk that threatens any asset for an organization. Of course, CS is a holistic process made up of procedures, techniques, and tools from heterogeneous nature such as scientific, technological, economic, financial, legal, or social, among others. CS is the natural response to current incidents that hit the organization assets, so much so CS transformation is still on progress. CS can be a challenge for any organization but generates efficiency advantages by eliminating duplicities and optimizing dedications and efforts.

Therefore, it seems necessary that those responsible for an organization security have a wider and solider knowledge baseline that covers the whole CS scope [7]. Thus, a cybersecurity professional (e.g., a telecommunication engineer responsible for cybersecurity for an organization) should be fluent in methodologies and techniques specific to his traditional academic education and work experience, but also he should know and consider other areas or disciplines that will be essential for decision making under his responsibilities to organization activities. This paper brings an education proposal that tries to contribute to develop CS via a master's degree.

The rest of this paper is organized as follows: Sect. 2 describes a brief background about security and its approaches in organizations that leads to CS and gives rise to the proposed CS master's degree. Across Sect. 3, this paper presents the master's degree objectives and competences to be achieved by students, which adjust to the generic needs of current organizations with sustainability development requirements. In addition, during Sect. 3, this paper also presents the master's degree curriculum and program in terms of subjects, contents, and credits. Finally, Sect. 4 brings the main conclusions about the proposed master's degree in CS.

## 2 Comprehensive Security Background

Well into the 21$^{st}$ century, it can be somewhat accessory to arise the main reasons why individuals require security in any of their daily facets (e.g., personal, social, family, work, health, etc.). A security that, in general, is innate to human nature itself and that is closely related to the protection of needs and aversion to damage, both accidental and deliberate [8].

In this sense, until the middle of the 20$^{th}$ century [9], security consisted of the aspects of its physical component, fundamentally related to the protection of people (e.g., their life, health, and well-being, etc.) and the tangible assets of companies (e.g., buildings, machinery, goods, etc.). However, from the moment that electronic systems begin to develop in a generalized way and technological means begin to be incorporated into business processes, this traditional security concept has been modified and has been extended to new fields of action. It is not only about protecting the technology that automates production processes and lowers costs, but this technology has led to the recognition that also logical aspects must be protected, that is, business data, processed or stored in information systems, are tremendously valuable and are part of the intangible assets of the company [10].

This change in the conception of security has increased, even more, since new technologies are a constant feature and become a real part of the personal world, where the cyber aspects are highly critical. Therefore, the physical and logical aspects of security cannot be analyzed individually but must be considered jointly, since there are relationships and interdependencies between them that require a wider and more cohesive vision of the concept of security [11]. That is CS.

While regulated education is focused on cybersecurity [12], security [13] or safety management [14] as isolated topics, no university master's degree has been found (at least in Spain, according to the information on official master's degrees provided by ANECA [15]) by the authors of this work with a greater approach like CS. The analysis performed on master's degrees compliant with the European Higher Education Area (EHEA) framework, reveals that three main modules that define competences in cyber defense, cyberattacks and cybersecurity management support all of them.

## 3 Master's Degree in Comprehensive Security

A master's degree in CS (MCS) is presented as a proposal to overcome the gap detected in current cybersecurity master's degrees in the path to CS. MCS is twofold oriented, inasmuch as it intends that the students who complete it can either develop a doctoral thesis or pursue a professional career.

The very nature of the CS concept makes it neither necessary nor advisable to establish itineraries or specialties in the MCS. In this way, the MCS aims to cover the main topics related to CS from a global perspective.

All the MCS attendants will acquire training in all subjects related to CS with no compartmentalization nor distinction and MCS will prepare students for achieving professional certifications such as Certified Information Security Manager (CISM) from ISACA [16] or Certified Information Systems Security Professional (CISSP) from ISC$^2$ [17]. The MCS constitutes an extraordinary access path for multidisciplinary training and research in the CS field from a critical, respectful, ethical, and informed perspective.

### 3.1 MCS Objectives

The MCS, attending to the needs and trends of the security sector and the markets [18, 19], sets as its main objective the complete training of professionals in the most relevant fields of CS. More specifically, this main objective of the MCS can be broken down into the following specific objectives:

OBJ1. To manage security processes and services in an efficiency and quality way.
OBJ2. To run products, processes and services in CS.
OBJ3. To lead, coordinate and supervise multidisciplinary CS teams.
OBJ4. To prepare, direct, coordinate and manage competitive projects in CS.
OBJ5. To research, develop and innovate in the fields related to CS.
OBJ6. To apply the acquired knowledge to solve problems in CS environments.
OBJ7. To know legislations, regulations, and standards around CS.
OBJ8. To communicate knowledge, reasoning and conclusions to general audience.

Every MCS graduated student must have achieved all of the previous stated objectives, so students will have developed and acquired the necessary skills to assume a series of competencies.

### 3.2 MCS Competences

The competences that students will acquire throughout the MCS are linked to the MCS objectives. These competences are divided into basic, general and specific ones.

**MCS Basic Competences.** MCS students will demonstrate that they have achieved the following basic competences:

BC1. To lead multidisciplinary working groups in international environment.
BC2. To apply solutions that solve the raised problems.
BC3. To work efficiently, knowing how to organize and plan personal work.
BC4. To deal with the information to use, identifying the sources and resources.
BC5. To issue opinions and reasoned judgments on CS based on knowledge.

**MCS General Competences.** In addition to the basic competences established in the previous section, students must have achieved the following general skills:

GC1. To acquire knowledge to be real in the development and/or application of ideas.
GC2. To apply knowledge and to solve problems in new or unfamiliar environments.
GC3. To integrate knowledge and to formulate judgments based on information.
GC4. To clearly communicate their conclusions and the supporting reasons.
GC5. To gain learning skills that allow them to self-directed or autonomous studying.

**Specific Competences.** All previous competences are complemented by some other specifically linked to MCS, to be precise, referring to each of the CS aspects. These are the following:

SC1. To manage and organize the CS of a company by providing tangible benefits.
SC2. To apply efficiently CS risk assessment.
SC3. To cope with CS legal, regulatory, and standards frameworks.
SC4. To understand the principles related to criminal acts.
SC5. To design decision-making processes according to different intelligence models.
SC6. To be aware of trends and challenges regarding CS threats and vulnerabilities.
SC7. To be able to design and apply business continuity and disaster recovery plans.
SC8. To plan and execute specific protection systems for critical infrastructures.
SC9. To know the best strategies to manage crisis and situations.
SC10. To prepare, manage, coordinate, and run CS projects.
SC11. To know and deploy different physical protection systems solutions.
SC12. To design and supervise plans for people protection.
SC13. To know the basic requirements and systems for fire protection.
SC14. To coordinate the different services for electronic protection.
SC15. To supervise the processes of occupational risks prevention and its legislation.
SC16. To organize and prepare evacuation and protection plans.
SC17. To know the criteria to ensure security for public and private entities.
SC18. To be aware of the importance of the information and its systems security.
SC19. To know different access control systems models.
SC20. To apply the appropriate controls to protect networks and telecommunications.
SC21. To know the advantages and requirements to apply cryptology solutions.
SC22. To be aware of the importance of complying with international standards.
SC23. To know methodologies for the acquisition and development of systems.
SC24. To plan and supervise applications and communications security.

### 3.3 MCS Curriculum and Program

The MCS proposal is a 60 credits degree and has a marked globalizing character as evidenced by its objectives, the competences that students must demonstrate, and the breadth of its contents. For these reasons, the syllabus does not contemplate the possibility of different specialties within the global framework of CS.

As shown in Table 1, the MCS splits into four modules. Module I, Module II and Module III have three subjects each, while Module IV entails a master degree's thesis.

**Table 1.** MCS structure. Subjects overview

| Module | Subjects | Credits |
|---|---|---|
| I: Strategic and Management Aspects within CS | I.1 – CS Management and Organization | 5 |
| | I.2 – CS Risk and Crisis Management | 5 |
| | I.3 – Law, Criminology and Crime | 4 |
| II: Basic Aspects of Physical Security | II.1 – Electronic and Physical Protection | 6 |
| | II.2 – Civil Protection, Fire Fighting and Hazards | 5 |
| | II.3 – People, Companies and Goods Protection | 6 |
| III: Basic Aspects of Cybersecurity | III.1 – Information Security and Cryptology | 6 |
| | III.2 – Networks Security and Access Control | 6 |
| | III.3 – Systems Architectures, Developments and Operations | 5 |
| IV: Master's Thesis | IV.1 – Master's Thesis | 12 |

The following subsections describes MCS modules and their subjects.

**Module I: Strategic and Management Aspects Within CS.** This is a 14 credits module made up of the following three subjects:

*I.1 – CS Management and Organization.* This is a five credits subject whose main characteristics are shown in Table 2.

**Table 2.** I.1 – CS management and organization. Description

| Brief description of the contents | Objectives | Competences |
|---|---|---|
| Value and contribution of CS; Corporate needs; Integration of CS in organizations; Organizational structures and their operation; Human resources, materials, and activities in CS; Strategies and Tactics in CS; Local CS vs. Global CS; CS and Defense; International organizations and CS trends; Intelligence cycle and models; Strategies and decisions around CS; CS Dashboards and indicators; Bases of business economics; and Financial asset management | OBJ1, OBJ2, OBJ3, OBJ4, OBJ5, OBJ6, and OBJ8 | BC1, BC2, BC3, BC4, and BC5; GC1, GC2, GC3, GC4, and GC5; SC1, SC2, SC3, SC5, SC6, SC9, and SC10 |

*I.2 – CS Risk and Crisis Management.* This is a five credits subject whose main characteristics are shown in Table 3.

**Table 3.** I.2 – CS risk and crisis management. Description

| Brief description of the contents | Objectives | Competences |
|---|---|---|
| CS Organizational principles; Planning; anticipation and psychological CS aspects; Communication management; International CS scenarios; Crisis management techniques and models; CS Trends; CS Organizational and functional structures; Business Continuity Plans; Business continuity management systems; Concept and bases of risk; CS Framework and criteria; CS Contextualization, analysis, and risk management; Making judicious decisions: objectives, risks, and costs; CS Continuous improvement and risk management; CS New challenges: threats and vulnerabilities; and Critical infrastructures | OBJ1, OBJ2, OBJ3, OBJ4, OBJ5, OBJ6, and OBJ8 | BC1, BC2, BC3, BC4, and BC5; GC1, GC2, GC3, GC4, and GC5; SC1, SC2, SC3, SC6, SC7, SC8, SC9, and SC10 |

*I.3 – Law, Criminology and Crime.* This is a four credits subject whose main characteristics are shown in Table 4.

**Table 4.** I.3 – Law, criminology and crime. Description

| Brief description of the contents | Objectives | Competences |
|---|---|---|
| The legal conception of security as a public service; Citizen security: security forces and bodies; The State for the operation of public security services; Police collaboration and international legal framework; Personal data privacy; Intellectual property and other intangible assets; The National Security and Cybersecurity Strategies; Criminal phenomenology; Forensic profiling; Digital evidences; Fraud prevention and investigation methods; and Professional deontology | OBJ1, OBJ2, OBJ3, OBJ4, OBJ5, OBJ6, and OBJ8 | BC1, BC2, BC3, BC4, and BC5; GC1, GC2, GC3, GC4, and GC5; SC1, SC2, SC3, SC4, SC6, SC9, and SC10 |

**Module II: Basic Aspects of Physical Security.** This is a 17 credits module made up of the following three subjects:

*II.1 – Electronic and Physical Protection.* This is a six credits subject whose main characteristics are shown in Table 5.

**Table 5.** II.1 – Electronic and physical protection. Description

| Brief description of the contents | Objectives | Competences |
|---|---|---|
| Fundamentals of physical and electronic security; Technical protection measures; External protection; Internal protection; Punctual protection; Specifications; Functions and staging of active protection; Intrusion detection; Closed circuit television; Access control; Other protection systems: communications, patrolling, etc.; Integration and centralization of protection systems; and Security approaches in buildings | OBJ1, OBJ2, OBJ3, OBJ4, OBJ5, OBJ6, and OBJ8 | BC1, BC2, BC3, BC4, and BC5; GC1, GC2, GC3, GC4, and GC5; SC1, SC2, SC3, SC6, SC9, SC10, SC11, and SC12 |

*II.2 – Civil Protection, Fire Fighting and Hazards.* This is a five credits subject whose main characteristics are shown in Table 6.

**Table 6.** II.2 – Civil protection, fire fighting and hazards. Description

| Brief description of the contents | Objectives | Competences |
|---|---|---|
| Civil protection; Civil Protection and Emergency Management; Risks in civil protection; Systems of Civil Protection and Emergencies; Fire safety; The risk of fire in buildings; Causes and consequences; The Fire Prevention Regulations; Passive fire protection; Active fire protection; Prevention of occupational hazards; CS against occupational hazards; General and specific risks; Risks related to the working environment; and Prevention and protection against occupational hazards | OBJ1, OBJ2, OBJ3, OBJ4, OBJ5, OBJ6, and OBJ8 | BC1, BC2, BC3, BC4, and BC5; GC1, GC2, GC3, GC4, and GC5; SC1, SC2, SC3, SC6, SC9, SC10, SC13, and SC15 |

*II.3 – People, Companies and Goods Protection.* This is a six credits subject whose main characteristics are shown in Table 7.

**Table 7.** II.3 – People, companies and goods protection. Description

| Brief description of the contents | Objectives | Competences |
|---|---|---|
| Protection and regulations; Protection Training; Security at large events; Coordination and command centers; Counter surveillance program; European Protection Network; Dynamic protection at international or large summits; Methodology and Implementation of a CS department; Self-protection plans; Permanent coordination centers in protection; CS plan; Access controls and accreditations; Airspace and CS; Private CS and the Private CS Law; Competences and public authorities; Institutional participation and social recognition; Private CS activities; Companies and detective offices; Private CS personnel; CS services. Private CS measures; Administrative control and sanctioning regime; and Collaboration between public and private CS | OBJ1, OBJ2, OBJ3, OBJ4, OBJ5, OBJ6, and OBJ8 | BC1, BC2, BC3, BC4, and BC5; GC1, GC2, GC3, GC4, and GC5; SC1, SC2, SC3, SC4, SC6, SC9, SC10, SC14, SC16, and SC17 |

**Module III: Basic Aspects of Cybersecurity.** This is a 17 credits module made up of the following three subjects:

*III.1 – Information Security and Cryptology.* This is a six credits subject whose main characteristics are shown in Table 8.

**Table 8.** III.1 – Information security and cryptology. Description

| Brief description of the contents | Objectives | Competences |
|---|---|---|
| Steganography; Mathematical foundations of CS; Symmetric cryptography: stream and block encryption; Random Number Generation (RNG); Key agreement protocols; Asymmetric cryptography: RSA; Elliptic curves; Hash functions; Digital signature schemes; Public key infrastructure: digital certificates; (Post)Quantum cryptography; Side channel and fault induction attacks; and Biometric systems and identification of people | OBJ1, OBJ2, OBJ3, OBJ4, OBJ5, OBJ6, and OBJ8 | BC1, BC2, BC3, BC4, and BC5; GC1, GC2, GC3, GC4, and GC5; SC1, SC2, SC3, SC6, SC9, SC10, SC20, and SC21 |

*III.2 – Networks Security and Access Control.* This is a six credits subject whose main characteristics are shown in Table 9.

**Table 9.** III.2 – networks security and access control. Description

| Brief description of the contents | Objectives | Competences |
|---|---|---|
| Confidentiality, Integrity and Availability Requirements; Protocols: PPTP, IPsec, L2TP, SSL/TLS, SSH; Network devices and definition of collision domains; Security protocols in WLAN networks; Security in Bluetooth, NFC and RFID interfaces; Security in WSN: ZigBee (IEEE 802.15.4); Security in VANETs networks; Security in GSM/UMTS/LTE mobile networks; Access control mechanisms through smart cards; Access control methodologies; Authentication mechanisms; Anonymity and privacy: main concepts and challenges; and Tor networks | OBJ1, OBJ2, OBJ3, OBJ4, OBJ5, OBJ6, and OBJ8 | BC1, BC2, BC3, BC4, and BC5; GC1, GC2, GC3, GC4, and GC5; SC1, SC2, SC3, SC6, SC9, SC10, SC18, and SC19 |

*III.3 – Systems Architectures, Developments and Operations.* This is a five credits subject whose main characteristics are shown in Table 10.

**Table 10.** III.3 – Systems architectures, developments and operations. Description

| Brief description of the contents | Objectives | Competences |
|---|---|---|
| Products and systems certifications; Common Criteria and Other certifications; Forensic analysis of computer systems; Evidence: obtaining, analysis, custody, and procedures; Forensic analysis tools; Exploitation of software vulnerabilities; Exploitation of web systems; Vulnerability management; Cyber defense systems; Audit and analysis of events; Attack detection systems; Security Information and Event Management Systems; Malicious code: analysis and engineering; and Cyber espionage | OBJ1, OBJ2, OBJ3, OBJ4, OBJ5, OBJ6, and OBJ8 | BC1, BC2, BC3, BC4, and BC5; GC1, GC2, GC3, GC4, and GC5; SC1, SC2, SC3, SC4, SC6, SC9, SC10, SC22, SC23, and SC24 |

**Module IV: Master's Thesis.** The Master's Thesis (MT) is the culmination of the acquisition of knowledge throughout the MCS. MT is a 12 credits original work that cannot be presented until all the 48 teaching credits of the MCS have been passed, although its subject must begin to be considered by students from the beginning of the MCS to guarantee a choice aligned with their preferences and interests.

The MT pursues the implementation and development of the methodological and theoretical knowledge acquired during the MCS through the preparation and defense of an individual and original research work, of a theoretical or empirical nature, that collects most of the aspects dealt with in the subjects of the MCS.

The MT would focus on any of the topics proposed by the academic committee of the MCS or any other proposed by the student and accepted. Some potential topics for the MT are the following: CS risk assessment for a company and its impact on business; definition and deployment of CS in an international organization; design of a CS and crisis management model in an organization; or analysis of the regulatory and legal framework in force in terms of CS and its impact on controls.

**MCS Schedule and Faculty.** Professors and professionals with wide experience in their corresponding MCS's subjects could lead the MCS classes and master classes. All the MCS's subjects would have, at least, one master class in charge of a well-recognized and prestigious professional o researcher, as well as several invited speakers from academic and organizations institutions, both public and private.

Every subject would have a responsible in charge of ensuring the correct development of that subject. In addition, he would safeguard the delivery of all content, the availability of teaching materials, the existence of a unified teaching criterion for other teachers of the same subject, as well as a clear and equitable method of evaluating the subject.

It is important to note that the MCS teaching staff would be composed by professionals with a big experience in different fields and academics (with PhD title).This fact would be strongly conditioned by the special contents that are contemplated in the first two modules. Module III contents are usually part of the study plans of many universities, mainly in Mathematics, Engineering, etc. For Module I and Module II subjects, there could be prestigious specialists with wide professional and teaching experience assigned to each of those subjects. Nevertheless, an academic commission to achieve MCS objectives would coordinate all MSC subjects.

## 4 Conclusions

Coordination and communication are key elements in order to keep assets protected in an organization. Although CS can be a challenge, it generates an advantage within organization, especially in terms of efficiency and, consequently, sustainability.

Since CS promotes synergies for the security function in organizations by eliminating duplicities and optimizing dedications and efforts around processes, requirements, resources, or budgets, among others. One of the strategic levers for CS deployment in an organization is education, training and the promotion of CS culture, facilitating the acquisition of knowledge and skills.

To date, the authors have not found any university master's degree or similar regulated education. Current university education in focused on cybersecurity or safety as isolated topics instead of a greater approach as CS.

To fill this gap and provide a unifying perspective, this paper shows a master's degree proposal, whose main purpose is the complete training of professionals in CS. On this premise, the MCS intends that all professionals can face any challenge in CS, being able to define, present, and/or develop effective and efficient CS strategies.

As a result, the students who graduated from the MCS will be able to provide a comprehensive approach to the CS risks of an organization. This constitutes a more accurate and solid perception than if risks were assessed separately. A detailed CS assessment provides better risk management strategies for an organization, thanks to wide perspectives of threats, vulnerabilities, and controls perspectives.

Additionally, MCS promotes sustainability through efficiency and rationalization on the general asset protection process, by eliminating functional gaps and promoting cost savings, either by reducing expenses, or by avoiding them. In this sense, redundant functions can be removed and processes can be optimized. This means an improvement in cost control and productivity for the organization.

Finally, MCS collaborates in improving the CS culture inside the organization. This has a direct impact on reducing the number of CS incidents, decreasing their intensity, and increasing the reputation and image of the organization, both internally and externally.

**Acknowledgement.** This research has been partially supported by the Department of Information and Communication Technologies (TIC) at Institute of Physical and Information Technologies (ITEFI), Spanish National Research Council (CSIC).

## References

1. Data Never Sleeps 8.0: https://www.domo.com/learn/data-never-sleeps-8. Accessed 18 May 2021
2. World Economic Forum: The global risks report 2021. http://www3.weforum.org/docs/WEF_The_Global_Risks_Report_2021.pdf. Accessed 18 May 2021
3. Kledt, F., Evers, P., Benson, D.: Section II.2 - Organizational Complexity. In: Contemporary Theory and Practice of Organizations. Part I. Understanding the Organization. Ibidem-Verlag, Stuttgart, pp. 95–116 (2016). https://cup.columbia.edu/book/contemporary-theory-and-practice-of-organizations/9783838267470
4. Rahman, S.M., Donahue, S.E.: Convergence of corporate and information security. Int. J. Comput. Sci. Inform. Sec. **7**(1), 63–68 (2010). https://arxiv.org/abs/1002.1950v1
5. Cabric, M.: From Corporate Security to Commercial Force: A Business Leader's Guide to Security Economics. Butterworth-Heinemann, Boston (2017). https://www.sciencedirect.com/book/9780128051498/from-corporate-security-to-commercial-force
6. Cabric, M.: Corporate Security Management. Challenges, Risks, and Strategies. Butterworth-Heinemann, Boston (2015). https://doi.org/10.1016/C2014-0-04011-7
7. Kinslow, J.: Physical and IT security: the case for convergence. J. Sec. Educ. 75–91 (2006). https://doi.org/10.1300/j460v02n01_06
8. Jovanovic, M., Burbiel, J.: The evolving concept of security: a critical evaluation across four dimensions. In: Proceedings of the 9th Security Research Conference Future Security, pp. 502–506 (2014). https://www.researchgate.net/publication/278033393_THE_EVOLVING_CONCEPT_OF_SECURITY_A_CRITICAL_EVALUATION_ACROSS_FOUR_DIMENSIONS
9. Purpura, P.P.: Security: An Introduction. CRC Press, Boca Raton (2016). https://doi.org/10.1201/9781439894569
10. Halibozek, E.P., Kovacich, G.L.: The Manager's Handbook for Corporate Security: Establishing and Managing a Successful Assets Protection Program, 2nd edn. Butterworth-Heinemann, Boston (2017). https://www.sciencedirect.com/book/9780128046043/the-managers-handbook-for-corporate-security\#book-info

11. Stackpole, B., Oksendahl, E.: Security Strategy: From Requirements to Reality. Auerbach Publications, New York (2010). https://doi.org/10.1201/EBK1439827338
12. Masters Programs in Cybersecurity: https://www.masterstudies.com/Masters-Degree/Cyber-Security/. Accessed 18 May 2021
13. Masters Programs in Security: https://www.masterstudies.com/Masters-Degree/Security/. Accessed 18 May 2021
14. Masters Programs in Safety Management: https://www.masterstudies.com/Masters-Degree/Safety-Management/. Accessed 18 May 2021
15. ANECA, Listado de titulaciones: https://srv.aneca.es/ListadoTitulos/busqueda-titulaciones. Accessed 7 Jun 2021
16. ISACA's Certified Information Security Manager – CISM. https://www.isaca.org/credentialing/cism. Accessed 18 May 2021
17. $ISC^2$'s Certified Information Systems Security Professional – CISSP: https://www.isc2.org/Certifications/CISSP. Accessed 18 May 2021
18. Gartner's Top 10 Security Predictions 2016: https://www.gartner.com/smarterwithgartner/top-10-security-predictions-2016/. Accessed 18 May 2021
19. Microsoft Corporation, Accenture LLP: Future of physical security whitepaper-Extending converged digital capabilities across logical & physical environments, 2018. https://www.accenture.com/_acnmedia/PDF-100/Accenture-Future-of-Physical-Security-Whitepaper.pdf

# Author Index




J. J. Gude Prego et al. (Eds.): CISIS 2021/ICEUTE 2021, AISC 1400, pp. 415–416, 2022.
https://doi.org/10.1007/978-3-030-87872-6